DIFFERENTIATION AND DEVELOPMENT

MIAMI WINTER SYMPOSIA—VOLUME 15

1. *W. J. Whelan and J. Schultz, editors:* HOMOLOGIES IN ENZYMES AND METABOLIC PATHWAYS *and* METABOLIC ALTERATIONS IN CANCER, 1970
2. *D. W. Ribbons, J. F. Woessner, Jr., and J. Schultz, editors:* NUCLEIC ACID-PROTEIN INTERACTIONS *and* NUCLEIC ACID SYNTHESIS IN VIRAL INFECTION, 1971
3. *J. F. Woessner, Jr. and F. Huijing, editors:* THE MOLECULAR BASIS OF BIOLOGICAL TRANSPORT, 1972
4. *J. Schultz and B. F. Cameron, editors:* THE MOLECULAR BASIS OF ELECTRON TRANSPORT, 1972
5. *F. Huijing and E. Y. C. Lee, editors:* PROTEIN PHOSPHORYLATION IN CONTROL MECHANISMS, 1973
6. *J. Schultz and H. G. Gratzner, editors:* THE ROLE OF CYCLIC NUCLEOTIDES IN CARCINOGENESIS, 1973
7. *E. Y. C. Lee and E. E. Smith, editors:* BIOLOGY AND CHEMISTRY OF EUCARYOTIC CELL SURFACES, 1974
8. *J. Schultz and R. Block, editors:* MEMBRANE TRANSFORMATION IN NEOPLASIA, 1974
9. *E. E. Smith and D. W. Ribbons, editors:* MOLECULAR APPROACHES TO IMMUNOLOGY, 1975
10. *J. Schultz and R. C. Leif, editors:* CRITICAL FACTORS IN CANCER IMMUNOLOGY, 1975
11. *D. W. Ribbons and K. Brew, editors:* PROTEOLYSIS AND PHYSIOLOGICAL REGULATION, 1976
12. *J. Schultz and F. Ahmad, editors:* CANCER ENZYMOLOGY, 1976
13. *W. A. Scott and R. Werner, editors:* MOLECULAR CLONING OF RECOMBINANT DNA, 1977
14. *J. Schultz and Z. Brada, editors:* GENETIC MANIPULATION AS IT AFFECTS THE CANCER PROBLEM, 1977
15. *F. Ahmad, T. R. Russell, J. Schultz, and R. Werner, editors:* DIFFERENTIATION AND DEVELOPMENT, 1978

MIAMI WINTER SYMPOSIA—VOLUME 15

DIFFERENTIATION AND DEVELOPMENT

edited by

F. Ahmad J. Schultz

The Papanicolaou Cancer Research Institute

T. R. Russell R. Werner

University of Miami School of Medicine

Proceedings of the Miami Winter Symposium, January 1978
Sponsored by The Department of Biochemistry
University of Miami School of Medicine, Miami, Florida
Symposium Director: W. J. Whelan

and by

The Papanicolaou Cancer Research Institute, Miami, Florida
Symposium Director: J. Schultz

ACADEMIC PRESS New York San Francisco London 1978
A Subsidiary of Harcourt Brace Jovanovich, Publishers

Academic Press Rapid Manuscript Reproduction

ACADEMIC PRESS, INC.
111 Fifth Avenue, New York, New York 10003

United Kingdom Edition published by
ACADEMIC PRESS, INC. (LONDON) LTD.
24/28 Oval Road, London NW1 7DX

LIBRARY OF CONGRESS CATALOG CARD NUMBER

ISBN 0-12-045450-5

PRINTED IN THE UNITED STATES OF AMERICA

CONTENTS

SPEAKERS, CHAIRMEN, AND DISCUSSANTS

E. A. Adelberg, Department of Human Genetics, Yale Medical School, New Haven, Connecticut

F. Ahmad, Papanicolaou Cancer Research Institute, Miami, Florida

V. G. Allfrey, The Rockefeller University, New York

E. E. Atikkan, Department of Molecular Biology, National Institutes of Health, Bethesda, Maryland

R. E. Block, Papanicolaou Cancer Research Institute, Miami, Florida

T. W. Borun, Wistar Institute, Philadelphia, Pennsylvania

Z. Brada, Papanicolaou Cancer Research Institute, Miami, Florida

H. Busch, Department of Pharmacology, Baylor College of Medicine, Houston, Texas

J. M. Cardenas, Department of Biochemistry and Biophysics, Oregon State University, Corvallis, Oregon

W. Cieplinski, Department of Medicine, University of Connecticut School of Medicine, Newington, Connecticut

P. P. Cohen, Department of Physiological Chemistry, University of Wisconsin, Madison, Wisconsin

S. Counce-Nicklas, Department of Anatomy, Duke University, Durham, North Carolina

S. P. Craig, Department of Biology, University of South Carolina, Columbia, South Carolina

S. R. Dienstman, Department of Pathology, New York University School of Medicine, New York

P. Feigelson, Department of Biochemistry, Columbia University, New York

G. Felsenfeld, Laboratory of Molecular Biology, National Institutes of Health, Bethesda, Maryland

I. B. Fritz, Banting and Best Department of Medical Research, University of Toronto, Toronto, Ontario, Canada

W. J. Gehring, Biozentrum, University of Basel, Basel, Switzerland

D. Gershon, Department on Biochemistry, Roche Institute of Molecular Biology, Nutley, New Jersey

N. Ghosh, Department of Medicine, New York University Medical Center, New York

A. Gierer, Max Planck-Institüt für Virusforschung, Tubingen, West Germany

H. Green, Department of Biology, Massachusets Institute of Technology, Cambridge, Massachusetts

S. B. Greer, Department of Microbiology, University of Miami School of Medicine, Miami, Florida

C. Grimmelikhujzen, European Molecular Biology Laboratory, Heidelberg, Germany

J. B. Gurdon, MRC Laboratory of Molecular Biology, Cambridge, England

F. Haurowitz, Department of Chemistry, Indiana University, Bloomington, Indiana

R. J. Hay, Department of Cell Culture, American Type Culture Collection, Rockville, Maryland

E. Hertzberg, Department of Cellular Biology, The Rockefeller University, New York

R. A. Hickie, Department of Pharmacology, University of Saskatchewan, Saskatoon, Canada

L. S. Hnilica, Department of Biochemistry, Vanderbilt University, Nashville, Tennessee

F. C. Kafatos, Department of Cellular and Developmental Biology, Harvard Biological Laboratories, Cambridge, Massachusetts

R. G. Kallen, Department of Biochemistry and Biophysics, University of Pennsylvania School of Medicine, Philadelphia, Pennsylvania

J. Kallos, Hospital for Joint Diseases, New York

D. G. Kerley, Department of Botany, University of Wisconsin, Madison, Wisconsin

E. Kohen, Papanicolaou Cancer Research Institute, Miami, Florida

A. Kootstra, Division of Biology, Oak Ridge National Laboratories, University of Tennessee, Oak Ridge, Tennessee

K. Latham, Department of Medicine, Uniformed Services University, Bethesda, Maryland

R. C. Leif, Papanicolaou Cancer Research Institute, Miami, Florida

C. C. Liew, Banting Institute, University of Toronto, Toronto, Ontario, Canada

I. C. Lo, University of Toronto, Toronto, Ontario, Canada

H. E. Lodish, Department of Biology, Massachusetts Institute of Technology, Cambridge, Massachusetts

W. R. Loewenstein, Department of Physiology and Biophysics, University of Miami School of Medicine, Miami, Florida

W. T. McAllister, Department of Microbiology, Rutgers Medical School, Piscataway, New Jersey

B. J. McCarthy, Department of Biochemistry and Biophysics, University of California, San Francisco, California

K. S. McCarty, Department of Biochemistry, Duke University, Durham, North Carolina

T. Maden, Department of Embryology, Carnegie Institute, Baltimore, Maryland

B. L. Malchy, Department of Biochemistry, Queen's University, Kingston, Ontario, Canada

A. Marks, Banting and Best Department of Medical Research, University of Toronto, Toronto, Ontario, Canada

H. P. Meloche, Papanicolaou Cancer Research Institute, Miami, Florida

R. E. Miller, Department of Medicine, Case Western Reserve University, Cleveland, Ohio

B. Mintz, The Institute for Cancer Research, Philadelphia, Pennsylvania

D. H. Mintz, Department of Medicine, University of Miami School of Medicine, Miami, Florida

J. Morrissey, Department of Biology, University of California, San Diego, La Jolla, California

B. W. O'Malley, Department of Cell Biology, Baylor College of Medicine, Houston, Texas

R. Oshima, Department of Pediatrics, University of California, San Diego, La Jolla, California

I. H. Pastan, National Cancer Institute, National Institutes of Health, Bethesda, Maryland

R. Patnaik, Department of Medicine, Sinai Hospital, Baltimore, Maryland

J. Poupko, Department of Pharmacology, University of Miami School of Medicine, Miami, Florida

M. M. Rasenick, Department of Pathology, Yale University School of Medicine, New Haven, Connecticut

R. H. Reeder, Department of Embryology, Carnegie Institution of Washington, Baltimore, Maryland

H. V. Rickenberg, Division of Molecular and Cellular Biology, National Jewish Hospital and Research Center, Denver, Colorado

T. R. Russell, Departments of Biochemistry and Medicine, University of Miami School of Medicine, Miami, Florida

U. Rutishauser, The Rockefeller University, New York

W. J. Rutter, Department of Biochemistry and Biophysics, University of California, San Francisco, California

I. B. Sabran, Institute for Cancer Research at Fox Chase, Philadelphia, Pennsylvania

P. Sarin, National Institutes of Health, Bethesda, Maryland

S. Seaver, Department of Molecular Biology, Vanderbilt University, Nashville, Tennessee

W. A. Scott, Department of Biochemistry, University of Miami School of Medicine, Miami, Florida

H. S. Shapiro, Department of Biochemistry, College of Dentistry and Medicine of New Jersey, Newark, New Jersey

E. Silverstein, Department of Medicine, SUNY Downstate Medical Center, Brooklyn

H. Smith, Department of Biological Sciences, University of New Orleans, New Orleans, Louisiana

A. C. Stoolmiller, Department of Biochemistry, Eunice Kennedy Shriver Center, Waltham, Maryland

H. K. Stanford, President, University of Miami, Coral Gables, Florida

G. S. Stein, Department of Biochemistry and Molecular Biology, University of Florida, Gainesville, Florida

M. Urban, Fox Chase Cancer Center, Institute for Cancer Research, Philadelphia, Pennsylvania

W. Verdeckis, Department of Cell Biology, Baylor College of Medicine, Houston, Texas

J. D. Watson, Cold Spring Harbor Laboratories, Cold Spring Harbor, New York

S. Weinhouse, Department of Biochemistry, Fels Research Institute, Temple University, Philadelphia, Pennsylvania

H. M. Weintraub, Department of Biochemistry, Princeton University, Princeton, New Jersey

R. Werner, Department of Biochemistry, University of Miami School of Medicine, Miami, Florida

C. West, Department of Anatomy, University of Pennsylvania, School of Medicine, Philadelphia, Pennsylvania

H. G. Wood, Department of Biochemistry, Case Western Reserve University, Cleveland, Ohio

A. A. Yunis, Department of Medicine, University of Miami School of Medicine, Miami, Florida

S. Yuspa, National Institutes of Health, Bethesda, Maryland

D. Zouzias, Department of Pathology, New York University, New York

A. Zweidler, The Institute for Cancer Research, Philadelphia, Pennsylvania

PREFACE

This volume is the fifteenth in the continuing series published under the title "Miami Winter Symposia." In January 1969, the Department of Biochemistry of the University of Miami and the University-affiliated Papanicolaou Cancer Research Institute organized the first two of these symposia. This is the tenth year in which the symposia have been held.

As topics for the Miami Winter Symposia we select areas of biochemistry in which recent progress offers new insights into the molecular basis of biological phenomena. In previous years we organized two symposia. The first, sponsored by the Department of Biochemistry, emphasized the basic science aspects of the chosen topic, while the second, sponsored by the Papanicolaou Cancer Research Institute, dealt with the application of this research to the cancer problem. The proceedings of the two symposia were published in separate volumes. With cancer research becoming increasingly concerned with basic cellular mechanisms, the division of the symposia into basic research and cancer-related research became rather academic. For this reason the 1978 meeting was organized as a single symposium and the proceedings are being published in a single volume. The topic of "Differentiation and Development" represents a logical sequel to last year's symposia on cloning and genetic manipulation of recombinant DNA. Many of the techniques described in last year's symposia are being employed to analyze the complex organization and regulation of eukaryotic genomes.

Associated with the symposia is the Feodor Lynen Lecture, named in honor of the Department of Biochemistry's distinguished visiting professor. Past speakers have been George Wald, Arthur Kornberg, Harland G. Wood, Earl W. Sutherland, Jr., Luis F. Leloir, Gerald M. Edelman, A. H. T. Theorell, and Paul Berg. This year the Lynen lecture was given by James D. Watson. These lectures have provided insights into the history of discovery, and have included the personal and scientific philosophies of our distinguished speakers. The Lynen lecturer for 1979 will be Francis Crick, and the symposium will deal with the regulation of gene expression and posttranslational processing of proteins. To bring forward as much of the recent work as possible, short communications are presented in a poster session. The abstracts of these short communications appear in this volume.

Our arrangement with the publishers is to achieve rapid publication of the symposium proceedings, and we thank the speakers for their prompt submission of manuscripts. Our thanks also go to the participants whose interest and discussions provided the interactions that bring a symposium to life and to the many local helpers, faculty, and administrative staff who have contributed to the success of the present symposium. Special gratitude should be accorded to the organizers and coordinators of the program: W. J. Whelan (joint director with J. Schultz), K. Brew, Sandra Black, and Olga F. Lopez and Virginia Salisbury who did the major job of assembling the typescripts for the Papanicolaou Cancer Research Institute.

The financial assistance of several Departments in the University of Miami School of Medicine, namely, Anesthesiology, Dermatology, Ophthalmology, and Pathology, as well as the Howard Hughes Medical Institute, Boehringer Mannheim Corporation, Eli Lilly and Company, Hoffmann-La Roche Incorporated, Smith Kline and French Laboratories, and the Upjohn Company, is gratefully acknowledged.

F. Ahmad
T. R. Russell
J. Schultz
R. Werner

IN FURTHER DEFENSE OF DNA

James D. Watson
Cold Spring Harbor Laboratory
Cold Spring Harbor, New York

I feel most honored to be giving the Lynen Lecture, for I am the first not trained as a biochemist - I take this to mean that my students have practiced it well. When Professor Whelan finally cornered me to say whether I would accept the invitation, I had to admit I was deeply tempted because it was to be autobiographical and that might mean a little gossip. But then he said I had to write it up, and I feared the lawyers would again be in my life. But finally I thought, "Well, I might as well give it, because if I don't they might ask Sinsheimer." So now I wish to relate why I got in my present position running a Lab with an interest in tumor viruses, and how in the process became a minor actor in our current drama about DNA.

The best time to start is in the late winter of 1958 when Salva Luria invited me to give the Miller lectures at the University of Illinois. It was cold, and living at the Union was not exactly fun. But I met Nomura, who then was a postdoc with Spiegelman, and that led to Masayasu's spending the subsequent summer at Harvard. Sol's notebooks were unbelievably neat, and classical music was played at all times. One evening Van Potter showed up and spoke on cancer before Gunsalus and his biochemists. He was thinking along the lines of the new ideas of Umbarger and Pardee on feedback inhibition, a concept to which I had not previously paid much attention. I got quite excited and returned to Harvard thinking I should someday give a course on cancer.

At that time tRNA was getting most of the attention, and Francis Crick's adapter theory was on everyone's tongue. Alfred Tissieres and I were studying ribosome subunits and about to publish that 30s+50s=70s. It was inherently dull, and whenever he saw me, Francis said that we should start working with in-vitro systems for protein synthesis. Paul Zamecnik, however, already was making proteins in *E. coli* extracts, and we thought it made more sense to take up work on the RNA and protein components of the ribosome. Then we still believed ribosomal RNA was the template for protein synthesis.

ISBN 0-12-045450-5

In February of 1959 Francis arrived with Odile to be a visiting professor in the Chemistry Department for the spring term. He did the DNA→RNA→protein story for graduate students, while I gave it at the introductory level in Biology II. This was a new course and was to provide beginning Harvard students with their first opportunity to learn the revolution in molecular genetics. I used my last lecture to talk about cancer, and soon afterward became manic when I realized that Seymour Cohen's recent work might be the key to how viruses make certain cells cancerous. He had just reported that the T-even phages had genes that code for enzymes involved in DNA metabolism. If this was also true for animal viruses, the proliferative response of cells to viral infection might fall out. Unlike bacteria, most cells in a higher animal are turned off for DNA synthesis. So many, if not all, animal viruses might need to carry genes that would ensure the turning on of the cellular apparatus for DNA synthesis.

The Cricks, having never found suitable housing, were living with the Richs. Besides giving his Harvard lectures, Francis was flying all over the country talking about his adapters, and by the end of the term collapsed with nervous exhaustion for several days of bed rest. All in all his visit was a tremendous success, and the Chemistry Department thought, why not make Francis an offer. They then sent out letters for recommendations that could be used to buttress their case with McGeorge Bundy. One went to Pauling, who wrote back that Crick had a fertile mind, but if they were looking for a first-rate protein crystallographer they should seriously consider Murray Vernon King, an almost unknown chemist who had also spent time in Brooklyn. Francis was happy to get the offer, but never seriously considered it, for the MRC promptly responded with the decision to build a major new lab which would bring Fred Sanger together with Francis, John Kendrew, Max Perutz, and Hugh Huxley, who by then wanted to leave University College to return to Cambridge.

Just before the Cricks left to spend the summer at Berkeley, Francis and I received the Warren Triennial Award from Massachusetts General Hospital. It was our first prize, and the $1,500 each looked big. Each of us was to give the lecture at the Museum of Science, and after Francis gave his adapter ideas, I followed giving a talk on Cancer that totally bombed. It took only a minute or two to get over the idea that animal viruses might carry genes that can turn on cellular DNA synthesis. The remaining thirty-five minutes were largely bull in which I tried to tie in both the Warburg effect and RNA tumor viruses. When the lecture was over, I felt awfully depressed. Many of my friends came over to say they had enjoyed it, but compared with Francis's clean presentation, I feared I had convinced almost no one.

I thus welcomed the chance to spend part of that summer over at Massachusetts General Hospital with John Littlefield who had started tumor virus work with the Shope Papilloma virus. Past results suggested that this wart virus might be the best source of a small homogeneous double-stranded DNA. Lots of rabbit warts could be bought from a Kansas trapper called Johnson, and after they were mashed up in a blender, the DNA was easily isolated. To our surprise, I found two components (21s and 28s) when I ran it in the Model E, and we suspected that we were dealing with a monomer-dimer situation. Upon storage the 28s form converted to the 21s species. Subsequent CsCl banding led to an estimated molecular weight of about five million, a figure surprisingly close to more direct measurements made recently with the electron microscope. Then over in Paul Doty's lab I found the sharp denaturation curves were rapidly reversible, and thought what beautiful proof that we were dealing with small homogeneous double helices. Never did we consider that we had circular molecules, and that the 28s were closed supercoils while 21s molecules were relaxed circles with single-stranded cuts.

The following term I devoted my formal teaching to a course on the Biochemistry of Cancer and for the first time tried to master the literature. The Warburg effect was hard going, and I went down to NCI to talk with Dean Burk, then the most prominent scientist arguing on Warburg's side. To my surprise he had only a modest-sized lab and had virtually stopped active experimentation on why most tumor cells consumed so much glucose.

The manuscript of our papilloma observations was submitted to the Journal of Molecular Biology in early 1960. Just then my student, Bob Risebourgh, was doing the first clean experiments showing that T2 RNA never became an integral part of any ribosomal subunit but was a separate species that bound to 70s ribosomes in high Mg^{++}. By contrast, Nomura, Hall and Spiegelman had just previously concluded that T2 RNA was a special type of ribosomal RNA. Bob's work, however, made us doubt that ribosomal RNA never carried any genetic specificity, and we came to think that T2 RNA was a hint of the real templates in uninfected cells. This then was a most unorthdox idea, and on my way in late March to a seminar at Oak Ridge, I stopped off at Memorial Hospital to see Leo Szilard who had just been found to have bladder cancer. Leo was as intellectually demanding as ever and kept busy calculating the radiation dose that he should receive. He was quite skeptical about the messenger concept, and soon turned our conversation over to his recent theory of antibody formation.

Early in May the Groses came over from Paris to live temporarily in my father's tiny house at 10 1/2 Appian Way. Immediately Francois started to see if messenger-like molecules were in uninfected E. coli cells. The first experiments gave positive hints, but we thought it wise not to talk prematurely. When in mid-June we drove up to New Hampton for the Nucleic Acid Gordon Conference, neither of us brought up the possibility of messenger-like RNA, even during drinks after the evening sessions. On the last day we talked by phone with Matt Meselson to learn that he, with Francois Jacob and Sydney Brenner, also had evidence of the messenger. Upon our return to Harvard, Wally Gilbert began helping us with sucrose gradients, and by the fall the messenger RNA concept became a solid base for all further work. In particular, we had a new outlook on making proteins in the test tube, and Alfred soon stimulated synthesis with semipurified mRNA preps. One day he asked Helga Doty for some Poly A to throw into his reaction mix, but it showed no effect.

The thought processes of the molecular biologists and our biochemist colleagues were now almost indistinguishable. There was no longer any need to argue our respective cases, and only Chargaff wanted to fight. A unique opportunity came at the 1961 Cold Spring Harbor Symposium, the first time there was massive talk about mRNA. Toward the end of a session, Sydney stood up to say that the term "messenger" was most appropriate for the new RNA form since it was a very mercurial substance and Mercury was the "God of Messengers". Chargaff arose and asked if Dr. Brenner knew that Mercury was also the "God of Thieves".

Over the next five years, I increasingly pushed much of my lab onto Norton Zinder's newly discovered RNA phages, reasoning we should always study the simplest systems. At the same time I was more and more aware that Arthur Kornberg's lab consistently did better biochemistry than the rest of us. And everyone knew what you would do if you were in Arthur's lab - get in by 9 a.m. and purify your enzyme. So I encouraged my student, John Richardson, to purify RNA polymerase better than anyone else, hoping that he would find evidence for specific binding to DNA. Though we learned much news about the physical properties of RNA polymerase, in retrospect all we observed were nonspecific interactions. That problem didn't break until 1969 when Andrew Travers arrived from Sydney's lab and he and Dick Burgess found σ. Most excitingly σ disappeared from polymerase purified from T4-infected cells, and we thought,

"At last we know the basis for the early vs late mRNA turn-ons; and even more important, maybe changes in RNA polymerase specificity will be important for embryology."

Ever since we had the double helix I had avoided thinking much about embryology. As a student I had a course on invertebrates with Paul Weiss, and despite being scared by his nasty frown, I enjoyed his "Principles of Development". And later in Europe, I temporarily knew enough French to read eagerly the first edition of Brachet's, "Embryologie Chimique". But after the double helix emerged, I began to realize how difficult it would be to meaningfully describe at the molecular level even the most simple embryological process. No one with sense should anymore get excited by the observation that one substance goes up and another down during differentiation. We all knew this must happen. You could go from here to eternity measuring enzyme levels, but nothing would come out. Yet silly nonsense kept appearing from prominent mouths that soon we were going to have real embryological breakthroughs. So starting in 1959 I annually gave a lecture in Biology II against embryology and what was wrong with Woods Hole. I minced no words that if we were to get anywhere, we should stick to soluble problems. The Harvard Biologists finally could not take me any longer and used my 1966 sabbatical to get me out of the course. Carroll Williams from then on gave my lectures. I didn't really mind, because "The Molecular Biology of the Gene" had appeared, and I already had a much larger audience than I could warn against phony optimism.

It is not true now, but until all too recently most biologists, at least those I grew up with, were not that deep. They too often threw out corn instead of sense. As Americans they had to be optimistic, so why not be optimistic about embryology? But I thought it was a dead subject and should stay in the deep freeze. Common sense told me it would never really move unless there was a path back to DNA. But then there was no way to get out the DNA pieces you wanted to study. The eucaryotic genome was too big. If you wanted to be sensible and yet have future embryological ambitions, you worked with the DNA animal viruses.

Strongly influencing me was the recent discovery of the mouse virus polyoma. Its chromosome was almost as small as those of the RNA phages and yet it would make cells cancerous. Ever since Stoker and Dulbecco got the field going in the early 1960's, I wanted to go into it, at least in a vicarious fashion. When the Directorship of the Cold Spring Harbor

Laboratory fell open, the opportunity arrived. John Cairns couldn'ttake the administrative chaos any more, and given the Lab's decrepit state finding a new Director was not going to be simple. Its Board of Trustees did not know what to do, and Gunsalus wanted to offer the job to a phage geneticist located in Dallas. But I couldn't see it falling in the hands of someone with no emotional attachment to Cold Spring Harbor whom I feared would use it as a stepping stone back to Germany. Late in 1967 I said I would take it over on a part-time basis with the idea that I would build up a group which would do DNA tumor-virus molecular biology.

A few months later Liz and I were married, and as soon as the Harvard spring term ended, we came down to Long Island to spend the summer. The Lab had no free cash that anyone might consider spending, and much of July and August I spent throwing out junk from the library so that we might have a decent space to read journals. John Cairns told me to go after Joe Sambrook if I wanted to start animal virus work, and luckily Joe was to be about during part of Phil Marcus' and Gordon Sato's summer course on animal cells and viruses. By September I had decided to put together a grant request, and Lionel Crawford, who had come over for the summer to collaborate with Ray Gesteland, helped me draw up plans for converting James' lab into space for tumor virus work. Joe by then seemed eager to come here from the Salk, and the application went off to NIH and eventually ended up in the Genetics Study Section. One of its members was Charles Thomas who knew that we necessarily would work with larger amounts of DNA than was handled by most non-molecular tumor virologists. So he asked that a committee be set up to ask whether there might be a safety problem.

Until then there had essentially been no public discussion about possible biohazards of tumor virus work. Those scientists who grew up in the tradition of medical microbiology naturally knew that animal viruses should be treated as potentially dangerous. They could reassure themselves that animal virologists didn't have shorter life spans than the average, and Dulbecco and Stoker, for example, did their experiments on the open bench. Perhaps they had moments of mild apprehension when they first got into tumor virus work, but soon realized it made sense to focus their worries on well-documented hazards.

Committees never move fast, and we got the grant to start SV40 work without any restriction as to the possible biohazard. Our immediate aim was to get experiments going as fast as possible, and during our first year Joe and Bill Sugden focused on purifying eucaryotic RNA polymerases with the hope of turning up σ -like specificity factors. Like others before them, they found evidence for the I and II forms, but never had success with either in observing specific transcription. In the meanwhile, Carel Mulder was trying to find a restriction enzyme which would break SV40 at a specific site so that a linear denaturation map could be worked out by Hajo Delius and maybe later the early and late mRNAs assigned to clean locations on the genome.

The potential biohazard problem surfaced again after Bob Pollack arrived here in 1971. Bob started out as a physicist and didn't react as his former mentor, the trained M.D., Howard Green, who thought such talk was unwarranted. While medical students start out worrying whether they might get sick from a patient, you can't become a doctor unless you lose that fear and start practicing your trade. Bob, however, was a product of a Brooklyn socialist heritage which told you that worker's lives are often lost in the shuffle of their boss's success. He was eager to take on the biohazard dilemma and was a prime mover behind the first Asilomar gathering that took place in February 1973. It focused on the potential health hazards of tumor virus research, and data were presented, for example, on the SV40 contamination of the early Salk and Sabin vaccines. No general conclusion could emerge. There was not a trace of evidence that what we did was dangerous and would lead to any new cases of cancer. Yet, maybe we hadn't waited long enough, for the incubation period of cancer can be very long. About the only solid impact was that NCI became likely to fund more elaborate containment facilities for work with tumor viruses and to sponsor courses on safer lab procedures. Mouth pipetting disappeared, and we tried to operate so that viruses would not go up in aerosols created by our experimental procedures. But it was impossible, and is still, to know whether our responses were adequate. No one had any idea of the magnitude of the possible dangers and maybe there was no risk at all.

We could by now press ahead at increasing speed because the first useful restriction enzymes had been found and many more were soon to come from Richard Roberts' group in Demerec's lab. Early and late SV40 mRNAs were found both here and at NIH to be coded by different DNA strands and each to occupy about one-half of the genome. The size assigned to the early region was just sufficient to code for the T (tumor) antigen which Peter Tegtymeyer was estimating to be in the 80,000 - 100,000 range. An obvious next step was to purify this protein, since it was the only viral protein expressed in SV40-transformed cells and must be the oncogenic product which converts normal cells into their cancerous equivalents. Klaus and Mary Weber came down on leave from Harvard with this objective, hoping subsequently to find out how it works. Though Klaus is about as talented a protein chemist as you find anywhere, he found the problem more than he bargained for. After a year of hard work, he gave up since there was no way to get enough cells. T antigen is made normally in small quantities, and he was limited to work with cells that grow only in monolayers. Even if he had grown an order of magnitude more cells, the effort would likely have failed, and the possibility never opened up to work at the Kornberg level.

This problem was only to break several years later when Eugene Lukanidin came here from Moscow and with John Hassel and Joe Sambrook isolated new SV40-adenovirus hybrid in which the SV40 T antigen gene came under control of a late adenovirus promoter. This leads to a tenfold-greater T antigen yield per cell. In addition, these hybrids, like wild-type adenoviruses, grow on cells that multiply in suspension to reach numbers far in excess of those limited to monolayers. Given these tricks, last year Bob Tjian used classical high-level biochemistry to show specific binding of T antigen to DNA fragments containing the origin of replication of SV40. I was very pleased, since our almost ancient ideas on how DNA tumor viruses made cells cancerous were looking more and more correct.

Already by 1973 restriction enzymes were very key tools for deep probing of the viral genome and seemed likely to be even more so as the recombinant technology was just beginning to come out of the Bay region. With plasmids to use as cloning vehicles, any piece of eucaryotic DNA eventually could be prepared in amounts suitable for detailed molecular characterization. We still then found preparing enough viral DNA for chemical analysis a major task, and Wally Gilbert's and Fred Sanger's powerful new DNA sequencing procedures were yet to appear. So we immediately thought about using recombinant DNA technology to amplify tumor virus genes. In this way we wouldn't have to grow large amounts of virus anymore, and if biohazards did exist, they now could be greatly lessened.

But first Bob Pollack and then many participants at the 1973 Nucleic Acid Gordon Conference began to question whether we might inadvertently, or conceivably quite intentionally, use recombinant DNA technology to produce bacteria with extended pathogenicity. Questioning began as to whether in fact certain experiments should never be done. A close vote in favor of possible restraints was taken at the Gordon Conference, and the organizers led by Maxine Singer and Dieter Soll sent off a letter to Science that asked that the matter be taken up by a body like the National Academy. Phil Handler agreed and Paul Berg was asked to come in with a report. Subsequently he asked David Baltimore, Herman Lewis, Dan Nathans, Sherman Weissman, and me to meet with him when he came East to MIT early in 1974.

This was a period when the rights of innocent third parties were being taken increasingly serious. We should be more than fair to the dishwashers and technicians on whom we might blow viruses, or they on us, and so on. Motivated primarily by a desire to be maximally socially conscious and without any evidence that recombinant DNA was dangerous, we called for a partial moratorium until we had a big meeting the following February. It became Asilomar II. In retrospect Asilomar I didn't produce anything except a small book, "Biohazards in Biological Research", which Cold Spring Harbor put out at an almost give-away price and kept reprinting and after several years made a tiny profit. I thought with Asilomar II we might have a book that would be a best seller. So to start with, I was quite enthusiastic about the whole affair. David Baltimore subsequently argued that people did not want the further hassle of still another meeting with manuscripts, and I got no book and stopped thinking about the matter. The thought never occurred to me that the new Asilomar could lead to any formal rules, since I found the subject too hypothetical for a reasonable response.

I could not have been more wrong. To start with the National Academy organized a press conference at which analogies were made to the war time decision of our nuclear physicists to desist from further publishing their experiments. This led to the press producing copy that we molecular biologists had a genie that we might not be able to contain, and that society might be at risk. None of us then thought it pertinent that we already take chances with DNA. Every time a new baby comes forth it is the product of new forms of DNA, and a priori we do not know whether the consequences will be for the good or bad. Recombinant DNA, per se, is not something first brought forth by science. It is an obligatory fact of life whose occurrence is far wider than generally perceived. Viruses, for example, have the potential to cross species barriers and to carry DNA between unrelated organisms. Then, however, it seemed more to the point to worry about what bad might come out of our labs rather than take on the even more complex happenings that must occur naturally. That myopia, of which I also was guilty, will haunt us for a long time.

As soon as Asilomar II started, it became apparent that a hectic rush into guidelines was foreordained. Sydney set the tone by saying we must be responsible to society and face up to what our restriction enzymes might yield. I was not convinced and rose to say that since we were unable to bring forth any rational guidelines for tumor virus research, how could we now react to a situation where we could not even guess the size of our potential opponent. As far as I was concerned, everyone might as well go home. That response generated stony silence, and I soon was the meeting's outcast to be later eaten up by the members of the press.

As the meeting progressed, its main preoccupation became the so-called safe strains. I am not sure who first lit on this device, but it was Sydney who orchestrated its star role. Why not take K12 and make it so enfeebled that it could never escape into the sewers of MIT? To most this seemed to be almost the perfect response to the dilemma of not knowing whether we had something to worry about. But how much do you want to spend making even safer a safe bug, and to be sure K12 is safe? Certainly twenty-five cents or even five dollars and maybe fifty thousand dollars. But would we do it if it were eventually to cost more than twenty-five million and potentially lead to so many mucked up memoranda of understanding that to cover our flanks we should go to law school before risking recombinant DNA experimentation?

Here there is no point in recounting the three years' travail that have elapsed since Asilomar II. So many thousands of man years have already been consumed that I would not further dwell on the matter were it now not so obvious that embryology acquired a future the moment we became able to clone specific pieces of eucaryotic DNA. The way is now open to find proteins that bind to specific pieces of DNA, and over the next decade we should come to understand key embryological steps the same way we now understand the lactose operon.

Today there is nothing to stop us except for the inability of biohazard experts to make rational responses to questions whose answers demand knowledge that we will never possess - or the agony of preparing countless memoranda of understanding that will no longer be valid once you advise a clearer way to do your experiment - or the inability of the ever growing NIH bureaucracy to know how to deal with novel experiments for which the guidelines have not yet been provided. All this is most maddening - especially as I now don't know of a single person who does recombinant DNA research who feels the tiniest apprehension. A little Dictyostelium DNA in *E. coli* I don't think would ever make Harvey Lodish even slightly tremble. If it did, I would think him cuckoo.

Unfortunately, lots of well intentioned outsiders today see recombinant DNA as a test case for the scientists' responsibility to society. Yet by now almost every molecular biologist wishes to expunge the guideline controversy from their consciousness. A return, however, to Asilomar II freedom will not be easy. I sensed that all too well as I made a short statement this past December before a public NIH hearing which Don Fredrickson called to consider relaxing certain of the more onerous guidelines. My message to his advisory body was that it was a national disgrace that we were wasting our time with untestable speculations, and that the National Institutes of Health already had enough to do dealing with real human diseases. In response, a lawyer who represented the Ford Foundation created environmental lobby, "The National Resources Defense Council", shrilly demanded that I prove that recombinant DNA research was safe. I looked across at him and said, "How do I know that you are not a paid killer sent by the Commies to do in our science?" He seemed stunned and did not reply. After the session ended, he came up to me and indignantly said, "How can you let me down? You scientists have created this issue, and you should keep it going". To which I replied, "Because I was a jackass is no reason for you to continue to be one."

Soon recombinant DNA is going to be back in front of Congress, and they will ask again whether there should be formal laws which will aim to reassure, say, the comfortable people of Princeton that its University's molecular biologists will not make them all sick. I focus upon the affluent because the more money you have the more likely you have been frightened by DNA. Poor people do not know about DNA. You have to have leisure to be able to read and worry whether your community should permit DNA research. Only then will you learn that P4 level experiments are said to be potentially more dangerous than P3 and then need much persuading that P2 should go forward.

There will be no way, however, that we can ever make clear to our non-scientific associates why Asilomar II was so frightened by human DNA. Future work with it was declared possibly the most risky, and hence it is still virtually impossible to do anywhere. This was a strange decision and totally disregarded the fact that we have DNA in our sperm and eggs. Countless tons of human DNA daily already gets spread about without our recombinant DNA technology, but this is not polite conversation. So laboratory rearrangements of human DNA are banned while normal and abnormal sex go on their uncontrollable ways.

Despite all this past confusion, the voices which the newspapers report still remain largely those who want to shut us down. To my dismay, the vast majority of our nation's biologists, the only group which has the competence to put its house back in order, stays largely mute. They seem not to remember that historically freedom is more easily lost than won, and that the Director of NIH needs to hear from them as well as from the do-nothing fringe of the environmental movement. He must necessarily act as if he is not playing free and fancy with our nation's health. The only way he can do this is for the responsible scientific community to come to his aid and firmly say that an overblown issue should not stay overblown forever.

Five years have passed, and we have more than respected the right of all sides to be heard. Now in the absence of the slightest evidence that any danger to society is involved, we should go forward at the maximal possible speed. No longer are we a pastoral society, and we shall not survive well, if at all, without more science in our future. Instead, these days we too often play timid, if not mildly guilty, and then wonder why we are increasingly impotent in carrying along the public. As long as we whine and not come to the heart of the matter, we just look self-serving. To say that Asilomar II was science at its best, and that all would be fine if it had not been for Mayor Vellucci and his Cambridge circus, is to miss the point. Our problem is not recombinant DNA, it is ourselves. I thank you.

THE ADIPOSE CONVERSION OF 3T3 CELLS

Howard Green
Department of Biology
Massachusetts Institute of Technology
Cambridge, Massachusetts 02139

Two factors commonly influencing the choice of systems for the study of differentiation are:

(1) The degree to which the system can be simplified. This depends on the number of cell types involved, and how well they function under the simplified conditions.

(2) The extent to which the system permits the study of change in the differentiated state. There are many cell types which, after a process of differentiation _in vivo_, continue to express their differentiated state under simple culture conditions, but they are of limited value for study of how the differentiated state was brought about. To investigate the actual process of differentiation, most culture work has employed interacting explants, commonly of mesenchyme and primitive epithelium. At the level of cell culture, perhaps myogenesis and erythroid differentiation have enjoyed the greatest attention. New possibilities have recently been created by the use of teratomas, which have lent themselves to interesting and important experiments at different levels of complexity.

Within the last few years, investigation of the process of fat cell formation has become relatively easy, owing to the development of a very simple culture system for its study (13, 14, 15, 16). The principal function of adipose tissue is very well defined; it is to act as an energy storage bank for the animal; for this purpose the adipose cell withdraws fatty acids from the circulation or synthesizes them from glucose, and stores them as triglyceride. At other times, the cell hydrolyzes the triglyceride and releases the fatty acids and glycerol for utilization elsewhere. This process is a very important part of the energy metabolism of higher animals and is under the control of numerous hormones, such as insulin, epinephrine, glucagon and ACTH. These chemical functions of the adipose cell do not require a complicated structure at the tissue level. While not completely homogeneous with respect to cell type, adipose tissue is simple

ISBN 0-12-045450-5

compared with other tissues; it consists of adipose cells, capillary endothelial cells and fibroblasts, of which at least some are also preadipose cells.

Though the adipose cell interacts with the endothelial cell, particularly by transfering lipoprotein lipase to it,so as to make the enzyme accessible to the plasma lipoproteins, the functions of the adipose cell do not seem to be controlled by any adjacent cells, but rather by chemicals reaching the cell from a distance.

Of course, the adipose cell does sometimes engage in morphogenetic functions having nothing to do with its chemical functions. For the most part, the adipose tissues are distributed in such a way as to fit body contours but, especially in the human, they may undertake to form the body contours on their own, in such a way as to make them more interesting or attractive or to make them more practical for such purposes as sitting down. These morphogenetic functions do not involve change in the construction of the tissue, but only changes in the number and size of the adipose cells. Since these variables are under chemical control, adipose cells at different body sites cannot have identical susceptibility to that control.

Adipose tissue does not begin to develop until late in embryonic life. This means that either precursor (preadipose) cells are not present earlier, or if they are present, the signal for their differentiation must be lacking. Most adipose cells form in early post natal life (19, 17). There is also evidence that some preadipose cells persist even in grown animals, and may convert to adipose cells under certain conditions (22, 35, 46, 50, 9).

Origin of the Tissue Adipose Cell

This is a problem investigated mainly by successive histological examination of developing adipose tissue. All investigators agree that the fat cell is of mesenchymal origin. Before the era of the electron microscope, some histologists believed that it originated from a reticuloendothelial cell (for a review of these studies see Wasserman, (52)). On the other hand, actual observation of fat cell formation in the living state with chambers inserted into the ear of rabbits led others to the conclusion that fat cells formed from highly elongated flattened cells with extended processes, a morphology typical of fibroblasts. These cells could be located at some distance from blood vessels (4).

Later work on developing fat pads in newborn mice by electron microscopy confirmed the view of Clark and Clark that the cell type of origin is a fibroblast, sometimes remote from a capillary, and separated from it by a basement membrane (30). However, allowance must be made for the possibility that fatty cells can sometimes arise from other cell types, especially in disease states. This can evidently occur under cell culture conditions (57).

The Origin of 3T3 Cells

The 3T3 line was evolved from the disaggregated cells of late fetuses of the mouse (48). Since that is the stage at which the adipose tissues develop or are about to develop, it seems reasonable that the original cultures would have contained preadipose cells. In normal development, preadipose cells lose their ability to grow when they mature; but if they are kept in the growing state under culture conditions, there is no obvious reason why they should not become established cell lines, like other mesenchymal cell types of rodents.

Though, under certain conditions, the appearance of 3T3 cells may resemble that of endothelial cells (34), they are fibroblasts by functional criteria. They synthesize abundant collagen, which assembles into fibers (10, 12). This collagen has been found to be mainly of Type I and partly of Type III; no collagen of Type IV has been detected (11).

The Conversion of 3T3 Cells into Adipose Cells

When the growth of 3T3 cells is arrested, a certain proportion will convert to adipose cells. This is true for each of a considerable number of randomly chosen clones isolated from a 3T3 stock stored frozen for over a decade (14, 15). Though all clones gave rise to some adipose cells, the clones differed markedly in the frequency with which each did so under the same standard conditions. For one clone, 3T3-C2, the frequency was so low that this clone served as a valuable control for studies on the adipose conversion in other clones. On the other hand, the quality of the adipose cells produced was not very different for the different clones. Hence the curious fact that the degree of differentiation could be described quantitatively by a frequency or probability of adipose conversion (15). The adipose conversion is not the only form of differentiation that has this property (32).

Earliest Stages of the Adipose Conversion

In living cultures or in the living tissues of animals, the earliest sign of the adipose conversion is the appearance of increasing numbers of fat droplets in the cytoplasm. This is probably a more sensitive criterion than enzymatic or chemical changes measured in whole cultures, because of the asynchrony of the participating cells and the easy microscopic identification of rare cells whose adipose conversion precedes that of the rest of the population. In fixed cultures the triglyceride droplets are stained bright red by Oil Red O, a fat soluble dye.

Cells with numerous brightly staining droplets generally do not possess highly extended processes. In cultures where the adipose conversion is slow, it is sometimes possible to see clusters of cells at an earlier stage (proadipocytes), before phase separation of the triglyceride has taken place. They have remarkably extended processes, sometimes quite numerous, and usually very flattened. The staining of these stages is useful in revealing the shape of resting 3T3 cells before it has been much changed by the adipose conversion. The outline of these extended cells cannot be seen without the lipid accumulation because of the extensive overlapping of neighboring cells. Though sections examined earlier by electron microscopy had shown cytoplasmic overlapping in confluent cultures of 3T3 cells (49), the amount of overlapping and the extent of the processes involved were revealed by the proadipocytes to be much greater than expected (15).

As the adipose conversion progresses, the 3T3 cells withdraw their processes and change their shape from highly extended to nearly spherical. This process is very characteristic, and is usually quite advanced by the time triglyceride droplets become numerous. It resembles what was found by direct observation of developing tissue fat cells (4).

In dealing with the 3T3 system as a culture model for the development of the tissue adipose cell, it is necessary to validate the fidelity of the model by as many criteria as possible. This problem has come up before in connection with studies of growth control and the effects of oncogenic viruses, a field in which 3T3 cells have been widely used. All established cell lines are abnormal to some degree, and the chromosomal constitution of 3T3 cells, while not as extensively rearranged as that of many other established lines, is definitely aneuploid (48, 29). Since so much is known about the metabolic behavior of adipose tissues, it is important in what follows to point out the extensive similarities so far demonstrated for the 3T3 system, so that studies of the adipose conversion, about which relatively little is known, can be undertaken with confidence in that system.

Enzymatic Changes During the Adipose Conversion

During the adipose conversion of 3T3 cells, the only lipid that accumulates is triglyceride; the cholesterol and phospholipid content of the cells does not change appreciably (13). As the triglyceride accumulates, the cytoplasm increases in protein content but to a much more modest extent (14), so that triglyceride accounts for an ever increasing proportion of the total cell mass.

A cell showing specialization of this type might be expected to increase its rate of synthesis of triglyceride. Fatty acids for triglyceride synthesis may be either synthesized by the cells or acquired from the medium. The rate of fatty acid synthesis accompanying the triglyceride accumulation can be measured by determining the rate of incorporation of ^{14}C-labeled acetate into the triglyceride. As shown in Fig. 1, this rate is very low in growing 3T3 cells and increases strikingly when the cells have stopped growing and begun to undergo the adipose conversion. The maximum increase in acetate incorporation in such experiments may exceed 100 fold. Similar results have been reported by Mackall et al, (27).

The rate of utilization of external fatty acids for triglyceride synthesis has been examined in the same way using ^{14}C-labeled palmitate. The rate of incorporation also increases sharply during the adipose conversion (Fig. 1).

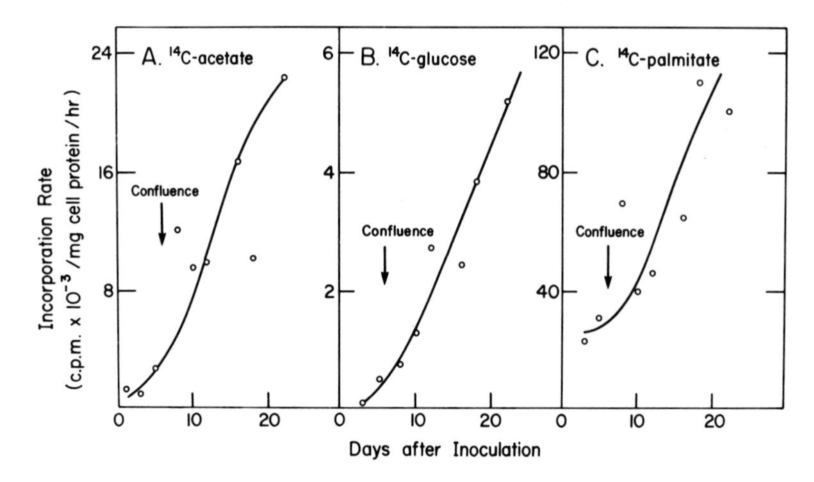

Figure 1. Increasing Rate of Triglyceride Synthesis from Precursors During the Adipose Conversion

Cultures of 3T3-L1 cells were grown to confluence and allowed to undergo adipose conversion. At intervals, the rate at which a culture could incorporate a labeled precursor into triglyceride was determined (14).

Whether the fatty acids are synthesized from acetyl CoA, or obtained from the medium, the glycerol to be acylated in adipose cells must be generated as glycerol 3-phosphate from glucose within the cells (51, 54). As in the case of tissue adipose cells, 3T3 cells undergoing adipose conversion have virtually no ability to utilize exogenous glycerol (56). The rate of incorporation of ^{14}C labeled glucose into triglyceride is therefore a measure of synthesis both of fatty acid and glycerol portions. This rate increases at least as much as the rate of ^{14}C acetate incorporation (Fig. 1).

Increases of such magnitude in the rate of a synthetic process clearly suggest increases in the amounts of the enzymes participating in the synthesis. The change in activity of a number of those enzymes has now been measured in different laboratories using extracts of 3T3 cells prepared before and after adipose conversion (Table 1). The enzymes of fatty acid synthesis, the acylating enzymes, and lipoprotein lipase, all increase very appreciably during the adipose conversion. Owing to the different conditions used, the relative increases of different enzymes is of doubtful significance, except within a given experiment. Within the class of fatty acid synthesizing enzymes at least some undergo the same relative increase (27).

So far, changes in the amounts of two of the enzymes of fatty acid synthesis have been measured immunochemically - acetyl CoA carboxylase (27) and fatty acid synthetase, (Ahmad, Russell and Ahmad, tnis symposium). In both cases, the amount of enzyme protein increased sufficiently during the adipose conversion to explain the increase in activity.

Another enzyme with less obvious relation to triglyceride synthesis, glutamine synthetase, also undergoes a marked increase in activity during the adipose conversion (Miller, Jorkasky and Gershman, this symposium). Enzymes of the glycolytic pathway undergo only relatively small increases (ibid, 28).

Influence of Hormones on the Lipogenic Process

The metabolism of adipose tissue is under chemical control, and many of the chemicals acting on the tissues are hormones. The nature of hormone action in the 3T3 system may be taken as an indication of the extent to which the culture system resembles intact adipose tissue.

Perhaps the most important hormone acting on adipose tissue is insulin. This hormone has many actions on the adipose cell (24), but whether they are described as lipogenic or anti-lipolytic, the result is to increase the triglyceride content of the cells.

Table 1 Enzymatic Changes in 3T3 Cells Undergoing the Adipose Conversion

Role of Enzyme	Enzyme	Increase*	
Fatty Acid Synthesis	ATP Citrate Lyase	11 X	Mackall et al, (27) Grimaldi et al, (18) Williams & Polakis, (55)
	Acetyl CoA carboxylase	11 X	Mackall et al, (27)
	Fatty acid synthetase	11 X	Mackall et al, (27) Grimaldi et al, (18)
	Malic Enzyme	8 - 15 X	Kuri-Harcuch & Green (26)
	Pyruvate carboxylase	18 X	Mackall & Lane, (28)
Acylation of Glycerol	Glycerol phosphate acyl transferase	80 X	Kuri-Harcuch & Green, (26)
	Diglyceride acyl transferase	6 X	Grimaldi et al, (18)
Utilization of external lipid	Lipoprotein Lipase	80 X - 180 X	Eckel et al, (58) Wise & Green, (56)
	Fatty acid: CoA ligase	9 X	Grimaldi et al, (18)

*Expressed per unit of cell protein, sometimes estimated by me from the author's data. Where more than one report exists, only the higher value is cited. As usually not all cells in a culture convert to fat cells, these values are likely to be under-estimates. Data obtained using 3T3-L1 and 3T3-F442A.

3T3-L1 cells respond to insulin at concentrations down to about 1 ng/ml. The effect is seen as an increase in the amount of triglyceride accumulated, either in the living cells examined under the microscope, or in fixed cell layers stained with Oil Red O and examined with the naked eye (14). The effect of insulin may be also seen in short term experiments on the rate of incorporation of labeled glucose into triglyceride. This rate is increased many fold in the presence of high insulin concentrations, and appreciably at concentrations below 10 ng/ml (14). During the adipose conversion, insulin receptors appear in increasing number on the cell surface (36). The sensitivity of the cells to insulin is itself affected by the prior presence of insulin, for the number of receptors increases much more when insulin is present than when it is absent (36). A number of enzymes important in the lipogenic process also increase much more in activity when the cultures are exposed to insulin; among these are malic enzyme (Kuri-Harcuch and Green, unpublished) and lipoprotein lipase (56).

The effect of epinephrine in reducing triglyceride accumulation can be demonstrated on stained cultures (13) or by a reduction in the rate of incorporation of labeled glucose into triglyceride (14). Sensitivity to the effect of lipolytic agents develops in the course of the adipose conversion. For example the cells become very much more sensitive to the effects of ACTH or β-adrenergic agonists; this was revealed by the much greater ability of these compounds to increase adenyl cyclase activity and the cAMP content of the cells after the adipose conversion took place (41). As sensitivity to the effects of ACTH develops when the adipose conversion takes place in the absence of the hormone, this sensitivity develops as part of the program of differentiation and is not induced by the hormone (41).

In nearly all cases, the effects of hormones on the adipose conversion of 3T3 cells were those to be expected for an adipose cell. Only in the case of glucagon were we surprised to find no obvious effect. This point was clarified by Rubin, Lai and Rosen,(41) who showed that the mouse is an unusual species in this respect, for the adenyl cyclase activity of its adipose tissue is relatively insensitive to the action of glucagon.

Effects of Drugs on the Adipose Conversion

Many drugs have already been shown to have important effects on the adipose conversion of 3T3 cells. In principle, these drugs can be divided into two classes:

1. Those which affect the basic process of differentiation, and
2. Those which modulate the rate of lipid accumulation, without affecting the primary differentiation.

One drug that seems to be firmly in the first class is bromodeoxyuridine. The effect of this compound in preventing differentiated functions was discovered by Wessells (53) and Stockdale et al (47), and since that time has been extended to many systems. If 3T3 cells were allowed to grow to confluence in the presence of bromodeoxyuridine at a concentration of about 10^{-7} Molar, the adipose conversion was sharply reduced. At 10^{-5} Molar, adipose conversion was abolished. This effect was demonstrated by the failure of cells grown in the presence of bromodeoxyuridine to accumulate any lipid after reaching confluence and entering the resting state (16). It was also demonstrated by the failure of the confluent cells to increase their ability to incorporate labeled glucose into cellular triglyceride (Fig. 2)(14).

In order to prevent adipose conversion, bromodeoxyuridine must be present during the previous growth of the cells and therefore is incorporated into cellular DNA. While it has seemed likely, in this form of differentiation as in others, that the effects of bromodeoxyuridine are exerted as a result of its presence in the DNA, recent studies on another system cast doubt on this idea (6) and suggest that other possibilities should be considered more seriously (43).

The mechanism of bromeoxyuridine action may be illuminated by the very interesting experiments on the effect of 1-methyl-3-isobutyl xanthine on the adipose conversion of 3T3 cells. As first described by Russell and Ho (42), this compound (and prostaglandin $F_{2\alpha}$) exerted a lipogenic effect scored as an increase in the number of fat cells at a given time. If methyl isobutyl xanthine was present for only 24 hours and then removed before adipose conversion became

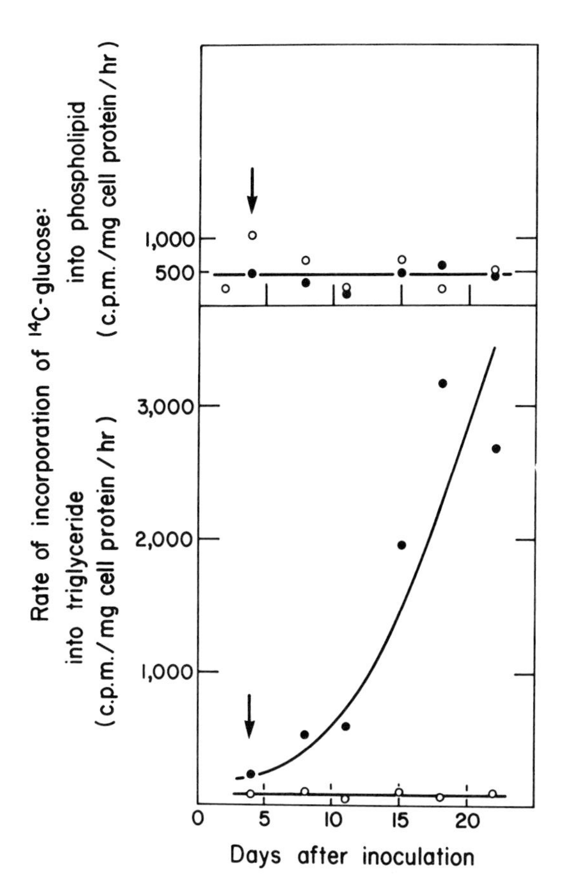

Figure 2 Effect of Incorporated Bromodeoxyuridine on the Rate of Triglyceride Synthesis from Glucose

Cultures of 3T3-L1 cells were grown to confluence (arrows) in the presence of bromodeoxyuridine (10^{-5} M) and deoxycytidine ($5x10^{-4}$M), or in the absence of both drugs. At intervals, cultures were washed free of medium and incubated with ^{14}C labeled glucose for 5 hours. The amount of label incorporated into cell triglyceride and cell phospholipid was then determined. Cells grown in the absence of bromodeoxyuridine increased their rate of triglyceride synthesis from glucose sharply after reaching confluence (solid circles). Those grown in the presence of bromodeoxyuridine did not (open circles). Phospholipid synthesis was not affected either by the adipose conversion or by bromodeoxyuridine. Data from (14).

prominent, its effect became obvious days later. This result might suggest that its effect is on an early stage of the differentiation process. Though cyclic nucleotides have well known effects on fat metabolism (45), and dibutyryl cyclic AMP diminishes fat accumulation by 3T3 cultures (13), the effects described by Russell and Ho could not be duplicated by any cyclic nucleotides. A determination of whether methyl isobutyl xanthine acts on the early stages of the differentiation or on the reaction rates leading to triglyceride accumulation will be important for analysis of the process. This question becomes particularly important if methyl isobutyl xanthine can reverse the effects of bromodeoxyuridine (Russell, this symposium).

Another class of substance affecting differentiated functions is that of tumor promotors, as exemplified by phorbol myristate acetate. This compound had been shown to interfere with myogenesis (5) and with differentiation in Friend erythroleukemia cells (40). More recently it has also been shown to inhibit the adipose conversion in 3T3 cells (7). Since phorbol ester is not likely to directly influence triglyceride metabolism, it may be that its effect is on some early stage of the differentiation.

Another drug recently demonstrated to have striking effect on the adipose conversion of 3T3 cells is indomethacin (55). This drug was shown to stimulate lipogenesis strongly, an effect that may in part be due to inhibition of prostaglandin synthesis.

The Lipoprotein Lipase of 3T3 Cells

One of the interesting features of the adipose cell is its use of lipoprotein lipase to obtain fatty acids from circulating lipoproteins (39, 44, 37). This enzyme is made by the adipose cell (38) and localizes to the cell surface. It can be discharged from that position in isolated adipose tissues (3, 21) or on disaggregated adipose cells (33) by the addition of heparin.

In the adipose tissues, the lipoprotein lipase of the adipose cells is transferred (presumably across basement membranes) to the capillary endothelial cell, where it localizes to the luminal surface. Here it acts on the plasma lipoproteins, particularly VLDL and chylomicrons. It can be discharged from this surface by heparin (25).

Lipoprotein lipase has been purified virtually to homogeneity (8, 2). This enzyme is probably multimeric with a subunit size of 64,000 daltons. Lipoprotein lipase has many properties that distinguish it from other lipases (31).

Lipoprotein lipase activity was found to be nearly undetectable in growing cultures of 3T3-L1 or 3T3-F442A; but when the cells were allowed to undergo adipose conversion, the enzyme activity rose sharply and reached levels approximately **2** orders of magnitude higher than in growing cells (56). The lipase had the properties of salt sensitivity, alkaline pH optimum and serum requirement characteristic of lipoprotein lipase, and was released from the surface of the cells by heparin. The lipase activity extractable from the cells was sharply increased by previous exposure of the cells to insulin. No lipoprotein lipase developed in confluent 3T3-C2 cells, a clone which, under these conditions, does not undergo adipose conversion.

The lipoprotein lipase of adipose tissue has been demonstrated to play an important role in the utilization of plasma triglycerides. This utilization depends on hydrolysis, since most of the glycerol portion of the triglyceride is not incorporated by the adipose tissue (39,44). Lipoprotein particles are not seen traversing the space between the capillary endothelial cell and the adipose cell (30). In contrast, in the numerous experiments carried out in cultured cells of different kinds, it has nearly always been found that triglycerides are taken up slowly and mainly in intact form, the glycerol portion being incorporated as efficiently as the fatty acid portion (1,23). Since all cultured cells appear to utilize fatty acids very well, it seems reasonable that their mode of handling triglyceride reflects their lack of lipoprotein lipase.

This view is supported by the fact that adipose 3T3-L1 cells utilize triglyceride much more efficiently than other cultured cells, and they utilize the fatty acid portion of external triglyceride much more efficiently than the glycerol portion (56). This was shown in experiments in which the cultured cells were exposed to labeled triolein emulsions (Fig. 3). 3T3-L1 utilized the fatty acid of the triolein about 2 orders of magnitude more quickly than 3T3-C2 cells, but they utilized the glycerol portion of the triolein at only about 1/20th the rate of the fatty acid.

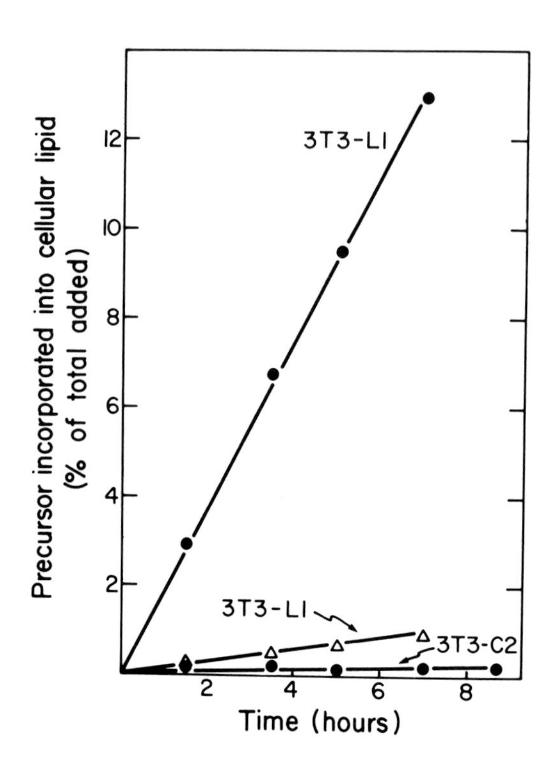

Figure 3 Triglyceride Utilization by Adipose 3T3 Cells

3T3-L1 cells having undergone adipose conversion were exposed to labelled triolein emulsions and the rate of incorporation of label into cellular triglyceride was determined. When the label was in the fatty acid part of the triolein, it was incorporated about 100 times more rapidly by 3T3-L1 cells than by 3T3-C2 cells (solid circles). When the label was in the glycerol portion of the triolein, it was incorporated at only a low rate, even by 3T3-L1 cells (triangles). Data from Wise and Green, (56).

Lipoprotein lipase, seems, then, to be an essential enzyme for the efficient utilization of external triglyceride and develops in 3T3 cells only in the course of their adipose conversion. It may be asked whether the presence of triglyceride in the form of lipoprotein is required for the development of the lipase i. e. whether the lipase is an inducible enzyme or whether it appears as part of the program of differentiation without regard to the presence of its substrate.

This question was examined by following the adipose conversion in medium whose serum lipids had been removed in various ways. These experiments showed that adipose conversion could occur in the absence of serum lipids and that under these conditions lipoprotein lipase could develop normally (56).

It is therefore clear that the differentiation program provides lipoprotein lipase whether the enzyme is able to function or not. The cells are able to accumulate triglyceride whether the fatty acids are obtained from exogenous sources or synthesized endogenously. The manner of obtaining the fatty acids seems to be optional, and to depend on the amount of exogenous lipid available. This strongly implies that the regulation of fatty acid synthesis is a secondary, rather than a primary part of the adipose conversion.

General Conclusions

The 3T3 system seems to be a valid model for the study of fat cell formation. Under simple culture conditions, clones of 3T3 cells develop the essential functions of adipose cells in the absence of any interacting cell type. In adipose tissues, the endothelial cells have an important function, but they are evidently not required for the onset and development of the adipose conversion. Possible areas for future investigation may be grouped as follows:

1. Intrinsic Susceptibility - this is clonally distributed among 3T3 lines and quasi-stably inherited. Genetic or epigenetic factors appear to determine the probability of adipose conversion under standard conditions (15). Perhaps such factors could be determinants of the differing degrees of adipose tissue development that take place in different body sites.

2. Effect of Chemicals and Hormones - as noted above, the adipose conversion can be influenced by many hormones and drugs. Some of these effects are relevant to the field of differentiation and others to the regulation of lipid metabolism. A third category may be seen in the effects of drugs which do not, at least in the concentrations used in animals, affect the adipose tissues but which, at the higher concentrations possible in culture, do have marked effects on the adipose conversion of 3T3 cells (55). This raises the interesting possibility of testing in the 3T3 system for pharmacological effects applicable to other cell types.

3. The Role of Growth - though there has been a great deal of controversy over the relation between growth and differentiation, the adipose conversion is a particularly clear example: growing cells do not convert and fully converted cells do not grow. 3T3 cells are an established line, and the adipose conversion can be postponed indefinitely by maintaining the cells under conditions in which they can grow. Cells begin to convert when their growth is slowed or arrested either in confluent culture or in suspension culture stabilized with methyl cellulose. Though the process can be reversed even after some triglyceride has accumulated, by changing the culture conditions, the cells become unable to

revert to the growing state when triglyceride accumulation passes a certain point.

4. The Program of Differentiation - we seem to be dealing with a change in program from one directed mainly to growth to one designed to maximize triglyceride synthesis. This form of differentiation is not one likely to lead to dominance of the cytoplasm by one or a small number of proteins. On the basis of the enzymes so far studied, the adipose conversion is likely to produce increases in the activity of all the enzymes involved in fatty acid synthesis, utilization of external lipids, the synthesis of glycerol phosphate and its acylation, and ancillary enzymes such as those of the phosphogluconate pathway which, like malic enzyme, produce NADPH for the reduction required in fatty acid synthesis. Perhaps as many as 50 enzymes may be involved. It is therefore obvious that in this "simple" form of differentiation an extensive regulatory system must coordinate the levels of different enzymes. This coordination will naturally depend on the nature of the substrates available for triglyceride synthesis, the hormones acting on the cell etc.

At this stage we cannot say exactly what the differentiation program consists of. The problem is that we cannot yet distinguish primary enzymes affected by the adipose conversion from possible secondary ones whose activity increases through their coupling by regulatory mechanisms to the activity of the primary enzymes. In order to analyze this problem, we will require methods of interfering specifically with secondary enzymes.

REFERENCES

1. Bailey, J. M., Howard, B. V. and Tillman, S. F., J. Biol. Chem., 248 (1973) 1240.

2. Bensadoun, A., Ehnholm, C., Steinberg, D. and Brown, W. V., J. Biol. Chem., 249 (1974) 2220.

3. Cherkes, A. and Gordon, R. S. Jr., J. Lipid Res., 1 (1959) 97.

4. Clark, E. F. and Clark, E. L., Am. J. Anat., 67 (1940) 255.

5. Cohen, R., Pacifici, M., Rubinstein, N., Biehl, J. and Holtzer, H., Nature, 266 (1977) 538.

6. Davidson, R. L. and Kaufman, E.R., Cell, 12 (1977) 923.

7. Diamond, L., O'Brien, T. G. and Rovera, G., Nature, 269 (1977) 247.

8. Englerud, T. and Olivecrona, T., J. Biol. Chem., 247 (1972) 6212.

9. Faust, I. M., Johnson, P. R. and Hirsch, J., Science, 197 (1977) 391.

10. Goldberg, B., Green, H. and Todaro, G., Exptl. Cell Res., 31 (1963) 444.

11. Goldberg, B., Cell, 11 (1977) 169.

12. Green, H. and Goldberg, B., Proc. Natl. Acad. Sci., 53, (1965) 1360.

13. Green, H. and Kehinde, O., Cell, 1 (1974) 113.

14. Green, H. and Kehinde, O., Cell, 5 (1975) 19.

15. Green, H. and Kehinde, O., Cell, 7 (1976) 105.

16. Green, H. and Meuth, M., Cell, 3 (1974) 127.

17. Greenwood, M. R. C. and Hirsch, J., J. Lipid Res., 15 (1974) 474.

18. Grimaldi, P., Negrel, R. and Ailhaud, G., Europ. J. Biochem. (1978) (in press).

19. Hirsch, J. and Knittle, J. L., Federation Proceedings, 29 (1970) 1516.

20. Hoffman, S. S. and Kolodny, G. M., Exp. Cell Res., 107 (1977) 293.

21. Hollenberg, C. H., Amer. J. Physiol., 197 (1959) 667.

22. Hollenberg, C. H. and Vost, A., J. Clin. Invest., 47 (1968) 2485.

23. Howard, B. V., Howard, W. J., Dela Llera, M. and Kefalides, N. A., Atherosclerosis, 23 (1976) 521.

24. Jungas, R. L. and Ball, E. G., Biochemistry, 3 (1964) 1696.

25. Korn, E. D., J. Biol. Chem., 215 (1955) 1.

26. Kuri-Harcuch, W. and Green, H., J. Biol. Chem, 252 (1977) 2158.

27. Mackall, J. C., Student, A. K., Polakis, S. E., Lane, M. D., J. Biol. Chem., 251 (1976) 6462.

28. Mackall, J. C. and Lane, M. D., Biochem. & Biophys. Res. Comm., 79 (1977) 720.

29. Miller, D. A., Miller, O. J., Dev, V. G., Hashmi, S., Tantravahi, R., Medrano, L. and Green, H., Cell, 1 (1974) 167.

30. Napolitano, L., J. Cell Biol., 18 (1963) 663.

31. Olivecrona, T., Bengtsson, G., Marklund, S-E., Lindahl, U. and Hook, M., Federation Proceedings, 36 (1977) 60.

32. Orkin, S. H., Harosi, F. I., and Leder, P., Proc. Natl. Acad. Sci., 72 (1975) 98.

33. Pokrajac, N., Lassow, W. J., and Chaikoff, I. L., Biochem. et Biophys. Acta, 139 (1967) 123.

34. Porter, K. R., Todaro, G. and Fonte, V., J. Cell Biol., 59 (1973) 633.

35. Poznanski, W. J., Waheed, I. and Van, R., Laboratory Investigation, 29 (1973) 570.

36. Reed, B. C., Kaufmann, S. H., Mackall, J. C., Student, A. K. and Lane, M. D., Proc. Natl. Acad. Sci., 74 (1977) 4876.

37. Robinson, D. S. and Wong, D. R., Horm. Metab. Res. Suppl., 2 (1970) 41.

38. Rodbell, M., J. Biol. Chem., 239 (1964) 753.

39. Rodbell, M. and Scow, R. O., Amer. J. Physiol., 208 (1965) 106.

40. Rovera, G., O'Brien, T. G. and Diamond, L., Proc. Natl. Acad. Sci., 74 (1977) 2894.

41. Rubin, C. S. and Lai, E. and Rosen, O. M., J. Biol. Chem., 252 (1977) 3554.

42. Russell, T. R. and Ho, R., Proc. Natl. Acad. Sci., 73 (1976) 4516.

43. Rutter, W. J., Pictet, R. L. and Morris, P. W., Ann. Rev. Biochem., 42 (1973) 601.

44. Schotz, M. C., Stewart, J. E., Garfinkel, A. S., Whelan, C. F., Baker, N., Cohen, M., Hensley, T. J. and Jacobson, M., _In_ Drugs Affecting Lipid Metabolism. W. L. Holmes, L. A. Carlson and R. Poaletti, eds. Plenum Press, (1969) New York. pp 161.

45. Steinberg, D., Mayer, S.E., Khoo, J.C., Miller, E.A., Miller, R.E., Fredholm, B. and Eichner, R., Adv. in Cyclic Nucleotide Res., 5 (1975) 549.

46. Stiles, J.W., Francendese, A.A. and Masoro, E.J., Amer. J. Physiol., 229 (1973) 1561.

47. Stockdale, F., Okazaki, K., Nameroff, M. and Holtzer, H., Science, 146 (1964) 533.

48. Todaro, G. and Green, H., J. Cell Biol., 17 (1963) 299.

49. Todaro, G., Green, H. and Goldberg, B., Proc. Natl. Acad. Sci., 51 (1964) 66.

50. Van, R.L.R., Bayliss, C.E. and Roncari, D.A.K., J. Clin. Invest., 58 (1976) 699.

51. Vaughan, M., J. Lipid Res., 2 (1961) 293.

52. Wasserman, F., in Handbook of Physiol., Section 5, Adipose Tissue, A.E. Renold and G.F. Cahill, Jr. Eds., Amer. Physiol. Soc., Wash. D.C. (1965) pp 87.

53. Wessells, N.K., J. Cell Biol., 20 (1964) 415.

54. Wieland, O. and Suyter, M., Biochem. Z., 329 (1957) 320.

55. Williams, I.H. and Polakis, S.E., Biochem. Biophys. Res. Comm., 77 (1977) 175.

56. Wise, L.S. and Green, H., Cell, 13 (1977) 233.

57. Zucker-Franklin, D. and Grusky, G., J. Cell Biol., 70 (1976) 410a.

58. Eckel, R.H., Fujimoto, W. Yand Brunzell, J.D., Biochem. and Biophys. Res. Comm., 78 (1977) 288.

DISCUSSION

T. BORUN: In December, in Science, there was a report that intravenous injections of glycerol had a pronounced inhibitory effect on the growth of rats and it also affected lipid metabolism. Have you had the opportunity to examine the effect of glycerol in the culture media on the transformation process?

H. GREEN: Yes. Adipose tissue does not utilize glycerol because it does not have glycerol kinase. 3T3 cells whether adipose or non-adipose also do not utilize glycerol appreciably, so I think it has no effect when added in rather high concentration. The glycerol for esterification of fatty acids in this cell type as in the case of the fat cell must be generated endogenously from glycerol phosphate.

T. BORUN: When you do these experiments, do you freeze the cells in glycerol or DMSO?

H. GREEN: Well, when we freeze them we freeze them in glycerol

W.J. RUTTER: I was wondering whether it is possible in your case to use one of the cell lines which forms adipocytes rather readily and co-culture it with another cell line and recruit adipocyte formation from that population by the more efficient converters.

H. GREEN: I can't answer that question satisfactorily. We tried such experiments some time ago but it became evident that the presence of another cell population in the same culture altered the composition of the medium in such a way to effect the adipose conversion in the first cells and so until we decide how that works we can't do that kind of reconstruction.

W.J. RUTTER: The other question is whether the process is reversible.

H. GREEN: Yes, it is reversible up to a certain point. A cell can accumulate a quite obvious amount of fat, and if the culture conditions are changed, let us say if it is transferred, the cell will begin to grow, dilute out the fat and revert in every way to a preadipose cell. However there is a certain point of accumulation of triglyceride beyond which the cell can not revert to the growing stage. That is then an end cell stage just like the adipose cell of mature tissues, and it is difficult to define just what amount of triglyceride accumulation puts the cell over into an end stage cell.

W.J. RUTTER: In the particular case of BrdU, which inhibits the process, is it possible to reverse by simply culturing the cells in the absence of BrdU for a certain number of generations?

H. GREEN: Yes, thats reversible.

R.E. MILLER: We have observed similar changes in glutamine synthetase as those you showed for lipoprotein lipase. We found that the enzyme level increases to a peak and then falls. I wonder if you have an explanation or suggestion for the reason for the fall in the enzyme specific activity?

H. GREEN: No, but I think it must have something to do with regulation. I don't think that it is due to deterioration of the system. I have a number of reasons for thinking that, which are too complicated to go into now, but I think that once the enzymes appear, there are probably compensating regulatory mechanisms which then push the enzyme back to some lower level.

R.E. MILLER: You said that you observed this with the two other enzymes as well, which ones specifically?

H. GREEN: It is true for the glycerol phosphate acyltransferase and lipoprotein lipase.

R.E. MILLER: Do they actually decrease down to the basal level?

H. GREEN: No, they stay higher than they were before the conversion.

C. WEST: In consideration of the question of why cells at low density don't seem to undergo this conversion, will conditioned medium from a culture which is undergoing adipose formation have an effect on low density cells?

H. GREEN: Well thats difficult to explain for the reason that I mentioned to Dr. Rutter.

C. WEST: But this was in the presence of other cells.

H. GREEN: Well, even the preadipose cells deplete the medium of certain important components but your question may be able to be answered in this way: If you take exponentially growing cells and suspend them in culture stabilized with methocellulose, they can't grow like most normal cells but convert to adipose cells. So under those conditions they are not receiving any kind of conditioning effect; the only thing that is common to both systems is that you stop their growth. Thus, it is clear that arresting growth is very closely related to the onset of the adipose conversion, no matter how you do it.

C. WEST: By artificially blocking growth with a drug inhibitor are you able to achieve similar effects?

H. GREEN: Yes, but those are a little more difficult to interpret.

S. WEINHOUSE: I wondered if any of those substances which promote transformation might work in these transformed cells.

H. GREEN: I think that Dr. Russell will be talking in much greater length about a number of these agents. Perhaps he has some information on that.

H.V. RICKENBERG: Regarding the agents which promote and inhibit the conversion of these cells, I believe you find that adrenergic agents and dibutyryl cAMP inhibit while $PGF_{2\alpha}$ and MIX, which is an inhibitor of cAMP phosphodiesterase promote the conversion. There seems to be an anomoly here.

H. GREEN: Well, the first part I can answer. The effect of beta adrenergic agents is quite what would be expected from their effect on fat cells and I would rather leave the other question to Dr. Russell whose experiments I was citing because he will no doubt talk about that and probably answer that question.

P. FEIGELSON: With respect to the insulin effect on the lipoprotein lipase do you believe that insulin is influencing a developmental process or is insulin acting on the cells as it would upon fat cells? For example, is the insulin effect reversible if you remove the insulin?

H. GREEN: Yes, it is reversible but it also does act on the developmental process.

REGULATION OF THE CONVERSION OF 3T3-FIBROBLASTS INTO ADIPOSE CELLS

Thomas R. Russell
Departments of Biochemistry and Medicine
University of Miami School of Medicine
Miami, Florida

3T3 fibroblasts convert into adipose cells and the process appears to be a type of differentiation which can be studied in cell culture. Dr. Green has presented an excellent review of this model system in the previous chapter and I refer interested readers to it. Briefly, 3T3-L fibroblasts differentiate into adipose cells when maintained in a non-growing state (1, 2). Upon conversion into adipose cells at least nine enzymes involved in triglyceride synthesis increase in activity (3-7). In the case of acetyl CoA carboxylase (3) and fatty acid synthetase (8), the increase in enzyme activity is the result of new enzyme synthesis. The number of insulin receptors per cell is 7 fold higher following conversion (9) and the mature adipose cells develop an ACTH-responsive adenylate cyclase (10). The adipose cells respond to hormones that stimulate triglyceride synthesis and decrease triglyceride content (11). Thus, the differentiation into adipocytes involves the activation of a new cellular program which includes the synthesis of new enzyme protein, the formation of a hormone receptor-enzyme complex and an altered cellular metabolism controlled by hormones known to regulate lipid metabolism in normal adipose cells.

Our studies are concerned with the regulation of the conversion of 3T3-L fibroblasts into adipose cells and this chapter will deal with this aspect of the differentiation process. Studies were initially undertaken to determine the role, if any, of the cyclic nucleotides, cyclic AMP and or cyclic GMP, in regulating the adipose conversion. These early studies (12) employed 1-methyl-3-isobutylxanthine (MeiBu-Xan), a synthetic caffeine analog. MeiBu-Xan inhibits cyclic nucleotide phosphodiesterase(s) and increases intracellular cyclic nucleotide levels (13). When confluent cultures of 3T3-L fibroblasts are treated with MeiBu-Xan for eight days, numerous adipose clusters develop while few are evident by day eight in nontreated cultures (12). This observation indicated that MeiBu-Xan is able to accelerate the differentiation process. Recent results (14) have shown that MeiBu-Xan promotes maximal adipose cluster formation by day seven, as no

ISBN 0-12-045450-5

change in cluster formation occurs between day 7 and 21. In control cultures, on the other hand, the number of clusters increase linearly reaching the level of the MeiBu-Xan treated cultures by day 21. A typical adipose cluster in MeiBu-Xan treated cultures on day 7 is similar in size to one on day 21 but the triglyceride content per cell is less (figure 1).

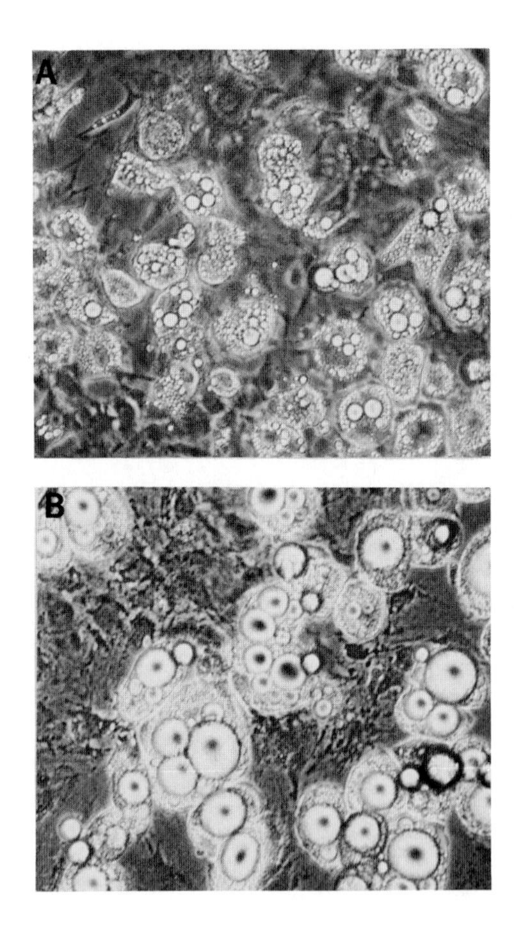

Figure 1. Phase Contrast Micrograph of a typical adipose cluster on day 7 (A) and day 21 (B) from confluent cells triggered to differentiate with MeiBu-Xan from day 0-4.

Since MeiBu-Xan was chosen for these studies because of its ability to increase intracellular cyclic nucleotide levels, it was assumed that other agents that function in a similar manner would also promote the adipose conversion. Yet, other standard agents employed to mimic cyclic nucleotide action were not effective in promoting the adipose conversion (12). These included 8-bromo-cAMP, caffeine, dibutyryl cAMP, 8-bromo-cGMP and dibutyryl cGMP. 8-Bromo-cAMP induced the cAMP phosphodiesterase in 3T3-L cells as did MeiBu-Xan indicating that the cells were able to respond to 8-bromo-cAMP, however, it was not an effective promoter of the adipose conversion under these experimental conditions. It is conceivable that cyclic nucleotides may be involved in controlling the differentiation of 3T3-L fibroblasts into adipose cells and that more refined experimental procedures now being developed in our lab may shed more light on this question. As of now the role, if any, of either cAMP or cGMP in the regulation of the differentiation is unclear. MeiBu-Xan, nevertheless, is a very potent stimulator of the process.

MeiBu-Xan may function as an inducer of differentiation independent of the cyclic nucleotides. It is of interest that numerous purine derivatives structurally similar to MeiBu-Xan have been shown to induce the differentiation of erythroleukemia cells into hemoglobin producing cells (15). We have recently shown that MeiBu-Xan is a potent stimulator of granulopoeisis (16) increasing the ratio of granulocytes to macrophages from 1.0 in control cultures to 4 in MeiBu-Xan treated cultures. Thus, in three experimental systems, the conversion of 3T3-L fibroblasts into adipose cells, of erythroleukemia cells into hemoglobin producing cells and of precursor stem cells into granulocytes structurally similar purine derivatives (all closely related to hypoxanthine) are able to accelerate a differentiation process. Delineation of the mechanism of action of these purine derivatives be it through the cyclic nucleotides or not should yield important information on early events that occur when a new cellular program is activated.

A terminally differentiated cell is a stable cell type that remains as such for its life cycle. Thus, agents that promote early events in differentiation may work irreversibly since once the inducer has switched the cell program from one type to another it is no longer needed. In the specific case of the conversion of 3T3-L fibroblasts into adipose cells our first observations (12) revealed that MeiBu-Xan accelerated the adipose conversion as numerous adipocytes developed in the presence of the drug by day eight. Studies were then undertaken to determine the effect of a two day treatment with inducer on the ultimate appearance of adipose cells. Cells treated for two days with MeiBu-Xan, rinsed to remove inducer and then fed with medium minus MeiBu-Xan still showed a maximal conversion by day 7-8 (12). Thus, the presence of MeiBu-Xan from days 0-2 was sufficient to trigger a process which morphologically was not apparent for 3-4 days later. At the biochemical level fatty acid synthetase is not altered during the MeiBu-Xan treatment, yet when the cells are ready to express fatty acid synthetase (around day 8) its activity increases 10-fold over that in cells not exposed to MeiBu-Xan (8). These results indicate that continuous addition of MeiBu-Xan to the culture medium is not required to elicit morphological and biochemical expression of differentiation which occurs days after the inducer treatment is discontinued.

3T3-L fibroblasts spontaneously differentiate into adipose cells in the absence of treatment with MeiBu-Xan. 3T3-4 cells, on the other hand, do not show a high degree of conversion in 3 weeks yet some cluster development is evident. It was thus of interest to ascertain the role of an inducer in such a low probability cell line. MeiBu-Xan is a very effective promoter of the conversion (figure 2) in 3T3-4 cells. The effect is very similar to that seen in highly probable lines such as 3T3-L i.e. short treatment with MeiBu-Xan induces differentiation which is manifest morphologically long after removal of MeiBu-Xan from the culture medium. The reason why some clones convert at a high probability and others at a low probability is not understood. Nevertheless, in the susceptible cells in either case it appears that the conversion is inducible in an identical manner by MeiBu-Xan. 3T3-C2 is a cell line with a zero probability of conversion. In these cells MeiBu-Xan is totally ineffective indicating that the drug can only function as an inducer in lines that are able to recognize it is a trigger.

The induction of differentiation by MeiBu-Xan is a phenomenon occuring at the level of the intact cell and thus information on its mechanism of action is lacking. Since the conversion of 3T3-L fibroblasts into adipose cells does involve an increase in the activity of at least nine enzymes involved in triglyceride synthesis two of which have been shown to involve new enzyme synthesis (3, 8), it is reasonable to assume that such a change in the cellular program requires an alteration in chromatin expression. It is experimentally possible to alter chromatin expression with the thymidine

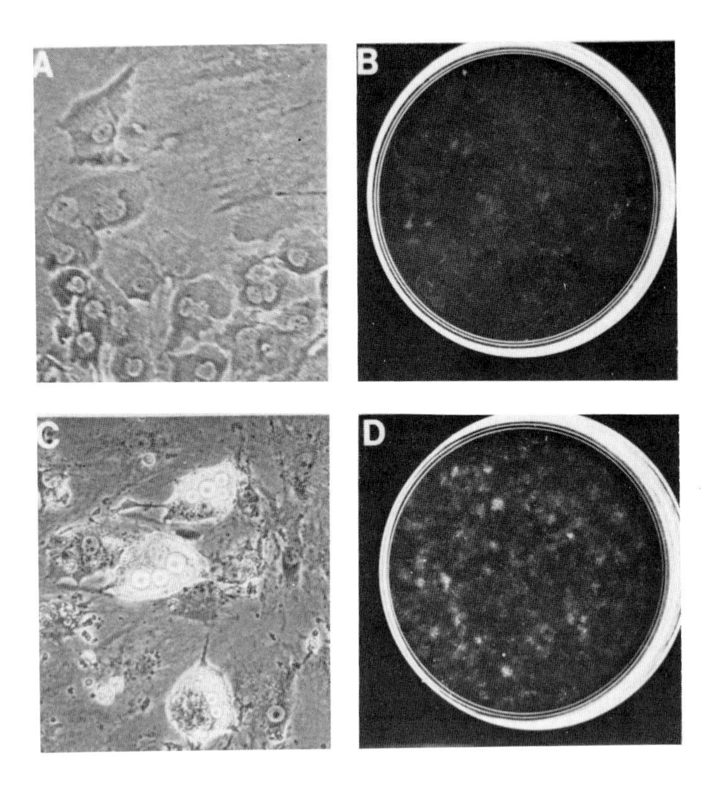

Figure 2. Effect of MeiBu-Xan on the conversion of 3T3-4 fibroblasts into adipose cells. Phase contrast micrograph and photograph respectively, of control (A, B) and MeiBu-Xan triggered cultures (C, D).

analog, 5-bromodeoxyuridine (BrdUrd) and then to test the effect of inducer on the altered chromatin. BrdUrd functions in procaryotes to cause the lac repressor to bind with higher affinity to the lac operator DNA than in unsubstituted DNA (17). However, an increase in the concentration of the inducer of the lac operon is sufficient to reverse the BrdUrd effect on repressor binding. Thus, BrdUrd-DNA and inducer can be thought of as competitors for repressor. BrdUrd leads to higher affinity of the repressor for DNA but the inducer at higher concentrations than required in unsubstituted DNA can compete with and remove the repressor from the DNA. It is apparent that the replacement of the methyl group in the 5 position of thymidine with bromine leads to higher affinity of a regulatory protein to DNA.

In eucaryotes, growth of cells in the presence of BrdUrd results in the suppression of cellular differentiation (18). BrdUrd also blocks the conversion of 3T3-L fibroblasts into adipose cells (2). Results from our laboratory (14) have recently shown that thymidine in the culture medium in excess of BrdUrd is able to prevent the effect of BrdUrd on the conversion which supports the contention that the analog must be incorporated into DNA to exert its block on cytodifferentiation. This is supported by the fact that the addition of BrdUrd to non-growing cells does not block cytodifferentiation (14). Thus, neither metabolites of BrdUrd nor BrdUrd itself can block differentiation simply by being present in the cells. The analog apparently must be incorporated into DNA. Recalling the procaryotic model it is of interest to determine if chromosomal proteins which have been postulated to be involved in the regulation of differentiation and gene expression bind more or less tightly to DNA substituted with BrdUrd. Recent evidence from two laboratories (19, 20) suggest that both histone and non-histone chromosomal proteins bind more tightly to BrdUrd substituted-DNA than non-substituted DNA. It has recently been postulated that BrdUrd blocks cytodifferentiation by causing putative regulatory molecules to bind more tightly to DNA. Since MeiBu-Xan is able to accelerate the converison of 3T3-L fibroblasts into adipose cells, studies were undertaken to determine the effect of BrdUrd and MeiBu-Xan on the process of differentiation in this model system. A time course showed that BrdUrd was able to block the spontaneous differentiation seen over 21 days in control cultures (14). Thus, the cells became apparent low probability cell lines similar to those already isolated. However, when the cells grown in BrdUrd were treated with inducer a partial recovery of the adipose conversion was obtained (14). In other words, an agent acting as an inducer can override a BrdUrd dependent block in cytodifferentiation. This is the first demonstration of an inducer functioning to override a BrdUrd dependent suppression of differentiation. As these results are a phenomena seen at the cellular level they have to be interpreted with caution but suggest that an understanding of MeiBu-Xan action may yield important information on the molecular mechanism of how cells differentiate. Based on these preliminary observations it is interesting to speculate that BrdUrd could block the adipose conversion by preventing the release of certain regulatory molecules that must dissociate (or rearrange) in order for a new cellular program to become activated. The different probability cell lines would then be manifestations of a different affinity of these regulatory molecules for DNA. BrdUrd functions to cause tight binding of the regulatory molecules to DNA yielding low probability lines even when starting with a highly probable cell line. The inducer then functions to release by competition the putative regulatory molecules thereby promoting differentiation. This is an attractive model based on the known mechanism of action of BrdUrd in procaryotes and the new information on BrdUrd in eucaryotes (19, 20) and the known action of MeiBu-Xan as an inducer of the adipose conversion (12, 14). Certainly other models including non-DNA linked action of BrdUrd cannot be completely ruled out at this stage. Further work is being directed toward determining the molecular mechanism of action of MeiBu-Xan in triggering differentiation.

REFERENCES

1. Green, H. and Kehinde, O. (1974) Cell 1 113-116.

2. Green, H. and Meuth, M. (1974) Cell 3 127-133.

3. Mackall, J.C., Student, A.K., Polakis, S.E. and Lane, M.D. (1976) J. Biol. Chem. 251 6462-6464.

4. Kuri-Harcuch, W. and Green, H. (1977) J. Biol. Chem. 252 2158-2160.

5. Mackall, J.C. and Lane, M.D. (1977) Biochem. Biophys. Res. Commun. 79 720-725.

6. Eckel, R.H., Fujimoto, W.Y. and Brunzell, J.D. (1977) Biochem. Biophys. Res. Commun. 78 288-293.

7. Wise, L.S. and Green, H. (1978) Cell 13 in press.

8. Ahmad, P., Russell, T.R. and Ahmad, F. manuscript in preparation.

9. Reed, B.C., Kaufmann, S.H., Mackall, J.C., Student, A.K. and Lane, M.D. Proc. Nat. Acad. Sci., U.S.A. 74 4976-4980.

10. Rubin, C.S., Lai, E. and Rosen, O. (1977) J. Biol. Chem. 252 3554-3557.

11. Green, H. and Kehinde, O. (1975) Cell 5 19-27.

12. Russell, T.R. and Ho, R.J. (1976) Proc. Nat. Acad. Sci., U.S.A. 73 4516-4520.

13. Schultz, G. , Hardman, J.G., Schultz, K., Davis, J.W. and Sutherland, E.W. (1973) Proc. Nat. Acad. Sci. U.S.A. 70, 1721-1725.

14. Russell, T.R. submitted for publication.

15. Gusella, J.F. and Housman, D. (1976) Cell 8 263-270.

16. Miller, A., Russell, T.R. and Yunis, A.A. submitted for publication.

17. Lin, S. and Riggs, A.D. (1971) Proc. Nat. Acad. Sci., U.S.A. 69 2574-2576.

18. Rutter, W.J., Pictet, R.L. and Morris, P.W. (1973) Ann. Rev. Biochem. 42 601-646.

19. Gordon, J.S., Bell, G.I., Mortinson, H.C. and Rutter, W.J. (1976) Biochemistry 15 4778-4786.

20. Schwartz, S. (1977) Biochemistry 16 4101-4108.

DISCUSSION

H.V. RICKENBERG: How rigorously have you excluded an involvement of cyclic AMP in your phenomenon, for example, have you measured the cellular concentration of cyclic AMP after the administration of MIX.

T.R. RUSSELL: Cyclic AMP may not be the mediator as dibutyryl cAMP or 8-bromo-cAMP do not work under our experimental conditions. We cannot look at cyclic nucleotide levels in these cells because we are dealing with a homogeneous population of cells and are changing it into a heterogeneous population and it is not possible to determine in which cell type cAMP levels may be changing.

H.V. RICKENBERG: You are saying that the percentage of cells that is converted is not sufficiently high to permit you to detect even relatively slight increases in cyclic AMP?

T.R. RUSSELL: The initial conversion occurs very early on, and I do not feel that we could detect changes in cAMP levels in the preadipose cells in the high background of fibroblasts present.

H.V. RICKENBERG: A related question - Have you tried to administer cyclic AMP or dibutyryl cyclic AMP repeatedly or in a pulsative manner?

T.R. RUSSELL: You mean adding and taking it off and then adding it again?

H.V. RICKENBERG: No, just re-adding it.

T.R. RUSSELL: No.

J.M. CARDENAS: Have you considered the possibility of MIX being incorporated into DNA and therefore, functioning in essentially an analogous but reverse way from BrdU?

T.R. RUSSELL: We have considered that, and we will be receiving radioactive MIX and hope to be able to attempt to study this. We should be able to see what MIX is metabolized into and whether or not it is incorporated into DNA.

H. GREEN: I think I can answer one of the questions that was just raised. We could not detect any effect of DMSO on the adipose conversion in 3T3 cells. But I would like to ask Tom Russell, what is the effect of MIX on mature adipose cells of adipose tissue?

T.R. RUSSELL: MIX is a lipolytic agent in adipose tissue.

R.A. HICKIE: One agent missing from your list of phosphodiesterase inhibitors is papaverine. Have you tried this agent?

T.R. RUSSELL: No we have not.

C. WEST: It seems to me that this process of conversion could be broken down into two processes, conversion itself, and then the expression of that conversion. Have you attempted to distinguish between those two possibilities and the action of MIX on either one or both of these?

T.R. RUSSELL: You mean is the MIX effect after the conversion rather than prior to the conversion?

C. WEST: Yes.

T.R. RUSSELL: MIX certainly programs the cells to express both morphological and biochemical changes that don't occur until long after removal of MIX from the cultures. However, at this time the exact point of MIX action is unknown.

D. GERSHON: If 3T3 cells have a heterogenous karyotype is there any correlation between the capacity to differentiate and chromosome constitution? This is asked also of Dr. Green. The second question is in view of the effect of insulin - could this stimulation of the various enzymes that you have shown be attributed to a decrease in the rate of degradation?

T.R. RUSSELL: There is no correlation between chromosome number and the ability of the cells to differentiate. I do not believe that the increase in enzyme activity could be due to a decrease in the rate of degradation but this has not been investigated. Certainly fatty acid synthetase is not detected immunologically before the conversion but is following the conversion.

M.M. RASENICK: You don't show data for the effects of dibutyryl cAMP plus MIX or for the sole addition of other methyl xanthines (e.g. theophylline). What effect might these other methyl xanthines have and might cAMP analogues inhibit this MIX induced transformation?

T.R. RUSSELL: Caffeine and hypoxanthine do not have any effect alone at least under our experimental conditions. cAMP analogues certainly inhibit the synthesis of fatty acid synthetase but this most likely is due to the effect of cAMP on inhibition of the induction of FAS and not necessarily due to an ability of cAMP to block differentiation.

M.M. RASENICK: Additionally, I suggest that in order to determine the effect of MIX on intracellular cAMP you might employ the immunocytochemical localization technique of Steiner and his colleagues. This should allow you to determine in a quantitative manner whether there is a MIX induced increase in intracellular cAMP levels and should indicate the levels of such a change.

T.R. RUSSELL: That is certainly possible. We have not done such experiments.

R.E. MILLER: Do you believe that the presence of insulin is required for the MIX mediated potentiation of adipocyte conversion?

T.R. RUSSELL: No, addition of insulin is not required for the MIX effect. However, low levels of insulin are present in serum and thus a role of insulin in the MIX effect has not been completely ruled out.

R.E. MILLER: I would like to comment, Mr.Chairman, if I may. We have measured insulin in commercially available calf serum by radioimmunoassay. We find that the concentration can vary among batches of calf serum by 10-fold or more.

T.R. RUSSELL: To what level?

R.E. MILLER: We find that the immunoreactive insulin concentration in calf serum varies from less than 0.2 ng/ml to a maximum of 2 to 3 ng/ml. Adipocyte conversion is markedly diminished or fails to occur in our confluent 3T3-L1 cultures maintained for 20 days in culture medium containing less than 0.02 ng/ml immunoreactive insulin (due to 10% calf serum). In contrast, there is significant adipocyte conversion in similar cultures maintained for 20 days after confluence in medium containing more than 0.1 ng/ml immunoreactive insulin (due to 10% calf serum). This suggests to us that insulin may be required for the conversion of 3T3-L1 cells to adipocytes.

T.R. RUSSELL: Insulin certainly does not function in the irreversible way that MIX does in promoting the conversion but it may be necessary for the conversion to occur.

THE USE OF TRANSFORMED FIBROBLASTIC CELLS AS A MODEL FOR DIFFERENTIATION

I. PASTAN, M. SOBEL, S. ADAMS, K.M. YAMADA, B. HOWARD
M. WILLINGHAM and B. DE CROMBRUGGHE
Laboratory of Molecular Biology
Division of Cancer Biology and Diagnosis
National Cancer Institute
National Institutes of Health
Bethesda, Maryland 20014 U.S.A.

Abstract: Cultured fibroblastic cells have a characteristic elongated and flattened appearance. Transformation of these cells, particularly by RNA tumor viruses, produces a marked and characteristic change in their appearance. This change is referred to as the morphologic phenotype of transformation. The cells become rounded and their surfaces become covered with microvilli. The microvilli probably account for the increased ability of plant lectins to agglutinate these transformed cells. Our evidence indicates that the initial step in leading to the morphologic phenotype of transformed cells is a decrease in adhesion to substratum. Decreased adhesion in RSV transformed CEF is in large part due to a decrease in the content of a major cell surface adhesive protein (CSP).

The mechanism of the decrease in CSP has been examined in detail and is due to a viral function (sarc) which decreases the level of translatable mRNA for the protein. The alteration in gene activity seen after transformation is closely related to changes that have been observed during some types of differentiation and is a simple model for studies on differentiation.

INTRODUCTION

During differentiation cells take on or lose the ability to make specific proteins, but the mechanism by which cells undergo this process is unknown. Thus information from simpler systems where detailed mechanistic studies can be carried out is obviously desirable. Some ideas on how differentiation could occur have been generated from observations on how gene activity is controlled in bacteria (1). In *Escherichia coli* specific proteins interact with single genes or with common sites on different genes to initiate or repress transcription. One such system is controlled by cyclic AMP. The cyclic AMP receptor protein (CRP) will bind to specific regions of DNA in the presence of cyclic AMP and enable RNA polymerase to initiate transcription at specific promoters (1).

Another system that promises to provide useful information about differentiation is the Rous sarcoma virus-transformed chick embryo fibroblast. Cultured chick embryo cells, like many other normal cultured fibroblastic cells, make large amounts of collagen and a high molecular weight surface protein (CSP) (2). This latter protein is also called LETS (large external transformation sensitive) protein or fibronectin (reviewed in 2). Upon transformation of chick embryo cells by Rous sarcoma virus, their content of collagen and CSP dramatically falls (reviewed in 2 & 3). Our experiments indicate that the low content of CSP is in part responsible for the altered appearance and behavior of RSV-transformed chick embryo cells. Our experiments also indicate that Rous sarcoma virus exerts its action at or close to the gene level to decrease collagen and CSP synthesis.

The Phenotype of the Transformed Cell

RSV transformed chick embryo cells can readily be distinguished from normal CEF by their rounded shape and low adhesion to substratum. Indeed many types of transformed cells have this type of altered appearance. When examined by scanning electron microscopy, the surfaces of the transformed cells are covered with microvilli, whereas the surfaces of normal cells are relatively smooth.

Because CSP is a surface protein and because it is one of the most prominent proteins that is diminished in transformed cells, Yamada *et. al* (4) added purified CSP to a culture of RSV transformed chick embryo cells. Within 24 hours the transformed cells had acquired the appearance of normal CEF (4). Further as growth proceeded, the CSP-treated transformed cells assumed the orderly growth pattern of parallel alignment of cells that characterizes normal cells. Experiments with [^{14}C]-amino acid labelled CSP indicated that the protein was actually reconstituted on the surface of the transformed cells. Other experiments with fluorescent CSP derivatives showed that the added CSP was present in the same fibrillar pattern that endogenously synthesized CSP assumed in normal cells (5).

A major role for CSP is that of an adhesive protein. When cellular adhesion is increased most transformed cells appear capable of assuming a normal flattened appearance (6). Microvilli are lost as the cells shift from a round to a flattened shape (reviewed in 6).

Olden and Yamada (7) used standard amino acid incorporation techniques to investigate the basis of the low CSP content of transformed chick embryo cells. They observed that such cells had a 4-6 fold decrease in the synthesis rate of CSP (7). Thus the major reason for low CSP levels in RSV-transformed cells is a decrease in synthesis rate.

Adams *et. al* (3) have used another approach to determine the basis of the low rate of CSP synthesis. They extracted RNA from normal and RSV-transformed chick embryo cells and compared the ability of these RNAs to promote CSP synthesis in a cell-free extract prepared from wheat germ. RSV-transformed CEF have about a 5 fold decrease in their capacity to translate two proteins. One is CSP; the other is collagen. In the studies of Adams *et. al* (3) the functional capacity of the RNA preparations to promote protein synthesis was measured. The diminished activity observed could have been due to a decreased rate of RNA synthesis or an increased rate of RNA degradation, or a failure in processing of the RNA.

To assess more directly the number of RNA molecules of collagen mRNA present, Howard *et. al* (8) transcribed preparations of collagen mRNA into double stranded DNA using purified reverse transcriptase. The preparations of DNA were then restricted with various restriction enzymes and the digestion products separated by polyacrylamide electrophoresis. DNA fragments that were characteristic of and derived from collagen mRNA were detected. These DNA fragments were decreased when mRNA from RSV-transformed chick embryo cells was examined. Thus we conclude that the decrease in functional collagen mRNA in RSV-transformed CEF is not due to an alteration that makes the RNA untranslatable. Rather it is more likely that the mRNA is structurally intact but diminished in quantity.

Current efforts are directed towards preparing large amounts of collagen and CSP DNA by gene cloning techniques. This should allow us to make precise measurements of the content of collagen and CSP mRNA in RSV-transformed cells and determine the molecular basis of this alteration as well as permitting study of the regulation of collagen and CSP in development.

REFERENCES

(1) I. Pastan and S. Adhya. Bacteriol. Rev. 40 (1976) 527.

(2) K.M. Yamada and I. Pastan. Trends in Biochemical Sciences 1 (1976) 222.

(3) S.L. Adams, M.E. Sobel, B.H. Howard, K. Olden, K.M. Yamada, B. de Crombrugghe and I. Pastan. Proc. Natl. Acad. Sci. USA 74 (1977) 3399.

(4) K.M. Yamada, S.S. Yamada and I. Pastan. Proc. Natl. Acad. Sci. USA 73 (1976) 1217.

(5) K.M. Yamada, submitted, 1977.

(6) M.C. Willingham, K.M. Yamada, S.S. Yamada, J. Pouyssegur and I. Pastan. Cell 10 (1977) 375.

(7) K. Olden and K.M. Yamada. Cell 11 (1977) 957.

(8) B.H. Howard, S.L. Adams, M.E. Sobel, I. Pastan and B. de Crombrugghe. J. Cell Biol. 75 (1977) 345a.

DISCUSSION

H. BUSCH: I think that the work you have reported is exceedingly interesting. It relates to some work that we have done on ribosomal protein mRNA which apparently turns over very rapidly. In the normal liver it is present in extremely small amounts, but in hepatomas it is present in very large amounts. With regenerating liver which represents a growing liver, in which cell division occurs at very high rates, there is a huge increment in the amount of mRNA for ribosomal proteins. What this means is that in G_0 there is a very low amount of mRNA for synthesis of ribosomal proteins and since your cells may be in G_0 as compared to fibrosarcoma cells which are clearly actively cycling, is it possible that your LETS or cell surface protein mRNA turnover is related to the phase of the cell cycle, rather than to differentiation *per se*?

I. PASTAN: I think it is not likely to be the case. Something that most people do not seem to realize, those who do not work with Rous sarcoma virus transformed chick cells, is that under standard conditions there is absolutely no difference in the growth rate in tissue culture, or the saturation density of Rous sarcoma chick cells relative to the untransformed cells. The situation is very different, say, from 3T3 cells, where the normal cells really stop growing when the cells get crowded together.

H. GREEN: Since the collagen molecule, at least the helical portion of it is so unusual, you ought to be able to easily show from the nucleotide sequence of DNA how it corresponds to the collagen molecule. Have you done that?

I. PASTAN: Yes, we have actually done sequencing of some of those fragments. We went into it with more enthusiasm than we should have. It is very likely the fragments we are generating are in the processed region of the molecule - collagen is processed at both ends, amino and carboxyl. The sequence of one of the fragments that we actually sequenced did not look like the structure of an RNA coding for glycine proline repeats you would expect from the center of the molecule. So that approach has not worked. We plan to clone fragments of the gene and by hybridization methods to show whether the fragment is derived from collagen mRNA.

P. FEIGELSON: Some of our recent studies have led to similar conclusions. We have shown that two proteins that are synthesized exclusively in the liver are not synthesized by any of the rat Morris minimal deviation hepatomas we have examined. We have explored the biochemical basis for the deletion of these proteins which are consequences of neoplastic transformation and the subsequent progression that ensues. Heterogeneous mRNA dependent translational assay systems enabled us to demonstrate that although mRNA isolated from hepatomas was active in directing total protein synthesis, these mRNAs are devoid of detectable levels of functional mRNA coding for tryptophan oxygenase or alpha 2u globulin. We have also recently prepared a pure cDNA probe for alpha 2u globulin. Using this probe we have shown that the gene for alpha 2u globulin is not due to generation of a functionally inactive form of this mRNA, nor to a defect in its processing but rather following neoplastic transformation the genome for alpha 2u globulin becomes transcriptionally silent (Ramanarayanan-Murthy, Colman, Morris, Feigelson, Cancer Res. 36: 3594-3599, 1976; Sippel, Kurtz, Morris, Feigelson, Cancer Res., 36:3588-3593, 1976).

I. PASTAN: Yes, it seems as if the old saw I learned in medical school that cancer is characterized by a failure in differentiation is probably correct.

E. ADELBERG: Would you comment on what you think is the relation between adhesion and lectin agglutinability? Is it the protein that does it, or are you saying that adherence to the substrate changes agglutinability?

I. PASTAN: Well, we have a view of lectin induced agglutinability that differs from others. I think no view is generally accepted but I think ours is correct. I think that the ability of transformed cells to agglutinate in a standard adhesion assay where you compare virus transformed cells to parent, is due to microvilli on the cell surface. These occur when a cell is round. It is true of mitotic cells, trypsinized cells and a mutant cell line isolated in my laboratory from 3T3 cells. Mechanically it is easy to stick two cells together that are covered with microvilli. Now we suggest that when CSP is present on cells, the cells tend to be flat and their surfaces tend to be smooth.

E. ADELBERG: Would they retain that state even in suspension?

I. PASTAN: The question would be that when you take a cell that is flat, such as a 3T3 cell and you do an agglutination assay, how long does it take for the surface to deform and either develop microvilli or blebs or other irregularities and what would be the agglutinability then? With 3T3 cells if you wait thirty minutes or so after removal and then do an agglutination assay you would find that you would not be able to distinguish between a normal and a transformed cell.

E. ADELBERG: Have you done that delayed experiment?

I. PASTAN: Yes, we have done that experiment with 3T3 cells. If you keep 3T3 cells around a while, before the agglutination test is performed, then their surfaces start to get rough and then they agglutinate more easily.

E. SILVERSTEIN: Would you clarify whether the Sarc protein is present normally in the host cell, and then what happens in the several types of transformants that you showed?

I. PASTAN: Well, the Sarc that I referred to would be a product of the Rous sarcoma viral genome, and would only be present in cells that were infected with transforming Rous sarcoma virus.

E. SILVERSTEIN: Is this strictly a Rous system?

I. PASTAN: The data that I spoke about today and the data presented by Erickson and others, are of the Rous system.

VIRAL TRANSFORMATION OF PANCREATIC BETA CELLS IN CULTURE

WALTER A. SCOTT, J. D. HARRISON and D.H. MINTZ
Departments of Biochemistry and Medicine
University of Miami School of Medicine
P.O. Box 875
Miami, Florida 33152 U.S.A.

The availability of beta cell lines capable of expression of differentiated function could contribute substantially to our understanding of growth and differentiation of pancreatic islet cells. Spontaneously occurring (1,2) and radiation-induced islet cell tumors (3) have been investigated in an attempt to develop such lines. It has not yet, however, been possible to sustain in continuous cell culture both proliferative and differentiated functions.

Viral transformation of neonatal rat pancreatic monolayer cultures has provided us an alternative experimental system. In these cultures cells are attached to a substratum and are therefore readily accessible to virus suspensions added in place of culture medium. Theoretically, the possibility for cloning and amplification of differentiated tissues by viral transformation is very attractive, and this approach has been successfully applied to mouse hypothalmic neurosecretory cells (4), hamster pineal cells (5), mouse astrocytic neuroglial cells (6), human vascular endothelium (7), and various differentiated chicken tissues (8,9). We have isolated a number of cell lines by transforming rat neonatal pancreatic cells with simian virus 40 (SV40) and are currently characterizing these cell lines in hopes that some of them will maintain differentiated endocrine cell functions or can be induced to express those functions.

Transformation of Neonatal Rat Pancreatic Endocrine Cells by Simian Virus 40.

Monolayer cultures of pancreas endocrine cells were established from trypsin, collagenase dispersed neonatal rat pancreas (10). The cultures were depleted of fibroblasts by seeding the dispersed tissue in culture medium (Medium 199 containing 10% calf serum and 16.7 mM glucose) and decanting into fresh dishes after 15 hours. During this period the fibroblasts attach more rapidly than the endocrine

ISBN 0-12-045450-5

cell clusters and the cells which attach to the plastic in the second set of dishes are enriched for endocrine tissue.

Two days after the decanting, the medium was changed to Dulbecco's modified Eagle's medium (DMEM) containing 10% calf serum and 5.6 mM glucose. After 24 hours, this medium was removed and simian virus 40 (10^8 plaque forming units per 60 mm dish) was added in the same medium except that the glucose concentration was adjusted to 16.7 mM. This protocol promotes exocytosis in pancreatic beta cells and secretion of insulin, and it is expected that this might provide favorable conditions for the uptake of virus particles which become attached to the surface of these cells. If this occurs, virus infection may be specifically targeted to the cell type which is active in secretion. It has not yet been possible to prove that such targeting occurs, but shortly after viral infection of a culture which contains fibroblasts, viral specific protein (tumor antigen) is seen in endocrine cell clusters but not in fibroblasts.

By 20 to 30 days after viral infection, there was clear evidence of cell proliferation in most infected cultures, but none in cultures that had been mock-infected. The majority of the cultures were subcultured by 45 days after infection and were subsequently subcultured as they approached confluence. Growth rate increased with repeated passage, eventually reaching a doubling time of 18 to 24 hours. These transformed cells will grow in Eagle's minimal essential medium, medium 199, and Ham's F12 medium (all supplemented with 10% calf serum and 16.7 mM glucose); but the growth rate is slightly faster in DMEM, the medium in which they were isolated.

Virus tumor antigen (T-antigen) was present in many cells at the time of first subculture passage and after several subculture passages the cells were 90 to 100% T-antigen positive. Clones of these cells have been isolated which are also T-antigen positive.

Infectious virus was rescued from one cell line and can presumably be rescued from other lines as well. A cell line previously purified by three successive single colony isolations was cocultivated with African green monkey kidney cells (BSC-1 cell line) in a ratio of one transformed cell to five monkey cells. Cell fusion was induced by addition of Sendai virus (ultraviolet-inactivated virus supplied by Mircobiological Associates) method of Davidson (11). After incubation for 1 month a lysate was prepared by freezing and thawing. The presence of infectious virus in this lysate was demonstrated by plaque assay on BSC-1 cell monolayers.

Further evidence that these cell lines are stably transformed includes: (a) injection of transformed cells into immune-deficient rats results in the production of tumors, (b) transformed cells proliferate under conditions where they cannot attach to a solid substrate (i.e., when they are suspended in medium containing 1.2% methyl cellulose), and (c) transformed cells have been serially passaged through fifty subcultures over a period of a year.

Production of Insulin-Immunoreactive Products by Transformed Cell Lines.

In spite of great increases in cell number shortly after viral transformation of pancreatic endocrine cell cultures, insulin content (measured by radioimmunoassay on acid-ethanol extracts of the cells) increases only slightly. After 42 days in culture (prior to the first subculture passage), content of five independently infected cultures ranged from 39 ng to 135 ng of insulin per 60 mm culture (average 99 ng per culture). A comparable mock-infected culture contained 41 ng of insulin. After the first subculture passage a comparable collection of cultures which had regrown to near confluence ranged from 6 ng to 25 ng of insulin per 60 mm culture (average 11 ng per culture). The mock-infected culture of this age could not be successfully subcultured for comparison.

At least part of this apparent retention of differentiated function after the first subculture passage may be explained as a "helper" effect by which cells which cannot be subcultured by themselves are capable of reattaching to a culture dish and retaining their differentiated functions in the presence of healthy cells which provide nutrients or other favorable conditions. To test this possibility, SV40 - transformed Balb 3T3 cells were added to a mock-infected rat pancreatic endocrine monolayer culture and the mixed culture was trypsinized and subcultured. The SV40-transformed Balb 3T3 cell line by itself contained undetectable insulin (less than 1.2 ng per 60 mm culture in an acid-ethanol extract), yet the mixed cell subculture (after it had grown to confluence) contained 16 ng of insulin in an acid-ethanol extract. This possible "helper" effect of transformed undifferentiated cells has not been further explored since, in any event, acid-ethanol extractable insulin was not detected after the second subculture passage in any of our cell lines.

Our belief that the transformed cells growing in these cultures are derived in at least some cases from pancreatic beta cells is due to two additional pieces of evidence.

1) Neisor et al. (12) who have described cell lines with properties similar to our transformed cell lines, have shown that tumors produced by the injection of these cells

into immune-deficient rats, regain the ability to produce acid-ethanol extractable insulin, even though the cells initiating the tumor had not contained acid-ethanol extractable insulin for many cell culture passages. In addition, when cell suspensions derived from these tumors were allowed to incorporate ^{3}H-leucine, a radioactive polypeptide was synthesized which bound to anti-insulin antibody (covalently attached to sepharose 4B) and which comigrated with authentic rat insulin when chromatographed on Sephadex G50.

2) If cultures of transformed cells are extracted by sonication in glycine buffer (0.2 M glycine buffer, pH 8.8 containing 0.125% (w/v) human serum albumin and 1% (v/v) normal lamb serum) rather than by extraction with acid-ethanol, a soluble immunoreactive insulin-like material is obtained. This material has been detected in nearly every transformed pancreatic cell line we have studied, but not in SV40-transformed Balb 3T3 fibroblasts.

Figure 1.

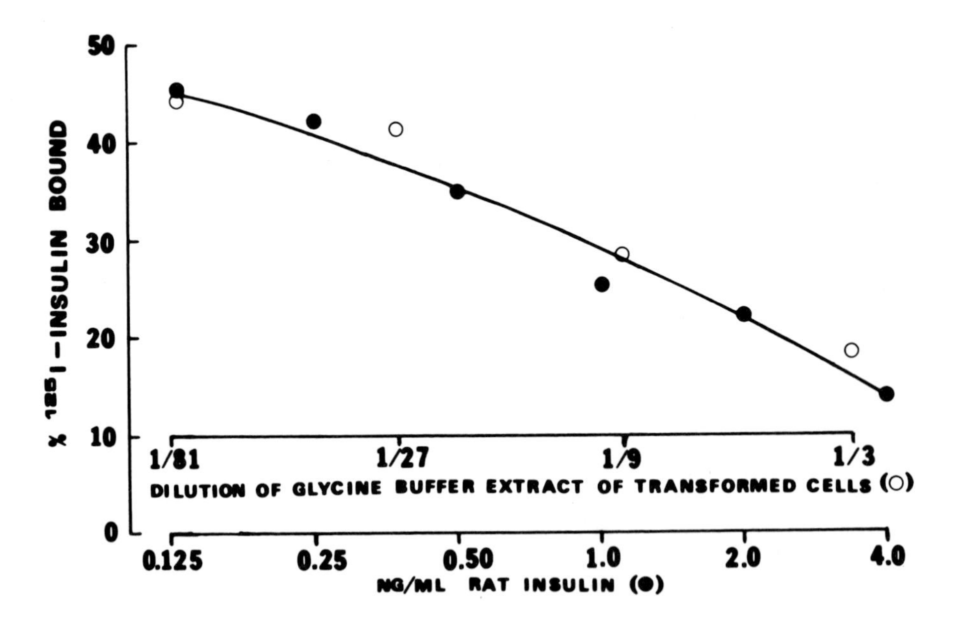

Inhibition of binding of ^{125}I-insulin to guinea pig anti-insulin antiserum by purified rat insulin and by glycine buffer extracts of SV40-transformed pancreatic endocrine cells. Binding was determined by the method of Herbert et al. (13) in a total incubation volume of 1.0 ml for each determination.

Figure 1 shows that increased concentration of glycine buffer extract from transformed cells in the radioimmunoassay results in increased displacement of iodinated insulin from the anti-insulin antibody in parallel with the similar displacement by authentic rat insulin (used in the standard curve). Although we do not yet know the affinity of this glycine buffer extractable immunoreactive insulin-like material for insulin antibody, the result in Figure 1 allows us to quantitate this material in "insulin weight equivalents". Various transformed cell lines we have isolated contain between 2 and 120 ng insulin equivalents per 10^7 cells. (This could correspond to a greater actual weight of immunoreactive material if the antigen has low affinity for insulin antibody). The immunoreactive material is contained within transformed cells and is not released into the culture medium unless the cells have been damaged.

Cell Lines Transformed by an SV40 Temperature Sensitive Mutant.

Uncontrolled cell proliferation associated with the transformed cell phenotype may in fact be incompatible with expression of the endocrine differentiated phenotype. If this is true, it may still be possible to manipulate transformed cell lines to reverse the transformed phenotype and obtain reexpression of the differentiated phenotype. Temperature sensitive mutants of the SV40 A cistron have been shown to yield transformed cell lines whose transformed features are temperature conditional (6,14-18). It is also possible that these mutants will display temperature conditional differentiated properties. Precedence for this prediction is obtained from the work of Boettiger et al. (8) and Pacifici et al. (9) who have demonstrated that temperature sensitive mutants in the src gene of Rous Sarcoma Virus are capable of producing transformed differentiated functions when the temperature is elevated.

Monolayer neonatal rat pancreas endocrine cell cultures were infected with SV40 mutant tsA28 (kindly supplied by Dr. P. Tegtmeyer) following the same infection protocol described for wild type SV40. After infection the cells were maintained at 32°C. The appearance of T-antigen positive cells and appearance of proliferating cells followed the pattern seen for wild type-infected cultures except that the process was slower, as expected for cultures maintained at 32°C. Proliferating cells appeared in substantial numbers only after 50 to 60 days. After the cells had been passaged in culture a number of times, growth rate increased (one cell line currently grows at 32°C with a doubling rate of 30 to 40 hours). In these cell lines, as with wild type SV40-

transformants, acid-ethanol extractable insulin was lost during the first few subculture passages, but immunoreactive insulin-like material is detected after many tissue culture passages when assayed in glycine buffer extracts of the cells.

Characterization of these cell lines is, at present, incomplete. We need to determine whether we can rescue temperature sensitive mutants of SV40 from these cells by fusion with permissive cells, whether the cell growth is, in fact, temperature sensitive, and whether elevating the temperature can cause differentiated function to be re-expressed. These investigations are currently underway.

Acknowledgements

This work was supported by funds from the Juvenile Diabetes Research Foundation and by NIH grant AM-21213. We are indebted to Mrs. Fermina Varacalli and Mrs. Carol Quigley for excellent technical assistance.

References

(1) M.R. Murray and C.F. Bradley, J. Cancer, 25 (1935) 98.

(2) W.L. Chick, V. Lauris, J.S. Soeldner, M.H. Tan and M. Grinbergs, Metabolism, 22 (1973) 1217.

(3) J.R. Duguid, D.F. Steiner and W.L. Chick, Proc. Natl. Acad. Sci. USA, 73 (1976) 3539.

(4) F. DeVitry, M. Camier, P. Czernichow, P. Benda, P. Cohen and A. Tixier-Vidal, Proc. Natl. Acad. Sci. USA, 71 (1974) 3575.

(5) S.A. Wells, R.J. Wiertman and A.S. Rabson, Science, 154 (1966) 278.

(6) J.L. Anderson and R.G. Martin, J. Cell. Physiol., 88 (1976) 65.

(7) M.A. Gimbrone and G.S. Fareed, Cell, 9 (1976) 685.

(8) D. Boettinger, K. Roby, J. Brumbaugh, J. Biehl and H. Holtzer, Cell, 11 (1977) 881.

(9) M. Pacifici, D. Boettinger, K. Roby and H. Holtzer, Cell, 11 (1977) 891.

(10) A.E. Lambert, B. Blondel, Y. Kanazawa, L. Orci and A.E. Renold, Endocrinology, 90 (1972) 239.

(11) R.L. Davidson, Exp. Cell. Res., 55 (1969) 424.

(12) E.J. Niesor, G.B. Wollheim, D.H. Mintz, B. Blondel, A.E. Renold and R. Weil, unpublished results.

(13) V. Herbert, K.S. Lau, C.W. Gottlieb and S.J. Bleicher, J. Clin. Endocr., 25 (1965) 1375.

(14) R.G. Martin and J.Y. Chou, J. Virol., 15 (1975) 599.

(15) P. Tegtmeyer, J. Virol., 15 (1975) 613.

(16) J.S. Brugge and J.S. Butel, J. Virol., 15 (1975) 619.

(17) M. Osborn and K. Weber, J. Virol., 15 (1975) 636.

(18) G. Kimura and A. Itagaki, Proc. Natl. Acad. Sci. USA, 72 (1975) 673.

DISCUSSION

R. HAY: Have you tried other feeder layer systems besides the 3T3?

D. MINTZ: We have not. Others have done so with co-cultured adrenal cells and more recently with hepatocyte co-culture.

R. HAY: We have done similar work with the pancreatic acinar cell system and I thought you might be interested to know that co-cultivation studies with irradiated human lung fibroblasts led to stimulation of both cell division and tritiated thymidine incorporation without any doubt whatsoever. The functional counterpart however was not effective. That is, we were unable to extend functional aspects of acinar cell metabolism in monolayer type culture with standard feeder layer co-cultivation.

C. WEST: In your experiment studying insulin synthesis in fetal cells was the rescuing effect of feeder cells tested in the absence of serum?

D. MINTZ: The most remarkable observation was increased survival of the endocrine pancreatic cells.

C. WEST: Is there an effect in the absence of serum?

D. MINTZ: Yes. Both long-term survival of pancreatic endocrine cells and enhanced hormone secretion can be detected, even in the absence of serum.

C. WEST: Is there a similar amount of enhancement in the presence of serum?

D. MINTZ: Both features, survival and release of hormone into the medium, are lower than in 10% serum, but absolutely increased over that observed in serum-less conditions.

C. WEST: In the tumors which were put in vivo, are the cells which are secreting insulin still transformed?

D. MINTZ: The cells that were dissociated from the tumor can be shown to be T antigen positive in subsequent culture and can be serially passed as transformed cells thereafter.

C. WEST: Do these serially passed cells still secrete insulin?

D. MINTZ: No, once the cells are removed from the animal, acid ethanol extractable insulin is no longer detectable.

W. RUTTER: Do the SV40 transformed cells secrete glucagon or somatostatin?

D. MINTZ: After the first tissue culture passage the transformed cells do not contain glucagon and we have not measured somatostatin.

W. RUTTER: Do the tumors which are formed produce somatostatin or glucagon?

D. MINTZ: We have not looked at somatostatin in the tumors and the tumors do not contain glucagon. Dr. Eric Niesson in Dr. Renold's and Dr. Wild's laboratories, in Geneva, initiated these studies and, I believe, he has similar findings.

W. RUTTER: Have you ever cultivated these tumors in animals which are diabetic and does that change the level of insulin production?

D. MINTZ: No, we have not done that yet. It will be tried, but the long term survival of irradiated diabetic animals may prove to be a problem.

P. COHEN: Do the animals that bear tumors show signs of hypoglycemia or any evidence of hyperinsulinemia?

D. MINTZ: The animals that bear tumors appear to gain weight at a more rapid rate than do non tumor-bearing animals that have been similarly irradiated, but we have not detected hypoglycemia in these animals.

PANCREAS SPECIFIC GENES AND THEIR EXPRESSION DURING DIFFERENTIATION

W.J. RUTTER, R.J. MACDONALD, G. VAN NEST, J.D. HARDING,*
A.E. PRZYBYLA,+ J.M. CHIRGWIN and R.L. PICTET
Department of Biochemistry and Biophysics
University of California, San Francisco
San Francisco, California 94143

Abstract: Pancreas-specific messenger RNAs, in particular that for amylase, have been measured during the course of development. In addition the modulation of specific gene expression during differentiation *in vitro* by bromodeoxyuridine and the glucocorticoid, dexamethasone, has been examined. BrdU depresses the accumulation of all of the tissue-specific mRNAs. Dexamethasone selectively increases the concentration of amylase mRNA. Changes in the levels of mRNAs correlate directly with changes in the levels of pancreas-specific proteins. Pancreatic rudiments release into the incubation medium a characteristic subset of cellular proteins. Some of these, including amylase, represent the secretory digestive (pro)enzymes. Another group of proteins are also released by immature rudiments which have not yet accumulated the digestive enzymes. Some of these proteins persist throughout development and may be diagnostic of and functionally associated with pancreatic differentiation.

INTRODUCTION

The developing embryonic pancreas is a useful paradigm for organogenesis and differentiation. The main objective of our studies is to elucidate the specific regulatory compounds that elicit the formation of a particular cell type and the specific molecular events associated with the differentiative process.

The development of the rat pancreas is schematically represented in Figure 1. The pancreas arises as a diverticu-

---

*Present Address: Dept of Biological Sciences, Columbia University, New York, N.Y. 10027
+Present Address: Dept of Biochemistry, University of Georgia, Athens, Ga 30602

ISBN 0-12-045450-5

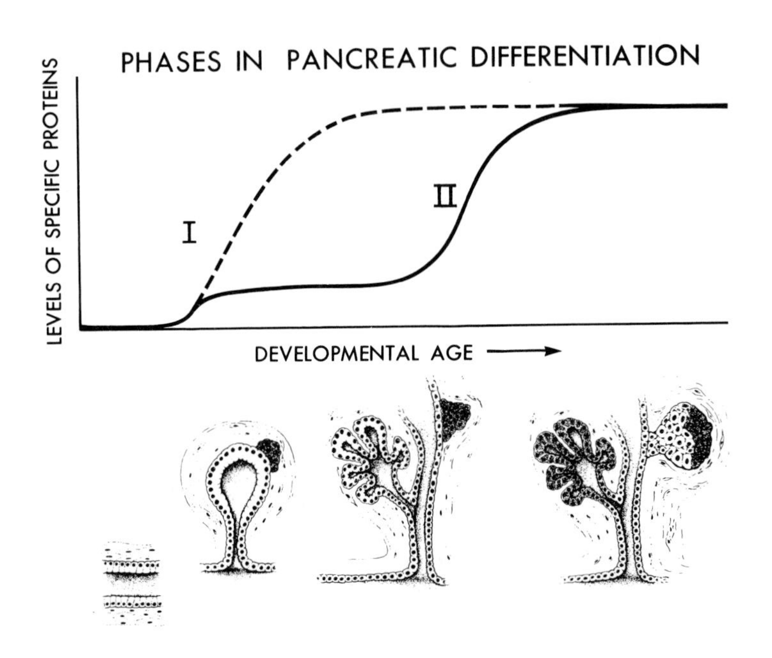

lum of the gut, at 11 days of gestation (parturition = 21 days). It is composed of a monolayer of epithelial cells that develops as a bulb surrounded by mesenchymal tissue (1,2). There follows a period of rapid proliferation and morphogenesis with the formation of both the islets which contain the endocrine cells and the branched structure of ducts and acini typical of the adult exocrine pancreas (1). As in most other embryonic epithelial organs (3,4), mesenchymal tissue is required for normal development of the epithelial pancreatic cells (5,6). We have shown that an extract from embryos can replace the pancreatic mesenchyme (6-9). This mesenchymal factor promotes synthesis of DNA and differentiation of exocrine cells which produce large quantities of specific products. Without the mesenchymal factor, the mesenchyme-deprived epithelium stays constant in size and differentiates almost exclusively into endocrine tissue, becoming in effect an islet (10,11).

Our studies on early morphogenesis and the pattern of the intracellular accumulation of the cell-specific products showed that the endocrine A cells which synthesize glucagon differentiate very early (2,12). Cells containing α granules are detected concomitant with the formation of the diverticulum; during the following 24 hours glucagon reaches a specific concentration similar to the adult state. We know little about the profile of accumulation of glucagon in the primitive A cells prior to 11 days.

The accumulation profile of insulin and of the exocrine enzymes is biphasic (12-14). The primary differentiative transition leads to the formation of the typical pancreatic morphological structures and the low accumulation of the various exocrine specific products and insulin. This level corresponds to a few thousand molecules per cell, and remains constant for several days as the cells proliferate actively. This period has been termed the protodifferentiated state. Then a secondary transition results in greatly enhanced levels of these proteins coincident with the development of the zymogen granules in the acinar cells and β granules in the B cells. The B cells and exocrine cells do not develop synchronously (the secondary transition appears delayed about one day in the exocrine cell). Furthermore the accumulation profiles of the exocrine enzymes (zymogens) are not coordinate; the half times vary by as much as two days. At the differentiated state (20 days) the two most abundant enzymes, amylase and chymotrypsin comprise 40% of the total protein of the cell. This represents more than three orders of magnitude increase in the concentration of these cell-specific proteins over those present in the protodifferentiated state. In the adult there is a substantial change of the relative proportion of the exocrine enzymes, over those present in the late embryo. This had initially been termed a "modulation" of the differentiated state.

The Synthesis During Development of Proteins Selectively Released (Secreted) Into the Incubation Medium

In addition to the accumulation of specific products, differentiation in secretory cells includes the development of the secretory mechanism. We have previously demonstrated that for insulin the capacity to secrete is present at least as early as the time when the hormone granules appear (15). If it is assumed that secretory capability develops early in the exocrine cells and that secretion is a selective process, then it is possible to biologically select a few presumably cell-specific proteins from among the many intracellular proteins. We have analyzed qualitatively the production of proteins released into the medium during the course of development. For these studies, pancreases are dissected from embryos at different stages of development and the radioactive products released during a 6 hr incubation period in the presence of ^{3}H-leucine were analyzed by two-dimensional gel electrophoresis. The labeled proteins released by the embryonic pancreases are clearly a limited subset of a large number of total cell proteins. Thus the accumulation of these proteins in the incubation medium is not due to random leakage or cell lysis.

Figure 2 shows the proteins released in vitro by freshly dissected adult pancreatic lobules. The two-dimensional gel pattern is quite simple with most of the labeled proteins appearing in about 10 major and a number of minor spots. In the two-dimensional system, modified proteins appear as molecules of similar molecular weight, but differing in charge. Thus a single protein may appear as a series of spots each differing by a unit charge (16). Such an array of spots is considered here as a single protein species. Protein 1 has been identified as amylase by immunoprecipita-

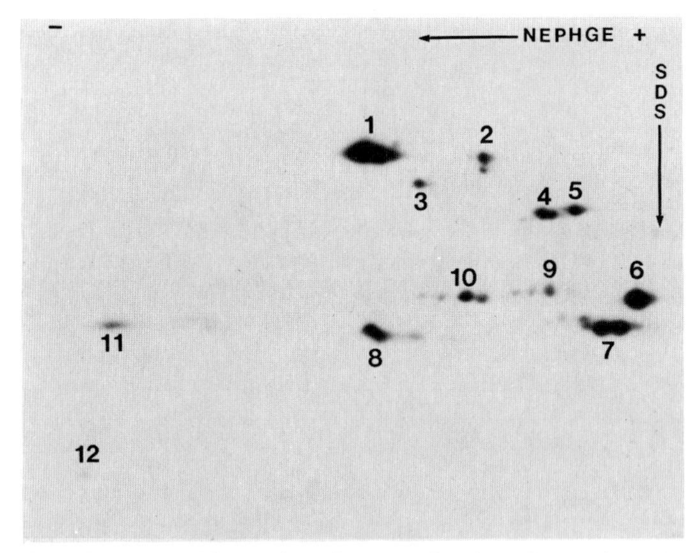

Fig. 2. Two-dimensional gel electrophoresis of adult pancreas secretions. Adult lobules were dissected and labeled in vitro with ^{3}H-leucine for 6 hr by the method of Scheele & Palade (22). The media containing the released material was collected, dialyzed, lyophilized and subjected to non-equilibrium pH gradient electrophoresis (NEPHGE)-SDS gel two-dimensional electrophoresis as described by O'Farrell et al. (16). Radioactive proteins were visualized using fluorography as described by Bonner & Lasky (23).

tion with rabbit anti-amylase antiserum and also by its ability to bind to glycogen, a property not shared by the other secretory products (17). The other proteins remain unidentified at this time but the size and isoelectric point of some of the rat enzymes (18) and comparisons with beef and guinea pig pancreas secretions (19-21) suggest that these are primarily digestive enzymes or proenzymes.

Amylase (protein 1) is a major product in the 18 day released material but is not present in earlier pancreatic preparations (Fig. 3). This result does not agree with our demonstration that amylase activity is present at 17 day and indeed even in the protodifferentiated state (13-14 days) (13). Amylase activity is also present in 17 day released material (data not shown). This suggests that there is another amylase protein present in the embryonic pancreas. Protein 2, a minor spot in adult secretions, has approximately the same molecular weight as adult amylase but is more acidic as indicated by its migration in the pH gradient of the first dimension electrophoresis. In the material released by 18 day and earlier rudiments protein 2 appears as a major protein group showing charge heterogeneity. This group is also precipitable by glycogen and therefore may represent a second amylase responsible for the activity measured in both the incubation medium and the pancreas. The presence of protein 2 but not protein 1 at day 17 when amylase activity is more than 10% that at day 18, supports this hypothesis. Preparative isoelectric focusing experiments have demonstrated that adult secretions show a major amylase activity in the basic region (isoelectric point about pH 8.5) and a very minor amylase activity with isoelectric point in the acidic region (isoelectric point about pH 5) that approximates the isoelectric point of protein 2. This putative amylase is not precipitated by antibodies prepared against purified adult amylase. The lack of cross-reactivity may indicate a distinctive gene product, or modification of the protein at the antigenic site(s).

In addition to amylase other major adult proteins (numbered 4, 5 and 7) are released by embryonic pancreases at different stages of development (Fig. 3). Proteins 7 is released from embryos 16 days and older, as well as in the adult, it is not detectable in the released material from 14 day embryos. Another group of three proteins just to the left of protein 7 is detected in material released from adult and 18 day cells but not earlier. Procarboxypeptidases A and B are known to have a size and charge approximating those of

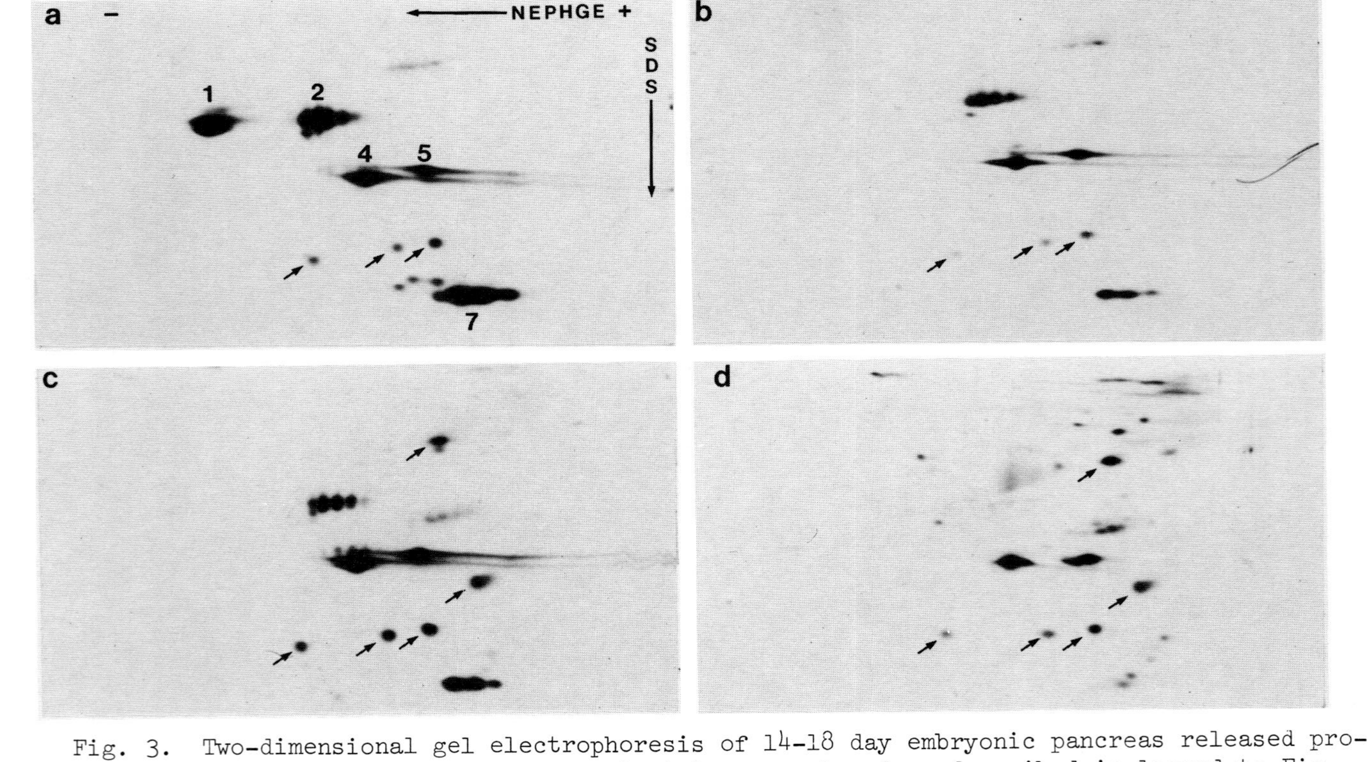

Fig. 3. Two-dimensional gel electrophoresis of 14-18 day embryonic pancreas released proteins. Organs were labeled and released material was analyzed as described in legend to Fig. 1. a, 18 day pancreas released material; b, 17 day released material; c, 16 day released material; d, 14 day released material. Arrows indicate embryonic specific proteins.

proteins 4 and 5 (21), major products in the material released by the pancreases but very minor components of whole cell protein preparations (24). However proteins 4 and 5 are prominent as early as day 14. Therefore they cannot be solely the procarboxypeptidases. These two embryonic proteins show a characteristic streaked pattern in the pH gradient electrophoresis dimension. Their identity is unknown. The prominence of proteins 4 and 5 could mask the presence of the procarboxypeptidases in that region.

Proteins 3, 6 and 8 are major components in the adult secretions but are not detectable in the released material of embryonic pancreases. These proteins are probably digestive enzymes or proenzymes appearing after birth. In the early embryo there are five prominent proteins released into the medium. These proteins are indicated by arrows in Fig. 3. All five are seen as major proteins released by 14 and 16 day embryonic pancreases. Three are major proteins in 17 and 18 day released material. Proteins 9 and 10 released by adult pancreas cells may be related to the two most basic proteins of the five proteins designated by arrows. Five similar or identical proteins are also released by 16 day embryonic liver (Fig. 4) and embryonic gut (data not shown) under the

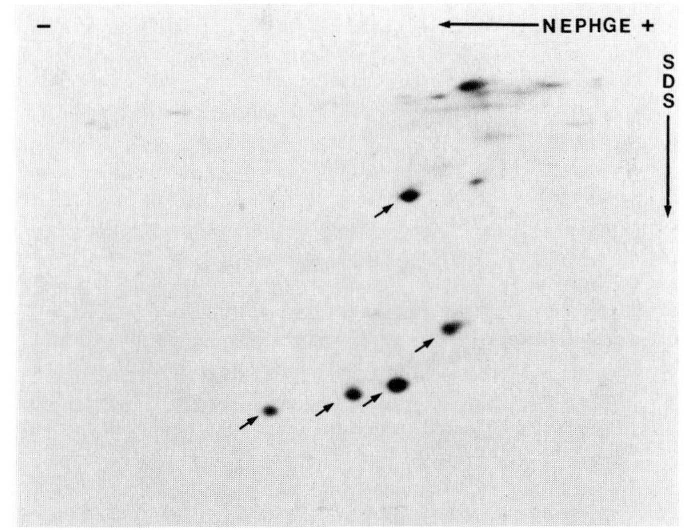

Fig. 4. Two-dimensional gel electrophoresis of 16 day embryonic liver released material. 16 day embryonic livers were labeled and released material was analyzed as described in legend to Fig. 2. Arrows indicate embryonic specific proteins.

same incubation conditions. They may represent a set of proteins common to several epithelial or secretory organs. While comprising a major portion of the material released by the embryonic organs, these five proteins are very minor components of whole cell protein preparations (not shown). They may be secreted products or perhaps more likely loosely bound surface proteins that are rapidly sloughed off in the fast growing embryonic cells. Further studies are needed to characterize these apparently non-secretory proteins, and to elucidate their function.

Analysis of mRNA by In Vitro Translation

We have extended our analysis of the pattern of accumulation of pancreatic specific products to the accumulation of their mRNAs. In the rat pancreas the high level of ribonuclease renders the isolation of intact RNA difficult. The homogenization of the tissue in the presence of high levels of multiple protein denaturants (guanidinium ion, thiocyanate ion, and sulfhydryl reducing compounds) has solved this problem by destroying RNAse activity (25). The mRNA species have been characterized by both *in vitro* translation and hybridization to complementary DNA.

In Figure 5, the translation products of mRNA obtained from the adult pancreas are compared with the proteins secreted by the pancreas. The majority of the total adult mRNA codes for a relatively few (approximately 12) proteins. Most of the adult products synthesized *in vitro* appear to be slightly larger precursor polypeptides of the pancreatic secretory proteins (26-28). For example, the primary translation product of amylase mRNA, identified by antibody precipitation and peptide analysis, is approximately 1500 daltons larger than the secreted amylase (27,28). The precursor proteins contain an additional amino acid sequence presumed to be the "signal peptide" hypothesized by Blobel and coworkers (29) to direct the vectorial discharge of nascent secretory polypeptides into the lumen of the endoplasmic reticulum, as the first step of their secretion.

Although individual mRNAs may not be translated with equal efficiency, changes in the relative *in vitro* synthetic rate of an individual translation product can be used to estimate changes in the level of the mRNA coding for that product. Comparison of results obtained from the translation product profiles for RNA isolated from adult and embryonic pancreases are also shown in Figure 5. Not all the secretory

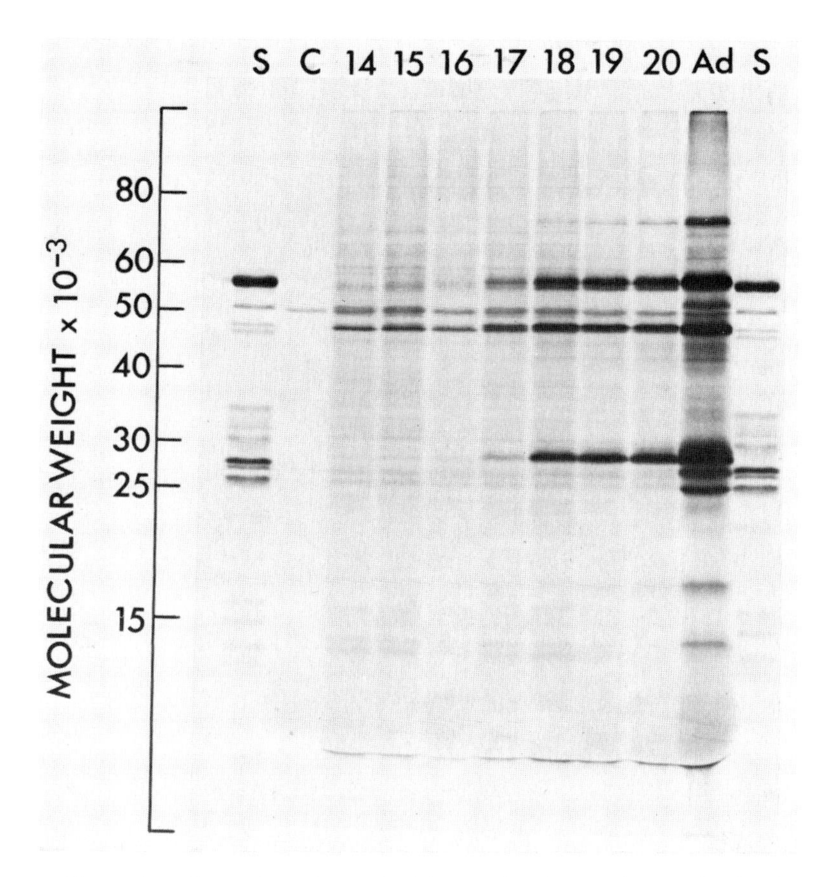

Fig. 5. *In vitro* translation of embryonic and adult rat pancreas RNAs in RNA-dependent reticulocyte lysates. *In vitro* protein synthesis was performed as previously described (27). Endogenous reticulocyte mRNA activity was eliminated as described by Pelham & Jackson (30). ^{35}S-methionine-labeled *in vitro* translation products were fractionated by SDS polyacrylamide gel electrophoresis and visualized by autoradiography. Slot S: Labeled secretion products of adult rat pancreas; Slot C: Control -- polypeptides synthesized by the mRNA-dependent reticulocyte lysate in the absence of exogenous RNA; Slots 14-20: Polypeptides synthesized in response to RNAs extracted from embryonic pancreases on the day of gestation indicated by the slot number; Slot Ad: Polypeptides synthesized in response to adult rat pancreas RNA. The asterisk indicates amylase in the secretion products.

proteins can be related to translation products. Approximately ten major translation products are especially evident in the later stages of development and in the adult and are superimposed over a relatively constant background of minor components. During development there is an increasing enrichment of the mRNAs coding for three major protein bands including amylase corresponding with the accumulation profile of the specific pancreatic secretory enzymes previously measured (13). Several translation products of the adult pancreas mRNA are not evident in the translation products of embryonic pancreas mRNA. This suggests that certain mRNA species appear only after birth.

The major translation products probably correspond to pancreas specific secretory proteins because of the similarity of size with the proteins released by embryonic pancreas (see Fig. 3). Further analysis by more discriminating separation techniques is required to determine whether a single molecular weight band contains a single translation product and whether a single band represents the same product throughout this developmental interval. A two dimensional gel electrophoretic analysis should resolve these uncertainties. By taking into account the time of appearance, the molecular weight and the isoelectric point of several exocrine enzymes, a tentative assignment can be proposed for several products seen in Figs. 2, 3 and 5. The assignment for amylase and procarboxypeptidases has already been discussed in relation to Figure 3. Proteins 7 and 11 could represent two isozymes of chymotrypsinogen (an acidic chymotrypsinogen has been described in the rat). Protein 8 is similar in size to trypsinogen, but it has a higher isoelectric point than expected. Finally, from its size and because rat ribonuclease runs as a basic protein in urea gels (32) protein 12 may be ribonuclease.

Analysis of mRNA by cDNA Hybridization

Qualitative changes in the abundance of mRNAs characteristic of the adult pancreas were also examined by means of cDNA-RNA hybridization analysis. Poly(A+) RNA was isolated from total adult pancreas RNA by means of oligo(dT) cellulose chromatography and was copied with viral RNA dependent DNA polymerase in the presence of an oligo(dT) primer. Approximately 90% of the cDNA product (total pancreas cDNA) is complementary to a population of frequent mRNAs having a total sequence complexity of about 20,000 nucleotides (sufficient to comprise 10-20 average size mRNAs); furthermore most of total pancreas cDNA is tissue specific (33). Thus, cDNA synthesized from total polyadenylated RNA is a probe for mRNAs coding largely for exocrine proteins. The kinetics of hybridization of that cDNA with total RNAs isolated from embryonic pancreases of increasing gestational age are shown in Figure 6. Between 14 and 20 days of gestation, the hybridization curves shift toward lower R_ot values, indicating that the concentration of mRNA sequences represented in the cDNA increases steadily during development. The shapes of the hybridization curves are clearly different, indicating that the various RNA sequences accumulate at different rates. At 14 and 16 days, some of these mRNAs may be absent or present at very low levels since not all of the cDNA is hybridized at the highest R_ot values tested. These results are in qualitative agreement with the *in vitro* translation previously described as well as with enzymatic activity data, viz, that the expression of pancreas specific genes during development is not coordinate.

Amylase mRNA Synthesis During Development

We have quantified changes in the abundance of the mRNA coding for one pancreatic secretory enzyme: amylase. Since amylase is a predominant pancreatic secretory product and has a higher molecular weight than most of the other secretory

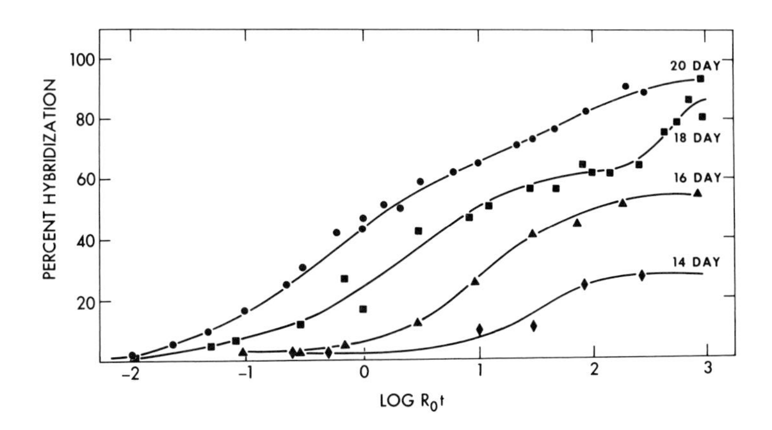

Fig. 6. Hybridization of total adult pancreas cDNA with embryonic rat pancreas RNAs. cDNA was synthesized and hybridized with total RNA isolated from embryonic pancreases of the given gestational age. Reprinted with permission from J.D. Harding et al. (1977). J. Biol. Chem. 252, 7391.

proteins, its mRNA can be resolved from other mRNAs by means of agarose-urea gel electrophoresis of total polyadenylated RNA. Amylase mRNA was eluted from the gel and amylase cDNA synthesized with viral RNA dependent DNA polymerase. This isolated amylase mRNA is approximately 90% pure as judged both by translation analysis *in vitro* and by the kinetics of hybridization of amylase cDNA with an excess of the mRNA template (data not shown).

In order to determine the tissue specificity of pancreatic amylase mRNA, amylase cDNA was hybridized with RNA from other rat tissues (Fig. 7). RNA from a tissue such as brain, or kidney does not hybridize. These tissues are not known to contain amylase. RNA from amylase-producing tissues (pancreas, parotid gland and liver) hybridizes with about 80% of the amylase cDNA. The relative positions of the liver and parotid curves on the R_ot scale imply that total RNA from the parotid contains about 1000 times more amylase mRNA than total RNA from liver. This result parallels differences in amylase specific activity in the two tissues. The fact that pancreas RNA hybridizes at lower R_ot values than parotid RNA implies a difference in amylase mRNA concentration between the two tissues. However the magnitude of this difference

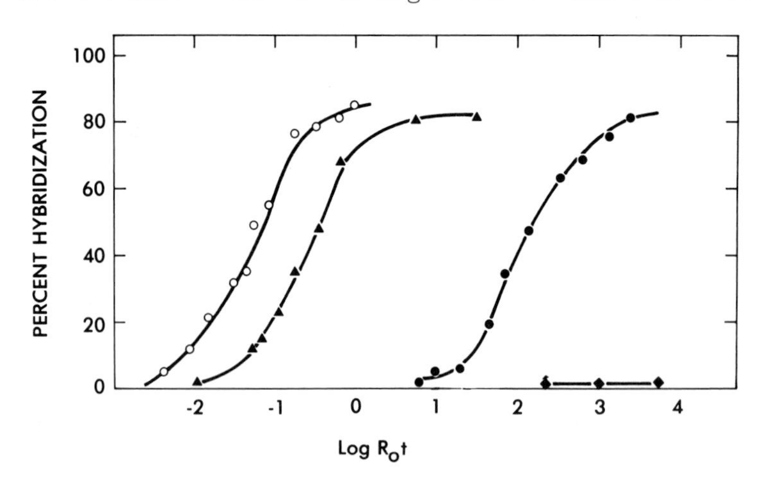

Fig. 7. Hybridization of amylase cDNA with adult rat tissue RNAs. Amylase cDNA was synthesized from amylase mRNA by viral RNA-dependent DNA polymerase. Total RNA was isolated from adult pancreas (O——O), parotid gland (▲——▲), liver (●——●) or brain (◆——◆) and hybridized with amylase cDNA as in ref. 33.

may not be quantified from the relative $R_ot_{1/2}$ values since the parotid RNA-pancreatic amylase cDNA hybrid is mismatched. Thus the levels of parotid amylase mRNA may be underestimated.

In order to determine whether amylase mRNAs in adult and embryonic rat tissues have a similar or identical nucleotide sequence we have determined the midpoint of the melting transition (Tm) of each of the RNA-cDNA hybrids shown in Figures 7 and 8. Adult or embryonic pancreas RNA hybrids have a Tm of about 76° (Table 1) and melt sharply over a temperature range of less than 10° (data not shown). The

TABLE 1

Thermal Stability of Amylase-cDNA-RNA Hybrids

RNA	Tm* (°C)
Adult Pancreas	76
18 Day Embryonic Pancreas	76
14 Day Embryonic Pancreas	75
Adult Liver	68
Adult Parotid Gland	69

*Each RNA was hybridized to a R_ot value at which maximum hybridization was obtained (see Figs. 2 & 3). Each value is the average of 3 independent determinations and is ±1°C.

parotid and liver hybrids have a Tm of about 69° (Table 1) and melt over at least a 20° range. The 7° difference in Tm implies that the nucleotide sequences of pancreas and either liver or parotid amylase mRNAs differ by about 5% (34). These results indicate that there are at least three different genes in the rat coding for amylase. These are expressed, respectively, in the pancreas, parotid and the liver. Embryonic pancreases either in the protodifferentiated state (14 days) or during the secondary transition (16-20 days) synthesize a pancreas-specific amylase mRNA as judged by Tm analysis. The two-dimensional gel analysis of the products secreted by pancreases of different developmental age as seen

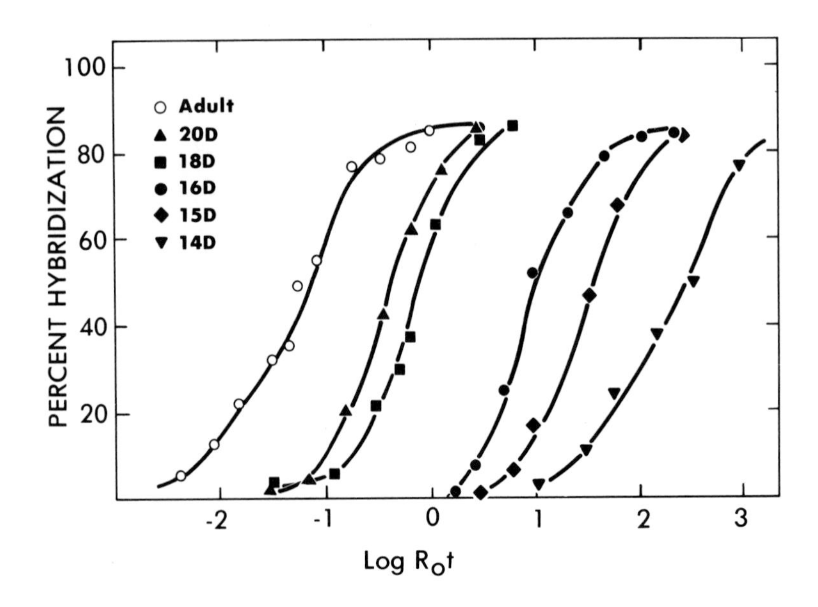

Fig. 8. Hybridization of amylase cDNA with embryonic rat pancreas RNAs. Total RNA was isolated from an embryonic rat pancreas of a given gestational age and hybridized with amylase cDNA as in ref. 33.

---

in Figure 3 suggest the synthesis of different amylase isozymes during development. If there is an embryonic amylase coded by a different gene than adult amylase, the two amylase mRNAs are very similar in nucleotide sequence.

In order to quantitate the changes in the abundance of amylase mRNA during development, total embryonic pancreas RNAs, isolated from pancreas at different ages, were hybridized with amylase cDNA (Fig. 8). Amylase mRNA is synthesized in the protodifferentiated pancreas as shown by the hybridization of 14 day RNA with the cDNA. As seen in Table 2, the concentration of amylase mRNA steadily increases during the secondary transition.

We have also estimated amylase mRNA levels during pancreatic development by in vitro translation. In this case amylase was detected by specific immunoprecipitation. Equal amounts of total pancreatic RNA from embryos at 14 through 20 days gestation were added to the reticulocyte cell-free system, and amylase immunoprecipitates were obtained. To correct for non-specific precipitated radioactivity, each

TABLE 2

Amylase mRNA Accumulation During Pancreatic Development

Embryonic age(days)	Amylase Accumulation* *in vivo*	Amylase mRNA Levels (% 20 day)	
		Translation Analysis[+]	Hybridization[++]
14	0.15	<1.0	0.2
15	0.27	2.8	1.2
16	1.8	9.4	4.1
17	7.8	17	--
18	33	78	54
19	70	100	--
20	100	100	100
Adult	74	550	630

*Data taken from Sanders & Rutter (24).

[+]Obtained by translation of RNAs in a message-dependent reticulocyte lysate, immunoprecipitation with anti-amylase immunoglobin, polyacrylamide gel electrophoresis of the immunoprecipitates and determination of the radioactivity in the amylase band.

[++]Obtained from hybridization of amylase cDNA with total RNA (Fig. 7). The $R_0t_{1/2}$ values are: 14 day, $2.0x10^{2}$; 15 day, $3.0x10^{1}$; 16 day, $8.5x10^{0}$; 18 day, $6.5x10^{-1}$; adult, $5.5x10^{-2}$.

immunoprecipitate was analyzed by SDS polyacrylamide gel electrophoresis (Fig. 9). Only the radioactivity found under the peak of completed amylase molecules (58,000 mol. wt.) was included in the estimate of amylase synthesis. The additional peaks of radioactivity in the profiles of anti-amylase precipitates were found to be predominantly incomplete amylase polypeptides generated in the reticulocyte system (28). Background levels were estimated by immunoprecipitation of the translation products of brain, a tissue deprived of amylase activity. This procedure permits a more precise

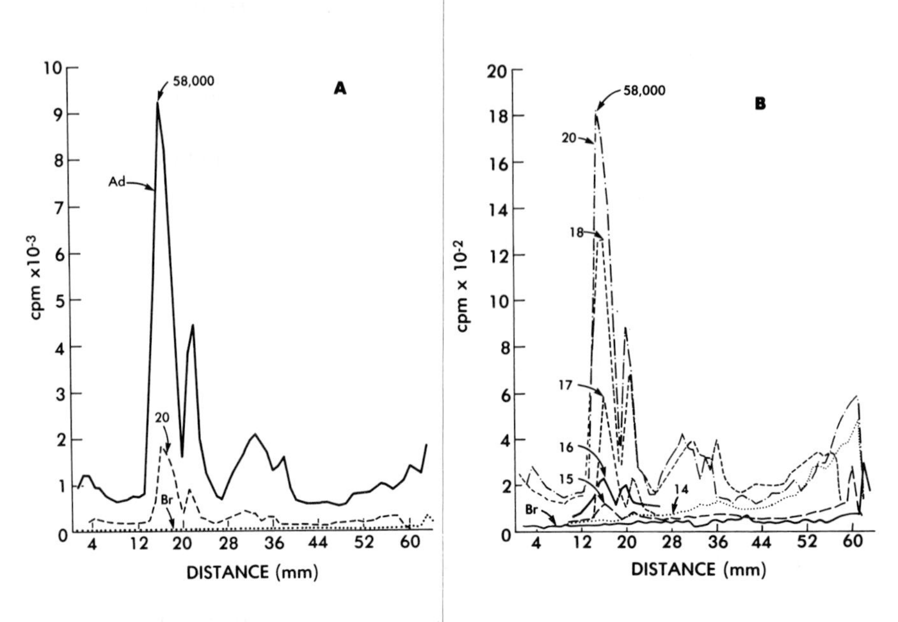

Fig. 9. Electrophoretic analysis of amylase immunoprecipitates. The amylase *in vitro* product was immunoprecipitated from reticulocyte cell-free assays containing RNA from adult pancreas, brain, and embryonic pancreases from days 14 and 20 of gestation. The immunoprecipitates were resolved by electrophoresis on cylindrical SDS polyacrylamide gels as described by Laemmli (31). Each profile is labeled according to the source of the RNA in the translation assay: *Ad*, adult pancreas; *Br*, rat brain; 14-20, the age in days of gestation for embryonic pancreases. The migration of the completed amylase translation product is marked by its molecular weight (58,000).

---

measurement of the synthesis of small amounts of amylase. By this method, pancreatic amylase mRNA levels were found to increase at least 36-fold from 15 to 20 days of gestation and 200-fold from 15 days of gestation to adult. No amylase mRNA was detectable in 14 day RNA by the translation assay. The limits of the assays indicate it is less than 1% of the level of 20 days. Thus the change in amylase mRNA is at least 550-fold from 14 days to adult. In general, the levels of amylase mRNA assayed in the translational system are in good agreement with the data obtained by hybridization with amylase cDNA (Table 2), except that the data obtained by cDNA hybridization are more reliable at low mRNA concentrations, i.e. at younger embryonic ages. The relative levels of amylase mRNA in 14 day pancreas RNA is 0.2% that present in 20 day pancreas RNA. Further the levels in adult are 6 times those in the 20 day pancreas. Thus there is a 3000-fold increase in the level of amylase mRNA over the entire developmental period. The collective data indicate that there is a dramatic increase in the steady-state concentration of amylase mRNA corresponding with or slightly preceding the increased synthesis of amylase. The rate of synthesis of amylase during this developmental interval correlates with the level of amylase mRNA.

Modulation of Pancreatic Development In Vitro

The development of the embryonic pancreas in organ culture is similar to its development *in vivo*. A cultured embryonic pancreas, explanted before overt cytodifferentiation of specific cell types has occurred, differentiates in the same temporal sequence as *in vivo*, into a normal proportion of exocrine and endocrine cells (1,2). We have thus used the system as a model of the developmental process *in vivo*. Glucocorticoids and the thymidine analog, 5-bromodeoxyuridine (BrdU) affect the synthesis of pancreas-specific proteins. Dexamethasone, a potent synthetic glucocorticoid, stimulates the synthesis of a restricted number of tissue-specific proteins (36). In contrast, BrdU blocks the accumulation of secretory proteins (36,37). We have recently sought to determine whether these compounds also affect the accumulation of specific mRNAs.

Effects of Dexamethasone.

In pancreatic rudiments treated with dexamethasone, tissue specific proteins increase from 40% in controls to 70% of total pancreatic protein. This is reflected by an increase in the number of zymogen granules in the cells (cf. Fig. 10 & 11). This increase is accounted for almost quantitatively by an ∿4-fold increase in amylase specific activity. Procarboxypeptidase B, a relatively minor component, is also increased, whereas the other exocrine proteins are relatively unaffected (35). Thus dexamethasone affects the expression of some pancreas specific genes, e.g., that coding for amylase, but not others.

The stimulatory effect of dexamethasone on the accumu-

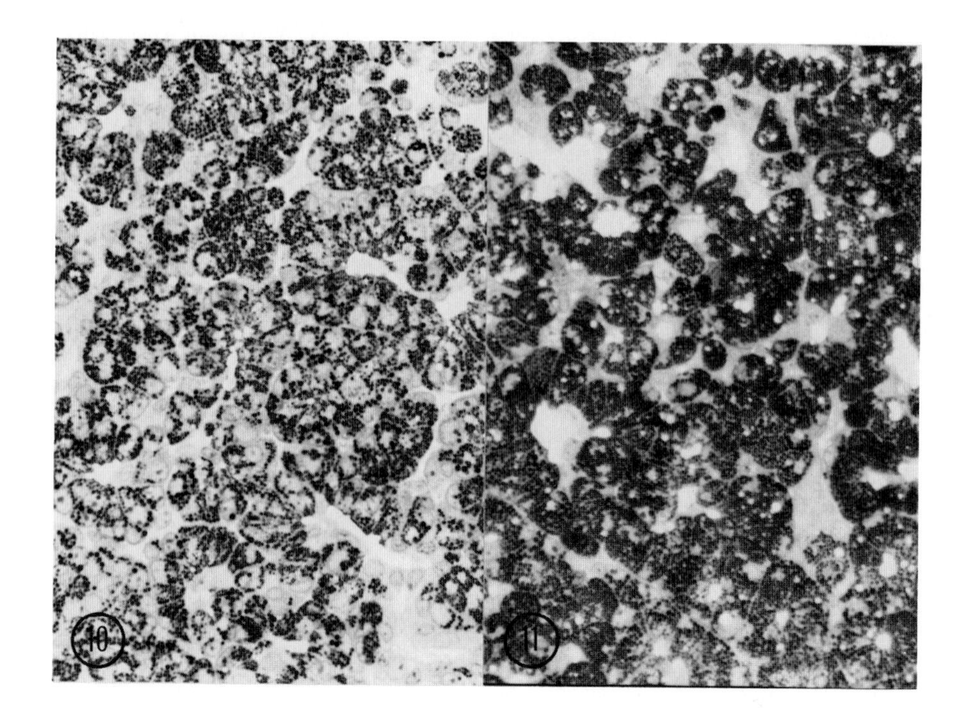

Figs. 10 & 11. Glucocorticoids increase the number of zymogen granules accumulating in the acinar cells. Pancreatic rudiments explanted at day 14 were cultured 6 days in the absence or presence of 10^{-7} M dexamethasone. In both cases the tissues develop normally but the acinar cells accumulate more zymogen granules in the presence of dexamethasone.

---

lation of amylase mRNA is apparent from both translation analysis and specific hybridization the results of which are summarized in Table 3. Equal amounts of total RNAs isolated from control and dexamethasone treated embryonic pancreases were translated in the reticulocyte lysate system. The amylase synthesized in the cell-free system was isolated from the total translation products by immunoprecipitation and resolved on a SDS cylindrical gel. RNA from dexamethasone treated pancreases coded for about 1.6 times more amylase than did control RNA. Estimation by hybridization with amylase cDNA, in an experiment analogous to that shown in Fig. 8 indicated that total RNA from dexamethasone treated pancreases contains about 2.4 times as much amylase mRNA as does RNA from controls. The significant difference between the 4-fold increase in amylase specific activity and the

TABLE 3

Effect of Dexamethasone and BrdU on Amylase Specific Activities and Amylase mRNA Levels in Cultured Embryonic Rat Pancreases

	Amylase Specific Activity (% Control)	$R_o t_{1/2}$ [a] (mol sec l^{-1})	cpm in Amylase immuno-precipitate [b]	Relative Amylase mRNA Levels: Hybridization (% Control)	Relative Amylase mRNA Levels: Translation (% Control)
Control	100	1.2×10^{0}	5030	100	100
+ Dex	440	5.0×10^{-1}	7960	240	160
+ BrdU	3	3.5×10^{1}	390	3	<8

a) $R_o t_{1/2}$ value of the hybridization of total embryonic pancreas RNA with amylase cDNA.

b) Embryonic RNA was translated in the reticulocyte lysate and the radioactivity in the amylase immunoprecipitate determined as in Figure 8.

approximate 2-fold increase in amylase mRNA accumulation (assayed by translation or hybridization) suggests that dexamethasone increases amylase specific activity only in part by increasing the steady state level of amylase mRNA; dexamethasone might also act on some other process such as the efficiency of translation of amylase mRNA.

Effects on BrdU

Growth of embryonic pancreases in the presence of BrdU results in the inhibition of the synthesis of the exocrine proteins and insulin, whereas DNA and overall protein synthesis are relatively unimpaired (36). Figures 12 and 13 show a pancreatic rudiment grown in BrdU; its morphology differs drastically from the normal rudiment (compare Fig. 10). The majority of cells in the BrdU treated rudiment have the morphological and biochemical features of duct cells. Moreover they line large fluid filled vacuoles (37). Thus BrdU treatment alters the proportion of different pancreatic cell types and may either block the further differentiation of a duct-like cell precursor into exocrine cells or change the differentiative program of a pluripotent stem cell which can differentiate into either a duct cell or exocrine cell.

SDS gel pattern of translation products coded by RNA isolated from BrdU treated embryonic pancreases demonstrates that synthesis of four of the five major control bands is clearly reduced by BrdU treatment. This effect is not due to a general inhibition of mRNA synthesis by BrdU since equal amounts of RNA from control and BrdU treated pancreases elicit the synthesis of similar amounts of TCA precipitable radioactivity in the reticulocyte system. That BrdU inhibits the synthesis of most, if not all, of the prominent pancreatic mRNAs is also evident from hybridization experiments with total pancreas cDNA (38). These experiments indicate that the mRNAs represented in total pancreas cDNA are present at about 3% of the control level in BrdU treated pancreases. In these experiments the time of initiation of treatment and the concentration of BrdU allows a small fraction of the cells to undergo the differentiative program.

The effect of BrdU on amylase mRNA levels in BrdU treated embryonic pancreases were also measured by the specific translation and hybridization assays already described above. As shown in Table 3, BrdU treated pancreases contained about 3% of the control level of amylase activity and amylase mRNA. Thus, BrdU blocks the synthesis of amylase and other secre-

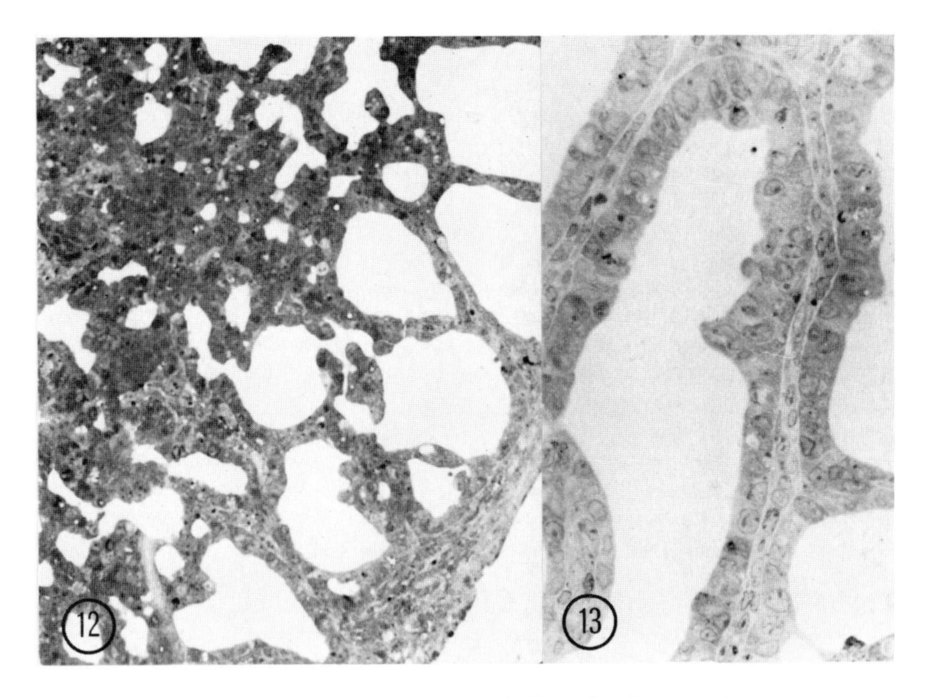

Figs. 12 & 13. Bromodeoxyuridine inhibits the accumulation of zymogen granules. Pancreatic rudiments explanted at day 14 were grown for 6 days in the presence of $2x10^{-5}$ M BrdU. The exocrine cells line large vacuoles and do not contain zymogen granules.

---

tory products by inhibiting the accumulation of their mRNA.

The mechanism by which BrdU inhibits the synthesis of tissue specific proteins (reviewed in 39) is still obscure. Inhibition of pancreatic acinar cell cytodifferentiation is related directly to the degree of incorporation of BrdU into DNA (36) suggesting a DNA-linked mechanism. This suggestion is further strengthened by the inhibition on the accumulation of the specific mRNA. Certain DNA binding proteins (40,41) and histones (42) have higher affinity for BrdU-substituted DNA than for normal DNA. We have shown that BrdU-substituted chromatin has greater thermal stability than normal chromatin (43). Thus, incorporation of BrdU into DNA during development could alter the affinity of DNA sites for specific regulatory proteins that effect the differentiative program. Alternatively, sites of incorporation of BrdU (in place of thymidine) may be specific targets for some enzymatic reaction crucial to the differentiative process.

CONCLUSION

The present investigations have provided new insights into the developmental program of the pancreas.

1. The development of the exocrine cells occurs in three phases: The primary regulatory transition (11 days, 20 somites) produces pancreatic acinar morphology and low level expression of pancreas-specific genes. The second transition (15 days of gestation) involves an increase in the synthesis of certain (largely secretory) proteins. The third transition, occurring postpartum, results in the expression of a new set of genes not detected in the embryo. This transition was unrecognized in previous analysis.

2. A set of proteins is selectively released by the early pancreatic rudiment. Some of these proteins are unlikely to be pancreatic zymogens and may have generalized epithelial functions. Others may be associated with general secretory functions. Some of these molecules may be associated with pancreas-specific functions; if so they will be useful markers for the primary developmental transition, the initiation of the pancreas-specific program.

3. The analysis of amylase-specific RNA via cDNA hybridization shows that pancreatic amylase mRNA is present in 14 day pancreatic rudiments, thus low level of amylase activity in these early pancreases is not due to expression of one of the other amylase genes (e.g. parotid or liver). These observations support the concept of low level cell-specific gene expression in the early phase of differentiation (the protodifferentiated state).

4. During the secondary transition the dramatic increase in synthesis of specific proteins is preceded by similar increase in the level of mRNA. The increase of the various mRNAs is not coordinate. Thus the secondary transition involves independent regulation of the expression of the genes for the various secretory proteins.

5. The glucocorticoid, dexamethasone, selectively enhances the accumulation of certain pancreas-specific genes, particularly amylase. We now show this effect is at least in part due to an increased level of amylase mRNA.

6. When grown in the presence of the thymidine analogue, 5-bromodeoxyuridine (BrdU), pancreatic rudiments fail to accumulate high levels of amylase mRNA or other pancreas-specific transcripts found in adult pancreas exocrine cells. This explains quantitatively the block in the expression of pancreas-specific proteins previously observed, and supports the view that BrdU alters the developmental program of the pancreas.

REFERENCES

(1) R.L. Pictet, W.R. Clark, R.H. Williams and W.J. Rutter, Develop. Biol. 29 (1972) 436.

(2) R.L. Pictet and W.J. Rutter, in: Handbook of Physiology, Section 7: Endocrinology, eds. D.F. Steiner and N. Freinkel, Vol. 1 (Williams and Wilkins, Baltimore, 1972) p. 25.

(3) C. Grobstein, Natl Cancer Res. Inst. Monogr. 76 (1964) 273.

(4) E. Wolff, Curr. Top. Devel. Biol. 3 (1968) 65.

(5) N. Golosow and C. Grobstein, Develop. Biol. 4 (1962) 242.

(6) W.J. Rutter, N.K. Wessells and C. Grobstein, J. Natl Cancer Inst. 13 (1964) 51.

(7) R.A. Ronzio and W.J. Rutter, Develop. Biol. 30 (1973) 307.

(8) S. Levine, R.L. Pictet and W.J. Rutter, Nature New Biol. 246 (1973) 49.

(9) S. Filosa, R. Pictet and W.J. Rutter, Nature 257 (1975) 702.

(10) R. Pictet, L.B. Rall, M. de Gasparo and W.J. Rutter, in: Early Diabetes in Early Life, eds. R.A. Camerini-Davalos and H.S. Cole (Academic Press, New York, 1975) p. 25.

(11) R.L. Pictet and W.J. Rutter, in: Cell Interactions in Differentiation, eds. M. Karkinen-Jaaskelainen, L.

Weiss (Academic Press, London, 1977) p. 339.

(12) L.B. Rall, R.L. Pictet, R.H. Williams and W.J. Rutter, Proc. Nat. Acad. Sci. USA 70 (1973) 3478.

(13) W.J. Rutter, J.D. Kemp, W.S. Bradshaw, W.R. Clark, R.A. Ronzio and T.G. Sanders, J. Cell Physiol. 72 (suppl. 1) (1968) 1.

(14) W.R. Clark and W.J. Rutter, Develop. Biol. 29 (1972) 468.

(15) M. de Gasparo, R. Pictet, L. Rall and W.J. Rutter, Develop. Biol. 47 (1975) 106.

(16) P.Z. O'Farrell, H.M. Goodman and P.H. O'Farrell, Cell 12 (1977) 1133.

(17) A. Loyter and M. Schramm, Biochim. Biophys. Acta 65 (1962) 200.

(18) T.G. Sanders, Doctoral Thesis, University of Illinois, Urbana (1970).

(19) L.J. Greene, C.H. Hirs and G.E. Palade, J. Biol. Chem. 238 (1963) 2054.

(20) A. Tartakoff, L.J. Greene and G.E. Palade, J. Biol. Chem. 249 (1974) 7420.

(21) G.A. Scheele, J. Biol. Chem. 250 (1975) 5375.

(22) G.A. Scheele and G.E. Palade, J. Biol. Chem. 250 (1975) 2660.

(23) W.M. Bonner and R.A. Lasky, Eur. J. Biochem. 46 (1974) 83.

(24) T.G. Sanders and W.J. Rutter, J. Biol. Chem. 249 (1974) 3500.

(25) J.M. Chirgwin, A.E. Przybyla and W.J. Rutter, in preparation.

(26) A. Devillers-Thiery, T. Kindt, G. Scheele and G. Blobel, Proc. Nat. Acad. Sci. USA 72 (1975) 5016.

(27) R.J. MacDonald, A.E. Przybyla and W.J. Rutter, J. Biol. Chem. 252 (1977) 5522.

(28) A.E. Przybyla, R.J. MacDonald, J.D. Harding, R.L. Pictet and W.J. Rutter, submitted.

(29) G. Blobel and D. Sabatini, in: Biomembranes, Vol. 2, ed. L.A. Manson (Plenum Press, New York, 1971) p. 193.

(30) H.R.B. Pelham and R.J. Jackson, Eur. J. Biochem. 67 (1976) 247.

(31) U.K. Laemmli, Nature 227 (1970) 680.

(32) B.T. Walther, R.L. Pictet, J.D. David and W.J. Rutter, J. Biol. Chem. 249 (1974) 1953.

(33) J.D. Harding, R.J. MacDonald, A.E. Przybyla, J.M. Chirgwin, R.L. Pictet and W.J. Rutter, J. Biol. Chem. 252 (1977) 7391.

(34) B.J. McCarthy and R.B. Church, Ann. Rev. Biochem. 39 (1970) 131.

(35) L. Rall, R. Pictet, S. Githens and W.J. Rutter, J. Cell Biol. 75 (1977) 398.

(36) B.T. Walther, R.L. Pictet, J.D. David and W.J. Rutter, J. Biol. Chem. 249 (1974) 1953.

(37) S. Githens, R. Pictet, P. Phelps and W.J. Rutter, J. Cell Biol. 71 (1976) 341.

(38) J.D. Harding, A.E. Przybyla, R.J. MacDonald, R.L. Pictet and W.J. Rutter, submitted.

(39) W.J. Rutter, R.L. Pictet and P.W. Morris, Ann. Rev. Biochem. 42 (1973) 601.

(40) S.Y. Lin and A.D. Riggs, Proc. Nat. Acad. Sci. USA 69 (1972) 2574.

(41) S. Lin and A.D. Riggs, Biochim. Biophys. Acta 432 (1976) 185.

(42) S.Y. Lin, D. Lin and A.D. Riggs, Nucl. Acids Res. 3 (1976) 2183.

(43) J. David, J.S. Gordon and W.J. Rutter, Proc. Nat. Acad. Sci. USA 71 (1974) 2808.

This research was supported by grants from the National Science Foundation (BMS72-02222), National Institutes of Health (AM21344), Juvenile Diabetes Foundation. R.P. was a recipient of a NIH Career Development Award; A.E.P. was a Helen Hay Whitney Foundation Fellow; G.V.N. is a NIH Post-doctoral Fellow. We would like to thank Jennifer Meek for technical assistance and Leslie Spector for preparing the manuscript.

DISCUSSION

R. OSHIMA: Are the insulin restriction maps of pancreatic DNA, using the insulin probe, the same as in other kinds of differentiated tissue?

W.J. RUTTER: We do not know that yet.

J. MORRISSEY: What is your measurement of purity of the cDNA probe that is supposed to be specific to amylase?

W.J. RUTTER: There are two measures of purity. The first comes from the translation of the mRNA preparation from which the cDNA was prepared. Amylase is almost the sole product, at least 90% of the product. The second is that rehybridization of that cDNA to the messenger RNA gives a complexity which is consistent with the molecule being pure.

S. SEAVER: You gave us a method for how you felt BrdU was influencing these cells but you did not really express your feelings on dexamethazone. Do you have ideas on how that is causing the altered increase in amylase accumulation?

W.J. RUTTER: I mentioned that I thought that about half of the increase was due to an increase in steady-state level of messenger RNA, and whether that increase is due to synthesis, be it transcription itself or turnover, is not defined by these studies; the other half is extremely interesting (the discrepancy of two-fold is beyond our error in the determinations). There must be some other dexamethazone-specific mechanism which increases the translation of amylase. There are three possibilities which are worthy of consideration: The first of these is that dexamethazone increases the translation efficiency of mRNA, perhaps by enhancing the capping of it (some fraction of it), it being again assumed that capping will promote efficiency of translation. The other method of enhancing translation might be a specific initiation factor for amylase. The final possibility is that a competition between the various messenger RNAs for the available ribosomes is somehow made more favorable for amylase mRNA in the presence of dexamethazone. For example, an increase of the steady-state level of ribosomes might enhance relative amylase synthesis. Preliminary experiments suggest the latter may be the case. Dexamethazone-treated rudiments contain about twice as much ribosomal RNA and total protein as control cells.

E.E. ATIKKAN: On dexamethazone addition, the alkaline phosphatase does not change?

W.J. RUTTER: Florence Muke had earlier shown that dexamethazone increases alkaline phosphatase in the gut. Dexamethazone has no influence whatsoever on alkaline phosphatase in our culture.

E.E. ATIKKAN: What about adding BrdU and dexamethazone together, what happens then?

W.J. RUTTER: We have not tried that exeriment.

S. COUNCE-NICKLAS: Do you have any information on pancreas specific proteins present in the mid-gut epithelium prior to day fourteen before the rudiment appears?

W.J. RUTTER: No. You will recall that prior to day fourteen there are about 5 to 10 thousand cells in the little rudiment and so it represents somewhat of tour de force for my colleagues to do these experiments at day fourteen.

MARROW COLONY STIMULATING FACTOR FROM HUMAN LUNG

M.-C. WU, S. S. FOJO and A. A. YUNIS
Departments of Medicine and Biochemistry
University of Miami School of Medicine and
the Howard Hughes Medical Institute,
Miami, Florida 33152, U.S.A.

Abstract: Colony stimulating factor (CSF) has been purified 2300-fold from human autopsy lung-conditioned medium yielding a specific activity of 2.7 x 10^6 u/mg protein. The CSF stimulates the growth of granulocytic and macrophage colonies from both mouse and human bone marrow. Purified human lung CSF (HLCM-CSF) is a glycoprotein with a MW of 41,000 D. It is stable at 50° C for 30' and over the pH range of 6.5 - 10.5. Its activity is destroyed by subtilisin and chymotrypsin, but is unaffected by neuraminidase. Antiserum prepared against purified HLCM-CSF in rabbits exerts strong inhibition on HLCM-CSF as well as on CSF from other human sources including serum, urine and placenta, but does not inhibit mouse lung CSF. Preliminary evidence is presented for a CSF-receptor site on the colony forming cell (CFU-C) membrane.

INTRODUCTION

The study of granulopoiesis and its control received its major impetus with the introduction in 1965-66 of a method for cloning bone marrow cells in soft agar (1, 2). Marrow cells cultured in soft agar under the appropriate conditions proliferate to form colonies composed of granulocytes and/or macrophages. The formation of these colonies has absolute dependence on the presence of an inducing substance which has been called colony stimulating factor (CSF). The committed stem cell giving rise to these colonies has been termed CFU-C (for colony-forming unit - CSF-dependent). In addition to its role in vitro, current evidence supports the concept of a role for CSF in the in vivo regulation of granulopoiesis (3).

ISBN 0-12-045450-5

In spite of the wide distribution of CSF in various tissues, the sources of human CSF have been limited. Thus, until recently most studies dealing with in vitro human bone marrow culture have utilized peripheral leukocyte feeder layer as a source of CSF. Human urinary CSF has been purified to homogeneity (4) but curiously this factor stimulates colony growth in mouse but not in human bone marrow. In our search for additional sources of human CSF we found that placenta (5) and lung (6) provide rich sources of activity which stimulate granulocytic and macrophage colony growth in both human and mouse marrow. In this report we describe the purification and characterization of human lung CSF, its immunological relationship to CSF from other sources, and some preliminary studies on its mode of action.

EXPERIMENTAL AND RESULTS

Assay of CSF. The method developed by Bradley and Metcalf (2) was used with slight modification (6). Mouse or human bone marrow cells were added to the medium to give 1 x 10^5 cells/ml or 2 x 10^5 cells/ml respectively and the mixture was plated in 1 ml aliquots. CSF was added directly to the plates, and was usually assayed at two concentrations. Plates were incubated at 37° C in a humidified incubator with 10% CO_2 in air. Colony counts were performed at 7 days (mouse) or 10 days (human) using a stereoscope at 25 x magnification. A colony is defined as an aggregate of 50 or more cells. A unit of CSF is arbitrarily defined as the amount of CSF which stimulates the formation of one colony under the specified conditions of the assay. Individual colonies were aspirated with a fine pasteur pipette and stained with 0.6% orcein in 60% acetic acid for morphologic identification.

Preparation of Human Lung Conditioned Medium (HLCM). Human autopsy lung was collected and incubated in serum-free medium as previously described (6). CSF activity in the medium became detectable in few hours reaching its maximum in 48 hrs. The appearance of CSF activity in the medium could be completely suppressed by the addition of cycloheximide or actinomycin D. After two days of incubation the medium was harvested by centrifugation to remove debris, followed by treatment at 56°C for 30' to inactivate complements, dialysis against distilled water and sterilization by filtration through millipore filters (0.45 μ) before assay. HLCM thus prepared had a CSF activity of 1000 - 2000 units/ml with a specific activity of 500 - 1200 u/mg of protein as assayed in mouse marrow and was similarly active in human marrow. The dose-dependent stimulation of mouse and human marrow colony growth by conditioned medium is illustrated in Fig. 1.

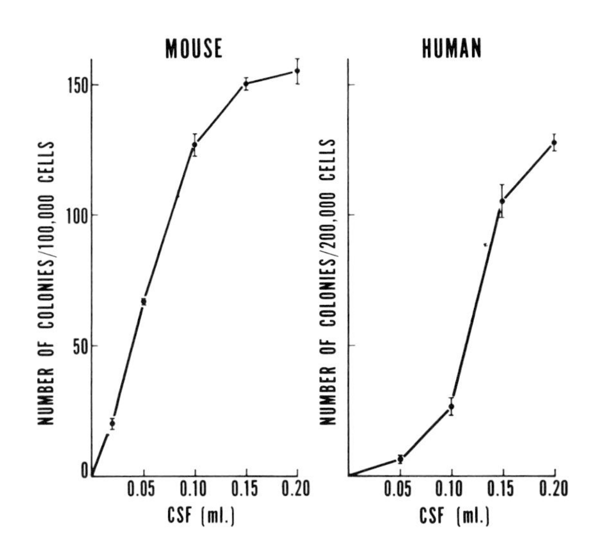

Fig. 1. Human lung CSF - dependent colony formation: Mouse and human bone marrows were cultured with increasing levels of HLCM-CSF as described in the assay.

A sigmoid dose response curve is noted as described for CSF from other sources (7). Fig. 2 shows the appearance of mouse and human marrow colonies as seen through the stereoscope at 25 x magnification.

Purification of Human Lung CSF

Using HLCM as the starting material, we have purified human lung CSF approximately 2300 fold. The purification steps included hydroxylapatite chromatography, preparative gel electrophoresis, preparative flat-bed isoelectrofocusing, and gel filtration on Ultrogel AcA44.

Hydroxylapatite chromatography. HLCM was applied to a hydroxylapatite column previously equilibrated with 0.01 M phosphate buffer pH 6.5. The column was washed with 0.06 M phosphate buffer and the "breakthrough" effluent and the buffer wash were combined and concentrated by ultrafiltration on

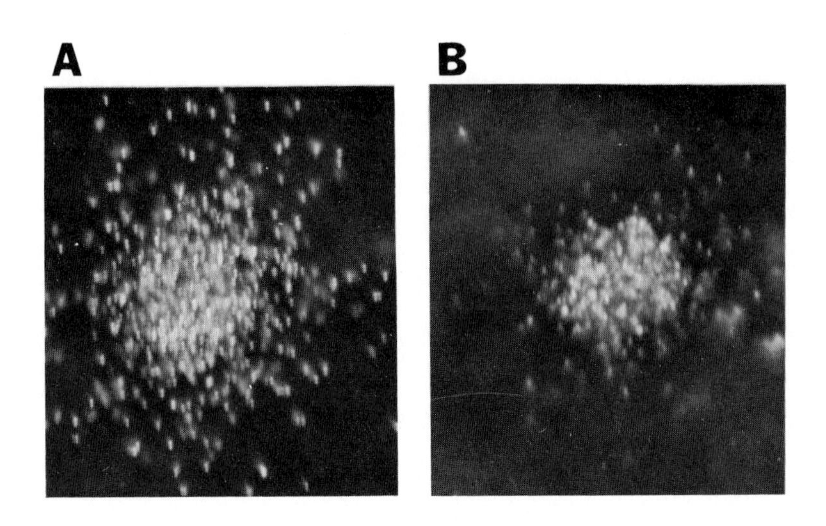

Fig. 2. Typical mouse (A) and human (B) colonies stimulated by HLCM-CSF as viewed in microscope (x 25).

Amicon (PM 10 membrane). This step resulted in 4-7 fold purification and 70-90 percent yield.

Preparative Gel Electrophoresis. The active concentrated pool from the hydroxylapatite column was dialyzed against stacking gel buffer and applied to the gel. Electrophoresis was carried out in 5 mM Tris-glycine buffer pH 8.6 at 30-40 mA for 15 hours. Eluates from the gel were collected in an automatic fraction collector (2.5 ml/fraction). A typical profile is shown in Fig. 3. Two major peaks of activity were consistently seen designated A and B. A third and minor activity peak was occasionally observed. Further characterization of CSF from peak A (CSF-A) and B (CSF-B) suggested that these two activities were distinct and this heterogeneity may have physiologic significance (vide infra).

Preparative Isoelectrofocusing. Although some fractions from the preparative gel electrophoresis had a high specific activity representing as much as 100-fold purification, it was necessary to sacrifice purity in order to maintain yield in this step. Thus, fractions with high and low specific activity included in peak A and B were all pooled, concentrated to 3 ml by Amicon ultrafiltration, dialyzed overnight against distil-

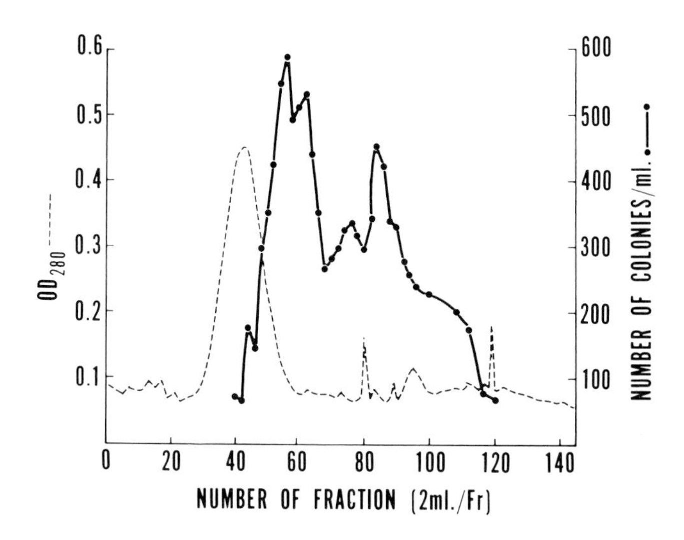

Fig. 3. Preparative gel electrophoresis: The pooled active fractions obtained from hydroxylapatite chromatography containing 103 mg and 6.1 x 10^5 units of CSF activity were concentrated to 15 ml, dialyzed against stacking gel buffer, then subjected to electrophoresis as described in Experimental Results. Fractions were dialyzed against distilled water, filtered and assayed using mouse marrow as described.

led water and then subjected to preparative flat bed isoelectrofocusing in granulated gel (8). The sample was applied to the Ultrodex gel bed as a narrow zone and isoelectrofocusing conducted overnight. Each fraction was eluted with 5 ml of 0.1 M NaCl containing 0.01% Tween 20. Eluates were dialyzed against distilled water, sterilized by filtration and assayed for CSF activity in mouse marrow.

A typical profile is shown in Fig. 4. The area of CSF activity fell in the pH range of 3.7 - 4.5 with partial overlap with a large protein peak which had the electrophoretic mobility of albumin. The active fractions excluding the "albumin" peak were pooled and concentrated resulting in 58-fold purification with an 18% yield.

Gel Filtration. One ml of concentrated material from the previous step was applied to an Ultrogel column (11 x 100 cm,

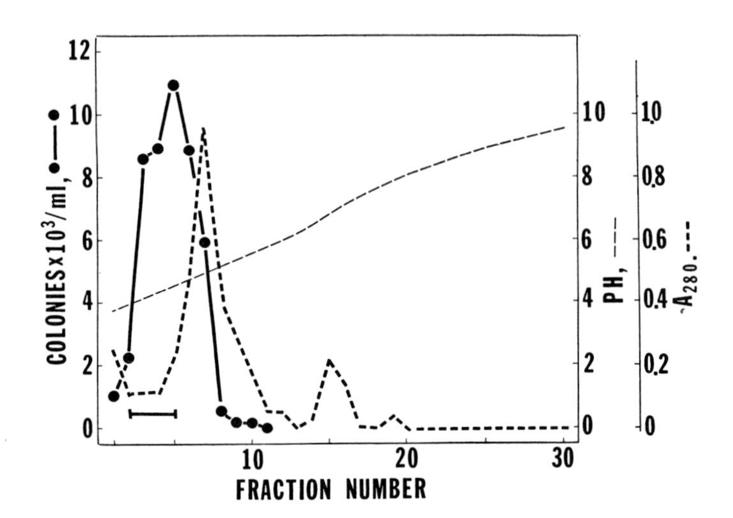

Fig. 4. Preparative isoelectrofocusing: The procedures are described in Experimental Results . One tenth of a ml aliquot from each fraction was diluted with 5% fetal calf serum in water to 1 ml, dialyzed against water, filtered and assayed.

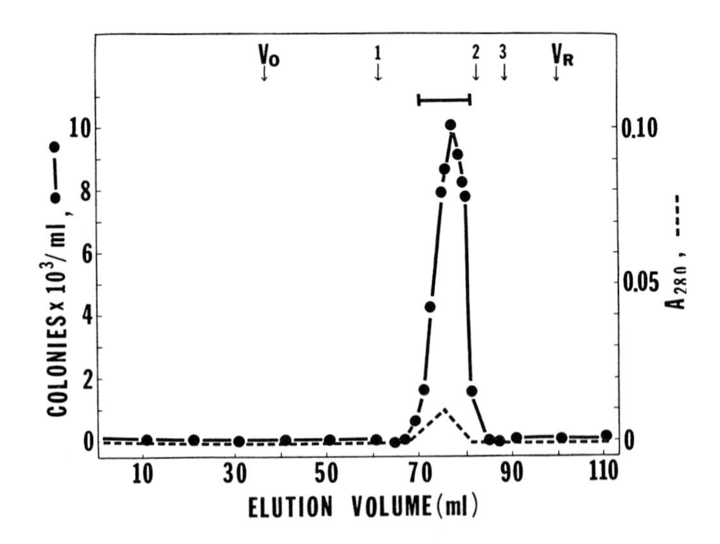

Fig. 5. Gel filtration: The procedures are described in Experimental Results. Fractions were diluted 1:100 with 5% fetal calf serum in water and assayed.

bed volume 100 ml), equilibrated with a solution of 0.1 M NaCl containing 0.01% Tween 20, 5 units of penicillin, and 5 ug streptomycin per ml. Elution was carried out with the same buffer and 1 ml fractions were collected and their CSF activity assayed. A typical gel filtration profile is shown in Fig. 5. This step resulted in approximately 3-fold purification yielding an overall purification of 2300 fold from the starting material with a final specific activity of 2.7 x 10^6 u/mg protein.

The purity of the final CSF preparation was examined by polyacrylamide gel electrophoresis after iodination with I^{125}. As shown in Fig. 6 three radioactivity peaks were observed with the major one superimposed on the CSF activity peak suggesting 40-60 percent purity of the CSF.

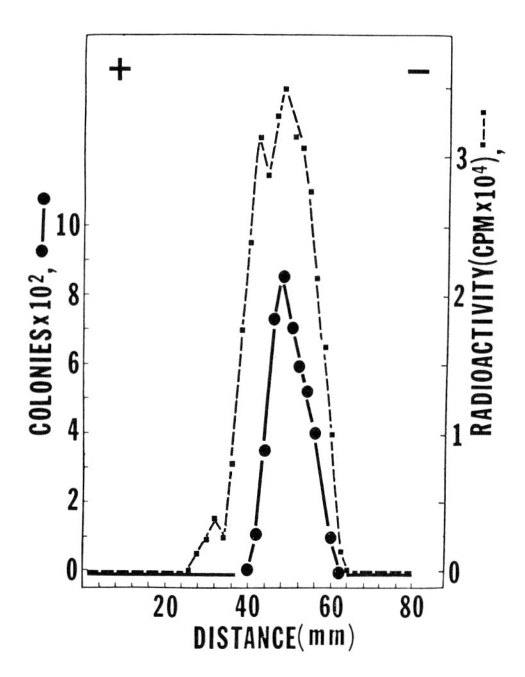

Fig. 6. Gel electrophoresis of I^{125} - CSF. The I^{125} - CSF (8600 u, 1.6 x 10^6 cpm) was subjected to polyacrylamide gel electrophoresis (7.5% gel). Gel was cut into 2 mm slices which were then counted in a Nuclear Chicago Gamma Counter for radioactivity and then extracted with 0.1 M NaCl containing 0.01% Tween 20, penicillin (40 u/ml) and streptomycin (40 µg/ml). The extracts were then dialyzed and assayed.

Some Properties of Purified Human Lung CSF (HLCM-CSF)

Like crude HLCM, purified HLCM-CSF stimulated the growth of both mouse and human CFU-C yielding a sigmoidal dose-response pattern. The colonies formed were of 3 types: granulocytic, macrophagic and mixed granulocyte-macrophage. Higher CSF levels stimulated the formation of predominantly granulocytic colonies while lower CSF concentration caused an increase in the proportion of macrophage colonies.

Molecular Weight. The MW of HLCM as calculated from gel filtration was 41,000 D. This value is in agreement with that of 42,000 D obtained from sucrose density sedimentation technique. It is also similar to the 45,000 D reported for human urinary CSF (9) but different from mouse lung CSF, MW 29,000 D and from mouse lung cell CSF, MW 70,000 D (10).

Temperature and pH Stability. Purified HLCM-CSF was stable at 50° C for 30' but lost 50% of its activity at 60° C and most of its activity at 70° C. This thermostability is similar to that of mouse lung CSF (11) and human urinary CSF (12). It was stable over the pH range of 6.5 - 10.5.

Effect of Proteolytic Enzymes, Neuraminidase, and Periodate. The activity was destroyed by subtilisin, chymotrypsin and periodate but was not affected by treatment with neuraminidase. The pattern of inactivation by proteolytic enzyme was similar to that observed for human urinary CSF. The susceptibility to periodate suggested but did not prove that HLCM-CSF is a glycoprotein. However, the glycoprotein nature of HLCM-CSF was indicated from the results of treatment with neuraminidase.

Isoelectric Focusing of Neuraminidase Treated HLCM-CSF. Purified CSF migrated in the pH range of 3.2 - 4.3 on isoelectrofocusing. However, after treatment with neuraminidase a shift in the location of activity was observed migrating as one sharp peak with an isoelectric point of 4.3 thus providing evidence that HLCM-CSF is a sialic acid containing glycoprotein. The sialic acid residues appeared to be nonessential for the in vitro activity of human lung CSF. Similar observations have been made with human urinary CSF (7), mouse lung CSF (11) and erythropoietin (13). It would appear from these results that the heterogeneity of CSF on isoelectrofocusing reflected differences in sialic acid content of the various peaks. This also appeared to be true for the heterogeneity observed on preparative gel electrophoresis. Thus, when CSF peak A and B obtained from preparative gel electrophoresis are each subjected to isoelectrofocusing a different migration pattern is observed (Fig. 7) but after neuraminidase treatment each moves to the same location as one sharp peak.

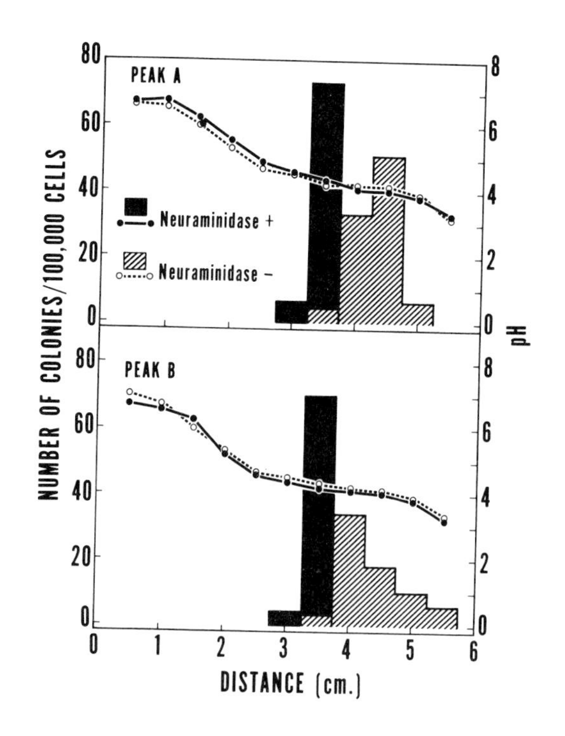

Fig. 7. Isoelectric focusing of neuraminidase treated HLCM-CSF. Three hundred microliters of peak A (1.1 mg/ml) and peak B (0.4 mg/ml) incubated in the presence or absence of neuraminidase (1 u) were run on isoelectric focusing as described in Methods. The pH gradients that were established during the run for the samples incubated in the presence (●——●) or absence (o------o) of neuraminidase are illustrated by the curves. The columns represent the CSF activity of peaks A and B incubated in the presence (■) or absence (▨) or neuraminidase.

While the sialic acid residue is not essential for the in vitro activity of CSF, we believe that the differences in sialic acid content of the various peaks may have physiologic significance. Thus, in work to be published, Miller, Gross and Yunis (14) fractionated human bone marrow cells by sedimentation in albumin gradient and assayed all fractions for CFU-C in response to CSF-peak A and B. Two distinct CFU-C populations were found: One rapidly sedimenting population exhibiting colonies after 7 days of culture in response to CSF peak A only and a more slowly sedimenting CFU-C peak did not exhibit colonies until 11 days of culture and did so in response to both CSF peak A and B. These results indicated that the two CSF peaks have different CFU-C specificities.

Immunological Relationship to CSF from other Sources

Partially purified HLCM-CSF obtained from the preparative gel electrophoresis step was used to immunize rabbits. Pre and postimmunization sera were collected and subjected to ammonium sulfate fractionation followed by chromatography on CEAE-cellulose to obtain an enriched IgG fraction free of non-specific inhibitors usually present in normal rabbit serum (15). The purified antibody thus prepared was capable of inhibiting HLCM-CSF activity by over 90 percent. Cross inhibition studies showed that anti-human lung CSF antibody was equally inhibitory to CSF from other human sources, namely serum, urine and placenta. (Fig. 8).

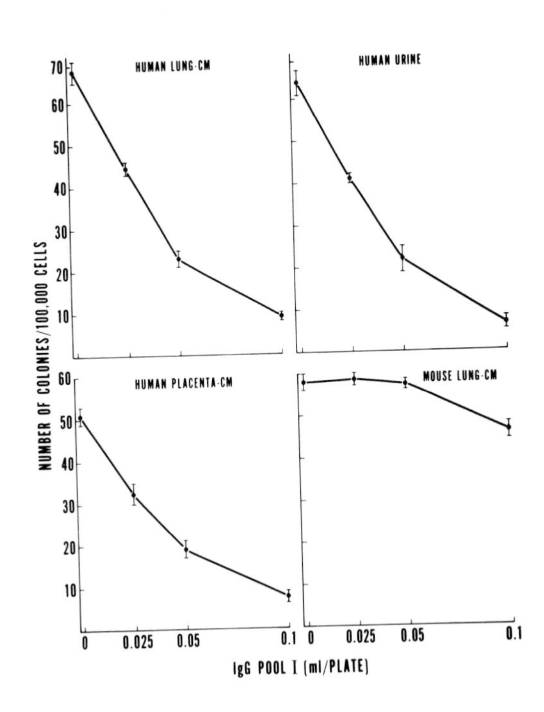

Fig. 8. Cross inhibition of anti-HLCM-CSF antiserum on CSF activity from other human sources.

This was true when either mouse or human marrow was used for assay.

In contrast the antibody showed little or no cross reactivity with mouse lung CSF.

Purification of CSF from Cultured Lung Cell Lines

More recently we have been successful in propagating "fibroblastic" cell lines from autopsy lung which secrete CSF into the culture medium. Serum free conditioned medium prepared from these cultures provided a distinct advantage over whole tissue conditioned medium in that it was virtually free of albumin and other tissue contaminating proteins having a starting specific activity of 15-20,000 units/mg protein or 10 fold that of tissue conditioned medium. Using a two step procedure involving preparative isoelectrofocusing and gel filtration we have purified cultured lung cell CSF about 1000 fold to a specific activity of 1.8×10^7 units/mg protein. The purified CSF is similar in its properties to CSF from whole lung tissue including MW, sensitivity to proteases, and isoelectric points before and after treatment with neuraminidase. It is inhibited by antiserum prepared against purified lung tissue CSF.

Mode of Action of Human Lung CSF

The mode of action of CSF on the CFU-C is not fully understood. Evidence has been provided for the existence of a hormonal receptor for erythropoietin on erythroid cell membrane (16) and the synthesis of a specific RNA as the initial step in the action of this hormone (13). Metcalf and Burgess have recently reported the stimulation by CSF of RNA synthesis in mouse marrow cells (17). We have made similar observations demonstrating the enhancement by CSF of the incorporation of H^3-uridine by rat marrow cells. We have further demonstrated that brief pretreatment of a CFU-C enriched rat marrow cell fraction with trypsin resulted in partial loss of response to CSF as evidenced by both decreased colony formation as well as reduced uridine uptake. These experiments were carried out as follows: Bone marrow cells (2×10^6 cells/ml) from Sprague Dawley rats were incubated with trypsin (0 - 5 mg/ml in Dulbecco's modified Eagle's medium (DME) at room temperature for 10 min. The cells were then washed 3 times in medium containing 10 percent fetal calf serum and 2.5% horse serum (DME-HSFC) to remove and inactivate the trypsin. They were then assayed for colony growth using mouse lung conditioned medium as a source of CSF. The results are shown in Table 1. Loss of CSF-induced colony growth is observed after treatment of CFU-Cs with trypsin.

TABLE 1

Effect of trypsin treatment on CSF-induced CFU-C growth

Trypsin mg/ml	Number of Colonies per 10^5 cell	Percent control
0 (Control)	104 ± 1	100
0.05	105 ± 3	100
0.50	86 ± 4	84
1.00	47 ± 1	45
5.00	27 ± 2	25

To study the effect of trypsin treatment on CSF-induced uridine incorporation by CFU-Cs, rat marrow cells were fractionated in Ficoll-Paque (Pharmacia) to obtain a CFU-C enriched cell preparation. This was then incubated with trypsin (2 mg/ml) at room temperature for 10 min followed by incubation without or with CSF (10^4 units/ml) for 3 hrs at 37° C. H^3-uridine (0.1 uCi) was then added to each sample (2 x 10^6 cells) and incubation continued for an additional 20 min. Radioactivity of the TCA-precipitable fraction was then determined. As can be seen from Table 2, the increment in uridine uptake induced by CSF was completely abolished by prior treatment of CFU-Cs with trypsin. These results suggested the removal by trypsin of a CSF-receptor site on the CFU-C membrane but further studies are needed for a definite conclusion.

CONCLUDING REMARKS

Progress in our knowledge of the mechanisms involved in the control of granulopoiesis has lagged behind that of erythropoiesis. The isolation, purification, and refinements in the method of assay of erythropoietin as well as the techniques developed for studying the various steps of heme synthesis have been largely responsible for the elucidation of the mechanisms involved in the control of erythropoiesis and the action of erythropoietin.

TABLE 2

Effect of trypsin treatment on CSF-induced stimulation of uridine incorporation by CFU-Cs.

	Preincubation	^{3}H-Uridine uptake CPM	Percent Control
Control Cells	H_2O	3922 ± 165	100
	CSF	5737 ± 50	146
Trypsinized Cells	H_2O	4160 ± 215	100
	CSF	4575 ± 112	110

Elucidation of the control of granulopoiesis is of utmost importance for a clear understanding of pathogenetic mechanisms involved in proliferative disorders of granulocytes including some leukemias. Although the role of CSF in vivo remains uncertain, there is evidence to suggest that it may function as a true granulopoietin. Accordingly, the isolation, characterization, availability, and the development of simplified methods of assay for human CSF are essential steps for further studies on CSF function in normal and abnormal granulopoiesis.

ACKNOWLEDGEMENTS

This work was supported in part by USPHS grants AM 09001 and AM 07114. A. A. Yunis is a Howard Hughes Investigator.

REFERENCES

(1) D. H. Pluznik and L. Sachs, J. Cell Physiol., 66 (1965) 319.

(2) T. R. Bradley and D. Metcalf, Aust. J. Exp. Biol. Med. Sci., 44 (1960) 287.

(3) D. Metcalf, in: Humoral Control of Growth and Differentiation Vol. I, eds. J. LoBue and A. S. Gordon (Academic Press, N.Y.) p. 91.

(4) E. R. Stanley, G. Hansen, J. Woodcock, and D. Metcalf, Fed. Proc., 34 (1975) 2272.

(5) R. J. Ratzan and A. A. Yunis, Clin. Res. 22 (1974) 402A.

(6) S. S. Fojo, M.-C., Wu, M. A. Gross, and A. A. Yunis, Biochim. Biophys. Acta., 394 (1977) 92.

(7) D. Metcalf, and E. R. Stanley, Aust. J. Exp. Biol. Med. Sci. 47 (1969) 453.

(8) A. Winter, S.-G. Hjalmarsson, and C. Karlsson, LKB Application Note 146 (1974).

(9) E. R. Stanley, and D. Metcalf, Proc. Soc. Exp. Biol. Med. 137 (1971) 1029.

(10) E. R. Stanley, and P. M. Heard, J. Biol. Chem., 252 (1977) 4305.

(11) J. W. Sheridan and D. Metcalf, J. Cell Physiol. 81 (1973) 11.

(12) E. R. Stanley, and D. Metcalf, Aust. J. Exp. Biol. Med. Sci 47 (1969) 467.

(13) E. Goldwasser, Fed. Proc. 34 (1975) 2285.

(14) A. M. Miller, M. A. Gross, S. S. Fojo, and A. A. Yunis Clin. Res. 25 (1977) 344A.

(15) E. R. Stanley, W. A. Robinson, and G. L. Ada, Aust. J. Exp. Biol. Med. Sci., 46 (1968) 715.

(16) C.-S. Chang, D. Sikkema, and E. Goldwasser, Biochem. Biophys. Res. Comm. 57 (1974) 399.

(17) A. W. Burgess and D. Metcalf, J. Cell Physiol. 90 (1977) 471.

DISCUSSION

S.R. DIENSTMAN: If I understood you correctly, you noted that the human lung CSF could stimulate granulopoiesis in mouse as well as human marrow. I was wondering, in view of your observation that the anti-human lung CSF does not neutralize mouse-lung CSF, if you have information as to whether the human lung CSF has the same target in the mouse marrow as the mouse lung CSF?

A.A. YUNIS: CFU-C fractionation has not been done to examine what peaks respond to which CSF. Is that what you mean?

S.R. DIENSTMAN: For example, an experiment could be done wherein mouse marrow cells could be stimulated with mouse and human lung CSF separately, followed by CFU-C assay after thymidine suiciding. If in either case the same number of CFU-Cs are eliminated by thymidine treatment it would suggest that both CSFs act on the same CFU-C population.

A.A. YUNIS: Yes, that might be a useful technique to answer the question.

S.R. DIENSTMAN: It appears from our own work and that of Naum at Seattle that the human lung macrophage growth factor (MGF) which stimulates alveolar macrophages is probably different from the MGF which stimulates peritoneal macrophages. Since many CSF and MGF preparations are similar, I wonder whether you have a lung macrophage MGF activity and where and when it separates out from the CSF activity.

A.A. YUNIS: We have not looked into that.

K. LATHAM: I was wondering whether the fairly high carbohydrate content that may have been in your CSF could have given you an anomalously high Stokes radius, and therefore an anomalously high molecular weight estimate. 40,000 would be too high. In fact if it was lower, it may be closer to the molecular weight that has been reported for other stimulatory factors.

A.A. YUNIS: That is a possibility. I suppose one should do molecular weight determinations after treatment with neuraminidase, but we have not done that.

W. VERDECKIS: In view of the fact that you have had some iodinated material which has CSF activity, have you tried binding that material to your CFU-C cells to see if it does bind?

A.A. YUNIS: Not yet.

THE ORGANIZATION AND EXPRESSION OF THE NATURAL OVALBUMIN GENE

B.W. O'Malley, S.L.C. Woo, A. Dugaiczyk, E.C. Lai,
D. Roop, S. Tsai and M.-J. Tsai
Department of Cell Biology
Baylor College of Medicine
Houston, Texas 77030

Abstract: The sequence organization of the structural ovalbumin gene and flanking sequences in native chick DNA was studied by restriction mapping and filter hybridization using a nick-translated probe generated from pOV230, a recombinant plasmid which contains a full-length ovalbumin DNA synthesized from ovalbumin mRNA. The structural sequences of the ovalbumin gene in native chick DNA were found to be non-contiguous because at least two restriction endonucleases that do not cut the structural sequence do cleave the natural gene into multiple fragments by cleaving within non-structural sequences interspersed between the structural sequences. The observation that all ovalbumin DNA-containing sequences were contained within a single DNA fragment generated by Bam HI digestion of total chick DNA has allowed the construction of an inclusive restriction map of the natural ovalbumin gene which contains at least two "insert sequences". Both "insert sequences" were located within the peptide-coding regions of the gene and the sizes of these "insert sequences" were estimated to be approximately 1.0 and 1.5 Kb, respectively. The same amazing ovalbumin gene organization is present in chick liver, oviduct and embryonic DNA. Pure single-stranded hybridization probes to each of the separate regions of the ovalbumin gene were prepared and hybridized to total chick DNA and oviduct nuclear RNA. Results indicate that transcription of the separated structural regions are coordinately regulated by estrogen.

INTRODUCTION

Recent studies have indicated that the regulation by steroid hormones of expression of the ovalbumin gene in the chick oviduct is via a "transcriptional control" mechanism (1-10). To further understand the detailed molecular

ISBN 0-12-045450-5

mechanism of this important regulatory event, large quantities of the natural ovalbumin gene are needed to study its interactions with hormone-receptor complexes, RNA polymerase and chromosomal components in vitro. In order to carry out these studies, the structure and organization of the gene coding for ovalbumin in native chick DNA must be established so that the gene may subsequently be purified and amplified by molecular cloning. Total chick DNA was thus digested exhaustively with a variety of restriction endonucleases and the DNA fragments were resolved by electrophoresis on agarose gels. To estimate the molecular weights of the DNA fragments containing ovalbumin DNA sequences, the DNA was transferred from the gels to nitrocellulose filters using the method of Southern (11). The nitrocellulose filters were subsequently employed for hybridization with specific probes generated from a recombinant plasmid pOV230 which contains a full-length $dsDNA_{OV}$ insert (12). The results of the present study have permitted us to construct a physical map of the natural ovalbumin gene and to conclude that the structural sequences coding for ovalbumin are not contiguous in native chick DNA. Pure single-stranded hybridization probes to regions of the structural ovalbumin gene separated by "insert sequences" were prepared from pOV230. These probes were employed to hybridize with excess total chick DNA or oviduct nuclear RNA. It was observed that all of the structural ovalbumin gene is a unique DNA sequence within the chick genome, and that the expression of these sequences in the chick oviduct are under coordinated control during hormonal stimulation.

EXPERIMENTAL

Chick DNA, free of protein and RNA, was prepared from liver and epithelial oviduct cells by a modification of the procedure of Marmur (13) as described previously (14). Fifty μg of total chick DNA were digested in a 100 μl reaction mixture containing 12 units of endonuclease in the appropriate buffer at 37°C for 10 hours. The endonuclease digests were heated at 68°C for 5 minutes before electrophoresis on slab agarose gels in Tris-acetate buffer as previously described (12). The DNA was subsequently transferred onto nitrocellulose filters by the method of Southern (11), and the filters were baked at 68° for 4 hours, followed by 16 hours in a 6 x SSC solution containing 0.02% of ficoll, polyvinylpyrollidone and bovine serum albumin according to the procedure of Denhardt (15). The filters were hybridized with [^{32}P]labeled DNA probes in the same solution containing 0.5% SDS and 1 mM EDTA at 68°C for 12 hours, washed three times with a 2 x SSC

+ 0.5% SDS solution at 68°C for a total of 3 hours and exposed to an X-ray film in the presence of a Dupont Cronex intensifying screen at -90°C for up to 3 days.

RESULTS AND DISCUSSION

pOV230 is a chimeric plasmid previously constructed in our laboratory that contains a full-length ovalbumin DNA insert (12). It was digested simultaneously with endonucleases *Hae* III + *Hind* III and DNA fragments were resolved by electrophoresis in agarose gels. This double enzymatic digest yielded two fragments containing the entire structural ovalbumin DNA sequences (Fig. 1A). The 1.15 Kb and the 1.45 Kb fragments contain DNA sequences corresponding to the 5'-terminus and the 3'-terminus of ovalbumin mRNA with respect to the single *Hae* III cleavage site present in ovalbumin cDNA, respectively (12,16). The sections of the gel containing the two DNA fragments were excised separately and the DNA recovered by diffusion (17). Subsequent electrophoretic analysis showed that up to 1 μg each of the DNA fragments migrated as single bands in agarose gels (Fig. 1B). The purified left and right DNA fragments were nick translated with [^{32}P]dGTP to about 8 x 10^7 cpm/μg by the method of Maniatis *et al*. (18). To test the purity and fidelity of the radioactive DNA fragments, various amounts of pOV230 digested with *Hae* III + *Hind* III were electrophoresed on an agarose gel. The DNA was transferred onto a nitrocellulose filter paper using the method of Southern (11) and subsequently allowed to hybridize with the nick translated-DNA preparations followed by extensive washing and radioautography. Both radioactive DNAs hybridized only to the proper DNA bands with at most a 1% cross-contamination (Fig. 1C). In addition, the bands were readily obtainable when as little as 100 pg of pOV230 DNA was used. Since ovalbumin sequences constitute only 2 Kb/7 Kb of the pOV230 DNA (12), the probes should be able to detect 2/7 x 100 pg = 30 pg of ovalbumin DNA sequences. It has previously been demonstrated that the ovalbumin gene is a unique DNA sequence and constitutes 2 Kb/10^6Kb of the chick genome, where 2 Kb is the approximate complexity of the structural ovalbumin gene and 10^6 Kb is the total number of nucleotides present per haploid chick genome (3-6). Thus, 15 μg of total chick DNA should contain 2 Kb/10^6Kb x 15 μg = 30 pg of the structural ovalbumin gene sequences, which would be detectable by the radioactive probes employed in the following experiments.

Total chick DNA was digested with *Hae* III and 15 μg of

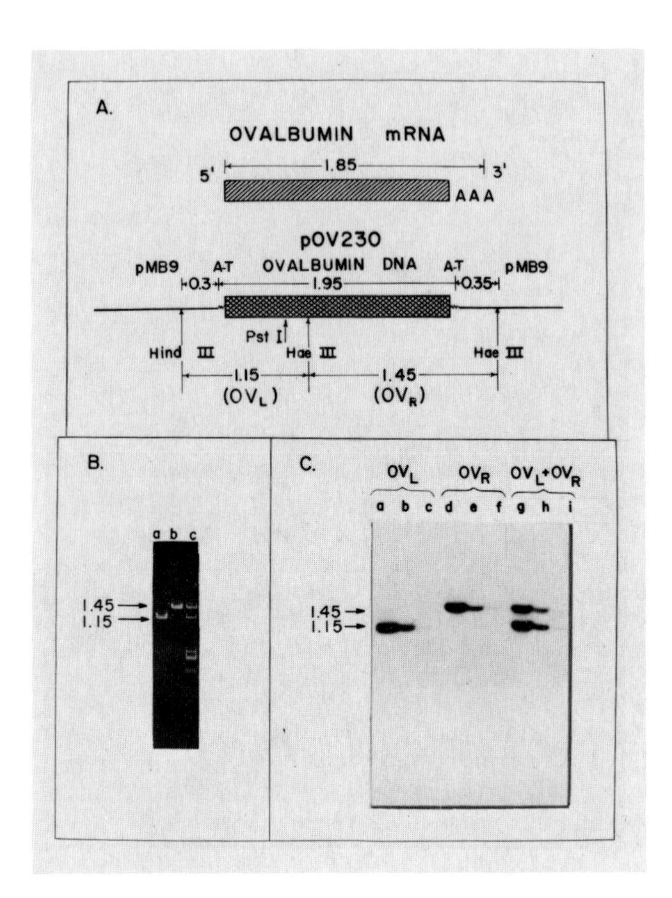

Fig. 1. The separation of OV_L and OV_R. Panel A illustrates the orientation of the full-length $dsDNA_{OV}$ insert in pOV230 (1). [cross-hatched bar], dsDNA insert; ———, pMB9 DNA and [wavy line], dA-dT linkers. The clevage sites of this recombinant DNA by *Pst* I, *Hae* III and *Hind* III are indicated by the arrows and the numbers represent the restriction fragment sizes in kilobases. Panel B is a 2% agarose gel of the isolated OV_L and OV_R after staining by ethidium bromide: a. 1 μg of OV_L; b. 1 μg of OV_R; and c. 1 μg of pOV230 cleaved with *Hae* III + *Hind* III. Panel C is a radioautogram of a 2% agarose gel containing various amounts of OV_L and OV_R after nick-translation. 10, 1 and 0.1 ng of OV_L and OV_R were applied separately to lanes a, b, c and d, e, f and together in lanes g, h, and i, respectively.

the DNA was electrophoresed on a 1% agarose gel. Since there is only one Hae III site in $dsDNA_{OV}$ (16), a total of two DNA bands was expected using both DNA probes, and only one band was expected when either one of the DNA probes was employed. OV_R yielded one DNA band at 2.5 Kb while OV_L yielded two bands at 1.3 and 1.8 Kb (Fig. 2A). When both

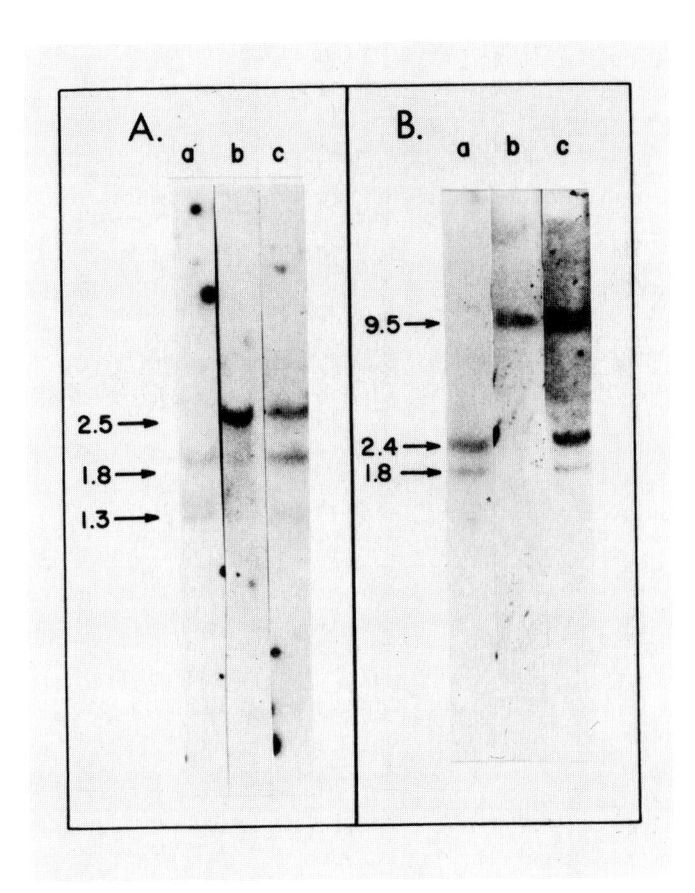

Fig. 2. Radioautogram of total chick DNA after restriction enzyme digestion and hybridization with the nick-translated probes. Probes used in lane a, b and c were OV_L alone, OV_R alone and OV_L with OV_R, respectively. Panel A: after Hae III digestion and Panel B: after EcoRI digestion. Upon prolonged exposure using some DNA fractions after RPC-5 column chromatography, a very faint 4th EcoRI fragment at 1.1 Kb was sometimes detectable.

probes were used in the hybridization reaction, all 3 bands were observed in the presence or absence of a large excess of added poly A. This experiment suggests that there exists a Hae III cleavage site in the left half of the structural ovalbumin gene in native chick DNA, which is missing in $dsDNA_{OV}$ synthesized from purified ovalbumin mRNA by viral reverse transcriptase. Furthermore, when EcoRI digested-DNA was electrophoresed on a 1% agarose gel and allowed to hybridize with the specific radioactive probes, a total of 3 bands was observed (Fig. 2B). Since EcoRI does not cleave $dsDNA_{OV}$ (16), this experiment suggests that there are two EcoRI sites within the natural ovalbumin structural sequence in chick DNA. Furthermore, when the EcoRI digested-DNA was allowed to hybridize with OV_L and OV_R separately, 1 band of approximately 9.5 Kb in length was generated with OV_R, and 2 bands of 2.4 Kb and 1.8 Kb in length were generated with OV_L (Fig. 2B). These bands could also be enriched from total chick DNA by "R-loop" formation using purified $mRNA_{OV}$ followed by oligo dT-cellulose column chromatography. Thus, both EcoRI cleavage sites seem to be located in the left half of the structural sequence of the natural ovalbumin gene.

To establish a preliminary restriction map of the natural ovalbumin gene within the chick genome, the correct orientation of the 3 EcoRI fragments must first be obtained. Since the 9.5 Kb EcoRI fragment hybridized only with OV_R while the 2.4 and 1.8 Kb fragments hybridized only with OV_L, the latter DNA fragments must be derived from the left side of the natural ovalbumin gene. When a probe was generated from OV_L after removing a 0.25 Kb fragment by Pst I digestion (Fig. 1A), only the 2.4 Kb fragment was obtained (data not shown). Thus, this fragment must be derived from the left-hand end of the ovalbumin gene, and the four EcoRI sites ($EcoRI_{a,b,c,d}$) can be ordered as shown in Fig. 3. Total chick DNA was then digested with various restriction endonucleases in the presence and absence of EcoRI. The DNA digests were again electrophoresed on agarose gels, hybridized with the specific probes and the results are summarized in Table 1. Bam HI is an enzyme that does not cleave $dsDNA_{OV}$ (16), and it has yielded a single DNA band of about 25 Kb with either OV_L or OV_R as expected. Codigestion with EcoRI resulted in 3 bands. The two small bands hybridized only with OV_L and were identical to those generated by EcoRI alone. The 9.5 Kb EcoRI fragment was reduced to a 5.2 Kb fragment, indicating that there exists a Bam HI cleavage site (Bam HI_b) 5.2 Kb to the right of $EcoRI_c$. The other Bam HI site (Bam HI_a) should then be approximately 25 Kb to

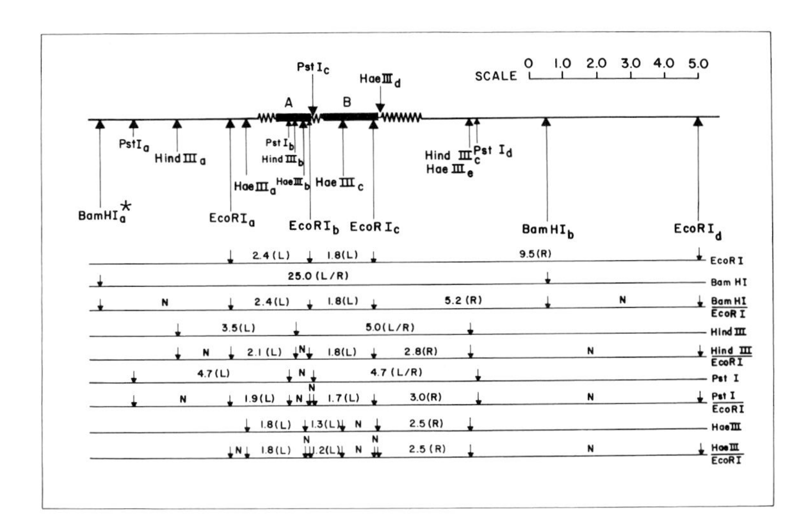

Fig. 3. A model for organization of the natural ovalbumin gene in chick DNA. DNA sequences present in $cDNA_{OV}$ are represented by and "insert" sequences are represented by , while flanking DNA sequences are represented by . Various restriction sites on the DNA are shown by the arrows. Those above the line are restriction sites present in $dsDNA_{OV}$ while those below are predicted by data summarized in Table 1. A scale is also shown in the upper right hand corner. An exception to the scale is the Bam HI^*_a site which should be 25 Kb to the left of the Bam HI_b site. The generation of DNA fragments of different sizes by various restriction digestions of this DNA is shown below the model: numbers represent the sizes of DNA fragments in kilobases; (L), (R) and (L,R) illustrates that the DNA fragment was either detected by hybridization with OV_L, OV_R or both OV_L and OV_R, respectively. (N) represents DNA fragments that should have resulted from various restriction digests but were not detected in the radioautograms because there was either no DNA sequence complementary to the probes employed, or insufficient lengths of the complementary sequences to form stable hybrids under the conditions employed.

TABLE 1

Sizes (Kb's) of Ovalbumin Sequence-Containing Chick DNA Fragments After Restriction Enzyme Digestions

	EcoRI	Bam HI	Hind III	Pst I	Hae III	Probe Used
-EcoRI	9.5	25.0	5.0	4.7	2.5	OV_R
+EcoRI	9.5	5.2	2.8	3.0	2.5	OV_R
-EcoRI	2.4 1.8	25.0	5.0 3.5	4.7	1.8 1.3	OV_L
+EcoRI	2.4 1.8	2.4* 1.8†	2.1* 1.8†	1.9* 1.7†	1.8* 1.3†	OV_L

*These fragments were derived from the 2.4 Kb EcoRI fragment.
†These fragments were derived from the 1.8 Kb EcoRI fragment.

the left of Bam HI_b (Fig. 3). Hind III is another enzyme that does not cleave $dsDNA_{ov}$ (16). Similar to EcoRI, it has produced more than one band from total chick DNA, indicating the existence of a Hind III site in native chick ovalbumin DNA. The sizes of the two DNA bands were 5.0 Kb and 3.5 Kb and both bands could be detected when hybridization was carried out with OV_L alone (Table 1). Hence the Hind III site must be located at the left half of the structural ovalbumin gene. Since only the 5.0 Kb fragment was observed using OV_R alone, there should be a Hind III site (Hind III_c) located at 5.0 Kb to the right of the Hind III_b. Furthermore, Hind III_c should be 2.8 Kb to the right of $EcoRI_c$ because the 9.5 Kb EcoRI fragment was reduced to 2.8 Kb by Hind III digestion. Consequently, Hind III_b is 2.2 Kb to the left of $EcoRI_c$, and the third Hind III site (Hind III_a) must then be located 3.5 Kb and 5.7 Kb to the left of Hind III_b and $EcoRI_c$, respectively (Fig. 3).

Although there is a Pst I site in the left half of $dsDNA_{ov}$, a Pst I digest of chick DNA has resulted in only a 4.7 Kb band using both OV_L and OV_R (Table 1), suggesting the presence of Pst I sites 4.7 Kb from both sides of this Pst I site (Pst I_c). Since the 9.5 Kb EcoRI fragment was digested to 3.0 Kb by Pst I, there should be a Pst I site (Pst I_d) 3.0 Kb to the right of $EcoRI_c$. Accordingly Pst I_c should be 1.7 Kb to the left of $EcoRI_c$. Indeed the 1.8 Kb EcoRI fragment detected by OV_L was digested to 1.7 Kb by Pst I. Hence $EcoRI_b$ should be 0.1 Kb to the left of Pst I_c. During the double digest, the 2.4 Kb EcoRI fragment was unexpectedly cleaved to a 1.9 Kb fragment (Table 1), suggesting the presence of an additional Pst I site (Pst I_b) between $EcoRI_a$ and $EcoRI_b$. This Pst I_b site should be 0.5 Kb to the left of $EcoRI_b$ since positioning this Pst I site 0.5 Kb to the right of $EcoRI_a$ would have produced a 2.0 Kb fragment instead of the observed 4.7 Kb fragment upon digestion of chick DNA by Pst I alone. The fourth Pst I site (Pst I_a) should then be 4.7 Kb to the left of Pst I_b (Fig. 3).

A Hae III digest of chick DNA has resulted in a single 2.5 Kb DNA band detected by OV_R (Table 1), indicating the presence of a Hae III site (Hae III_e) 2.5 Kb to the right of the Hae III site (Hae III_d) present in $dsDNA_{ov}$. Since this band was not cleaved by EcoRI, the $EcoRI_c$ site should be positioned to the left of Hae III_d. In the same Hae III digest, fragments of 1.8 Kb and 1.3 Kb were also detected by OV_L (Table 1). The same 1.8 Kb fragment could be obtained by treating the 2.4 Kb EcoRI fragment with Hae III (Table 1). Hence there should be two Hae III sites (Hae III_a and Hae III_b) within the 2.4 Kb EcoRI fragment (Fig. 3). In addition, the 1.3 Kb Hae III fragment was digested to 1.2 Kb by EcoRI (Table 1) indicating Hae III_b should be 0.1 Kb to the left of $EcoRI_b$. The same 1.2 Kb fragment could be produced by treating the 1.8 Kb EcoRI fragment of Hae III. Thus, an additional Hae III site 1.2 Kb to the right of $EcoRI_b$ should be present in the natural ovalbumin DNA (Fig. 3).

Based on these data, a preliminary restriction map of the natural ovalbumin gene was constructed (Fig. 3). Since only one hybridizable band was observed after digestion with Bam HI, all ovalbumin DNA-containing fragments are probably not derived from different parts of the chick genome and the multiple bands generated by various restriction enzymes must be due to the presence of respective cleavage sites in the "inserts" or "spacers" located within the structural ovalbumin gene. We thus propose in the preliminary model that there are a minimum of two "inserts" located within the left

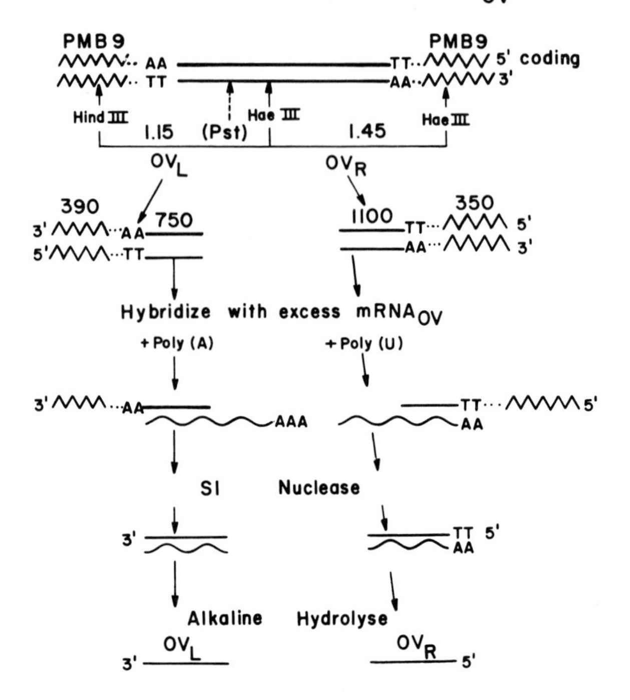

Fig. 4. Schematic representation of the preparation of pure hybridization probes to the left and right halves of $mRNA_{OV}$. Fragments corresponding to the left (OV_L) and right (OV_R) were labeled with [^3H]dCTP by nick translation to a specific activity of 5-10 x 10^6 cpm/µg. These fragments were hybridized to completion with a 100-fold excess of ovalbumin mRNA and either poly A or poly U. The reaction mixtures were then treated with S1 nuclease to digest the plasmid pMB9 sequences and anticoding sequences which remain single-stranded under the conditions employed. These steps were repeated with the hybridized [^3H]DNA to ensure the complete removal of sequences not contained in $mRNA_{OV}$. Pure single-stranded probes to the left and right halves of $mRNA_{OV}$ were then obtained by alkaline hydrolysis of the mRNA.

half of the structural ovalbumin gene. "Insert A" is located rather close to the left of the single *Pst* I_c site present in $dsDNA_{ov}$, and it contains at least one cleavage site each for *Pst* I, *Hind* III, *Hae* III and *Eco*RI. Since *Pst* I_b is 0.5 Kb to the left of *Eco*RI_b and both sites are contained in this "insert", its size must be in excess of 0.5 Kb. On the other hand, the entire insert is contained within the 1.9 Kb *Hae* III_a/*Pst* I_c fragment in which 0.5 Kb is present in the $dsDNA_{ov}$. Thus, the maximum size of "insert A" must be smaller than 1.4 Kb. In addition, "insert B" is located to the right of the *Pst* I_c site and very close to the left of the *Hae* III_d site present in $dsDNA_{ov}$, and should contain at least one *Hae* III site and one *Eco*RI site. To accommodate all the fragment sizes generated by *Hae* III, *Pst* I and *Eco*RI within the 0.25 Kb span between *Pst* I_c and *Hae* III_d in $dsDNA_{ov}$, the length of "insert B" should be in the neighborhood of 1.5 Kb. Finally, we wish to emphasize that it is entirely possible for the natural ovalbumin gene to contain additional "inserts" that do not contain any restriction sites for the restriction enzymes employed in the present study.

To study the regulation of transcription of the interspersed ovalbumin structural gene sequences, it was first necessary to prepare pure single-stranded hybridization probes to the left (5') and right (3') regions of the ovalbumin structural gene. This was accomplished by a procedure represented schematically in Figure 4. OV_L and OV_R were labeled with [^{3}H]dCTP by nick translation to a specific activity of 10^7 cpm/μg. The labeled fragments were hybridized with excess ovalbumin mRNA in the presence of poly A or poly U. Subsequent nuclease treatment of these reaction mixtures resulted in the digestion of plasmid pMB9 sequences and anticoding ovalbumin sequences which remain single-stranded under the conditions employed. Only the coding strand sequence remains hybridized to $mRNA_{ov}$ and undigested. Pure single-stranded left and right probes were then obtained after alkaline hydrolysis of the RNA. To answer the question whether the expression of the interspersed regions of the structural ovalbumin gene are regulated coordinately, the specific hybridization probes to the left and right halves of the structural ovalbumin genes were allowed to hybridize with total nuclear RNA extracted from estrogen-stimulated chick oviduct, estrogen-withdrawn oviduct and liver. Equimolar amounts of nuclear RNA corresponding to the left and right regions of the ovalbumin structural gene were detected in all three tissues (Fig. 5). These results

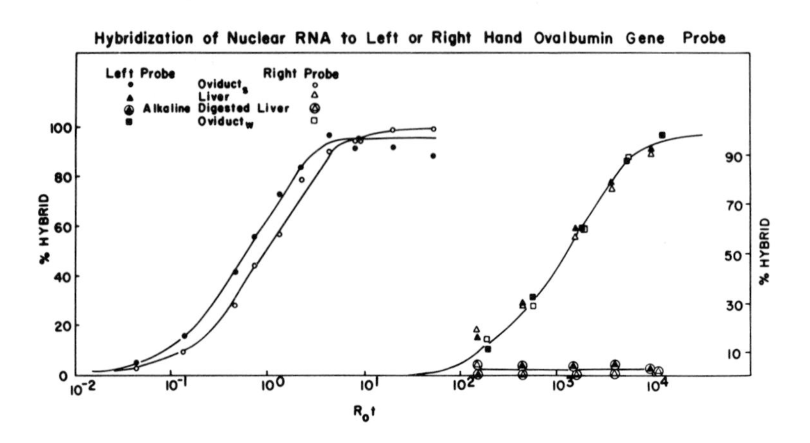

Fig. 5. Hybridization of pure OV_L and OV_R probes to nuclear RNA isolated from chick liver hormonally stimulated and withdrawn oviduct. Nuclei were prepared from diethylstilbestrol (DES) stimulated chick oviduct, DES-withdrawn chick oviduct and chick liver by a modification of the citric acid procedure described by Busch (19). A triton X-100 (0.1%) wash was included in the preparation of stimulated chick oviduct nuclei. RNA was isolated from these nuclei according to the procedure of Holmes and Bonner (20) except that two DNase (RNase-free) digestion steps were included to eliminate the DNA contamination in the final RNA preparation. Isolated RNAs were then hybridized to the [^{3}H]DNA probes (0.2 ng) in 0.6 M NaCl for the R_ot values indicated. The percentage of [^{3}H]DNA hybridized was determined by S1 nuclease digestion. The probes used were OV_L -●-●-, -▲-▲-, -■-■- and -Ⓐ-Ⓐ-; and OV_R -O-O-, -△-△-, -□-□- and -ⓐ-ⓐ-. The RNAs used were stimulated oviduct, -●-●- and -O-O-; liver, -▲-▲- and -△-△-; alkaline hydrolyzed liver RNA, -Ⓐ-Ⓐ- and -ⓐ-ⓐ-; and withdrawn oviduct, -■-■- and -□-□-.

suggest that these structural regions are under coordinate control in these tissues and that the non-structural insert sequences do not act as terminators in estrogen-withdrawn oviduct or nontarget cells.

We have recently cloned the *Eco*RI fragments of the natural ovalbumin gene prepared from chick DNA (21). Specific hybridization probes to the "insert sequences" can thus be prepared to answer such important questions as whether these sequences are repeated and whether they are transcribed *in vivo*.

REFERENCES

(1) B.W. O'Malley and A.R. Means, Science 182 (1974) 610.
(2) S.L.C. Woo and B.W. O'Malley, Life Sci. 7 (1975) 1039.
(3) J.J. Monahan, S.E. Harris, S.L.C. Woo, D.L. Robberson and B.W. O'Malley, Biochemistry 15 (1976) 223.
(4) S.E. Harris, A.R. Means, W.M. Mitchell and B.W. O'Malley, Proc. Natl. Acad. Sci. 70 (1973) 3776.
(5) D. Sullivan, R. Palacios, J. Stavnezer, J.M. Taylor, A.J. Faras, M.L. Kiely, N.M. Summers, J.M. Bishop and R.T. Schimke, J. Biol. Chem. 248 (1973) 7530.
(6) S.L.C. Woo, J.M. Rosen, C.D. Liarakos, D.L. Robberson and B.W. O'Malley, J. Biol. Chem. 250 (1975) 7027.
(7) S.Y. Tsai, M.-J. Tsai, R. Schwartz, M. Kalimi, J.H. Clark and B.W. O'Malley, Proc. Natl. Acad. Sci. 72 (1975) 4228.
(8) R.J. Schwartz, M.-J. Tsai, S.Y. Tsai and B.W. O'Malley, J. Biol. Chem. 250 (1975) 5175.
(9) R.D. Palmiter, Cell 4 (1975) 189.
(10) G.S. McKnight and R.T. Schimke, Proc. Natl. Acad. Sci. 71 (1974) 4327.
(11) E.M. Southern, J. Mol. Biol. 98 (1975) 503.
(12) L.A. McReynolds, J. Catterall and B.W. O'Malley, Gene (1977) in press.
(13) J. Marmur, J. Mol. Biol. 3 (1961) 208.
(14) S.L.C. Woo, T. Chandra, A.R. Means and B.W. O'Malley, Biochemistry (1977) in press.
(15) D. Denhardt, Biochem. Biophys. Res. Commun. 23 (1966) 641.
(16) J.J. Monahan, S.L.C. Woo, C.D. Liarakos and B.W. O'Malley, J. Biol. Chem. 252 (1977) 4722.
(17) P.A. Sharp, P.H. Gallimore and S.T. Flint, Cold Spring Harbor Symp. Quant. Biol. 39 (1974) 457.
(18) T. Maniatis, A. Jeffrey and D.G. Kleid, Proc. Natl. Acad. Sci. 72 (1975) 1184.
(19) H. Busch, Methods in Enzymology, Vol. XII part A, eds. L. Grossman and K. Moldave, (Academic Press, New York,

(1967) p. 434.
(20) D.S. Holmes and J. Bonner, Biochemistry 12 (1973) 2330.
(21) Woo, S.L.C., A. Dugaiczyk, E. Lai, J. Catterall and B.W. O'Malley, Proc. Natl. Acad. Sci. (in press).

We wish to thank Ms. Serlina Robinson, Ms. Celine Yu and Ms. Martha Ray for excellent technical assistance. This work was supported by NIH grant HD-08188, the Baylor Center for Population Research and Reproductive Biology and the Howard Hughes Medical Institute (S.L.C.W.).

DISCUSSION

J.B. GURDON: I think you described at the beginning some more stringent criteria for recognizing transcripts from chromatin than had been used before. Did I understand that this work is conducted with natural chromatin and if so, have you applied the same criteria to the reconstituted chromatin?

B.W. O'MALLEY: The answer to the first question is yes, the answer to the second is no. We have not had sufficient time to do the reconstitution experiments using the new radioactive technique.

B.L. MALCHY: I wonder if you could comment on the fact that you don't find the gene for ovalbumin or a homologous gene in the calf thymus genome.

B.W. O'MALLEY: Well, after all it is a chicken egg-white protein.

B.L. MALCHY: Wouldn't you expect some sort of homology that might cause cross hybridization?

B.W. O'MALLEY: If one takes these specific messages, prepares cDNAs and does the hybridization under stringent conditions with total DNA of other species, one really doesn't observe significant hybridization.

B.L. MALCHY: You know that calves and chickens are closely related, so unless the gene has been removed from the genome you would expect to get some cross hybridization.

B.W. O'MALLEY: Well, I can't say anything about whether it should or shouldn't be there, but direct hybridization of the cDNA to calf thymus DNA does not give the appropriate hybridization.

R.G. KALLEN: In your transcription experiment have you detected the insertion sequences in your primary RNA transcripts?

B.W. O'MALLEY: We are just now doing the experiments to determine whether the inserts are transcribed.

R.G. KALLEN: Is there any information available regarding RNA-processing enzymes?

B.W. O'MALLEY: We have done nothing with RNA-processing enzymes, but of course that will be something to consider if one wishes to reconstitute an appropriate transcription system - one would have to pay attention to enzymes that cut and religate the RNA sequences to affect the mature mRNA.

R.G. KALLEN: Do you have any information on palindromes in the DNA sequence of the insertions?

B.W. O'MALLEY: No. We are particularly interested in the sequence linking the message and the insert and also whether the sequence promotes folding into a secondary structure.

S.P. CRAIG: In your previous studies you have shown that the ovalbumin mRNA hybridizes to DNA with kinetics indicating that it is coded for by a unique gene. I was wondering if the complete cloned DNA sequence hybridizes with kinetics that would indicate that the inserted sequences could be redundant?

B.W. O'MALLEY: That experiment will be done shortly.

K.S. McCARTY: Have you made any attempt to determine if the steroid receptor proteins have the capacity to recognize the insert region?

B.W. O'MALLEY: No, but you can well imagine that we will be looking at that possibility.

K.S. McCARTY: The possibility that the insert DNA specifically interacts with the steroid receptor protein complex is obvious. We anxiously await the results of these experiments.

B.W. O'MALLEY: We have not forgotten about the receptor which is our first love. Eventually we will get back to that when we better understand the sequence-structure of the natural gene.

REGULATION OF GENE EXPRESSION IN NORMAL AND NEOPLASTIC CELLS

G. S. STEIN[1], J. L. STEIN[1], P. J. LAIPIS[1], S. K. CHATTOPADHYAY[2], A. C. LICHTLER[1], S. DETKE[1], J. A. THOMSON[1], I. R. PHILLIPS[1] and E. A. SHEPHARD[1]

[1]University of Florida
College of Medicine
Gainesville, Florida 32610

[2]National Cancer Institute
National Institutes of Health
Bethesda, Maryland 20014

Abstract: We have been studying the regulation of two sets of genetic sequences -- histone genes in human cells and AKR C-type viral sequences in mouse cell lines. Histone genes: Recently we observed the presence of two species of H4 histone mRNA on the polysomes of S phase HeLa S3 cells. Electrophoresis under denaturing and non-denaturing conditions indicate that the two H4 mRNAs differ in size. Both mRNAs translate H4 histones *in vitro*, lack poly A at their 3' termini and are capped at their 5' termini. Polyacrylamide gel electrophoresis and tryptic peptide analysis suggest that the polypeptides synthesized by the two mRNAs are similar. T1 digests of the two H4 mRNAs and two-dimensional fractionation of the resulting oligonucleotides suggest some differences in the nucleotide sequences. The possible biological significance of the two H4 histone mRNAs is discussed. RNA tumor viruses: The expression of AKR C-type RNA tumor virus has been studied in cell lines derived from AKR mouse embryos. Although all AKR cell lines examined contain the proviral sequences integrated into their genomes, viral producer and non-producer lines have been identified. Using a ^{3}H-labeled DNA complementary to AKR viral RNA sequences, the presence of viral RNA sequences in the nucleus and cytoplasm of viral producer cells has been detected. Chromatin from viral producer cells is an effective template for transcription of AKR viral RNA sequences. A significantly reduced level of hybridization is observed between AKR viral cDNA and nuclear RNA, cytoplasmic RNA and *in vitro* chromatin transcripts from non-producer AKR cells. Regulation of viral gene expression mediated at least in part at the transcriptional level is therefore suggested. The low level of hybridization of viral cDNA with in vivo and in vitro transcribed RNAs from non-producer cells may reflect either partial AKR viral transcripts or other endogenous, closely related C-type viruses. Although chromatins of viral producer and non-producer cells differ in their ability to transcribe AKR viral sequences, high resolution two-dimensional analysis of chromosomal proteins from both cell types does not reveal significant differences.

HISTONE GENES: INTRODUCTION

It is well established that histones play an important role in the structural and transcriptional properties of the eukaryotic genome (reviewed in 1-7). Hence over the past several years a considerable amount of attention has been focused on histone genes and on the regulation of histone gene expression. Histone genes are intriguing from a biological standpoint for a number of reasons. Of particular importance are the presence of five defined species of histone polypeptides and transcription of histone mRNA from reiterated genes. Multiple copies of histone genes raise an important consideration regarding the identity of the histone sequences. Have all copies of the genes coding for individual histones been conserved? Recently while fractionating histone mRNAs, with the objective of isolating mRNAs for individual histones, we isolated from the polysomes of S phase HeLa cells two mRNAs which code for the arginine-rich H4 histone. We have therefore been addressing the questions of whether the two mRNAs are distinctly different species and whether there are variations in the proteins coded by the H4 histone mRNAs.

HISTONE GENES: CHARACTERIZATION OF H4 HISTONE mRNAS

Preparative scale polyacrylamide gel electrophoresis of unlabeled 5-18S polysomal RNA from S phase HeLa cells in the presence of ^{32}P-labeled 5-18S tracer RNA gives the pattern shown in Figure 1. The individual bands were excised and the RNAs were eluted (8). RNAs were then translated in a wheat germ cell-free protein synthesizing system and the products of translation were electrophoresed with unlabeled marker histones on acetic acid-urea polyacrylamide gels. No preliminary purification to separate the histones from other translation products was carried out prior to electrophoresis.

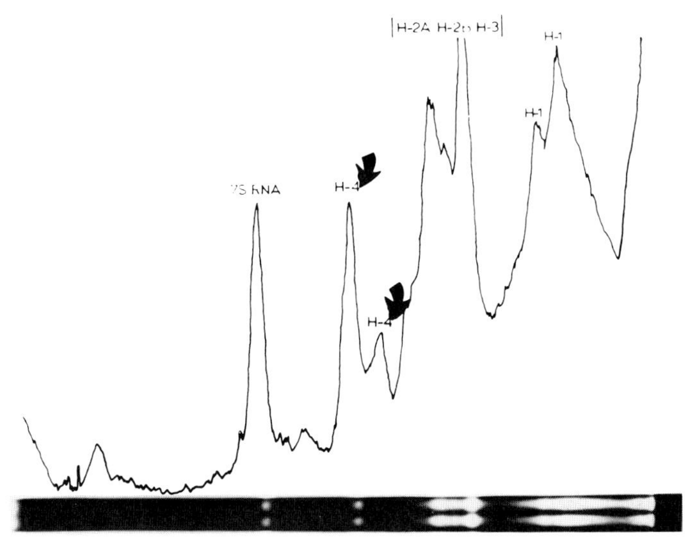

Fig. 1. Preparative acrylamide gel electrophoretic fractionation of 5-18S polysomal RNA from S phase HeLa S_3 cells. 75 μg of unlabeled RNA were combined with 700,000 cpm of ^{32}P-labeled, 5-18S RNA from S phase HeLa cells, loaded on 0.3 x 0.4 cm wells of a 6% acrylamide gel and eletrophoresed as previously described (8). The gel was analyzed autoradiographically and bands were excised as described (8). A densitometric tracing of one of the wells is shown along with a positive contact print of the negative showing two of the wells.

The gel was prepared for fluorography and used to expose Kodak RP-54 X-ray film which had been pre-exposed according to the method of Laskey and Mills (9) so that the darkening was approximately proportional to the amount of radioactivity in the gel. Exposed film was then scanned using a Joyce-Loebel densitometer. The results for RNAs from the faster (H4(1)) and slower (H4(2)) migrating H4 bands of Figure 1 are shown in Figure 2. It should be noted that the amount of ^{3}H-lysine label incorporated into hot acid resistant, trichloroacetic acid precipitable material was proportional to the amount of RNA added to the translation mixture. A similar procedure was followed to assign the coding properties for other bands shown in Figure 1. It can be seen that for both bands designated H4 (Figure 1), the majority of the labeled translation product comigrates with marker H4 histone (Figure 2A and 2B). Based on the assumption that each of the protein products has a similar specific activity and that the area of each peak is proportional to the amount of radioactivity in the gel band, planimetric integration of the peaks indicates that greater than 95% of the translation product of the faster migrating H4 band (Figure 1) and approximately 85% of the translation product of the slower migrating H4 (Figure 1) constitute H4 histone (8).

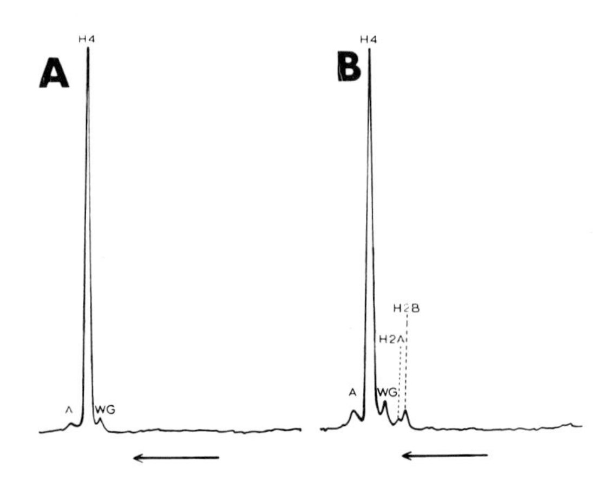

Fig. 2. Acetic acid-urea acrylamide gel electrophoretic analysis of *in vitro* translation products of RNA extracted from the H4 bands shown in Fig. 1. 15 μl of wheat germ *in vitro* translation products were electrophoresed on an acetic acid-urea acrylamide gel in the presence of marker histones, and fluorography was performed as described (8). The migration of marker histones is indicated. A. Translation products of faster migrating H4 band (H4(1)). B. Translation products of slower migrating H4 band (H4(2)).

Peak WG (Figure 2A and 2B) is a wheat germ protein (based on its translation in the absence of added mRNA); therefore its contribution to the total amount of protein has been subtracted in the above calculations. The fast-migrating material in band A may represent incomplete polypeptides, but has been included in the purification calculations. Our estimations of purity for the mRNAs are therefore somewhat conservative.

The question to which we initially addressed ourselves was the nature of the differences between the two H4 histone mRNAs. Since the buffer system used in the electrophoretic fractionation shown in Figure 1 contained 0.1 M Na^+ which would allow RNA to have considerable secondary structure, it is possible that the two mRNA species had the same molecular weight but different secondary structure. It is also possible that the slower migrating H4 band (Figure 1) is a complex of the faster migrating band with some other RNA species. We therefore denatured the two H4 histone mRNAs in hot 98% formamide and compared their electrophoretic mobilities in 8% polyacrylamide gels containing 98% formamide (10), conditions which promote migration of RNA solely as a function of molecular weight. If the two RNA species are the same molecular weight but are separated in aqueous gels because of differences in secondary structure or because of aggregation with smaller RNA species, they should comigrate in formamide. When the two RNA bands were eluted from an aqueous gel and rerun on parallel wells of an 8% acrylamide - 98% formamide gel, it can be seen (Figure 3) that the two bands have distinctly different mobilities. Further support for the contention that the two H4 mRNAs differ in molecular weight rather than in degree of secondary structure can be gleaned from their distinctly different electrophoretic migration in 6% acrylamide gels containing 10 mM glyoxal (Figure 4).

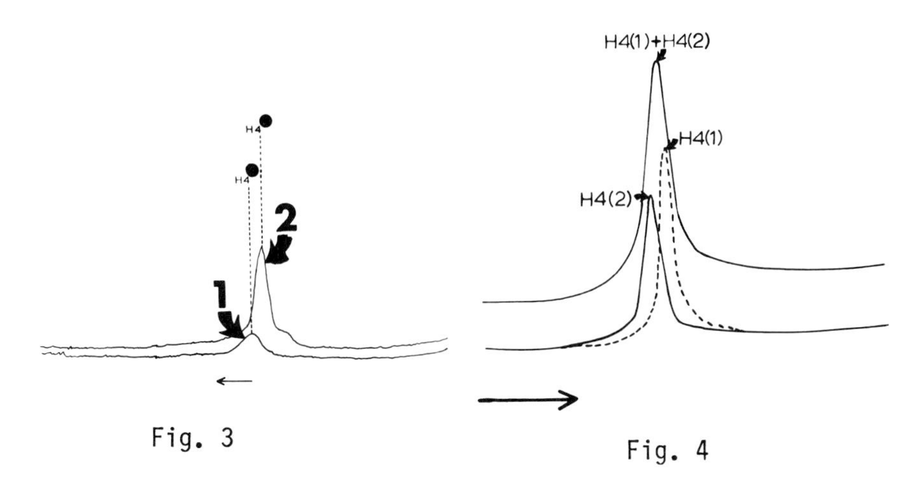

Fig. 3. Electrophoretic analysis under denaturing conditions of RNAs coding for histone H4. ^{32}P-labeled RNAs extracted from an acrylamide gel similar to that shown in Fig. 1 were electrophoresed on an 8% acrylamide-98% formamide gel as described (8). H4 coding bands were electrophoresed on adjacent wells. Densitometric scans were superimposed to facilitate comparison.

Fig. 4. Electrophoretic analysis of the two H4 histone mRNAs in a 6% acrylamide gel containing 10 mM glyoxal as described by McMaster and Carmichael (11).

Note that in a sample containing a mixture of the two H4 histone mRNAs, each species maintains its distinct mobility. Glyoxal interacts preferentially with the guanosine residue of the RNA molecule and the glyoxalated nucleic acid, essentially irreversibly denatured, migrates as a function of molecular weight alone (11).

The 5' and 3' termini of the two H4 histone mRNAs have been characterized. When chromatographed twice on oligo dT cellulose, both H4 mRNAs elute with the high salt wash, are quantitatively recovered and maintain unaltered electrophoretic mobilities (8). The latter results indicate that neither RNA species contains a poly A region long enough to allow it to be retained on an oligo dT cellulose column. Furthermore, these findings suggest that the differences in electrophoretic mobilities of the two H4 histone mRNAs are not attributable to poly A at the 3'-terminus of one of the mRNA species. Based on electrophoretic mobilities on Whatman DEAE paper in pH 4.0 citrate buffer of oligonucleotides resulting from digestion of RNAs with T1, T2 and pancreatic ribonuclease, it appears that both H4 histone mRNAs have capped 5'-termini (data not shown).

To assess further any differences in the nucleotide sequences of the two H4 histone mRNAs, both RNAs (^{32}P-labeled) were fractionated using the mini fingerprint technique of Fiers (12). The RNAs were digested with T1 ribonuclease and fractionated first by high voltage electrophoresis on cellulose acetate in pH 3.5 pyridine-acetate buffer containing 2 mM EDTA and 5 M urea. The oligonucleotides were transferred to polyethyleneimine plates using the glass rod transfer technique of Southern (13) and chromatographed in the second dimension at 64° using homomixture B (alkaline-digested yeast RNA in 7 M urea adjusted to pH 7.5) of Volckaert et al. (12). Figure 5A is a photograph of the T1 fingerprints of the two H4 histone mRNAs and the oligonucleotides are depicted schematically in Figure 5B. The identity of oligonucleotides was confirmed by base composition analysis (14) and alignment along vectors drawn between three or more oligonucleotides. While it is evident that most of the oligonucleotides of the two H4 mRNAs migrate similarly, some variations are observed in the lower regions of the fingerprints, the regions where larger and hence unique sequences are located. Subtle differences in the nucleotide sequences of the two H4 histone mRNAs have been confirmed by two-dimensional polyacrylamide gel electrophoretic analysis (first dimension: 25 mM citric acid (pH 3.5) - 6 M urea at 4°C; second dimension: Tris-borate buffer (pH 8.3)) of T1 ribonuclease-digested mRNAs as described by Fiers (12).

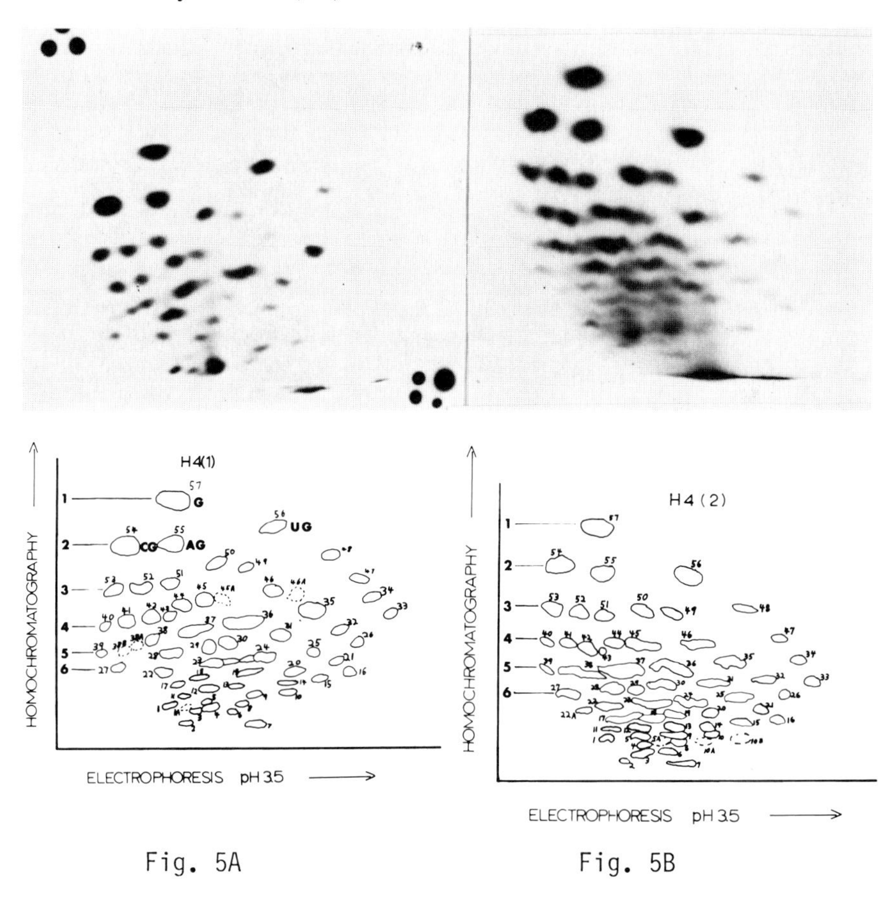

Fig. 5. A. Photograph of T1 fingerprints of the two H4 histone mRNAs. B. Schematic illustration of the fingerprints shown in Figure 5A.

HISTONE GENES: CHARACTERIZATION OF *IN VITRO* TRANSLATED PRODUCTS OF H4 HISTONE mRNAS

As shown in Figure 2, the polypeptides translated *in vitro* from both H4 histone mRNAs fractionated identically in acetic acid-urea polyacrylamide gels and electrophoresed coincidentally with H4 marker. Under these conditions fractionation is by charge and by molecular weight. The similarity of the histones translated *in vitro* was confirmed by digesting ^{3}H-leucine (Figure 6A), ^{3}H-lysine (Figure 6B) and ^{3}H-alanine (Figure 6C) -labeled polypeptides with trypsin and then carrying out one-dimensional peptide mapping (15); significant differences in in vitro translation products of the two H4 histone mRNAs are not evident. This conclusion was further substantiated by a two-dimensional procedure (first dimension: paper chromatography with butanol-acetic acid-H_2O; second dimension: electrophoresis in pyridine-acetic acid-H_2O) (15).

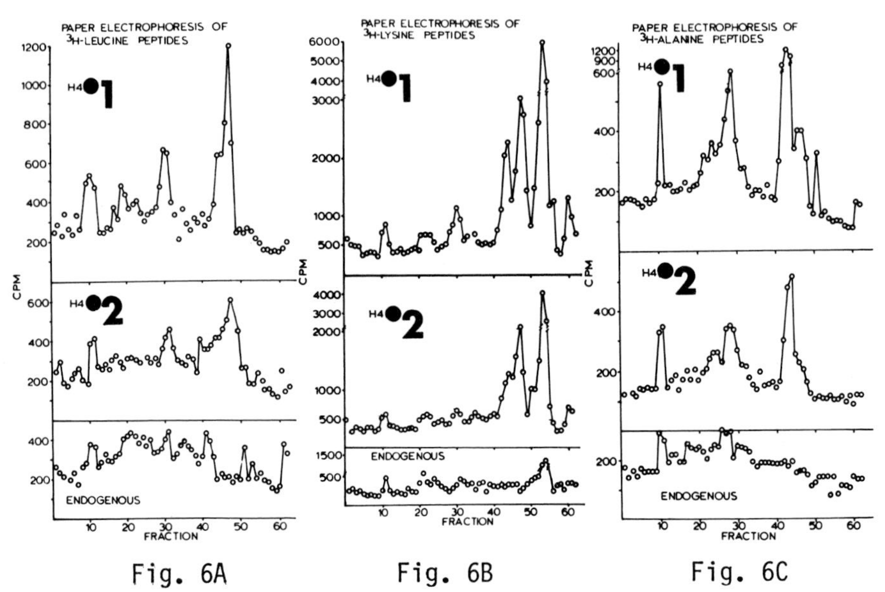

Fig. 6A Fig. 6B Fig. 6C

Fig. 6. One-dimensional tryptic peptide map of polypeptides translated in vitro from the two H4 histone mRNAs. A. ^{3}H-Leucine-labeled tryptic peptides. B. ^{3}H-Lysine-labeled tryptic peptides. C. ^{3}H-Alanine-labeled tryptic peptides. Electrophoresis was carried out on Whatman 3 MM paper in n butanol-acetic acid-H_2O as described by Katz et al. (15).

HISTONE GENES: CONCLUSIONS

Evidence has been presented which suggests that two species of mRNA coding for H4 histones are present on the polysomes of S phase HeLa cells. The H4 mRNAs differ in size, neither mRNA contains significant amounts of poly A at its 3'-terminus, both mRNAs have capped 5'-termini and there are differences in the nucleotide sequences of the two H4 mRNAs. The proteins translated in vitro by the two H4 mRNAs appear to be similar , although the possibility cannot be eliminated that there exist subtle differences in the amino acid sequences of the polypeptides. The presence of multiple forms of H4 mRNA in HeLa cells is consistent with findings of Grunstein et al. (16) that more than one species of H4 mRNA is present in sea urchins. One possible interpretation of our results is that the small H4 mRNA may represent a cleavage product of the large mRNA, thereby reflecting a processing step. Since there are multiple copies of histone genes, whether or not the proteins translated are similar the H4 mRNAs may represent transcripts of different genes, perhaps present in adult and embryonic histone operons such as those Zweidler has postulated (17). The differences in the two H4 mRNAs may also reflect a rearrangement of the H4 histone DNA sequences or post-transcriptional splicing of RNA transcripts outside the coding region. Sequence analysis of the two H4 mRNAs and the DNA sequences which serve as their templates should further elucidate the functional significance of the two H4 histone messages.

RNA TUMOR VIRUS: INTRODUCTION

The presence of endogenous C-type RNA tumor virus is associated with an increased incidence of leukemia in mice (18), but our knowledge of the regulation of expression of these viruses is limited. Inbred mouse strains such as AKR, C58 and C3H/Fg have been shown to contain, integrated into their genomes, 3-4 copies of sequences specifying ecotropic C-type murine leukemia viruses (MuLV) (19-21). In these mice expression of C-type viral sequences occurs shortly after birth and a high viral titer is present throughout life (22). These animals characteristically develop leukemia and die at an early age (6-12 months). Other inbred strains of mice (BALB/c, DBA/2, C2H/He) contain only 1-2 copies of integrated ecotropic C-type MuLV sequences, become viremic only late in life, and then only at low titer and in a small percentage of the population (21). Such mice develop leukemia at much lower frequencies than mice of the first category. A third class of inbred mice and most outbred mice (NZB, 129, C57/Bl6, NIH Swiss) apparently do not contain an entire copy of the ecotropic MuLV sequences, do not produce intact murine leukemia viruses and only rarely develop leukemia (19-21,23). In addition, all mouse strains so far examined contain multiple (7 or more) copies of sequences specifying at least two classes of murine xenotropic C-type viruses (19-21,24). Xenotropic viruses are unable to infect mouse cells but do grow in cells of other species in culture (25-27). Expression of these xenotropic MuLVs is variable; certain inbred mouse strains such as NZB produce large amounts of xenotropic virus. Although the expression of xenotropic viruses does not appear to be correlated with the incidence of leukemia, virus production is implicated in the etiology of nonmalignant diseases such as glomerular nephritis.

Rowe and his collaborators have shown that although newborn and adult AKR mice and cells derived from these animals express the AKR C-type ecotropic MuLV, cells derived from 15-17 day AKR mouse embryos do not produce the AKR MuLV (28, 29). However, sequences complementary to the AKR MuLV are present in the DNA of these early mouse embryo cells and virus production can be induced by treatment with various chemicals or ionizing radiations (28). Occasionally spontaneous virus production is observed in these early AKR embryo non-producer cells with frequencies estimated to be no more that 10^{-8} - 10^{-9} (28,29). It has been postulated that the virus is transmitted vertically, with genetic studies revealing the presence of two chromosomal loci for the integrated AKR MuLV proviral sequences - one of which has been mapped to chromosome seven (30,31).

The question therefore arises as to how expression of these integrated proviral sequences is controlled in the viral non-producer cells derived from AKR mouse embryos. We have examined the transcription of the AKR sequences *in vivo* and *in vitro* using cDNA prepared from AKR MuLV as a probe for detection of the viral RNA sequences in non-producer, spontaneously induced, and IUDR-induced 15-17 day AKR mouse embryo cells. Evidence is presented which suggests that regulation of AKR virus expression is mediated at least in part at the transcriptional level.

RNA TUMOR VIRUS: REPRESENTATION OF AKR VIRAL RNA SEQUENCES IN INTRACELLULAR FRACTIONS OF VIRUS PRODUCER AND NON-PRODUCER CELLS

To assess the levels at which regulation of AKR viral sequences occurs, we initially examined the nucleus and cytoplasm of viral producer and non-producer cells for the presence of AKR viral RNA sequences. RNA was isolated from the nuclear and cytoplasmic cell fractions by an SDS-phenol-chloroform procedure (32) and hybridized in RNA excess with a ^{3}H-labeled AKR cDNA probe (19). Such RNA-driven hybridization permits quantitative as well as qualitative measurements of AKR sequences. The presence of AKR viral RNA sequences was detected in both nuclear and cytoplasmic fractions of viral producer cells (Figures 7A and 7B). The maximal levels of hybridization of AKR cDNA with nuclear RNA (Figure 7A), cytoplasmic RNA (Figure 7B) and AKR viral RNA (Figures 7A and 7B) are similar (81%, 78% and 80%, respectively) suggesting that, as expected, the entire viral RNA sequence is present in the nucleus and cytoplasm of viral producer cells. By comparing the kinetics of the AKR cDNA-nuclear RNA ($Cr_ot_{1/2}$ = 3.3 x 10^1) and the AKR cDNA-cytoplasmic RNA ($Cr_ot_{1/2}$ = 20 x 10^1) hybridization reactions with that of the AKR cDNA-AKR viral RNA hybridization reaction, it can be calculated that AKR viral RNA sequences account for 0.04% and 0.007% of the nuclear and cytoplasmic RNAs, respectively. In contrast the maximal levels of hybridization observed between the AKR cDNA probe and RNAs isolated from the nucleus or cytoplasm of non-producer cells are distinctly lower than the levels of hybridization seen between the probe and either viral RNA or RNAs from producer cells. Nuclear RNA from non-producer cells protects 29-51% of the cDNA probe from S1 nuclease (Figure 7A) and cytoplasmic RNA from non-producer cells protects 21-39% of the probe (Figure 7B).

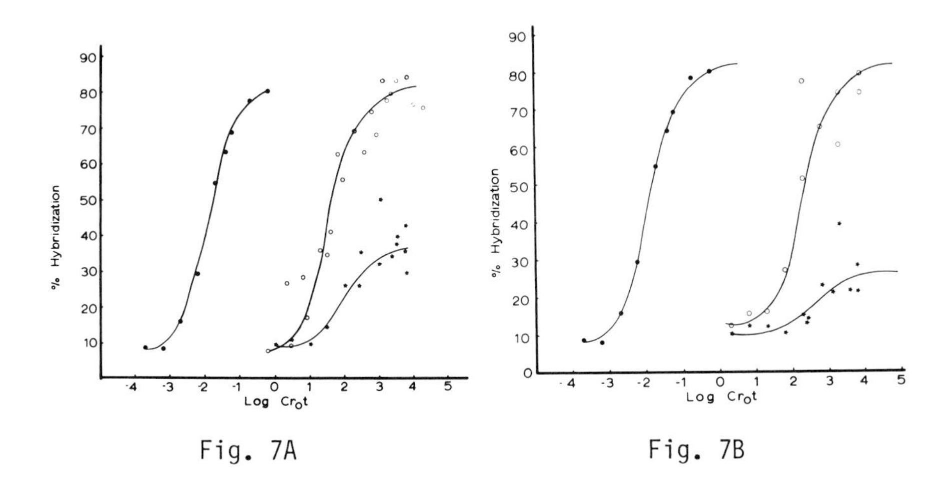

Fig. 7. Kinetics of hybridization of nuclear and cytoplasmic RNAs from viral producer and non-producer cells with AKR cDNA (assayed by S1 nuclease). A. Nuclear RNAs. B. Cytoplasmic RNAs. (●) 70S viral RNA; (○) producer; (★) non-producer.

In other experiments where the cytoplasm was further fractionated prior to RNA extraction, a similarly reduced maximal level of hybridization was observed for RNAs from non-producer compared with viral producer cells. The reduced levels of hybridization seen for nuclear and cytoplasmic RNAs in non-producer cells were obtained by determining the resistance of hybrids to S1 nuclease followed by precipitation of the hybrids with trichloroacetic acid. To confirm these results, hybrids were analyzed by hydroxylapatite chromatography and similar reduced maximal levels of hybridization of the AKR cDNA to non-producer RNAs were observed.

The data presented in Figures 7A and B represent hybridization experiments using RNAs extracted from different cell preparations. The maximal levels of hybridization observed for RNAs from non-producer cells varied among different batches of cells but were consistent for multiple extractions of RNA from the same cell preparation. From the kinetics of hybridization of the AKR cDNA probes to the nuclear and cytoplasmic RNAs from non-producer cells ($Cr_ot_{1/2} = 1.0 \times 10^2$ and 4.0×10^2, respectively), it can be calculated that these hybridizable sequences account for 0.014% and 0.004% of the nuclear and cytoplasmic RNAs.

There was a small (2-3 fold) but reproducible decrease in the rates of hybridization of RNAs from the nuclear as well as cytoplasmic cellular fractions of non-producer compared with producer cells. This decrease indicates that there are fewer copies of the hybridizable sequences in a non-producer compared with a producer cell. However, since the annealing reactions for producer and non-producer RNAs were carried out with similar amounts of RNA and probe, and since the $Cr_ot_{1/2}$ values differ only by a factor of 2-3, the observed differences in the maximal levels of hybridization of RNAs from producer and non-producer cells represent a true difference in the complexity of the sequence population which is hybridizable to the AKR cDNA probe.

It should be noted that the maximal levels of hybridization of nuclear RNAs from non-producer cells are consistently higher than those of cytoplasmic RNAs. In addition, in both producer and non-producer cells the representation of AKR hybridizable sequences is lower in the cytoplasmic fraction than in the nuclear fraction. These data could suggest post-transcriptional processing has occurred.

RNA TUMOR VIRUSES: IN VITRO TRANSCRIPTION OF AKR VIRUS RNA SEQUENCES FROM CHROMATIN OF VIRUS PRODUCER AND NON-PRODUCER CELLS

To examine further the regulation of AKR viral gene expression, we compared the abilities of chromatin from the AKR non-producer cells and spontaneous AKR viral producer cells to transcribe AKR viral RNA sequences *in vitro*. Chromatin was isolated from both cell lines and transcribed with *E. coli* RNA polymerase. The RNA transcripts were isolated and annealed to AKR viral cDNA. The data in Figure 8 indicate that there is a tenfold increase in representation of RNA sequences which hybridize with AKR viral cDNA in transcripts from chromatin of spontaneous producer as compared with non-producer cells. The $Cr_ot_{1/2}$ values of the RNA-cDNA hybridization reactions are 1×10^1 and 1×10^2 for transcripts from chromatin of producer and non-producer cells, respectively.

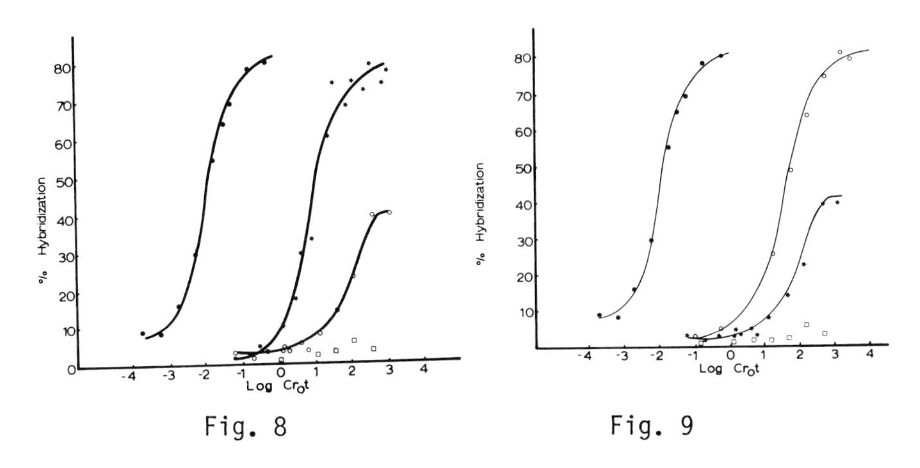

Fig. 8

Fig. 9

Fig. 8. Kinetics of hybridization of AKR cDNA with *in vitro* chromatin transcripts from spontaneous viral producer and non-producer cells. (●) 70S viral RNA; (★) producer; (○) non-producer; (□) endogenous RNAs from producer cell chromatin.

Fig. 9. Kinetics of hybridization of AKR cDNA with *in vitro* chromatin transcripts from IUDR-induced viral producer cells and non-producer cells. (●) 70S viral RNA; (○) IUDR-induced producer; (★) non-producer; (□) endogenous RNA from producer cell chromatin.

By comparing these $Cr_ot_{1/2}$ values with the kinetics of AKR cDNA-AKR viral RNA annealing, it can be calculated that 0.14% and 0.014% of the RNA transcripts from producer and non-producer chromatin hybridized with the AKR viral cDNA probe. Consistent with results from the hybridization analysis of cellular RNAs (Figure 7), the maximal level of annealing (23-39%) is significantly decreased for chromatin transcripts from non-producer compared with producer cells.

The template activities of chromatin preparations from producer and non-producer cells are similar, indicating that AKR sequences are not masked by a dilution effect in the chromatin transcripts from non-producer cells. It should also be noted that the size distribution of the *in vitro* RNA transcripts from chromatin of virus producer and non-producer cells are similar - more than 90% electrophoresing in the 4-14S region of polyacrylamide gels. Although endogenous RNAs are associated with our chromatin preparations, no hybridization above background was observed when endogenous RNAs were isolated and annealed with AKR viral cDNA probe (Figure 8), suggesting that chromatin transcripts which formed hybrids with the AKR cDNA were transcribed *in vitro*. However, further controls are necessary to establish that RNA transcription under our *in vitro* conditions reflects RNA synthesis as it occurs in intact cells.

We also analyzed *in vitro* RNA transcripts from chromatin of IUDR-induced AKR viral producer cells for the presence of AKR viral RNA sequences (Figure 9). A comparison of the kinetics of hybridization of the AKR cDNA probe to IUDR-induced cell chromatin transcripts ($Cr_ot_{1/2}$ = 40) with that of the probe to non-producer cell chromatin transcripts shows a 2.5-fold increase in transcription of AKR sequences. As was observed with chromatin from spontaneous viral producer cells, endogenous RNAs do not hybridize with the AKR cDNA probe above background levels (Figure 9).

It is interesting to note that although the chromatins of viral producer and non-producer AKR cells differ in their abilities to transcribe AKR viral sequences, high resolution one-dimensional and two-dimensional analysis of chromosomal proteins from both cell types does not reveal significant differences (Figure 10).

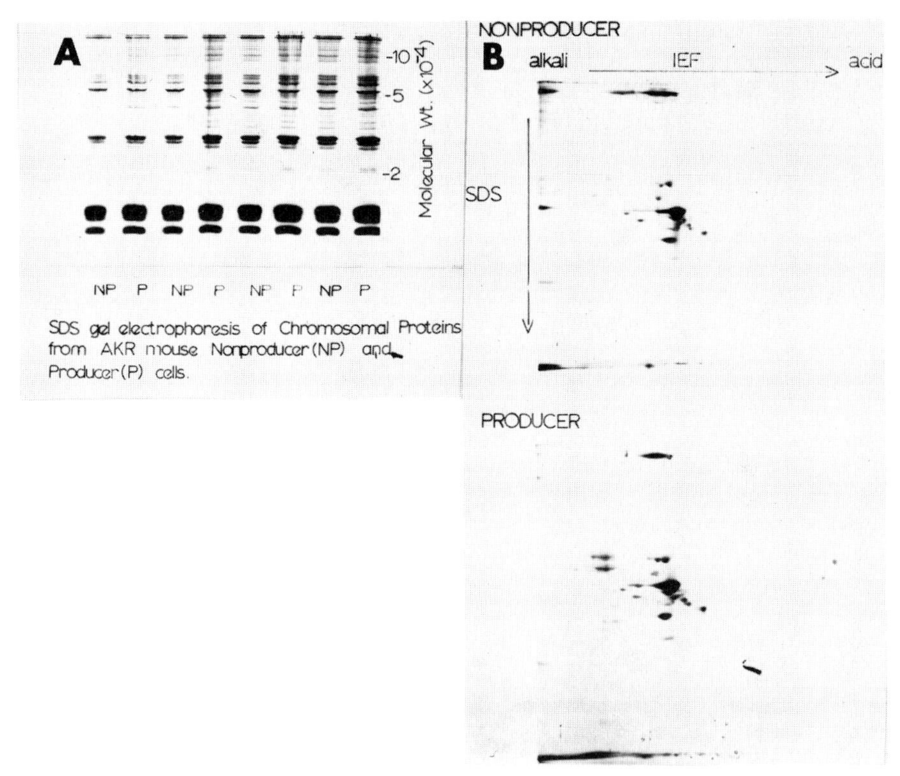

Fig. 10. Electrophoretic fractionation of chromosomal proteins from viral producer and non-producer AKR cells. A. One-dimensional SDS acrylamide gels. B. Two-dimensional fractionation according to the method of O'Farrell (42).

RNA TUMOR VIRUSES: CONCLUSIONS

We have begun to examine the regulation of gene expression in cells derived from embryos of AKR mice. Although both producer and non-producer AKR cell lines contain DNA encoding the complete AKR virus, expression of the sequence does not occur readily in the non-producer cell line. In order to examine the biochemical mechanisms which are responsible for this regulation of gene expression, we have examined the transcription of AKR viral sequences both in intact cells and in isolated chromatin. As expected, RNA molecules isolated from producer cells or transcribed *in vitro* from chromatin of these cells contain sequences complementary to the entire cDNA probe derived from the AKR MuLV. In contrast, RNA molecules isolated from non-producer cells or transcribed in vitro from chromatin of such cells presented a different, and somewhat ambiguous result. Only a portion of the AKR cDNA probe was protected by these RNAs (21-51%); the representation of these sequences was two to ten-fold less than observed in RNA from producer cells. These results suggest that regulation of expression of the AKR proviral sequences occurs at least in part at the transcriptional level. Whether or not regulation resides solely at the transcriptional level is predicated upon definitive identification of the RNA sequences which partially protect the AKR cDNA probe. These sequences may represent readout of a specific portion of the AKR proviral information, in which case we are dealing with either transcriptional regulation, with only a portion of the proviral genome "activated" or "derepressed", or a post-transcriptional mode of regulation, with a portion of the primary transcript lost during a processing step. If the latter mechanism prevails, it is most likely that such a processing step occurs within the nucleus. A loss of sequences detectable in the nucleus but not in the cyotoplasm is observed in our analysis of RNA sequences from non-producer cells, i.e., 27-51% maximal hybridization of nuclear RNA versus 21-39% maximal hybridization of cytoplasmic RNA.

Alternatively, the limited hybridization observed between RNAs from non-producer cells and the AKR cDNA probe may reflect annealing of the probe to closely related xenotropic MuLVs, which contain considerable sequence homology to AKR. Such xenotropic viruses have been detected in all mouse strains (25-27). Multiple subclasses of xenotropic viruses probably exist. Callahan et al. (24) have shown at least two distinguishable subclasses of xenotropic viruses which by nucleic acid hybridization exhibit 27-66% sequence homology between the two classes. These xenotropic viruses have approximately 50% of their sequences in common with the ecotropic viruses including AKR. Chattopadhyay et al. (19,20) have demonstrated that the AKR mouse genome, in addition to the 3-4 copies of the AKR proviral genome present per haploid cellular genome, contains seven or more copies of a second set of proviral sequences. Gelb et al. (33) reported 9-14 copies of sequences homologous to Kirsten MuLV in uninfected cells of several mouse strains. These multiple proviral sequences probably represent the genomes of xenotropic and perhaps other C-type viruses such as amphotropic or xenotropic-ecotropic recombinants (34-36). If these other viral sequences are transcribed in the AKR non-producer cells, they could hybridize to a limited extent with the AKR probe and thus account for the partial hybridization we have detected.

In an attempt to resolve whether sequences in viral non-producer cells which hybridize with AKR cDNA are partial AKR transcripts or represent other MuLV sequences, we are carrying out thermal denaturation studies. Hybridization analysis of *in vivo* and *in vitro* transcribed RNAs will also be carried out with a cDNA probe to total C-type virus from which AKR sequences have been removed.

In vitro transcription of C-type RNA viral sequences in nuclei and chromatin is not without precedence. Weissman's and Hurwitz's laboratories have demonstrated that in the avian system transcription of both AMV and RSV sequences can be detected *in vitro* (37,38). Transcription of the AMV sequences is sensitive to low doses of α-amanitin, suggesting that these proviral sequences are transcribed by RNA polymerase II (37). Janowski et al. (39) have shown that chromatin from Rauscher leukemia virus-infected BALB/c mouse spleen cells can be transcribed by *E. coli* RNA polymerase to yield sequences complementary to the RLV cDNA probe. Their observation that spleen cells from uninfected BALB/c mice contain some transcripts which hybridize to a limited extent with the RLV cDNA probe has a direct bearing on the interpretation of our results. Similar results were obtained by Benveniste et al. (40) who examined BALB/c cell lines for cytoplasmic RNA sequences related to an endogenous BALB/c MuLV or Kirsten MuLV. Normal, uninfected cells contain sequences which hybridize to a limited extent with both cDNA probes. These data are consistent with expression of endogenous viral sequences in non-producing cells. Varmus et al. (41) have also presented evidence for transcriptional control of the regulation of integrated murine mammary tumor viruses. They observed a complex dependence of viral expression on both viral and host genetic loci.

REFERENCES

(1) J. Bonner, M.E. Dahmus, D. Fambrough, R.-C. Huang, K. Marushige and D.Y.H. Tuan, Science, 159 (1968) 47.

(2) G.S. Stein, T.C. Spelsberg and L.J. Kleinsmith, Science, 183 (1974) 817.

(3) L.S. Hnilica, The Structure and Biological Function of Histones (CRC Press, Cleveland, Ohio, 1972).

(4) S.C.R. Elgin and H. Weintraub, Ann. Rev. Biochem., 44 (1975) 725.

(5) H. Busch, The Cell Nucleus, Vols. I-III (Academic Press, New York).

(6) R.D. Kornberg, Ann. Rev. Biochem., 46 (1977) 931.

(7) G. Felsenfeld, Nature, 271 (1978) 115.

(8) A.C. Lichtler, G.S. Stein and J.L. Stein, Biochem. Biophys. Res. Comm., 77 (1977) 845.

(9) R.A. Laskey and A.D. Mills, Eur. J. Biochem., 56 (1975) 335.

(10) J.C. Pinder, D.Z. Staynov and W.B. Gratzer, Biochemistry, 13 (1974) 5373.

(11) G.K. McMaster and G.G. Carmichael, Proc. Natl. Acad. Sci., 74 (1977) 4835.

(12) G. Volckaert, W. Min Jon and W. Fiers, Anal. Biochem., 72 (1976) 433.

(13) E.M. Southern, Anal. Biochem., 62 (1974) 317.

(14) G.G. Barrell, in: Procedures in Nucleic Acid Research, Vol. 2, eds G.L. Cantoni and D.R. Davies (Harper and Rowe, New York, 1971) p. 751.

(15) A.M. Katz, W.J. Dreyer and C.B. Anfinsen, J. Biol. Chem., 234 (1959) 2897.

(16) M. Grunstein, S. Levy, P. Schedl and L. Kedes, Cold Spring Harbor Symp. Quant. Biol., 38 (1973) 717.

(17) A. Zweidler, M. Urban and P. Goldman, Abstracts of the Tenth Miami Winter Symposia (Academic Press, New York, in press).

(18) L. Gross, Proc. Soc. Exp. Biol. Med., 78 (1951) 342.

(19) S.K. Chattopadhyay, D.R. Lowy, N.M. Teich, A.S. Levine and W.P. Rowe, Proc. Natl. Acad. Sci., 71 (1974a) 167.

(20) S.K. Chattopadhyay, D.R. Lowy, N.M. Teich, A. S. Levine and W.P. Rowe, Cold Spring Harbor Symp. Quant. Biol., 34 (1974b) 1085.

(21) D.R. Lowy, S.K. Chattopadhyay, N.M. Teich, W.P. Rowe and A.S. Levine, Proc. Natl. Acad. Sci., 71 (1974) 3555.

(22) W.P. Rowe and T. Pincus, J. Exptl. Med., 135 (1972) 429.

(23) E.M. Scolnick, W. Parks, T. Kawakami, D. Kohue, H. Okabe, R. Gilden and M. Hatanoka, J. Virol., 13 (1974) 363.

(24) R. Callahan, M.M. Leiber and G.J. Todaro, J. Virol., 15 (1975) 1378.

(25) J.A. Levy, Science, 182 (1973) 1151.

(26) S.A. Aaronson and J.R. Stephenson, Proc. Natl. Acad. Sci., 70 (1973) 2055.

(27) R.E. Benveniste, M.M. Leiber and G.J. Todaro, Proc. Natl. Acad. Sci., 71 (1974) 602.

(28) D.R. Lowy, W.P. Rowe, N. Teich and J.W. Hartley, Science, 174 (1971) 155.

(29) W.P. Rowe, J. W. Hartley, M.R. Lander, W.E. Pugh and N. Teich, Virology, 46 (1971) 876.

(30) W.P. Rowe, J.W. Hartley and T. Bremner, Science, 178 (1972) 860.

(31) S.K. Chattopadhyay, W.P. Rowe, N.M. Teich and D.R. Lowy, Proc. Natl. Acad. Sci., 72 (1975) 906.

(32) G. Stein, W. Park, C. Thrall, R. Mans and J. Stein, Nature, 257 (1975) 764.

(33) L.D. Gelb, J.B. Milstein, M.A. Martin and S.A. Aaronson, Nature New Biol., 244 (1973) 76.

(34) S. Rasheed, M.B. Gardner and E. Chan, J. Virol., 19 (1976) 13.

(35) J.W. Hartley and W.P. Rowe, J. Virol., 19 (1976) 19.

(36) J.W. Hartley, N.K. Wolford, L.J. Old and W.P. Rowe, Proc. Natl. Acad. Sci., 74 (1977) 789.

(37) L. Rymo, J.T. Parsons, J.M. Coffin and C. Weissman, Proc. Natl. Acad. Sci., 71 (1974) 2782.

(38) M. Jacquet, Y. Groner, G. Monroy and J. Hurwitz, Proc. Natl. Acad. Sci., 71 (1974) 3045.

(39) M. Janowski, L. Baugnet-Mahieu and A. Sassen, Nature, 251 (1974) 347.

(40) R.E. Benveniste, G.J. Todaro, E.M. Scolnick and W.P. Parks, J. Virol., 12 (1973) 711.

(41) H.E. Varmus, N. Quintrell, E. Medeiros, J.M. Bishop, R.C. Nowinski and N.H. Sarkar, J. Mol. Biol., 79 (1973) 663.

(42) P. O'Farrell, J. Biol. Chem., 250 (1975) 4007.

ACKNOWLEDGEMENT

These studies were supported by grants GM 20535 and CA 18874 from the National Institutes of Health and PCM 77-15947 from the National Science Foundation.

DISCUSSION

H. LODISH: I was wondering if you would care to comment on the recent paper by Melli in Cell which indicated that transcription of histone genes actually occurred more or less uniformly throughout the cell cycle and that histone mRNA was processed into the cytoplasm only during S phase.

G.S. STEIN: There are several points really with regard to interpretation of those experiments. First of all, the hybridization analysis carried out by Melli et al. was not with a homologous probe, but rather was carried out with cloned sea urchin histone genes. Therefore they are looking probably at best at about ten percent sequence homology. I think what is particularly more important is shown on my last slide. The problem is really the biological situation and that is, how one synchronizes the cells - what one is calling different phases of the cell cycle. The technique which they are using for obtaining G1 cells is double thymidine block. If one monitors the cell cycle after a double thymidine block, this is using HeLa cells, at between 9-11 hours, which is what one would call G1, at least 20-25% of the cells have nuclei incorporating ^{3}H-labelled thymidine (i.e. 20-25% of the cells are actually S phase cells). I think that is an important point with regard to what is the G1 and what is the S phase of the cell cycle. So based on that, I think one has to have reservations with regard to differences in histone mRNA synthesis at points during the cell cycle when double thymidine block is used for synchronization.

The other point that they make in support of their interpretation is that if they take S phase cells and block them with cytosine arabinoside they still get transcription of histone sequences. That is not particularly surprising in the sense that, in fact, we published about a year ago a paper which indicated, in agreement with their data, that if you block DNA replication with either cytosine arabinoside or hydroxyurea, you do in fact lose the histone message from the polysomes of the cells, but you still find that chromatin is effective as a template for transcription of histone sequences. The histone sequences are still present in the nucleus and one finds in fact an increased representation of the histone sequences in the post polysomal cytoplasm. So the coupling does not appear to be mediated at the transcriptional level and their observation that blocking DNA synthesis does not inhibit histone mRNA synthesis does not support the argument that histone genes are transcribed throughout the cell cycle. Those are essentially the reservations that we have with regard to the data of Melli et al. Have I answered your question?

H. LODISH: Partly - just one other point -in the experiments you have published and that you mention in your abstract, you really saw no synthesis of histone message at all during G1?

G.S. STEIN: In HeLa cells?

H. LODISH: In HeLa cells.

G.S. STEIN: That is right. We are using a mitotic selective detachment procedure where we find less than 1% S phase cells in the G1 population.

H.LODISH: Are you saying that this is evidence that the cells really are properly synchronized in G1 and therefore the transcription of the histone genes really is valid?

G. STEIN: Well, yes, that is what I am saying. It is not clear that what they are calling G1 is a G1 cell. It is a population with at least 20% S phase cells.

K. McCARTY: Your observations on sequence differences in peptide 17 of the H4 histone are very interesting. Do you have additional information to place the amino acid substitutions in either the hydrophobic or amino terminal region of the histone H4?

G. STEIN: No, that is a very good point. All that we have done in terms of characterization of the proteins are the one-dimensional and the two-dimensional peptide maps that we showed.

K. McCARTY: In view of the evolutionary conservation of H4 suggesting conservative amino acid substitutions limited to the hydrophobic region and its key role in nucleosome structure, these two histones have the potential to provide a tool for nucleosome structure analysis.

G. STEIN: It would be obviously interesting to find out.

T. BORUN: Are the two species of H4 message translated with equal efficiency?

G. STEIN: Yes, they are.

T. BORUN: In other words, X micrograms of H4(1) and H4(2) mRNA produce the same net stimulation?

G. STEIN: Yes, within a factor of two.

T. BORUN: Well, that is a pretty big factor. In other words, it is hard to get a sufficient number of micrograms of purified stuff from the gel free of the acrylamide, such that you can accurately determine the amounts that you are putting into the system.

G. STEIN: The quantitation that we are doing is based on labelling. On ^{32}P radioactivity - not on OD amounts.

T. BORUN: So you do not really know how much?

G. STEIN: That is why I am saying- within a factor of two.

T. BORUN: You work with WI-38, so have you been able to perceive the two different kinds of H4 in the WI-38 cells?

G. STEIN: We have not looked yet.

E. ATIKKAN: The chromosome distribution of HeLa S3 is not uniform. I am wondering if you could obtain, by random cloning, populations that have either one or the other of the histone messages exclusively, rather than having the two together? The HeLa R cells that we use, do not respond uniformly to various agents. We have been trying to clone them by a single cell cloning method to obtain a uniform culture. In the same way, it should be possible to clone S3 to obtain a uniform cell line that is derived from a single cell.

G. STEIN: I guess it is possible. We have not done it.

E. ATIKKAN: You have not done it?

G. STEIN: No. These are not cloned cells, or at least it has been too long since they were cloned to call them a clone.

E. ATIKKAN: So they come from a mixed, random population of cells.

G. STEIN: Yes, that is right.

A. ZWEIDLER: There are three whole sets of different histone variants that one can resolve by Triton electrophoresis. There is at least one fetal operon in mammals, one adult operon and one spermatocyte operon. I think you would look in the HeLa cells at a mixture of fetal and adult operon expression which would give you two different sets of nucleic acid sequences in the message because the operons have been evolving independently for quite a long time and because there is more divergence in the nucleic acid sequence than in the protein sequence. You could also have differences, of course, in the non-translated regions of the message. Actually you should use the Triton method to look at the translation products of your H4 mRNA fractions.

G. STEIN: That is why I would really like you to try to translate them.

A. ZWEIDLER: With respect to the fingerprint interpretation - it is very dangerous to look just at the pattern, because histones are notorious in partial cleavages of runs of basic residues that do not cleave the same every time. You have to isolate those spots to identify what they actually are in order to show if there are really sequence differences as opposed to the differences resulting from partial cleavage of the protein.

H. SMITH: You have recently shown and you mentioned it in your abstract that the coupling of DNA synthesis with histone synthesis is not mediated at the transcriptional level and one of your points has to do with post polysomal RNA. I was wondering if you would care to comment on your criteria for measuring post polysomal RNA?

G. STEIN: To be frank - not very stringent criteria. What we are calling post-polysomal RNA is simply - I would call it a crude post-polysomal RNA preparation. We spin down the nuclei, mitochondria, microsomes, and the polysomes. The supernatant solution which remains is what we are calling post-polysomal RNA. It is just an operational definition.

A. KOOTSTRA: Have you actually done the experiment whereby you label total histones, isolate chromatin from the cells and purify H4 protein, perform tryptic digestion on H4 followed by your finger printing technique to see if the differences persist under these conditions? In other words, do you see the differences of H4 in chromatin itself?

G. STEIN: We really have not looked at it. We have never done a peptide analysis of the chromatin.

BUSCH: We sequenced the H4 some time ago and did not find convincing evidence for two species. We were not troubled by a problem of complete digestion with trypsin; that is very easily controlled as we described the conditions some time ago. Unless another species is present in a rather small amount in the final purified histone population, my presumption is that you may have some intermediate on the way to final processing. Evidence for a pre-histone or a pre-pro-histone has not been demonstrated until now, but that may be part of what you are finding.

G. STEIN: Yes. My feeling is that what we are looking at are similar histones but different messages.

THE ROLE OF CYCLIC NUCLEOTIDES IN DEVELOPMENT

H. V. RICKENBERG
Division of Molecular and Cellular Biology
National Jewish Hospital and Research Center
and
Department of Biophysics and Genetics
University of Colorado School of Medicine
Denver, Colorado 80206

INTRODUCTION

The organism which we employ in most of our experiments, Dictyostelium discoideum, requires no detailed introduction. The cellular slime mold is used widely in the study of both morphogenesis and differentiation and its biology is the subject of a recent monograph (1). Very briefly, the organisms grow as unicellular amoebae; upon starvation the amoebae form multicellular aggregates of up to 10^5 cells per aggregate, which takes on the form of a pseudoplasmodium. This behaves in many respects as an integrated tissue and consists of two (possibly three) types of cells. The ratio of the two major classes of cells, i.e. of spores and of stalk cells, is fixed and arises, presumably, as a result of the interaction of the cells in the pseudoplasmodium. The two types of cells differ in their enzymic repertoire as well as morphologically. The spores, under proper environmental conditions, will germinate to form new amoebae whereas the cells which constitute the stalk are moribund.

In D. discoideum the chemotactic agent is cyclic AMP (2) which is released in a pulsatile manner (3) by the starving amoebae. It appears that the cyclic AMP is secreted initially by a few cells at the centers of aggregation territories and is detected by adjacent cells (4). The latter then respond by migrating toward the source of the cyclic nucleotide and by the release of their own pulse of cyclic AMP. The identity of cyclic AMP as the chemotactic agent during the development of D. discoideum is firmly established. The possibility that the cyclic nucleotide is also the agent which determines, whether a given cell will become a spore or a stalk cell is suggested by the, as yet unconfirmed claim (5), that prestalk cells contain about 10 times as much cyclic AMP (and, incidentally twice as much calcium) as do prespores. In other species of cellular slime molds,

ISBN 0-12-045450-5

taxonomically closely related to D. discoideum, a small polypeptide, rather than cyclic AMP, acts as chemotactic agent (6).

The immediate interest of our group is the role of cyclic nucleotides in the development of D. discoideum. More generally, however, we should like to explore the hypothesis of a causal relationship between starvation, the synthesis of cyclic nucleotides, the occurrence of multicellularity and, eventually, of differentiation. For this reason we shall discuss not only our current work with D. discoideum, but also certain older, published experiments (7) on the prokaryote Escherichia coli.

STARVATION AND THE ACCUMULATION OF CYCLIC AMP

A causal relationship between starvation and an increase in cellular cyclic AMP has been demonstrated in both prokaryotes and eukaryotes. The situation in the bacterium Escherichia coli may serve as an example of that relationship. Several years ago, in the course of a study of the phenomenon of catabolite repression, we examined the effect of starvation for carbon on cyclic AMP levels in the bacteria (7). In the experiment shown in Fig. 1, E. coli was cultured on a mineral salts medium with a growth-limiting concentration of glucose. When the glucose was exhausted, the cellular concentration of cyclic AMP rose rapidly to approximately 30 times the basal level; the rise was followed by the release of cyclic AMP into the medium. The cyclic nucleotide continued to be formed and secreted into the medium during prolonged starvation and at a time when its intracellular accumulation had ceased. The readdition of glucose to a starving culture of bacteria led to a very rapid drop in the cellular concentration of cyclic AMP. Table 1 shows cellular concentrations of cyclic AMP in E. coli during growth on different sources of carbon. It may be seen that there was an inverse relationship between the effectiveness of a given compound as source of carbon, as reflected in the generation time of the culture, and the concentration of cyclic AMP. It may also be seen that the ability of the culture to synthesize β-galactosidase increased as the cellular concentration of cyclic AMP rose.

A relationship between starvation and the formation of cyclic AMP in D. discoideum was first demonstrated by Bonner and his associates (2). Whereas in E. coli it is starvation, specifically for carbon, that induces the accumulation of

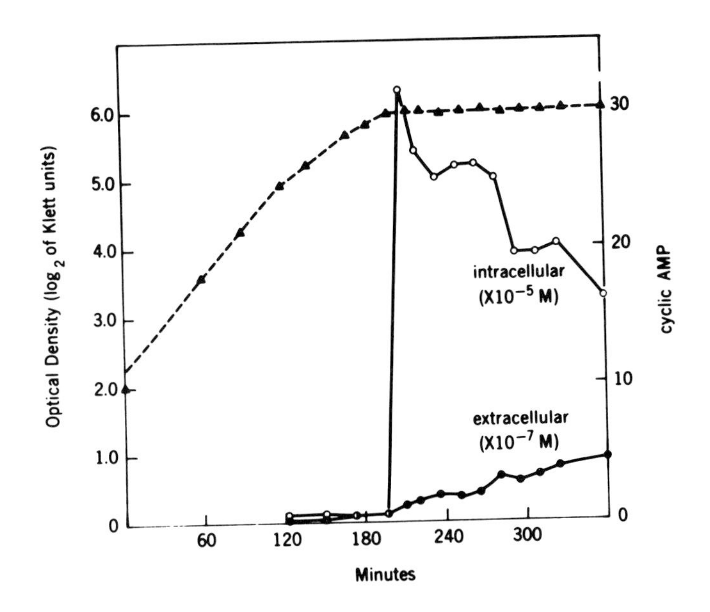

Fig. 1. The intracellular and extracellular concentrations of cAMP in a culture of AB257. Growth was limited by 3.5 mM glucose. Symbols: ▲, bacterial growth; o, intracellular cAMP; ●, extracellular cAMP.

cyclic AMP, the situation is less clear in D. discoideum. Our own observations (8) suggest that it is indeed starvation for carbon that triggers the synthesis of cyclic AMP in D. discoideum. We found that the presence of a metabolizable sugar sufficed to inhibit the synthesis of the cyclic nucleotide which occurred in the absence of that sugar. This is shown in Figs. 2 and 3. In Fig. 2 the effect of maltose on the synthesis of cyclic AMP by amoebae, plated and aggregating (in the absence of the sugar) on petri dishes, is shown. In the experiment described in Fig. 3 amoebae were incubated in suspension (9) in either the presence or absence of maltose and cellular concentrations of cyclic AMP were determined. α-Methylglucoside and arabinose are not metabolized by D. discoideum and were included in the experiments as controls for non-specific inhibitory, possibly osmotic effects of the sugars. The detailed procedures for the harvesting of the amoebae and the determination of cyclic AMP

TABLE 1

Effect of source of carbon on cellular concentration of cyclic AMP and ability to form β-galactosidase in *Escherichia coli*

Source of carbon	Generation time (min)	Intracellular cyclic AMP (μM)	β-Galactosidase specific activity*
Glucose, 10 mM	56	10	200
Glycerol, 20 mM	96	26	12,500
Succinate, 15 mM	170	50	14,500
Acetate, 30 mM	270	55	19,800
L-proline, 30 mM	218	200	43,800

E. coli strain AB257 was employed; procedures are described in Reference 7. *Specific β-galactosidase activity is given as nanomoles of o-nitrophenol formed per minute per milligram of bacterial protein at 37 C. Methyl-thio-β-D-galactoside at a concentration of 0.5 mM served as inducer of β-galactosidase.

is described in Reference 8. Clearly the presence of a metabolizable sugar sufficed to inhibit the synthesis of cyclic AMP. This suggests, but does not prove, that, conversely, it is the absence of a metabolizable source of carbon which stimulates the accumulation of the cyclic nucleotide in a manner analogous to the effects of starvation for carbon in *E. coli*. It should be pointed out, however, that work by Marin (10) indicates that starvation for amino acids triggers development in *D. discoideum*. The interpretation of this observation is complicated by the fact that a number of amino acids serve as an excellent source of carbon for *D. discoideum*; to our knowledge the effect of starvation for amino acids on cyclic AMP levels in the slime mold has not been examined.

That cyclic AMP is formed in the cells of higher

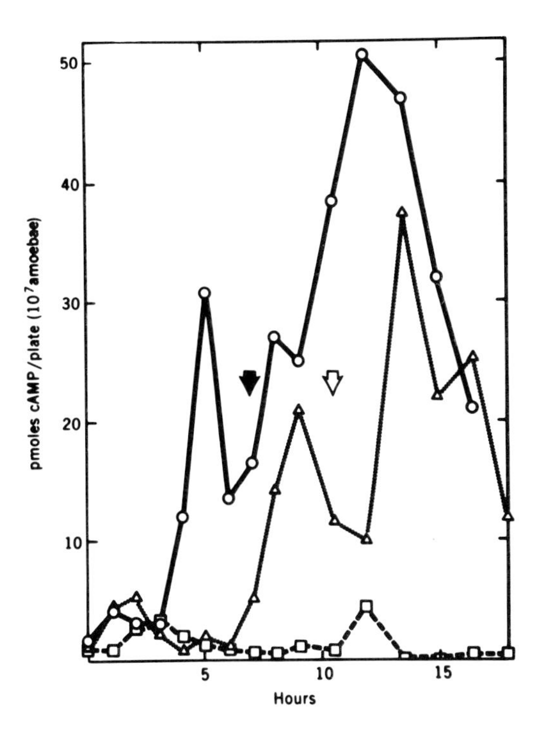

Fig. 2. Effect of sugars on total cAMP formed by amoebae aggregating on petri dishes. Unsupplemented buffer (o); buffer containing 0.1 M maltose (□) or 0.1 M α-methylglucoside (Δ). The black arrow marks the onset of streaming in the control preparation and the open arrow the onset of streaming in the preparation containing α-methylglucoside. See Reference 8 for details.

organisms in response to starvation is a matter of common knowledge; in the intact animal this response is mediated by hormones such as glucagon and epinephrine in the liver and in lipocytes, by epinephrine in muscle and kidney, etc. Little is known, however, about the effects of starvation on the accumulation of cyclic AMP in metazoan cells grown in the absence of hormones, i.e. in tissue culture. It appears that the removal of serum from the medium leads to an increase in cellular cyclic AMP (e.g., 11, 12). Whether serum deprivation is in any manner analogous to starvation, as described above for *E. coli* and *D. discoideum*, or whether serum acts as a source of growth factors which normally inhibit the accumulation of cyclic AMP, is not clear.

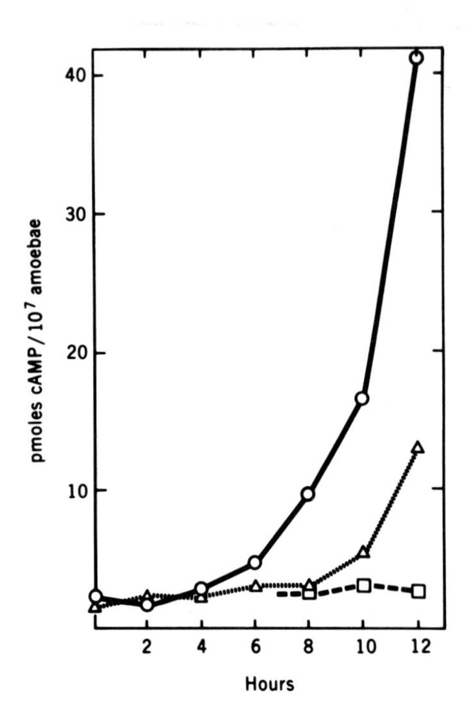

Fig. 3. Cellular cAMP in suspensions of amoebae incubated in phosphate buffer. Unsupplemented buffer (o), buffer supplemented with 0.1M maltose (□) or with 0.1 M arabinose (Δ). See Reference 8 for details.

Summarizing this section, we may conclude that starvation in prokaryotes, the lower eukaryotes, and in metazoa leads to the cellular accumulation of cyclic AMP. The precise mechanism by which the deprivation of nutrients brings about this effect is still not known; in the prokaryotes and lower eukaryotes it must, _ipso facto_, be cellular. In higher organisms a hormonal mechanism of cyclic AMP formation has been demonstrated; whether or not this hormonal regulation is superimposed on an underlying cellular mechanism of control based, for example, on "effectors" such as intermediates of metabolism, on the energy charge of the cell, or its plasma membrane potential, etc., remains to be seen.

THE EFFECTS OF CYCLIC AMP

Again, we shall start our inquiry at the bottom of the evolutionary ladder. The availability of viable mutants in

E. coli, defective in their cyclic AMP metabolism, has made it possible to elucidate both the physiological function and the mode of action of cyclic AMP at the molecular level in this organism. The subject has been reviewed extensively (13,14,15) and we shall be brief. It appears that in E. coli cyclic AMP facilitates the transcription of perhaps 50 to 200 genes (our estimate) which, typically, code for proteins not required under all conditions of growth. In a majority of cases the genes so affected are under dual control in that their expression also requires a second, operon-specific, signal. The case of the lac operon may serve as an example; effective transcription of this cluster of genes requires the exposure of the bacteria to both a β-galactoside which neutralizes the inhibition exercised by the operon-specific, proteinaceous repressor and a threshold concentration of cyclic AMP. This means that the operon is transcribed at a maximal rate only, if bacterial growth is limited by the availability of an effective source of carbon or if cyclic AMP is furnished exogenously. Referral to Table 1 illustrates this point.

Cyclic AMP forms a non-covalent complex with a protein of molecular weight 45,000; the complex interacts with the deoxyribonucleotide sequences which constitute the sites at which RNA polymerase binds to a given operon prior to its transcription. The interaction between the protein-cyclic AMP complex and the deoxyribonucleotide sequences in some manner facilitates the binding of the RNA polymerase and hence the transcription of the genes.

As already stated, cyclic AMP in E. coli modulates the transcription of genes which code for proteins not essential under all conditions of growth. These include enzymes which catalyze reactions that make available to the bacteria sources of carbon not attacked prior to the increase in the cellular concentration of cyclic AMP. It might be pointed out here that this energy source-mobilizing function of cyclic AMP in E. coli has its physiological counterpart in the activation of glycogenolysis by cyclic AMP in mammalian liver and muscle (16). In that case, though, the cyclic nucleotide stimulates the activity of an already existing protein. A closer parallelism to the situation in E. coli is found in the stimulation by adenyl cyclase-activating hormones of the de novo formation of enzymes which catalyze the synthesis of intermediates of energy metabolism from amino acids. An early example of this was the observation (17) that glucagon, epinephrine, and dibutyryl cyclic AMP stimulated the formation of serine dehydratase in rat liver. Unlike in the case of prokaryotes, the mechanism by which cyclic AMP regulates the synthesis of proteins in eukaryotes is not known, and this is one of several reasons why we decided to work with the relatively simple, differentiating eukaryote D. discoideum.

As stated in the Introduction, D. discoideum synthesizes cyclic AMP in response to starvation; the cyclic nucleotide is released by the starving amoebae in a pulsatile manner and brings about the formation of the multicellular pseudoplasmodium. As also pointed out earlier, we were able to demonstrate that metabolizable sugars inhibited the formation of cyclic AMP and, not surprisingly, blocked development (18). This is shown in Figs. 4 and 5. In one case

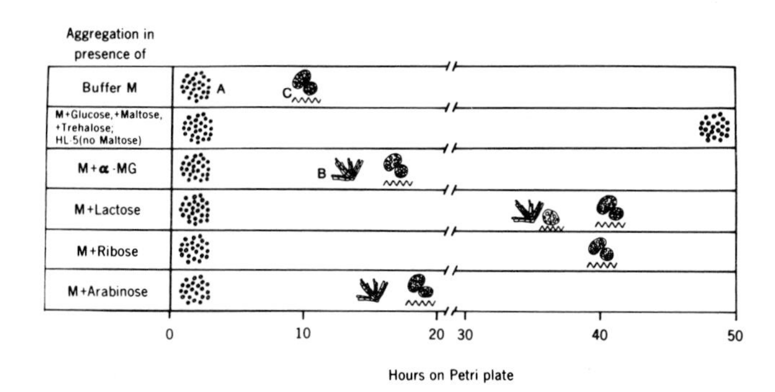

Fig. 4. Effect of sugars on aggregation. Washed amoebae were resuspended to 3×10^6 cells/ml in either medium HL-5 or in buffer containing substances as indicated. The final concentration of the sugars was 0.1 M. Cells were plated in 55 mm plastic petri dishes and development was followed microscopically. α-MG, α-Methylglucoside. (A) Single amoebae attached to bottom of petri plate; (B) aggregation streams; (C) mature, floating aggregates. For details see Reference 18.

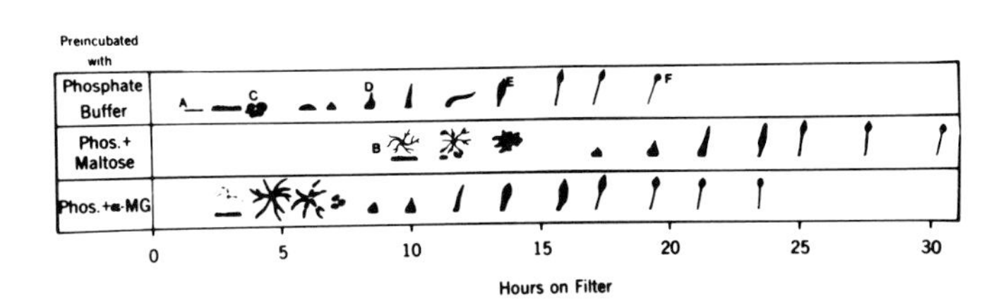

Fig. 5. Development on membrane filters after preincubation in phosphate buffer. Washed amoebae were resuspended to a density of 3×10^6 cells/ml in phosphate buffer, in buffer containing 0.1 M maltose, and in buffer containing 0.1 M α-methylglucoside (α-MG), respectively. The suspensions were then shaken for 8 h at 22 C, harvested, washed, and amoebae (2.5×10^7 per membrane filter) plated. Development was followed microscopically. (A) Rippling; (B) aggregation streams; (C) aggregates; (D) fingers; (E) immature fruiting bodies; (F) mature fruiting bodies. For details see Reference 18.

(Fig. 4) the sugars were added to amoebae plated in petri dishes and, as may be seen, aggregation did not occur in the course of the experiment. In the other case (Fig. 5) amoebae were preincubated in suspension with the sugar for 8 hours; the sugar was then removed and the amoebae plated on millipore filters in the absence of the sugar. The amoebae which had been preincubated with the sugar were retarded by approximately 8 hours in their subsequent development. The fact that certain "aggregateless" developmental mutants, arrested early in development, can be "rescued" by the application of exogenous cyclic AMP also suggests a role for the cyclic nucleotide in development (19).

We decided to explore what some of these functions might be. It is self-evident that the development of D. discoideum in the absence of exogenous nutrients entails the catabolism of preexisting macromolecules as well as the synthesis of new species which constitute the differentiated structure. Our original approach to the problem was indirect; we studied the effects of sugars, which inhibited development, on the degradation of glycogen, RNA, and protein, which had been formed during vegetative growth. Figs. 6, 7, 8 show that the sugars inhibited severely the degradation of the three classes of macromolecules. We are not sure, however, that

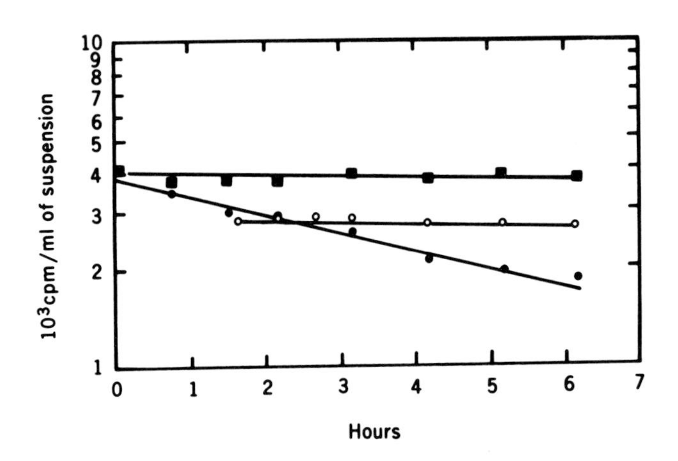

Fig. 6. Effect of glucose on degradation of preformed glycogen during incubation in phosphate buffer. The amoebae were grown in medium HL-5-glucose containing 390 μCi of [U-^{14}C]glucose/liter for approximately eight generations and resuspended in 0.017 M phosphate buffer. They were then incubated at 22 C with shaking, samples were taken at hourly intervals, the cellular glycogen extracted, and the radioactivity determined. Symbols: ●, no addition; ■, glucose added to final concentration of 0.1 M at 0 time; o, glucose added to final concentration of 0.1 M at 1.7 h after resuspension of the amoebae. For details see Reference 18.

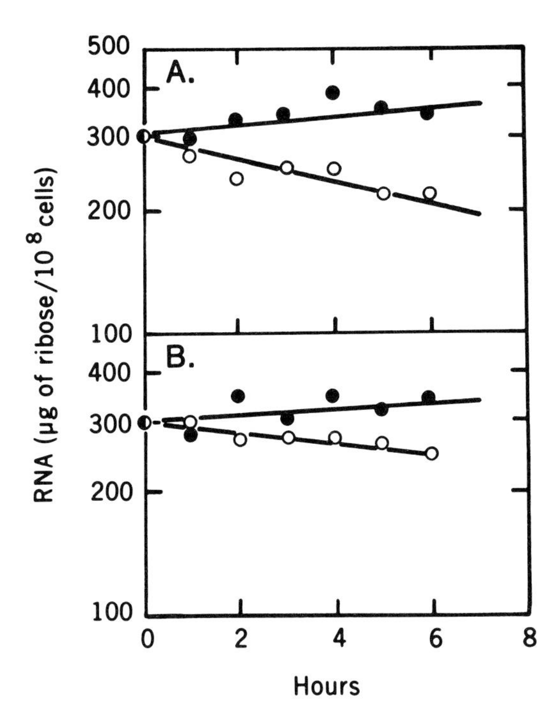

Fig. 7. Effect of maltose on degradation of amoebal RNA. Washed amoebae were resuspended in buffer or in buffer containing 0.1 M maltose and shaken at 22 C. RNA was determined on cellular TCA-precipitable material. (A) Incubation in the absence of cycloheximide: o, control; •, 0.1 M maltose. (B) Incubation in presence of 250 µg of cycloheximide per ml: o, no maltose added; •, 0.1 M maltose.

cyclic AMP is involved in these early degradative processes since, on the one hand, endogenous cyclic AMP appears to increase significantly only after approximately two to four hours of starvation and, on the other hand, starving amoebae respond to exogenous pulses of cyclic AMP also only after two hours of starvation (8).

We next examined the effects of exogenous cyclic AMP on the formation of several enzymes which had been shown earlier (1) to either increase or decrease during development. That exogenous cyclic AMP brings about the precocious formation of cyclic AMP phosphodiesterase (20) and of cell contact sites (21) had already been demonstrated. Here we report on the behavior of two enzymes; alkaline phosphatase activity increases late in normal development, at the stage

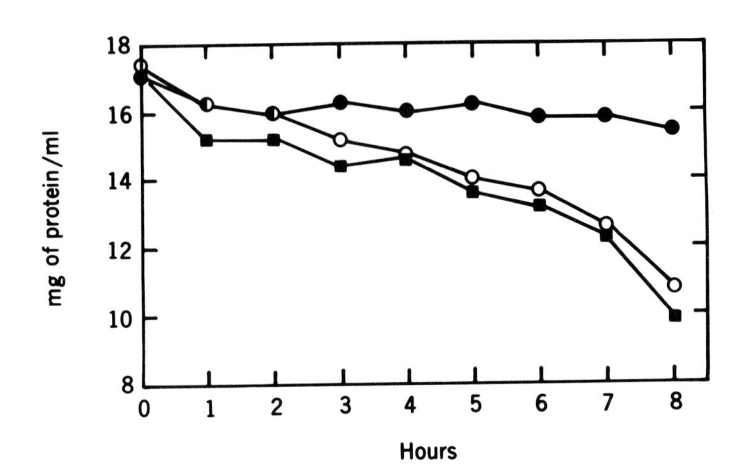

Fig. 8. Effect of maltose on degradation of amoebal protein. Conditions as described in legend to Fig. 7 (A). ■, control; ●, 0.1 M maltose; o, 0.1 M α-methylglucoside.

of culmination, when amoebae form fruiting bodies on a solid surface (1); the enzyme occurs primarily in stalk cells (22). The second enzyme, β-glucosidase-2, is also a "late" enzyme, formed normally during culmination; it however occurs predominantly in spores (22).

Amoebae suspended in dilute buffer to approximately 10^7 cells per ml were pulsed with cyclic AMP at 7 minute intervals such that its instantaneous concentration after each addition was 0.5 μmolar. The detailed procedure for the pulsing of the amoebae with cyclic AMP, or 5' AMP as a control, the sampling, and the enzyme assays are described elsewhere (23). Fig. 9 shows that alkaline phosphatase was induced by cyclic AMP; the inhibition by actinomycin and cycloheximide suggests that the increase in activity represented the _de novo_ synthesis of the enzyme. Cyclic AMP had the opposite effect on the synthesis of β-glucosidase. The findings shown in Fig. 10 demonstrate that the addition of cyclic AMP inhibited the increase in β-glucosidase activity which occurred in the absence of added cyclic nucleotide and which also seemed to represent the _de novo_ synthesis of the enzyme.

It may be concluded from these observations that the pulsatile addition of cyclic AMP brings about changes in the

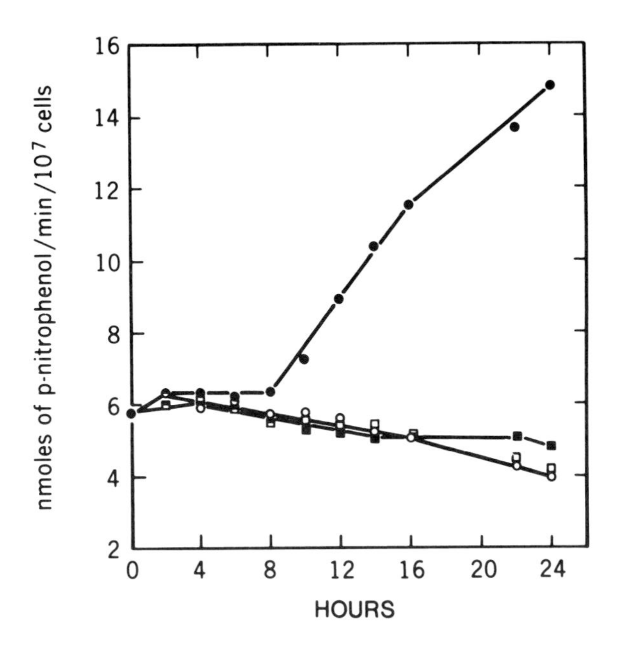

Fig. 9. Alkaline phosphatase activity. o——o—— control; ——●——●—— cyclic AMP; ■——■—— cyclic AMP and actinomycin D; ——□——□—— cyclic AMP and cycloheximide. For details see Reference 23.

enzymic composition of the amoebae which occur normally at a much later developmental stage when a pseudoplasmodium has been formed and close cell-cell contact has been established. It appears that with respect to the formation of alkaline phosphatase, the pulsatile addition of cyclic AMP can substitute for cell-cell contact. One caveat is in order here; we were not able to avoid the formation of small clumps of amoebae (between 4 and 12 cells per clump) in the shaken suspensions. It seems unlikely that the clumping plays a role in the formation of the alkaline phosphatase, since we obtained no larger increases in alkaline phosphatase activity when the suspensions were shaken more slowly and significantly larger clumps were formed. Reducing the density of amoebae in the shaken suspensions from 10^7 to 10^6 cells per ml also had no effect on their ability to form alkaline phosphatase. We added cyclic AMP in a pulsatile manner for two reasons. First, we wanted to mimic physiological conditions (3) and, secondly, the presence of cyclic AMP phosphodiesterase on the amoebal surface and in the buffer led to the rapid hydrolysis of the added cyclic AMP.

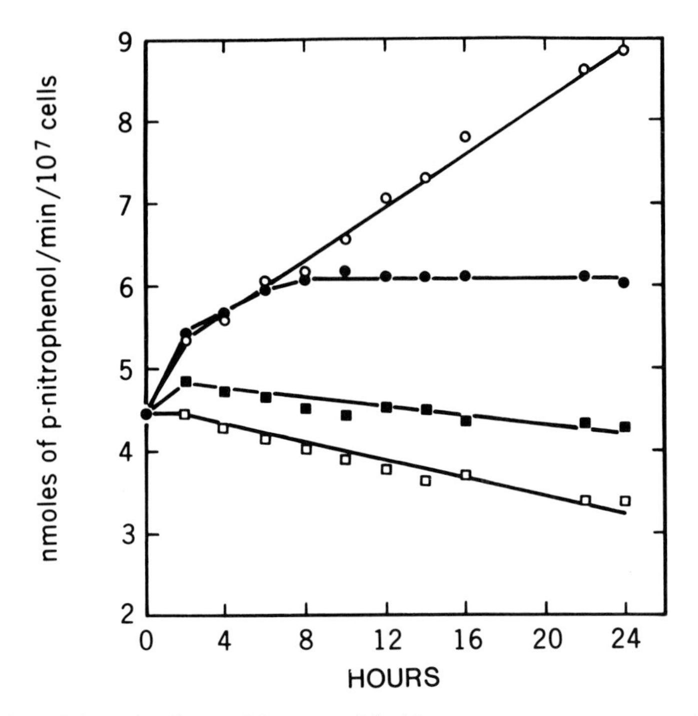

Fig. 10. β-glucosidase activity. —o—o— control; —●—●— cyclic AMP; —■—■— actinomycin D; —□—□— cycloheximide. For details see Reference 23.

The mechanism by which cyclic AMP effects these changes in enzyme content in D. discoideum is not understood; its elucidation is our major current concern.

Summarizing this section, we may conclude that cyclic AMP plays a dual role in the slime mold. The cyclic nucleotide acts as chemotactic morphogen which brings about the formation of a primitive tissue, the pseudoplasmodium. Cyclic AMP also has an intracellular function; it modulates the synthesis of, undoubtedly, a large number of proteins by a mechanism not yet understood.

CONCLUSIONS AND SPECULATION

The somewhat fragmentary and fragmented findings presented here, considered in conjunction with a large and amorphous mass of data in the literature, lead to certain generalizations. Thus it appears that cyclic AMP in eukaryotes as well as in prokaryotes is formed in response to starvation. The molecular mechanism of this response to starvation is not understood. The interesting data which will be presented at this meeting by Dr. Coe suggest that the synthesis of cyclic AMP may be linked to oscillations in the concentrations of intermediates of energy metabolism. However that may be, in both E. coli and D. discoideum starvation leads not only to the enhanced synthesis of cyclic AMP but also to its release by the cells. Investigators who have studied the hormonal stimulation of the formation of cyclic AMP in cultured mammalian cells are well aware of the appearance of the cyclic nucleotide in the medium. The physiological function, if any, of this release of cyclic AMP by cultured mammalian cells is not known. In E. coli the extracellular cyclic AMP plays no demonstrable role. In another prokaryote, Myxococcus xanthus (24), exogenously added cyclic AMP greatly enhanced the formation of fruiting bodies (ADP, however, was more effective than cyclic AMP). The intercellular morphogenetic role of cyclic AMP in the lower eukaryote Dictyostelium has been discussed. Further evidence for a morphogenetic role of the cyclic nucleotide comes from observations on organisms as widely separated evolutionarily as the moss Fumaria hygrometica where, it is claimed, cyclic AMP stimulated the formation of chloronema filaments (25) and the fresh water sponge Spongilla lacustris where the cyclic nucleotide enhanced the formation of gemmules (26). That cyclic AMP may play a morphogenetic role in the higher metazoa is suggested by the work of Robertson and Gingle (27,28). They showed that, when one-day old chick embryos were exposed to pulses of cyclic AMP, migration of the cells was affected and bending of the primitive streak toward the source of the cyclic nucleotide occurred. Cyclic AMP also enhanced the ability of the one-day old embryo cells to aggregate and it appears that cyclic AMP served as a relay signal, much as it does in the case of D. discoideum. (To what extent starvation plays a role in the synthesis and release of cyclic AMP by the chick embryo cells has not been investigated to our knowledge).

An intercellular, morphogenetic role of cyclic AMP then appears to be widespread. An intracellular function of cyclic AMP as modulator of protein synthesis also appears to be established in prokaryotes and in the higher as well as the lower eukaryotes. The mechanism of this effect is well understood only in the case of the prokaryote, E. coli. Furthermore, it is not clear to what extent the intracellular regulatory functions of cyclic AMP in the multicellular organisms depend on preceding morphogenesis and cell-cell interactions. Our own, albeit quite limited, data suggest that in *Dictyostelium* the intercellular and intracellular functions of the cyclic nucleotide were not obligatorily linked under our experimental conditions, i.e. in amoebae kept separated from one another by suspension in a liquid. It appears likely that under physiological conditions, when the amoebae are in close contact, the extra- and intracellular functions of cyclic AMP constitute a regulatory mechanism that may well integrate differentiation with morphogenesis.

The synthesis and function of cyclic AMP have been the focus of our experiments and their discussion in this presentation. We are quite aware, however, of the possibility that the cyclic nucleotide may not be the ultimate effector of the reactions which we study. Time and space do not permit us to dwell on the complex of relationships between cyclic nucleotides and inorganic ions, particularly calcium. However that may be, the findings discussed here are compatible with a role for cyclic AMP as a prehormone in both the phylogenetic and the ontogenetic sense. The possibility, that starvation was the *primum movens* for multicellularity, and cyclic AMP the mediator, is an intriguing one.

REFERENCES

(1) W. F. Loomis, *Dictyostelium discoideum*. A Developmental System (Academic Press, New York, San Francisco, London, 1975) 214 p.

(2) J. T. Bonner, D. S. Barkley, E. M. Hall, T. M. Konijn, J. W. Mason, G. O'Keefe, III and P. B. Wolfe, Develop. Biol., 20 (1969) 72.

(3) G. Gerisch and U. Wick, Biochem. Biophys. Res. Commun., 65 (1975) 364.

(4) R. K. Raman, Y. Hashimoto, M. H. Cohen and A. Robertson, J. Cell Sci., 21 (1976) 243.

(5) Y. Maeda and M. Maeda, Exptl. Cell Res., 84 (1974) 88.

(6) B. Wurster, P. Pan, G.-G. Tyan and J. T. Bonner, Proc. Nat. Acad. Sci. USA, 73 (1976) 795.

(7) M. J. Buettner, E. Spitz and H. V. Rickenberg, J. Bacteriol., 14 (1973) 1068.

(8) H. J. Rahmsdorf, H. L. Cailla, E. Spitz, M. J. Moran and H. V. Rickenberg, Proc. Nat. Acad. Sci. USA, 73 (1976) 3183.

(9) G. Gerisch, Naturwissenschaften, 46 (1959) 654.

(10) F. T. Marin, Develop. Biol., 60 (1977) 389.

(11) R. Kram, P. Mamont and G. M. Tomkins, Proc. Nat. Acad. Sci. USA, 70 (1973) 1432.

(12) J. Oey, A. Vogel and R. Pollack, Proc. Nat. Acad. Sci. USA, 71 (1974) 694.

(13) H. V. Rickenberg, Ann. Rev. Microbiol., 28 (1974) 353.

(14) A. Peterkofsky, in: Advances in Cyclic Nucleotide Research, Vol. 7, eds. P. Greengard and G. A. Robison (Raven Press, New York, 1976) p. 1.

(15) I. Pastan and S. Adhya, Bacteriol. Rev., 40 (1976) 527.

(16) G. A. Robison, R. W. Butcher and E. W. Sutherland, Cyclic AMP (Academic Press, New York, London, 1971) 531 p.

(17) J.-P. Jost, A. W. Hsie and H. V. Rickenberg, Biochem. Biophys. Res. Commun., 34 (1969) 748.

(18) H. V. Rickenberg, H. J. Rahmsdorf, A. Campbell, M. J. North, J. Kwasniak and J. M. Ashworth, J. Bacteriol., 124 (1975) 212.

(19) M. Darmon, P. Brachet and L. H. Pereira Da Silva, Proc. Nat. Acad. Sci. USA, 72 (1975) 3163.

(20) C. Klein, J. Biol. Chem., 250 (1975) 7134.

(21) G. Gerisch, H. Fromm, A. Huesgen and U. Wick, Nature, 255 (1975) 547.

(22) I. D. Hamilton and W. K. Chia, J. Gen. Microbiol., 91 (1975) 295.

(23) H. V. Rickenberg, C. Tihon and O. Güzel, in: Development and Differentiation in the Cellular Slime Moulds, eds. P. Cappuccinelli and J. M. Ashworth (Elsevier/North-Holland Biomedical Press, Amsterdam, New York, 1977) p. 173.

(24) J. M. Campos and D. R. Zusman, Proc. Nat. Acad. Sci. USA, 72 (1975) 518.

(25) A. K. Handa and M. M. Johri, Nature, 259 (1976) 480.

(26) T. L. Simpson and G. A. Rodan, in: Aspects of Sponge Biology, eds. F. W. Harrison and R. R. Cowden (Academic Press, New York, San Francisco, London, 1976) p. 83.

(27) A. Robertson and A. R. Gingle, Science, 197 (1977) 1078.

(28) A. R. Gingle, Develop. Biol., 58, (1977) 394.

The author holds an Ida and Cecil Green Investigatorship in Developmental Biochemistry and his work is supported by USPHS Grant AM-11046 and NSF Grant PCM 76-10272. Drs. M. J. Buettner, H. J. Rahmsdorf, Claude Tihon and Hélène Cailla participated in different aspects of the work. The author thanks Eva Spitz, Diann Miller, and Ömer Güzel for their excellent technical assistance.

DISCUSSION

P. FEIGELSON: I am not certain whether I understood you to state that cyclic AMP "activated" tyrosine aminotransferase. I am sure that you did not mean to leave the impression that cyclic AMP does not increase the rate of synthesis of this enzyme *in vivo*. In addition to pre-existing evidence our recent studies indicate that cyclic AMP and glucagon act on isolated hepatocytes to induce elevations in the catalytic activity of tyrosine aminotransferase with parallel selective elevation in pulse-incorporation of labeled amino acids into this enzyme protein and a parallel elevation in the immunochemical titer of this enzyme. We have some studies which are "in press" which demonstrate that glucagon and cyclic AMP also act *in vivo* to induce elevated levels of the tyrosine aminotransferase mRNA and that alpha-amanitin prevents the cyclic AMP mediated induction of this specific mRNA species. What remains unknown is whether this is a direct effect of cyclic AMP upon the tyrosine aminotransferase genome or whether this is an RNA processing phenomenon. What is certain is that some effects of cyclic AMP are pretranslational (Ernest and Feigelson, J. Biol. Chem. 253:319-322, 1978; Ernest, Chen and Feigelson, J. Biol. Chem. 252:67836791, 1977).

H.V. RICKENBERG: Did I suggest that the formation of tyrosine aminotransferase does not constitute *de novo* synthesis? I certainly was not suggesting that, but what I was pointing out is that the activation of glycogen phosphorylase kinase is the activation of a pre-existing enzyme. By contrast I pointed out that the formation of serine dehydratase and, of course, of tyrosine aminotransferase indeed constitute *de novo* synthesis. Your remark, if I understood you correctly, that the effect is indeed at the level of transcription is exciting and the evidence should settle arguments about the mechanism by which the synthesis of tyrosine aminotransferase is regulated by cylic AMP.

P. FEIGELSON: The experimental fact is that there is an induction in the level of the messenger RNA for tyrosine aminotransferase. This is compatible with transcriptional control by cyclic AMP. However, we cannot exclude any mechanism by which cyclic AMP selectively increased the rate of processing of the transcriptional precursor of tyrosine aminotransferase mRNA to functionally active message.

TRANSCRIPTIONAL AND TRANSLATIONAL CONTROL OF PROTEIN SYNTHESIS DURING DIFFERENTIATION OF DICTYOSTELIUM DISCOIDEUM

H.F. LODISH, J.P. MARGOLSKEE, and D.D. BLUMBERG
Department of Biology
Massachusetts Institute of Technology
Cambridge, Massachusetts 02139

INTRODUCTION

Dictyostelium discoideum is a good system in which to study developmental regulation of protein synthesis. The life cycle has been studied extensively by morphological, biochemical, and mutational techniques, and large amounts of cells can be induced to undergo synchronized differentiation (1-3). Development is initiated by starvation for one or more amino acids (4). Beginning at 6 hr, cells aggregate together in a cAMP-mediated process (5); by 10 hr, the aggregates of about 10^5 cells contain tight cell-cell contacts. These differentiate into a slug which is capable of directed migration; already at this stage there is separation of pre-spore and pre-stalk cells (6). The culmination stage, beginning at about 19 hr, is induced by overhead light and high ionic strength. The stalk cells elongate and vacuolate, extending upward in a cellulose-enclosed column. The mass of spore cells climbs and is pulled to the top of the stalk (3, 7). *Dictyostelium* is haploid, and mutants exist which are blocked at many developmental stages or which are defective in specific developmentally regulated enzymes (8-13).

In order to study the regulation of gene expression in such a system, one must identify genes whose activity increases or decreases at prescribed times of differentiation (reviewed in ref. 14). Analysis of morphological mutants suggests a programmed genetic control of sequential induction of groups of enzymes (11). In two cases (glycogen phosphorylase and UDP glucose pyrophosphorylase), it has been shown that appearance of enzyme activity is accompanied by an increase in the rate of synthesis of the enzyme protein (15, 16), but the results on the latter enzyme remain controversial (17). In the case of one enzyme, β-glucosidase-I, the increase in enzyme-specific activity is due to preferential stabilization of the enzyme to proteases (18); and the increase in another, trehalose-6-phosphate synthetase, may be due to activation of a masked enzyme (19). During the pre-aggregation stage, a number of new antigens and proteins appear on the cell surface, but again, it is not clear whether this is due to increased synthesis of the protein or to some other factor (20, 21).

ISBN 0-12-045450-5

Hybridization studies suggested changes in the sequence representation of single-copy DNA, but it is unclear whether some or all of these DNA transcripts are mRNA (22).

Because of these problems in relating changes in protein or RNA patterns to changes in gene expression, we have turned to analysis by single-dimensional and two-dimensional gel electrophoresis of labeled proteins synthesized at discrete developmental stages. In an effort to understand the regulation of synthesis of these proteins, we have studied translation of Dictyostelium mRNA isolated from different developmental stages in a wheat germ cell-free system and fractionated the products on similar gels. In parallel, we have investigated developmental changes in mRNA by hybridization experiments using cDNA templated with mRNAs isolated from different stages. Our results, summarized here, indicate that, at different times of differentiation control is at the translation or transcriptional level.

Pre-aggregation stage of differentiation. Differentiation is induced by starvation; deprivation of any of several amino acids is essential for initiation of the developmental process. During the first 15-20 min of differentiation, synthesis of no new proteins can be detected on 1D or 2D gels (23). However, there exist 6 polypeptides, prominent in the spectrum of proteins made by growing cells, whose synthesis, relative to other cell proteins, decreases precipitously during early differentiation. This transition is apparently mediated at the translational level: the total amount of translatable mRNA which can be extracted from growing cells or from cells at early stages of differentiation is the same. Synthesis of 5 of these 6 proteins can be achieved in the wheat germ cell-free system, and the amount of translatable mRNA for these polypeptides is unchanged in early differentiation. Comcomitant with these changes is a rapid reduction (about 3-fold) in the overall rate of polypeptide chain initiation, as could be seen by analysis of polysome profiles. To interrelate these phenomena, we propose that the activity of one initiation factor is reduced in early differentiation, thus reducing the overall rate of chain initiation (24). Translation of the aforementioned 6 proteins would then require higher levels of this factor than would other mRNAs.

Synthesis of two new proteins (mw of about 100,000) can be detected very early in differentiation (Fig. 1). Synthesis of these reaches a peak at 45 min and then declines; its synthesis cannot be detected after 90 min. Quantitation of these autoradiograms shows that peak synthesis of these polypeptides, relative to that of other cellular proteins, is at least 10-fold that characteristic of vegetative cells.

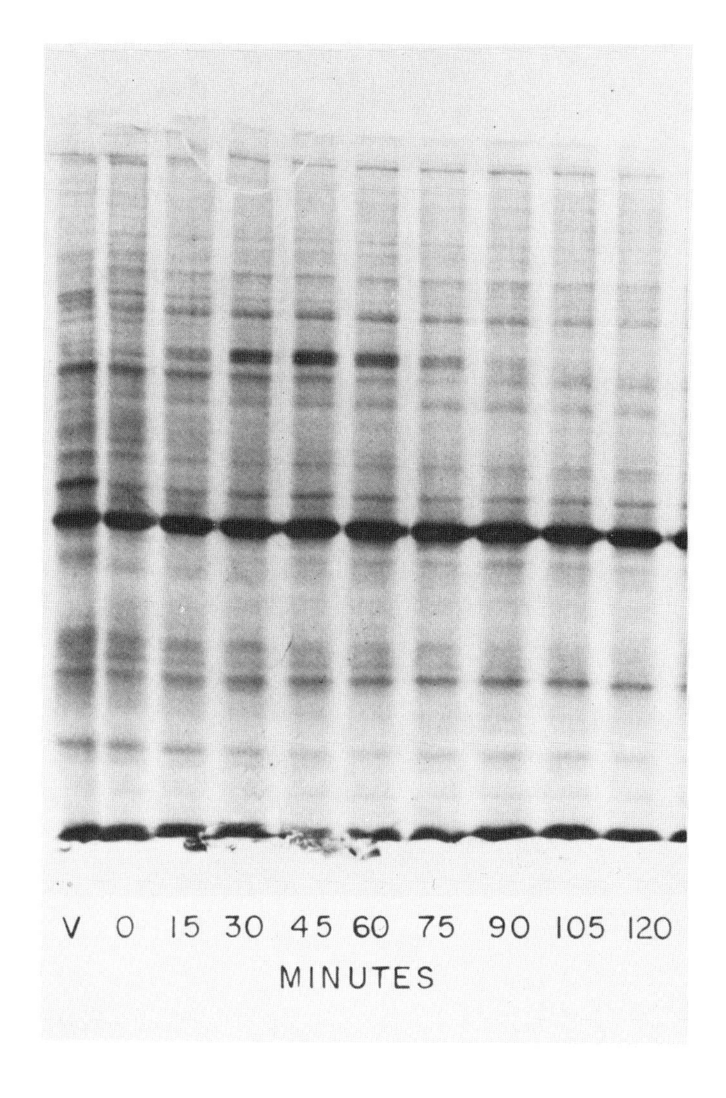

Fig. 1. Protein synthesis during early development. Vegetative cells (V) were harvested at a density of 2 x 10^6 cells/ml after growth for two generations in the presence of 60 μCi of [^{35}S]methionine. During development filters 13 mm in diameter containing 5 x 10^6 cells were placed on 10 μl drops of PDF containing 1 mCi/ml [^{35}S]methionine for 15 min periods and then harvested. Samples were labeled at 15 min intervals after starvation. Samples containing equal amounts of incorporated radioactivity were applied to each well of an SDS polyacrylamide gel.

Synthesis of these polypeptides is closely mirrored by the appearance and disappearance of the corresponding translatable mRNAs for these two polypeptides as assayed in vitro in a wheat germ protein synthesizing system (Fig. 2); translatable mRNA reaches a maximum at 45 min and declines with the same rate as observed in vivo. Therefore, the synthesis of these proteins is regulated by the transcription and degradation of their mRNAs. It is of interest that the apparent half-life of these mRNAs — less than 20 min — is much less than that characteristic of growing cells or of cells at 4 hr of differentiation — about 4 hr (Margolskee and Lodish, unpublished). [For technical reasons, it is not possible to measure accurately the half-life of total mRNA at very early stages of differentiation.]

Synthesis of these 2 proteins (and mRNAs) is developmentally regulated. We have studied the properties of a number of Dictyostelium mutants which grow normally yet exhibit no morphological changes characteristic of aggregation. Some of these agg⁻ mutants do not induce synthesis of these 2 polypeptides, while some do so at the normal rates and times (Fig. 3). The role of these polypeptides in differentiation is, of course, unknown; because they are the earliest known regulated polypeptides, they might be involved in regulation. It is of interest that some mutants which do not make these polypeptides do induce appearance of the cAMP binding protein, a cell-surface component involved in cell signalling during aggregation. Also, some mutants which do make these polypeptides do not accumulate normal levels of cAMP receptor, while others do. Early differentiation thus appears to be more complex than was first apparent.

Protein synthesis and cell aggregation. The most dramatic changes in the pattern of protein synthesis in differentiating cells occur during the time of late aggregation (23). As analyzed by 2D gel electrophoresis, there are about 40 proteins made by these (10 hr) cells which are not synthesized by pre-aggregation cells. In addition, there are about 10 proteins, including actin, whose synthesis is high during the pre-aggregation stages but which is markedly reduced after aggregation. There are qualitative and quantitative changes in the pattern of synthesis at later stages, but none so dramatic as those which occur during formation of tight cell-cell aggregates.

Insofar as synthesis of these 50 proteins can be achieved in a wheat germ cell-free system primed with Dictyostelium RNA, it could be shown that the initiation and cessation of synthesis of these polypeptides during differentiation was

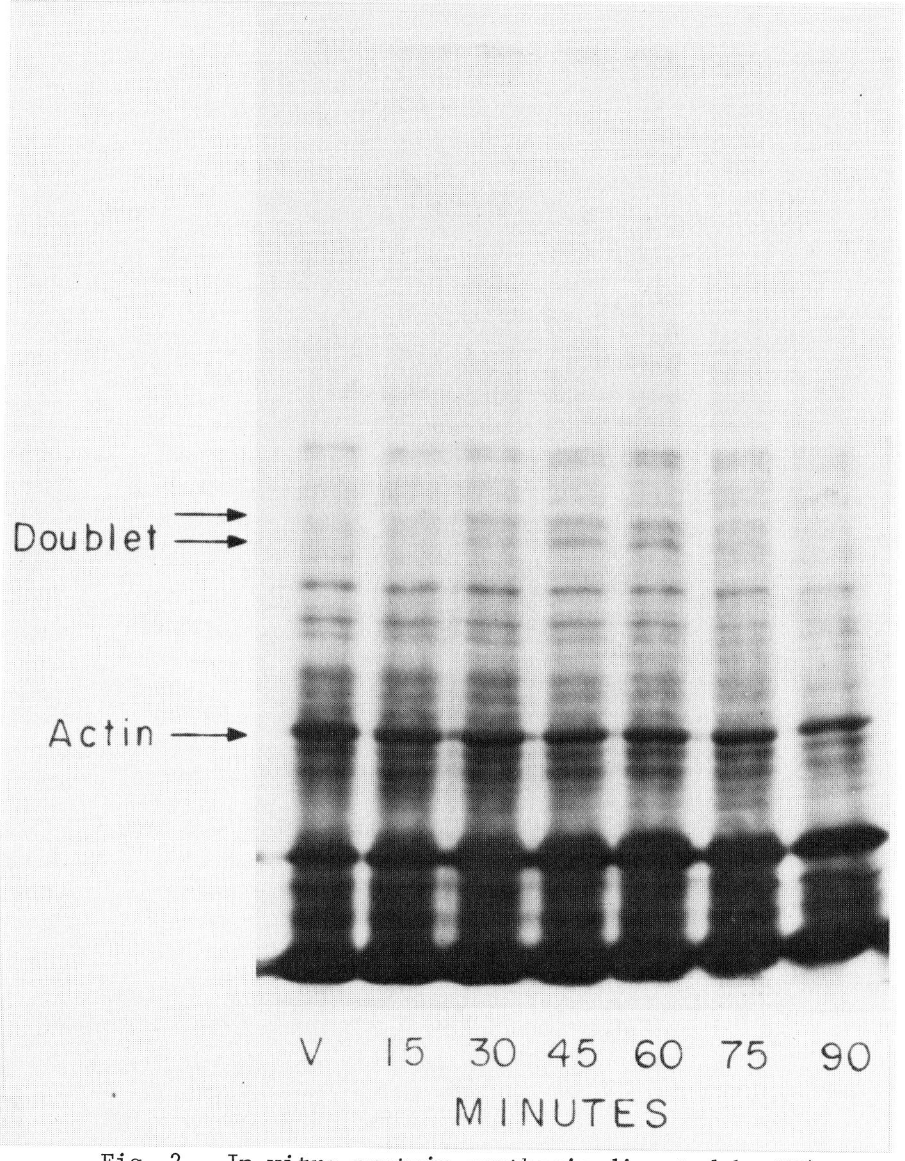

Fig. 2. *In vitro* protein synthesis directed by RNA extracted during early development. During the experiment shown in Fig. 1, total cytoplasmic RNA was extracted from 2×10^8 cells at 15 min intervals during development as well as from vegetatively growing cells. RNA (4 μg) from each time point was translated in a wheat germ cell-free protein synthesizing system in a 50 μl reaction volume containing [^{35}S]methionine. Equal reaction volumes were applied to each well of the gel.

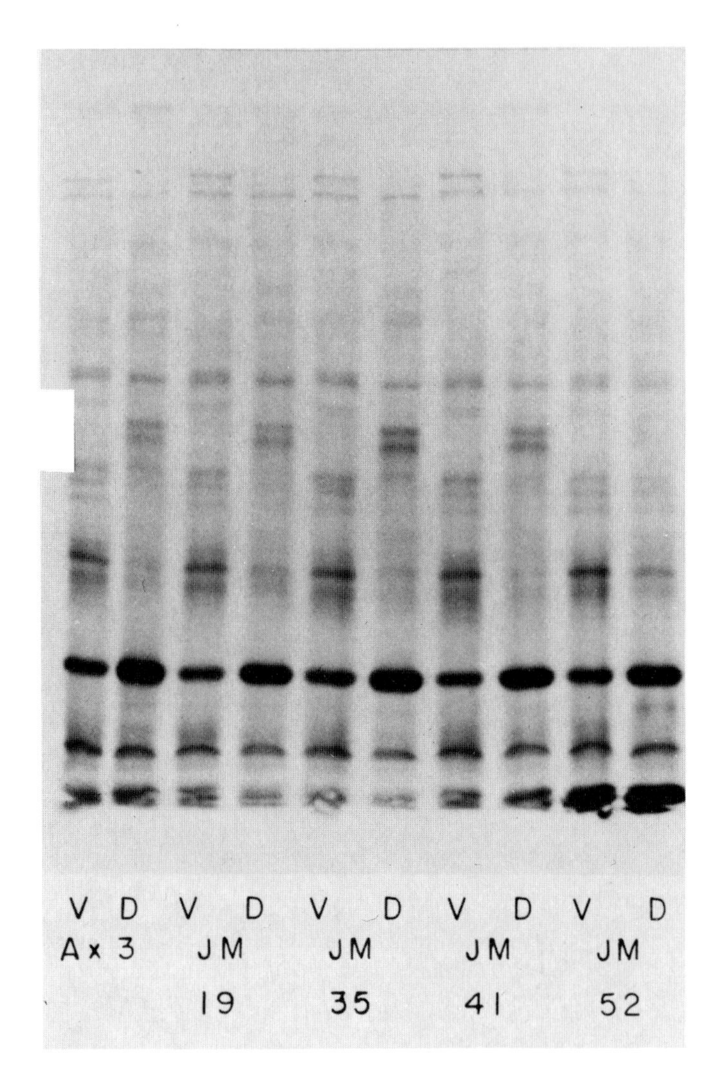

Fig. 3. Synthesis of "45 min" doublet in mutants. Cells were labeled as described in Fig. 1: (V) during vegetative growth, or (D) between 40 and 60 min after initiation of development. AX 3 is the wild type parent strain; the other strains are morphological mutants. JM 19 aggregates and JM 35 forms slugs. JM 41 and Agg 2 do not aggregate.

closely paralleled by the increase or decrease in the amount of the homologous mRNA. mRNAs do not appear in translatable form until synthesis of the homologous protein begins. We conclude that regulation of these polypeptides was also at the level of synthesis and destruction of the corresponding mRNA.

The changes in the pattern of protein synthesis (and translatable mRNAs) occurring during aggregation required continued cell-cell contact. These changes did not occur if otherwise aggregation-competent cells were prevented by rapid shaking from establishing large cell aggregates; they continued synthesis of polypeptides characteristic of early differentiation (25). Furthermore, agg$^-$ mutants which show no morphological signs of aggregation do not induce synthesis of most proteins characteristic of aggregating cells; they continue synthesis of polypeptides characteristic of pre-aggregating cells (Margolskee and Lodish, unpublished) (26). How cell-cell contact influences gene expression is not clear. In particular, we do not know whether a secreted "hormone" is responsible or whether the change is mediated by actual cell-cell contact.

The number of mRNA sequences present in differentiating cells. The sensitivity of the 2D polyacrylamide gels is about 0.01% of the cells' total protein synthesis. Proteins made below this level could not have been detected. Thus our estimate of about 100-200 developmentally regulated polypeptides is certainly low. In an attempt to analyze the overall changes in the mRNA populations during the developmental cycle, we prepared DNA complementary to mRNA (cDNA) using AMV polymerase to reverse transcribe poly(A)-containing cytoplasmic RNA into DNA. The cDNA is used as a radioactive probe in DNA/RNA hybridizations designed to determine the complexity of the different mRNA populations. Since the rate of the hybridization reaction between the cDNA and a vast excess of its mRNA template is determined by the concentration of unique nucleotide sequences in the driver mRNA population, it is possible to estimate not only the total number of distinct average size mRNA molecules in the population but also the relative concentration or abundance of the mRNAs within the cell.

Figure 4 shows a graph of the hybridization reaction between cDNA transcribed from the cytoplasmic poly(A)$^+$ vegetative mRNAs and a vast excess of its own vegetative poly(A)$^+$ cytoplasmic mRNA template (●———●). The % cDNA in S1 nuclease (single strand specific nuclease) resistant hybrid is plotted versus the log of the rate of the hybridization reaction expressed as the concentration of driver RNA in moles

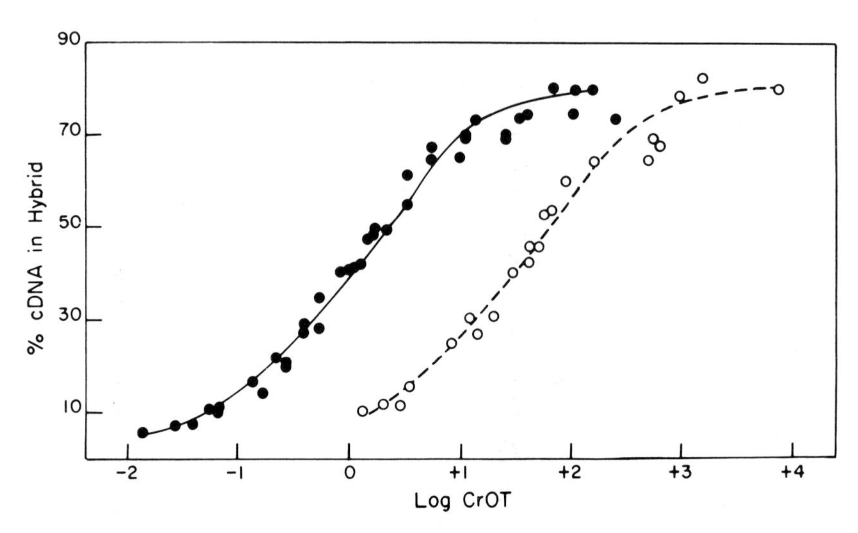

Fig. 4. Hybridization of vegetative, cytoplasmic poly(A)$^+$ RNA and total polysomal RNA to cDNA transcribed from vegetative cytoplasmic poly(A)$^+$ RNA.

Dictyostelium cytoplasmic RNA was prepared according to Batts-Young, *et al*. (27). The polyadenylated fraction was prepared by two cycles of binding and elution from an oligo(dT) cellulose column. Polysomes were prepared from the postnuclear supernatant as described by Alton and Lodish (24). Complementary DNA was prepared from the polyadenylated cytoplasmic RNA fractions according to the procedures of Kacian and Myers (28), except that unlabeled dNTPs were at 1mM, while the labeled dNTP concentration was 0.2 mM and actinomycin D was present at a concentration of 100 µg/ml. The cDNA was purified from the reactions according to Rothenberg and Baltimore (29). The fragment size was 600-1,000 nucleotides, and the specific activity was between 6×10^6 and 1×10^7 cpm/µg. Hybridizations were carried out in 0.24 M phosphate buffer, 0.5% SDS at 61°. Rates are normalized to 0.12 M phosphate (30). The concentration of driver RNA was between 2-100 µg/ml for reactions using poly(A)$^+$RNA and 100 µg/ml for reactions using unfractionated polysomal RNA. The cDNA concentration was 0.02 µg/ml. At various times after the initiation of the reactions, 10 µl aliquots were removed to 1.2 ml of S1 nuclease buffer (31). Aliquots (0.5 ml) were incubated for 30 min at 37° with or without 100 units of S1 nuclease. S1 nuclease resistant material was determined as described by Hereford and Roshbash (31). (●——●) Vegetative cytoplasmic poly(A)$^+$ hybridized to its own cDNA. (O----O) Vegetative polysomal RNA hybridized to cDNA templated by vegetative cytoplasmic poly(A)$^+$ RNA. The solid line represents a theoretical curve drawn using the relationship described by Bishop, *et al*. (32). In generating the curve, 26%, 35%, and 39% have been taken as the values for the amoung of cDNA in hybrid for three transitions with half maximal rates of 0.02, 0.31, and 1.93 (moles of nucleotides per liter per sec), respectively.

of nucleotides per liter times the time of the reaction in seconds (log C_rot).

Knowing the average size of Dictyostelium mRNA (1330 nucleotides) and the amount of mRNA per cell, one can calculate that vegetative cells contain about 3500 unique nucleotide sequences the size of an mRNA molecule. These sequences can be grouped into three classes on the basis of their abundance in the cell. Twenty-six per cent of cDNA transcribed from vegetative cytoplasmic poly(A)$^+$ mRNA contains approximately 31 sequences reiterated 1,000 times per cell. 35% of the cDNA is complementary to approximately 500 sequences present in vegetative mRNA at 100 copies per cell, and 39% of the vegetative cDNA is complementary to 3,000 sequences present between 10-20 copies per cell. The total number of unique sequences in the mRNA population present during vegetative growth represents transcription of 12-13% of the single copy portion of the Dictyostelium genome.

Also plotted in Fig. 4 is the progress of the reaction of cDNA transcribed from poly(A)$^+$ cytoplasmic RNA isolated from growing cells to a vast excess of RNA extracted from the polysomes of growing cells (O———O). This RNA has not been purified by passage through oligo(dT) cellulose and so contains both rRNA and mRNA sequences. The poly(A)$^+$ sequences only represent 2% of the total RNA concentration, and thus the rate of the hybridization reaction is 50-fold slower than the rate of the reaction between cytoplasmic poly(A)$^+$ and its cDNA ($C_rot_{1/2}$ 28.8 / $C_rot_{1/2}$ 0.63). It is important to observe that the cDNA reverse transcribed from the vegetative cytoplasmic poly(A)$^+$ RNA hybridizes completely to the vegetative polysomal RNA, indicating that all of the cytoplasmic poly(A)$^+$ RNA sequences detected in these experiments are present on the polysomes of growing cells and presumably represent actual mRNAs.

We conclude that growing cells contain about 3,500 different mRNA sequences. However, only about 500 of these are in high enough abundance that their polypeptide translation products could be resolved on our 2D gels.

Figures 5 and 6 show the results of a similar but more preliminary characterization of cytoplasmic poly(A)$^+$ RNA from cells at 13 hr (post-aggregation) and at 22 hr (culmination) of differentiation. Two important points can be made. First a comparison of mRNA from growing cells and cells at 13 hr or 22 hr of differentiation reveals major differences (Fig. 5). This figure shows that most of the mRNA sequences present in the cell during growth are retained in the population of RNAs

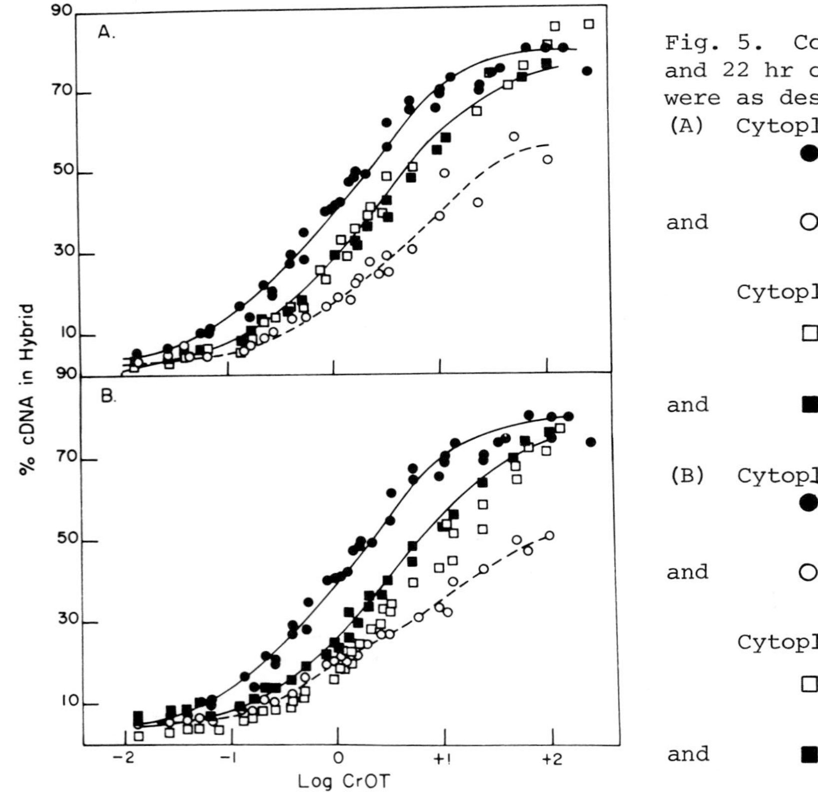

Fig. 5. Comparison of the complexity of vegetative, 13 hr, and 22 hr cytoplasmic, poly(A)+ RNA populations. Reactions were as described in Fig. 4

(A) Cytoplasmic poly(A)+ vegetative RNA hybridized to

● ^{3}H-labeled cDNA templated by vegetative cytoplasmic poly(A)+ RNA

and ○ ^{32}P-labeled cDNA templated by 13 hr cytoplasmic poly(A)+ RNA.

Cytoplasmic poly(A)+ 13 hr RNA hybridized to

□ [^{3}H]cDNA templated by vegetative cytoplasmic poly(A)+ RNA

and ■ [^{32}P]cDNA templated by 13 hr cytoplasmic poly(A)+ RNA.

(B) Cytoplasmic poly(A)+ vegetative RNA hybridized to

● ^{3}H-labeled cDNA templated by vegetative cytoplasmic poly(A)+ RNA

and ○ ^{32}P-labeled cDNA templated by 22 hr cytoplasmic poly(A)+ RNA.

Cytoplasmic poly(A)+ 22 hr RNA hybridized to

□ [^{3}H]cDNA templated by vegetative cytoplasmic poly(A)+ RNA

and ■ [^{32}P]cDNA templated by 22 hr cytoplasmic poly(A)+ RNA.

present in the cytoplasm at 13 hr or 22 hr of differentiation since cDNA templated by mRNA from growing cells hybridizes as completely to cytoplasmic poly(A)$^+$ RNA from cells at 13 or 22 hr of differentiation as it does to its own mRNA template (75-80%, □ vs. ●). By contrast, only 73% (55% / 75%) of cDNA templated by cytoplasmic poly(A)$^+$ RNA from cells at 13 or 22 hr of differentiation will hybridize to mRNA from growing cells (○----○). This indicates that the remaining 27% of the cDNA is complementary to RNA sequences which, to the limit of our resolution, are not present in vegetative RNA. This suggests that 25-30% of the cytoplasmic poly(A)$^+$ RNA from 13 and 22 hr of differentiation is transcribed from DNA expressed only during development. Whether these RNAs include a few abundantly transcribed sequences or many sequences present at a few copies per cell still remains to be determined, as does the question of whether these sequences are on the polysomes and represent actual mRNAs.

Since 13 hr and 22 hr cytoplasmic RNAs contain the majority of sequences expressed during growth (although not necessarily at the same abundance) as well as new sequences specific to development, it may be speculated that the total complexity of the late developmental RNA populations may exceed the complexity of the vegetative cytoplasmic RNA populations. Some indication of this is evident in Fig. 5 where the rate of the hybridization reaction between 13 hr and 22 hr RNAs and their own cDNAs is 3-4x slower than the rate of the hybridization reaction between vegetative RNA and its cDNA ($C_rot_{1/2}$ 2.5 [for 13 hr] versus $C_rot_{1/2}$ 0.63 [for vegetative cells]). Additional experiments are currently in progress to determine how much of this difference in rate represents an actual increase in the number of unique sequences expressed during differentiation.

With this reservation in mind, it is possible to make only a very preliminary estimate of the number of unique sequences present in 13 hr or 22 hr RNA. The hybridization reaction takes place over 5 logs and a theoretical curve which fits the 22 hr data assumes 3 abundance classes containing 30 sequences present at 300-400 copies per cell, 700 sequences present at 20-30 copies/cell, and 10,500 sequences present at 1-3 copies/cell. This population of RNA which contains most of the sequences expressed during vegetative growth as well as the sequences new to development represents transcription of about 50% of the single copy portion of the genome in agreement with results of Firtel (22) which show transcription of 56% of the single copy genome using saturation hybridizations. The numbers take into account the fact that there is 1/2 to 1/3 the amount of RNA in late developing cells (3). As a

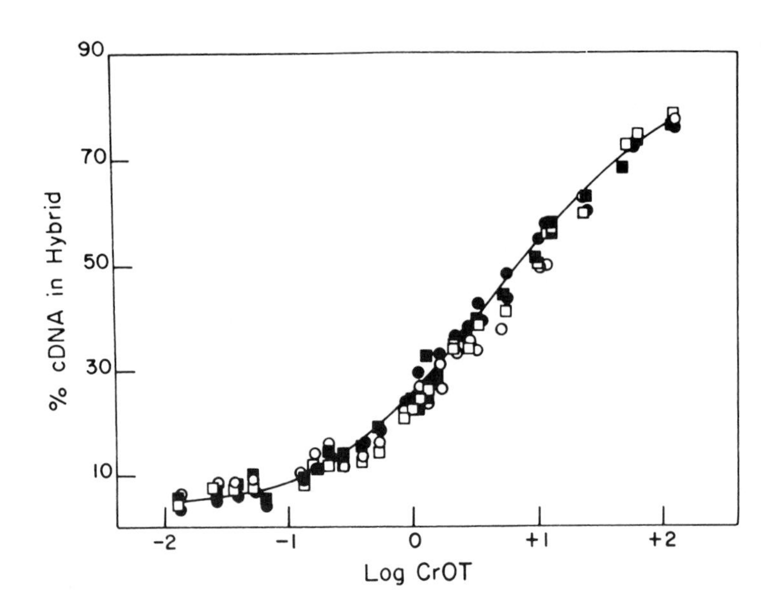

Fig. 6. Comparison of the complexity of RNA populations from 13 hr and 22 hr of development. Reactions are as described in Fig. 4.

13 hr cytoplasmic poly(A)$^+$ RNA hybridized to

● cDNA templated by 13 hr cytoplasmic poly(A)$^+$ RNA

○ cDNA templated by 22 hr cytoplasmic poly(A)$^+$ RNA

22 hr cytoplasmic poly(A)$^+$ RNA hybridized to

□ cDNA templated by 13 hr cytoplasmic poly(A)$^+$ RNA

■ cDNA templated by 22 hr cytoplasmic poly(A)$^+$ RNA.

result, the most complex class of cytoplasmic poly(A)$^+$ RNA sequences is present in the developing cells at 1/10 the level of the least abundant RNA in the cells during vegetative growth. Experiments are currently in progress to determine whether all of these RNAs are present on the polysomes and represent actual mRNAs.

A second point of interest is indicated in Fig. 6. Poly(A)$^+$ RNAs isolated from the cytoplasm of cells at 13 hr and at 22 hr of differentiation appear (to the resolution of this technique) to be identical. cDNA templated from 13 hr RNA hybridizes to RNA from 13 hr or 22 hr of differentiation at the same rate and to the same extent. Likewise, cDNA templated from 22 hr RNA hybridizes in a similar fashion to both 13 hr and 22 hr RNAs.

We conclude that changes in the transcriptional patterns are consistent with changes seen in the smaller population of proteins detected by 2D gel electrophoresis. Concomitant with aggregation there is the appearance in the cytoplasm of poly(A)$^+$ RNA (25-30% of the total cytoplasmic RNA) corresponding to new developmentally specific transcription. After 13 hr, however, there are no further major changes detectable in the cytoplasmic poly(A)$^+$ RNA populations.

REFERENCES

(1) J.T. Bonner, The Cellular Slime Molds, 2nd edition (Princeton University Press, Princeton, 1967).

(2) A. Jacobson and H. Lodish, Ann. Rev. Genetics, 9 (1975) 145.

(3) W.F. Loomis, Jr., Dictyostelium discoideum: A Developmental System (Academic Press, N.Y., 1975).

(4) Marin, F.T. Devel. Biol. 48 (1976) 110.

(5) J.T. Bonner, D.S. Barkley, E.M. Hall, I.M. Konijn, J.W. Mason, G. O'Keefe III, and P.B. Wolfe, Devel. Biol., 20 (1969) 72.

(6) K.B. Raper, J.E. Mitchell Sci. Soc., 56 (1940) 241.

(7) A.J. Durston, M.H. Cohen, D.J. Prage, M.J. Potel, A. Robertson, and D. Wonio, Devel. Biol. 52 (1976) 173.

(8) R.L. Dimond, M. Brenner, and W.F. Loomis, Jr., Proc. Nat. Acad. Sci. USA, 70 (1973) 3356.

(9) R.L. Dimond, P.A. Farnsworth, and W.F. Loomis, Jr., Devel. Biol., 50 (1976) 169.

(10) S. Free and W.F. Loomis, Jr., Biochemie 56 (1975) 1525.

(11) W.F. Loomis, Jr., R.L. Dimond, S.J. Free, and S. White, in: Eukaryotic Microbes as Model Developmental Systems, eds. D. O'Day and P. Horgen) (Marcel Dekker, New York, 1977) pp. 177-194.

(12) A.J. Warren, W.D. Warren, and E.C. Cox, Proc. Nat. Acad. Sci. USA, 72 (1975) 1041.

(13) K.L. Williams and P.C. Newell, Genetics, 82 (1976) 287.

(14) W.F. Loomis, S. White, and R.L. Dimond, Devel. Biol., 55 (1976) 171.

(15) D.A. Thomas and B.E. Wright, J. Biol. Chem., 251 (1976) 1258.

(16) J. Franke and M. Sussman, J. Mol. Biol. (1973) 173.

(17) K. Killick and B. Wright, Ann. Rev. Microbiol., 28 (1974) 139.

(18) M.B. Coston and W.F. Loomis, J. Bacteriol. 100 (1969) 1208.

(19) K. Killick and B. Wright, J. Biol. Chem., 247 (1972) 2967.

(20) H. Beug, F.E. Katz, and G. Gerisch, J. Cell Biol., 56 (1973) 647.

(21) S. Rosen, J.A. Kafka, D.L. Simpson, and S.H. Barondes, Proc. Nat. Acad. Sci. USA 70 (1973) 2554.

(22) R. Firtel, J. Mol. Biol., 66 (1972) 363.

(23) T.H. Alton and H.F. Lodish, Devel. Biol. 60 (1977) 180.

(24) T.H. Alton and H.F. Lodish, Cell 12 (1977) 301.

(25) T.H. Alton and H.F. Lodish, Devel. Biol., 60 (1977) 207.

(26) H.F. Lodish, T.Alton, R.P. Dottin, A.M. Weiner, and J.P. Margolskee, in: The Molecular Biology of Hormone Action, ed. J. Papaconstantinou (Academic Press, New York, 1976) pp. 75-103.

(27) B. Batts-Young, N. Maizels, and H.F. Lodish, J. Biol. Chem., 252 (1977) 3952.

(28) D.L. Kacian and J.C. Myers, Proc. Nat. Acad. Sci. USA, 73 (1976) 2191.

(29) E. Rothenberg and D. Baltimore, J. Virol., 17 (1976) 168.

(30) R.J. Britten, D.E. Grahn, and B.R. Neufeld, in: Methods in Enzymology, Vol. 29, part E, eds. J. Grossman and K. Moldave (Academic Press, New York, 1974) pp. 363-348.

(31) L. Hereford and M. Rosbash, Cell, 10 (1977) 453.

(32) J.O. Bishop, J.G. Morton, M. Rosbash and M. Richardson, Nature, 250 (1974) 199.

(33) G.A. Galau, W.H. Klein, R.J. Britten, and E.H. Davidson, Arch. Biochem. Biophys., 179 (1977) 584.

This research was supported by grant PCM-04869 from the National Science Foundation.

DISCUSSION

R. KALLEN: Have you developed homologous systems for the assay of potential regulatory proteins either (1) as isolated from *Dictyostelium* or (2) from the *in vitro* translation of mRNA populations obtained from early stage *Dictyostelium* that may be influencing mRNA translation?

H. LODISH: Yes and no. We can isolate cytoplasmic extracts from *Dictyostelium* which are very active in *in vitro* protein synthesis. Unfortunately they do not seem to initiate synthesis of chains. Thus, a number of experiments that we would like to do, we cannot do. But we certainly can translate *Dictyostelium* messages, as you have seen, in heterologous cell-free systems. There are a large number of experiments which could give us some idea of the nature of translational control which we can do in these extracts, and we are, in fact, trying to do them now.

T. MADEN: Is anything known about the changes in the surface of the cells at the time of aggregation, either from direct analysis of proteins or from immunological analysis for example?

H. LODISH: There clearly are such changes. There are proteins which appear on the cell surface before aggregation, for instance, the cAMP phosphodiesterase. There is also at least one lectin which appears on the surface and which seems to be involved in aggregation. Once the cells aggregate the analysis becomes more complicated because the cells begin to synthesize a slime sheath around them (which gives *Dictyostelium* the nickname "slime mold"). I am not sure that one can reliably detect changes thereafter. There have been a few surface proteins which people have noticed changing during development, for instance by iodination.

W. CIEPLINSKI: For how long do you label your cells, when you are labeling for one-dimensional gels?

H. LODISH: All of the one-dimensional and two-dimensional gels that I showed were done with cells after a 15 minute label.

W. CIEPLINSKI: I see, the question goes to the point of whether your early proteins are actually precursors of some of the late appearing proteins.

H. LODISH: You are speaking of the proteins whose synthesis we believe is under translational control?

W. CIEPLINSKI: Yes.

H. LODISH: I do not think so because the protein products of the wheat germ cell-free translation system are identical in their positions on the two-dimensional gels to the product which we get out of the labeled cell. If these proteins were glycosylated, phosphorylated, and so on, the mobility would change.

W. CIEPLINSKI: Or they would undergo some proteolytic cleavage.

H. LODISH: No. I said these are almost certainly cytoplasmic proteins which are not modified, so I do not think that is a problem in terms of the cessation of synthesis of these proteins.

D. GERSHON: Do you have any estimate on how much post-translational modifications (for example cleavage of small peptides of proteins, glycosylation or some other modifications) can account for the change in patterns on your gels?

H. LODISH: Well, I can give you a certain estimate. As I said before, those proteins which we can identify as products of the wheat germ cell-free protein synthesizing system, which then co-migrate with the proteins made by whole cells, almost certainly are not modified in any obvious way. As I mentioned, by two-dimensional gel electrophoresis, we can only detect synthesis in the wheat germ extracts of about 50% of the proteins which are made by growing cells or developing cells. Almost certainly that is because those proteins are modified in some way after their synthesis, so that the product accumulated in the cell is different from the product made in the cell-free system. But I am confident of the ones which we can see as cell-free products - those are almost certainly bona fide cases.

D. GERSHON: What would be the physiological significance of messengers that are found in cells in one to three copies?

H. LODISH: I think it could have great significance. There is a rather nice paper published by Eric Davidson and his associates (Archives of Biochemistry and Biophysics 179:584, 1977). He analyzed a large number of proteins, unique to liver, for which one could estimate the rate of synthesis from their turnover rates. By assuming reasonable parameters for rates of initiation of protein synthesis one could get an order of magnitude estimate of the number of copies of messenger RNAs for these proteins in a typical liver cell. He looked at about 30-40 liver specific proteins - various enzymes of amino acid metabolism and so forth. The messages for some of them were present in very low amounts per cell - 1 to 3 copies, whereas many more copies of mRNA per cell were present for other proteins. Specifically, proteins with a very high turnover rate would be synthesized at a very high rate in a liver cell and would be expected to have messenger RNAs in very high abundance, but a number of enzymes in liver cells - organ-specific enzymes - are present in very low levels, and their calculations would indicate that the messengers for these need be present in very small amounts. So I think one class of RNAs is easily as significant as the other. Obviously, RNAs for particular regulatory proteins may also be present in very small amounts.

W. CIEPLINSKI: I was wondering if you would care to speculate on the mechanism of translational control? Do you think it is accessibility of the message or that it is much more poorly initiated in the presence of new messages that have a much higher initiation rate - like what you showed in globin?

H. LODISH: Well, I think in this case that the simplest interpretation of the translational controls, which I think we mentioned in the Cell publication, is that something becomes very rate-limiting for chain initiation early in development. Say a particular initiation factor becomes inactivated in some mysterious way. Under these conditions messenger RNAs compete with each other for this particular factor.

W. CIEPLINSKI: Have you looked into the rate of initiation of early and late messages in your cell-free system?

H. LODISH: We are beginning to do precisely these experiments in heterologous cell-free systems as well as attempting to measure the number of ribosomes on each of these messenger RNAs, both in growing cells and developing cells. In these studies we are isolating different sized polysomes and translating the RNAs in each of these. These are complicated experiments, but we are trying them. Obviously when and if we get these sequences cloned these studies will be much easier to do.

HIERARCHICAL CONTROLS OF NUCLEOLAR SYNTHETIC FUNCTIONS

HARRIS BUSCH, N. R. BALLAL, R. K. BUSCH, B. SAMAL, Y. C. CHOI, F. M. DAVIS, H. M. KUNKLE, K. NALL, D. PARKER, M. S. RAO, L. I. ROTHBLUM and H. TAKAMI
Department of Pharmacology
Baylor College of Medicine
1200 Moursund Avenue
Houston, Texas 77030

Abstract: Nucleoli of cancer cells have high rates of rRNA synthesis. A variety of controls affect the fidelity and rates of rRNA synthesis as well as the availability of rDNA. Studies on the roles of nucleolar proteins show that proteins maintaining fidelity of nucleolar transcription are soluble in 0.6 M NaCl. Of the proteins decreased with increased nucleolar activity, nucleosomal protein A24 is an interesting conjugate of ubiquitin and histone 2A susceptible to specific cleavage by a specific proteinase (Eickbush _et al_, Cell _9_, 785, 1977). Moment-to-moment control of rates of nucleolar rDNA readouts may be related to cytoplasmic factors such as elongation factor EF-1 which we recently found in nuclei and nucleoli.

INTRODUCTION

Cancer is a disease in which a dysplastic phenotype, represented by uncontrolled cell growth and division, invasiveness and metastasis is transmitted genetically or epigenetically to the daughter cells (1).

Genetic analysis of cancer cells - Although aberrations have been reported in chromosomes of individual cancers, adequate evidence that there are differences in either the specific chromosomes or the DNA of cancer cells has not yet been reported (2). Other than the translocated Philadelphia chromosomes which may appear in myelogenous leukemia (3), no common clinical case of chromosome aberration exists. If there were structural aberrations in DNA, mutant proteins or specific gene

ISBN 0-12-045450-5

deletions could have been found unless such aberrations were in promoter or operator sequences. At present, either because of limitations of methods or because such events are very subtle, there is no convincing evidence for DNA aberrations in cancer cells (despite the mutagenic effects of carcinogens).

Where to search other than the genes? - In a sense, the cancer problem resembles a "detective story" in which the investigator makes an intensive search for a very elusive villain. What is it in the cancer cell that gives it the properties of (a) growth, (b) invasiveness, and (c) metastasis that permit destruction of the host?

Support for the idea that gene control of cancer cells is aberrant (4,5) emerged from recent studies which suggest that aberrations in gene control may be rectifiable. In essence, Illmensee and Mintz (6) showed that combined teratoma cells and normal blastocyst cells could produce normal heterozygous guinea pigs. Normal controls produced in normal sequence in the biological clock were apparently capable of converting teratoma cells to normal cells (6).

Assuming the genes of cancer cells are basically unchanged as evidenced by the lack of evidence for abnormal proteins in cancer cells, the normal chromosome complement of a number of hepatomas and the reversibility of the neoplastic phenotype during fetal development, the series of questions that arise are: (a) What are gene controls? (b) Why do they become aberrant in cancer cells? (c) Why does this aberration persist "permanently" in cancer cells (1)?

The nuclear proteins - Evidence that cytoplasmic elements rather than DNA controlled the cell phenotype emerged from studies of Gurdon (7) who showed that the nucleus was totipotent even up to stages of development of intestinal epithelial cells of frog tadpoles. Thus, the full information for development and differentiation of a whole frog is present in the nucleus of each cell. The direct controls of gene function are now generally considered to be the nonhistone nuclear proteins (derived from cytoplasmic synthesis) which were shown years ago to be heterogeneous (4) and later were directly implicated by Gilmour and Paul (8) as specifying RNA transcripts. The histones, which at one time were thought to control genes, are now being largely relegated to structural roles in the "nu-bodies" or nucleosomes.

Some uncertainties exist in the "dogma" that the nonhistone proteins control gene function. A definitive experiment has not yet been reported that shows the role of a specific nonhistone protein in the production of a single special gene readout although mixtures of nonhistone proteins have been reported to serve that function.

Numbers of gene control molecules - Of the nuclear proteins, histones are present in greatest amounts, i.e., each major histone species is present in amounts of approximately 2 pg (10^{-12} gram)/nucleus (9). Proteins C23-25, which are among the most abundant of the more than 200 species of nucleolar nonhistone proteins, are present in amounts of approximately 10-100 fg (10^{-15} gram) in nuclei and nucleoli. Such proteins are now readily visualized by staining on 2D gels, particularly after fractionation and purification. When ^{32}P and immunological methods are used, the levels of detection are 10-100 fold greater, i.e., they approach 500 attagrams (10^{-18} gram)/nucleus. Such amounts begin to approximate the levels probably important in gene control. For molecules of approximately 50,000 MW, amounts of 100 attagrams would equal 1200 molecules per nucleus. The increasing information on nuclear enzymes and immunologically active nuclear elements suggests that many nuclear proteins are present in such small amounts. Clearly, the task of purification and functional analysis of these proteins is a large one.

Specific functions of nonhistone proteins - The roles of these proteins have been difficult to define because the systems involved are highly complex. There are many important problems in evaluation of gene readouts, one of which relates to the crucial question of "fidelity" of the product produced. For *in vitro* systems, difficulties become apparent in (a) the enzyme, (b) the measurements, and (c) the template.

(a) The first problem is the RNA polymerase systems. Most RNA polymerases are "elongation" enzymes and the critical controls of the "initiation" reactions have not yet been clarified. RNA polymerases from eukaryotic cells are much more unstable than the *Escherichia coli* RNA polymerase and accordingly most workers have elected to use the *E. coli* enzyme. Unfortunately, there are many problems associated with the *E. coli* RNA polymerase, not the least of which is the uncertainty about its sensitivity to chromatin controls that are recognized by homologous RNA polymerases. Along with earlier studies (10), our studies on nucleolar gene readouts (11) showed that this enzyme not only makes wrong transcripts but also reads through "gene restriction proteins" that limit normal homologous enzymes. Accordingly, for nucleolar gene readouts, the results obtained by *E. coli* polymerases are not meaningful for analysis of initiation or other transcriptional controls.

Unfortunately, the other side of the coin is another problem. Eukaryotic RNA polymerases are unstable and obtainable only in relatively low yields. Other than the characterization of many subunits on one-dimensional SDS gels, little is known about their structures and functional conformations. Virtually nothing is definite about factors involved in their control like the sigma or rho factors defined for *E. coli* RNA polymerase.

(b) To evaluate transcriptional validity, each investigator has to establish that the results obtained are specific and meaningful for the system being analyzed. Hybridization of labeled RNA alone is not adequate because simple elongation of pre-existing chains provides molecular species capable of hybridizing to DNA or cDNA. The question of fidelity of synthesis may not be answered by such experiments, i.e., even if some hybridization may occur, large amounts of the synthetic reactions may produce wrong readouts. Recently, serious questions have been raised about "reconstituted systems" because proof is lacking in many instances that the mRNA transcript was not derived from pre-existing small fragments that were simply elongated.

(c) The template - It is most important to

define the DNA recognition sites which affect the activities of RNA polymerases and their limitations with respect to gene readouts. The mechanism of restriction of readouts to specific gene segments is a most intriguing and exciting area for future research (11).

What are the nonhistone nuclear proteins? - The demonstration in this and other laboratories that there are hundreds of nuclear proteins composed of many species of enzymes, structural proteins and other polypeptides (12) has led to concerted efforts to isolate, purify and determine the functions of these many protein species. First, methods were needed to separate and classify these proteins. Two systems are in use in our laboratory (Fig. 1) for two-dimensional analyses of these proteins, namely the "Orrick" system (Fig. 1A) (13) and the "O'Farrell" system (Fig. 1B) (14). In the former system, the acid-urea gel (first dimension) separates the proteins by charge, i.e., the proteins with the greatest positive charge migrate toward the cathode; in the latter system, the first dimension is isoelectric focusing which separates the proteins on the basis of migration into regions in which their overall net charge is essentially zero. In both systems, the second dimension employs SDS gel electrophoresis.

The second problem of nomenclature of this largely undefined group of proteins has resulted in variations from laboratory to laboratory. In our laboratory, the "Orrick" gel spots were arbitrarily divided into regions A, B and C representing fastest to slowest migration (A being the fastest in the first dimension). Within each region, the lower the number (assigned to nucleolar proteins first), the faster the migration. In this system, the histones migrated the fastest in both dimensions. Some of the most interesting proteins analyzed thus far are A11, A24, BA, B23, C14, C18 and C23 (the second letters were assigned in the same way to chromatin proteins); their functions remain to be defined.

In the "O'Farrell" system, numbers are assigned in our laboratory in terms of molecular weight estimates and isoelectric points. For example, protein 64/7.2 has a molecular weight of

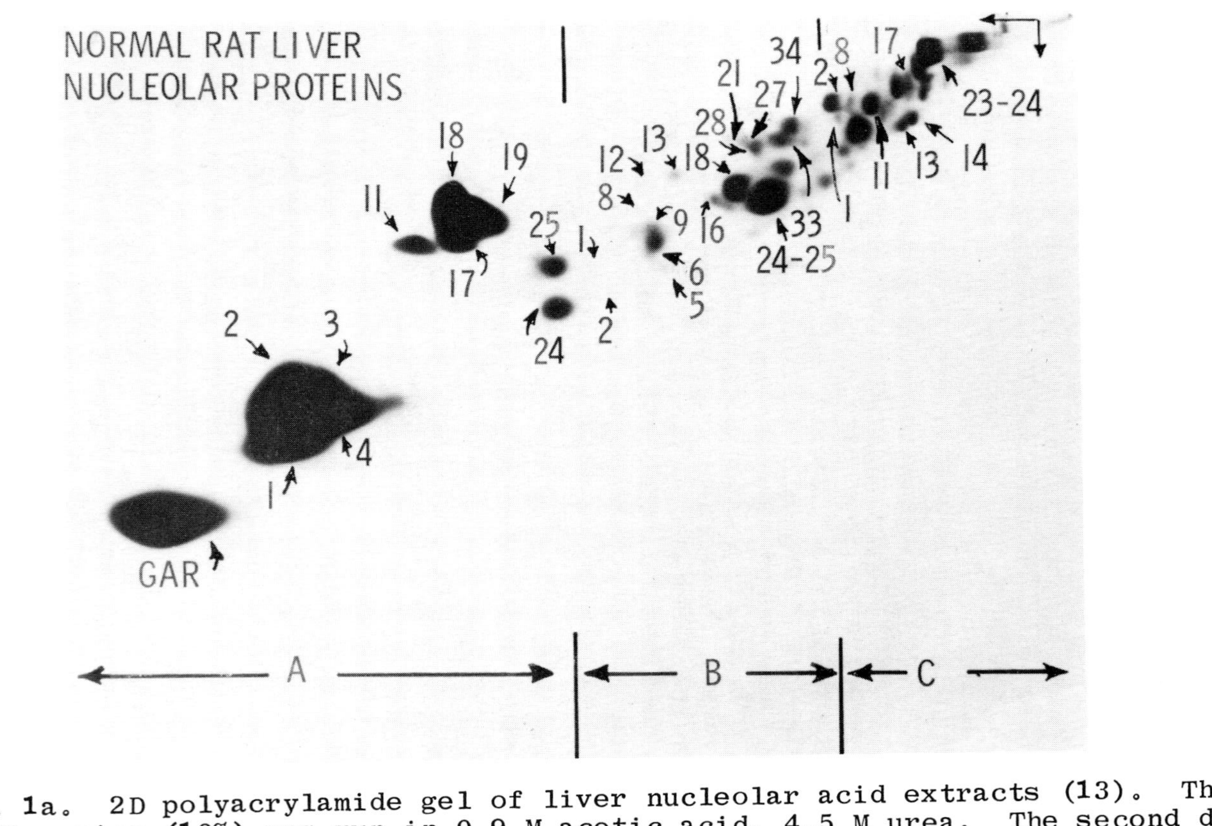

Fig. 1a. 2D polyacrylamide gel of liver nucleolar acid extracts (13). The first dimension (10%) was run in 0.9 M acetic acid, 4.5 M urea. The second dimension (12%) was run in 0.1% SDS, 0.1 M phosphate, pH 7.1. Proteins of interest are A11, A24, C13, C14, C23-24 (13).

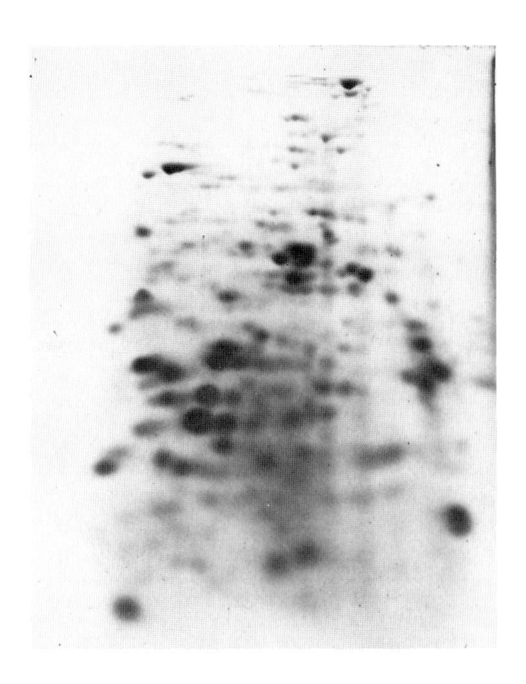

Fig. 1b. Two-dimensional "O'Farrell" (14) gel of the 0.075 M NaCl, 0.025 M EDTA extract of liver nuclei. The nonhistone nuclear proteins shown on this gel number approximately 150. The first dimension of this gel involves an isoelectric focusing system in the pH range of 3.5-10 and the second dimension is an 8% SDS gel, pH 7.5.

64,000 (estimated by standards) and an isoelectric point of 7.2 (defined by pH).

The nucleolus, an intranuclear organelle (10), is a specific representative of a highly integrated feedback system of gene function and control. THE NUCLEOLUS(15) was written to provide a general basis for understanding of the structure and function of this important cellular organelle. It is clear that the nucleolus primarily subserves a single role in cell function, i.e., the production of preribosomal particles that are subsequently transported to the cytoplasm to form mature ribosomes. The ribosomes (16) are essential polysomal elements for translation of mRNA. In most eukaryotic cells, they account for 80% to 90% of the cell RNA; the four species are 28S, 18S, 5.8S and 5.8S rRNA. The nucleolus, which is the site of synthesis of 28S, 18S and 5.8S rRNA, is the locus of synthesis of the vast bulk of cellular RNA.

The rDNA template for the synthesis of rRNA is contained in discrete chromosome segments which are visualized as "secondary constrictions" in metaphase or as NOR (nucleolus organizer regions) in specific types of cells. These templates for synthesis of rRNA have several interesting properties (Table 1).

The nucleolus responds with alacrity to cellular demands for growth products, hormonal stimuli and toxic substances. Thus far, the intrinsic mechanisms involved are still unclear. The three units of its activity are: (a) The rDNA genes (which code for 28S, 18S, 5.8S rRNA and their 45S precursors) and their adjacent promoter regions (Table 1); (b) the output arm (which includes RNA polymerase I) which forms the nucleolar 45S pre-rRNA and cleavage and assembly systems for rRNP particles destined to ultimately become the ribosomes; and (c) the input arm which includes all the vectors that operate to control the number of active genes and to influence the rates of their activity.

<u>The rDNA template</u> - The intensive research on the nucleolus in the last 15 years has established that it produces the rRNA (80-90% of the total RNA) of the eukaryotic cell. The genes for synthesis of rRNA (rDNA) are virtually all located in the nucleolus. The highly exciting topic of nucleolar gene control is just now beginning to unfold and an enormous amount remains to be learned not only about controls of nucleolar activity and the integration of its function with synthesis of ribosomal proteins (r-prot) and synthesis of mRNA. Table 1 is a summary of current analyses of the topography of the rDNA of several species by restriction endonucleases.

As noted earlier (15), a variety of species contain highly accessible rDNA which either sediments as a satellite or with sufficient difference in buoyant density that it separates from the

TABLE I

rDNA MAPS

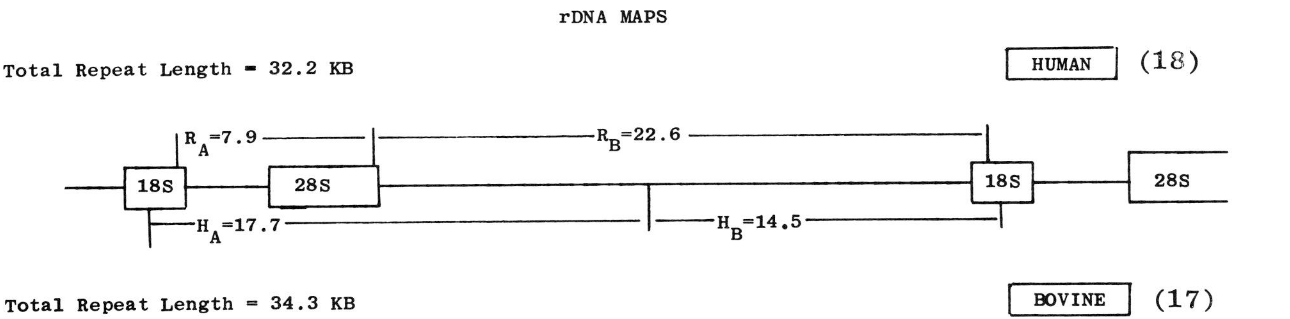

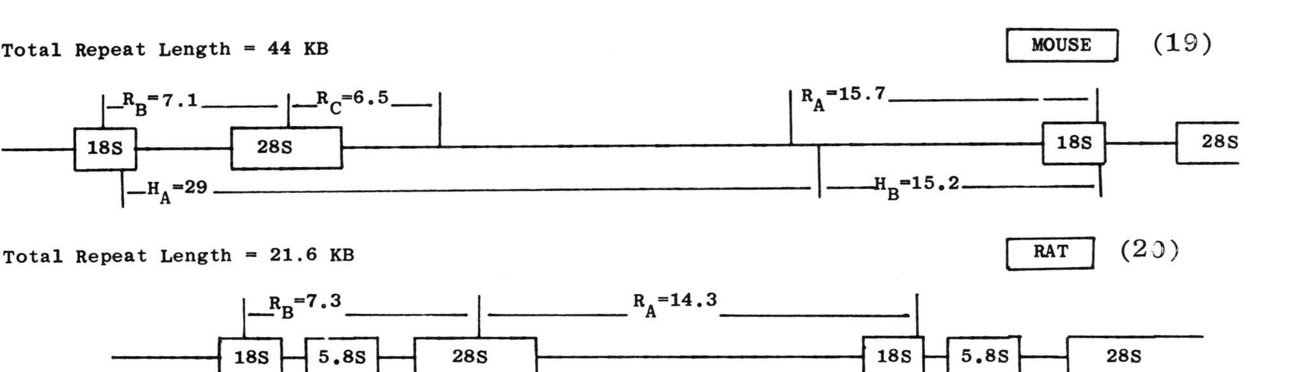

remainder of the DNA. However, in a number of species this rDNA is much less easily separable from the remainder of the DNA and, accordingly, it has been the subject of a long and frequently frustrating search. Needless to say, for the analysis of the input and output arms, the most direct analysis could be made by the search for a specific effect on promoter or quasi-promoter regions on the rDNA or its upstream structures.

Isolation of nucleolar chromatin - Recently, two methods have been employed in our laboratory for the isolation of rDNA of tumor and nontumor cells. The first approach to the isolation of rDNA containing chromatin, complete with the associated proteins, has been effected by Dr. M. Kunkle who utilized the procedure of Bachellerie et al (21) to obtain a fraction in which the rDNA and its associated nontranscribed spacers constitute 30% or more of the total DNA following initial isolation of nucleoli by methods similar to those worked out in our laboratory (22), followed by the disaggregation of the nucleoli in dilute phosphate (0.75 mM, 0.2 mM Mg^{++}) and resonication (21). Sedimentation of this fraction on Metrizamide gradients results in the separation of two bands, the lower one of which was enriched in 18S rDNA up to 1.1%. This chromatin fraction is now being analyzed for both its composition and for its ultrastructure. It contains the rDNA and hopefully its promoter and terminator regions. It will be of great interest to ascertain whether the constituents of these fractions are the same or differ in tumors and other tissues.

A parallel study has been made by D. Parker and L. Rothblum on the purification of rDNA itself from the chromatin fraction of nucleoli of Novikoff hepatoma and normal liver. In essence, the method employed is based upon the known high melting characteristics of rDNA by comparison with most of the other DNA, scission of the single-stranded DNA by S1 nuclease, rebinding of the rDNA to actinomycin D and sedimentation of the rDNA on CsCl gradients. Purification of the fractions has been detected with the ^{125}I-18S rRNA probe. The 18S rDNA was 1.3% of the total DNA obtained in this fraction and accordingly the overall purity of the rDNA is estimated to be 39%. These studies are being extended with the aid of restriction enzymes not only to verify the structures in Table 1 but also to determine whether it is possible to direct activity of this template with proteins isolated from the chromatin proteins associated with the rDNA in the Bachellerie preparation.

Nucleoli of tumor cells - A dominant theme of our research has been the study of mechanisms involved in the development of the enlarged, pleomorphic nucleolus of cancer cells. The observation that nucleoli of cancer cells are enlarged and irregular compared to those of growing or nongrowing normal cells of human adult tissues dates back more than 40 years to studies by MacCarty and his associates at the Mayo Clinic (23-25) and of Caspersson and his colleagues at the Karolinska Institute (26, 27). These observations were made long before it was recognized that the primary role of the nucleolus was the synthesis of preribosomal RNA and preribosomal RNP particles which are transported to the cytoplasm where they form the mature ribosomes.

Although at one time it was thought that the structure of rDNA which is transcribed for rRNA synthesis might have undergone mutations in cancer cells (15), there have now been many studies with a variety of techniques that have provided evidence that there are no marked differences in the ribosomal RNA products of cancer cells and other cells. If any differences do exist, they are very subtle and require even more elegant methods for their demonstration. Also, there is no compelling evidence that any of the elements of the protein synthesizing machinery of cancer cells differ from that of nontumor tissues although a large amount of evidence is required for this conclusion to be finalized.

What then are the differences that explain the pleomorphism of cancer cell nucleoli? To search for the factors involved, studies are now in progress on the controls of nucleolar function; these are not yet understood. They may include both the sensitive regions on or adjacent to the rDNA (promoter sites) as well as soluble elements of the cytoplasm such as receptor-stimulus complexes that interact with such sites (28).

Numbers and types of nucleolar proteins - In the nucleolus, a ratio of protein to RNA or DNA is approximately 8. The proteins in the nucleolus are of many types and include the precursors of ribosomal proteins as well as enzymes, histones and nonhistone proteins. The histones are present in an equal amount to DNA and account for 1/8 of the total nucleolar proteins (12).

Until the first demonstration by Orrick *et al* (13), it was generally assumed that there were relatively few species of nucleolar acid soluble proteins; however, the current evidence shows that there are approximately 200. Included in this group of proteins, C23-25 and C26 and C27 have been most intensively studied (12). They are typical nonhistone proteins with large quantities of glutamic and aspartic acid and, in addition, they are the first proteins in which clusters of acidic residues have been found associated with the phosphorylated sites (29).

During the nucleolar hypertrophy produced by thioacetamide treatment, a series of rapid changes occur in nucleolar proteins, some of which are markedly decreased and others, associated with synthesis of preribosomal ribonucleoprotein particles, are markedly increased in amount (30,31). These changes were studied by one- and two-dimensional polyacrylamide gel electrophoresis of the 0.4 N H_2SO_4 soluble nucleolar proteins which showed an overall increase in the ratio of nonhistone proteins to histones following thioacetamide treatment. Most spots that increased in size and density had electrophoretic mobilities of proteins of the preribosomal ribonucleoprotein particles. One spot (A25) remained constant in size and density during the course of the treatment. Interestingly, marked decreases were found very early in two protein spots (A11 and A24) while two other protein spots (C13 and C14) decreased slowly with time. These results indicate that the nucleolus rapidly exhibits multifaceted changes during alterations in cell function.

Protein A24 - Isolation and initial chemical characterization indicated that protein A24 was a nonhistone chromosomal protein with approximately equal amounts of acidic and basic amino acids; it constituted approximately 1.9% of the sum of histones 2A, 2B, 3, and 4 (32). Protein A24 was found to contain the tryptic and chymotryptic peptides of histone 2A as well as additional peptides and, accordingly, it was suggested that protein A24 contained a "nonhistone-like" polypeptide linked to a histone 2A molecule (32-37). More recently (33,34), protein A24 was found to have the structure:

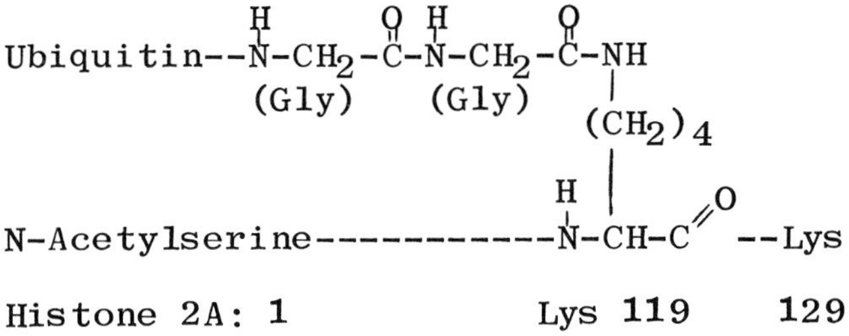

```
                   H      O  H       O
Ubiquitin--N-CH2-C-N-CH2-C-NH
                 (Gly)     (Gly)       |
                                      (CH2)4
                                         |
                                     H   |    //O
N-Acetylserine-----------N-CH-C      --Lys

Histone 2A: 1                  Lys 119     129
```

The reasons why it, like protein A11, C13 and C14, decrease in conditions which increase nucleolar function are not clear. We have postulated a "nu body" structure in which the ubiquitin of this protein might serve as a special recognition structure. One question was, how could this ubiquitin be cleaved off? Eickbush *et al* (38) in Moudrianakis' laboratory recently found a specific H2A protease which cleaves position 114-115 and thereby would remove the C-terminal ubiquitin containing sequence. In turn, this could decondense the rDNA containing chromatin.

Restriction Proteins - 3 M NaCl - 7 M urea extract - The proteins that associate with DNA are a broad group that include structural and enzyme proteins. Proof of the hypotheses that by association with DNA or histone, the nonhistone proteins specifically control the genes has not yet been achieved although the logic and evidence are overwhelming. The proteins that associate with DNA must be of at least two types, i.e., loosely and tightly binding proteins although logic dictates that they have a spectrum of binding characteristics rather than two classes of binding constants.

Our attention has been focused recently on the tightly binding proteins because of the remarkable characteristics they have. The dilute salt solutions (Zubay-Doty buffer 0.075 M NaCl-0.025 M EDTA, pH 7.2; Tris 0.01 M, pH 7.2) and 0.35 and 0.6 M NaCl 0.01 M Tris, pH 7.2, extract a broad spectrum of proteins with respect to both size and isoelectric point, pI. When the proteins were extracted with 0.6 M NaCl, the fidelity of nucleolar transcription was markedly reduced.

In both chromatin and nucleoli, the fraction that remains behind after 0.6 M NaCl extraction of the chromatin residue and is then extracted with 3 M NaCl-7 M urea is somewhat enriched in proteins with pI ranging from 4.5 to 6. Such highly acidic proteins probably best represent the "nuclear acidic proteins" which possess capability of DNA binding. Recently, Mamrack _et al_ (29) analyzed the composition of some of the tryptic peptides of the acidic proteins, C23-C25 in our old numbering system. These peptides have novel structures containing many acidic residues, i.e., up to 12 glutamic and aspartic residues are in series with 3 phosphoserine residues. This extremely acidic portion of these molecules is much more remarkable than the older "clusters of basic amino acids" in the histones and not only represents a truly novel structure but also the type of region that could exhibit great specificity for base binding in DNA. Needless to say, this fraction was only observed thus far in proteins C23-25 but it has important implication for the vast array of protein in the 3 M-7 M NaCl extract and indeed for another group of proteins whose structure is quite unknown, the proteins of the fraction of low pI in the more soluble nuclear protein fractions.

Nucleolar Antigens - Antibodies to nucleoli in this laboratory were prepared by immunization of rabbits with whole nucleoli of Novikoff hepatoma and normal rat liver (15,39). Immunofluorescence analysis studies showed specificity of localization of the antibodies to the nucleolus (39).

Immunological studies on chromatin proteins of tumors and other tissues demonstrated the presence of an antigen, NAg-1 (40), in tumors and fetal liver. These results are consistent with the presence in tumor nuclei of fetal proteins that were not detected in normal growing or nongrowing tissues. With the techniques developed in studies on NAg-1 (40), a reinvestigation was made of the antigens in nucleoli and nucleolar chromatin. In addition to NAg-1, tumor nucleoli contain another antigen, noAg-1, which is different from NAg-1 and appears to be more limited in localization.

Nucleolar chromatin of Novikoff hepatoma ascites cells contains an antigen (noAg-1) detected with antinucleolar antibodies by the immunodiffusion technique. This antigen was distinguished from the previously reported nuclear chromatin antigen NAg-1 (40) by the findings that tumor nucleolar antibodies which formed immunoprecipitin bands with noAg-1 did not do so with NAg-1 and that tumor cytosol, which contains NAg-1, formed immunoprecipitin bands with tumor chromatin antibodies but not with antibodies to tumor nucleoli. Tumor nucleolar chromatin contains both NAg-1 and noAg-1, but only noAg-1 formed bands with tumor nucleolar antibodies, noAg-1 is a component of tumor nucleolar chromatin that was not soluble in 0.075 M NaCl-0.025 M EDTA, pH 8, and only slightly soluble in 0.01 M Tris-HCl, pH 8. noAg-1 was not found in liver nucleoli. Antibodies to liver nucleoli formed immunoprecipitin bands with liver nucleolar antigens but none were confluent with those formed between tumor nucleolar antibodies and antigens of tumor nucleolar chromatin. Absorption of the tumor nucleolar antibodies with whole tumor cells or whole liver pressate did not alter band formation with noAg-1. Three antigens in liver nucleoli were not found in tumor nucleoli.

The antinucleolar antisera have recently been used to compare the nucleolar antigens which were partially fractionated by differential solubilization from nucleoli. Fourteen antigens were detected by these antisera; ten of these antigens were detected by both antisera. Ouchterlony double-diffusion analysis of soluble extracts from normal rat liver and Novikoff hepatoma ascites nucleoli and fetal rat liver nuclei provided evidence for antigens found only in liver extracts, only in tumor extracts, or only in tumor and fetal extracts. Antisera preabsorbed to remove antibodies to common antigens of liver and tumor provided confirmatory evidence for one nucleolar antigen in liver that was not found in tumor or fetal rat liver, one antigen in tumor that was not found in adult or fetal rat liver, and three antigens in both tumor and fetal rat liver that were not found in adult rat liver.

What is the role of the nucleolar antigens? Although many possible functions can be projected for these interesting molecular species, it is clear that they cannot serve structural roles inasmuch as the amounts of these proteins are too small to be equimolar with any of the major components of the nucleolar RNP granules or preribosomal products. In addition, they do not seem to be in amounts sufficient for elements of the transport systems such as proteins C23-25.

The most logical candidates for roles of these proteins would be either as elements of the gene control systems or elements of nucleolar enzymes. Inasmuch as the field of enzyme chemistry of tumors is replete with fetal enzymes which do not appear to exert any special functional variation from the corresponding nonfetal components. The possibility of their being one or more nucleolar fetal enzyme certainly seems reasonable although the alternative possibility of their being gene regulators exists, it cannot yet be established.

The key experiments now in progress are the purification and analysis of these nucleolar proteins. Basically, there is no doubt that the absorption systems in which affinity binding of these proteins is effected and from which they are eluted in purified states is working well at present. Thus far, four components of these antigens have been eluted and currently they are being purified further. The roles of these proteins will be evaluated in both the functional and expanded systems of normal liver nucleoli, thioacetamide treated liver nucleoli and tumor nucleoli and their rDNA containing fractions to determine whether they exert a stimulatory or regulatory role.

Reconstitution of nucleolar chromatin - When nucleolar chromatin was extracted with increasing concentrations of NaCl, it was found that the specificity of preribosomal RNA synthesis as determined by homochromatography fingerprinting was lost at salt concentrations greater than 0.6 M. Attempts have been made to re-establish fidelity of transcription by reassociating the 0.6 M NaCl soluble proteins with the residual chromatin. Preliminary results indicate that the RNA transcribed from the chromatin reconstituted with an excess of 0.6 M NaCl extract had an enrichment of ribosomal marker oligonucleotides, as visualized by homochromatography fingerprinting. Control experiments performed in the absence of 0.6 M NaCl extract showed that the RNA transcribed was nonspecific.

Elongation factor 1 (EF-1) in nuclei and nucleoli - Liver nuclei isolated by the hypertonic sucrose method contained 11% of the total cellular elongation factor activity. Of the total nuclear EF activity, 90% was in the nucleoplasm and 10% in the nucleoli. The nuclear and cytosol EF activity were the same with respect to dependence on GTP, ribosomes and polyuridylate; both were inhibited by heparin. The presence of EF activity in the nuclei is not due to cytoplasmic contamination since (a) only 0.14% of cellular lactate dehydrogenase was present in these purified nuclei and (b) less than 1% of radioiodinated EF-1 was found in the nuclei when added to the liver before homogenization. The percentage of EF activity in the nuclei was not appreciably altered when nuclei were isolated in the presence of rRNA (41).

Although nuclear EF may simply represent a protein in transit for integration into preribosomal particles, EF may serve a regulatory function in a nucleocytoplasmic feedback system or may be involved in nuclear transcription reactions.

REFERENCES

(1) H. Busch, Molecular Biology of Cancer (Academic Press, New York, 1974).

(2) A. A. Sandberg and M. Sakurai, in: The Molecular Biology of Cancer, ed. H. Busch (Academic Press, New York, 1974) p. 81.

(3) P. C. Nowell, H. P. Morris and V. R. Potter, Cancer Res. 27 (1967) 1565.

(4) H. Busch, Histones and Other Nuclear Proteins (Academic Press, New York, 1965).

(5) H. Busch, W. J. Steele, H. Mavioglu, C. W. Taylor and L. Hnilica, J. Cell. Comp. Physiol. (Suppl. 1) 62 (1963) 95.

(6) K. Illmensee and B. Mintz, Proc. Nat. Acad. Sci. USA 73 (1976) 549.

(7) J. B. Gurdon, in: The Cell Nucleus, Vol. I, ed. H. Busch (Academic Press, New York, 1974) p. 471.

(8) R. S. Gilmour and J. Paul, J. Mol. Biol. 40 (1969) 137.

(9) K. Smetana and H. Busch, in: The Cell Nucleus, Vol. I, ed. H. Busch (Academic Press, New York, 1974) p. 73.

(10) R. H. Reeder, J. Mol. Biol. 80 (1973) 229.

(11) N. R. Ballal, Y. C. Choi, R. Mouche and H. Busch, Proc. Nat. Acad. Sci. USA 74 (1977) 2446.

(12) H. Busch, N. R. Ballal, M. O. J. Olson and L. C. Yeoman, in: Methods in Cancer Research, Vol. XI, ed. H. Busch (Academic Press, New York, 1975) p. 43.

(13) L. R. Orrick, M. O. J. Olson and H. Busch, Proc. Nat. Acad. Sci. USA 70 (1973) 1316.

(14) P. H. O'Farrell, *J.* Biol. Chem. 250 (1975) 4007.

(15) H. Busch and K. Smetana, The Nucleolus (Academic Press, New York, 1970).

(16) M. Nomura, A. Tissieres and P. Lengyel, in: The Ribosomes (Cold Spring Harbor Laboratory, 1974).

(17) N. Blin, E. C. Stephenson and D. W. Stafford, Chromosome 58 (1976) 41.

(18) N. Arnheim and E. M. Southern, Cell 11 (1977) 363.

(19) S. Cory and J. M. Adams, Cell 11 (1977) 795.

(20) D. L. Parker, L. I. Rothblum and H. Busch. Unpublished data.

(21) J. P. Bachellerie, M. Nicoloso and J. P. Zalta, Eur. J. Biochem. 79 (1977) 23.

(22) H. Busch and Y. Daskal, in: Isolation of Nuclei and Preparation of Chromatin I., Vol. XVI, eds. G. Stein, J. Stein and L. J. Kleinsmith (Academic Press, New York, 1977) p. 1.

(23) W. C. MacCarty, Amer. J. Cancer 26 (1936) 529.

(24) W. C. MacCarty, Amer. J. Cancer 31 (1937) 104.

(25) W. C. MacCarty and E. Haumeder, Amer. J. Cancer 20 (1934) 403.

(26) T. O. Caspersson, Cell Growth and Cell Function. A Cytochemical Study (Norton, New York, 1950).

(27) T. O. Caspersson and L. Santesson, Acta Radiol. Suppl. 46 (1942) 1.

(28) H. Busch, Cancer Res. 36 (1976) 4291.

(29) M. D. Mamrack, M. O. J. Olson and H. Busch, Biochem. Biophys. Res. Commun. 76 (1977) 150.

(30) N. R. Ballal, I. L. Goldknopf, D. A. Goldberg and H. Busch, Life Sci. 14 (1974) 1835.

(31) N. R. Ballal, Y.-J. Kang, M. O. J. Olson and H. Busch, J. Biol. Chem. 250 (1975) 5921.

(32) I. L. Goldknopf and H. Busch, Biochem. Biophys. Res. Commun. 65 (1975) 951.

(33) M. O. J. Olson, I. L. Goldknopf, K. A. Guetzow, G. T. James, T. C. Hawkins, C. J. Mays-Rothberg and H. Busch, J. Biol. Chem. 251 (1976) 5901.

(34) I. L. Goldknopf and H. Busch, Proc. Nat. Acad. Sci. USA 74 (1977) 864.

(35) D. H. Schlesinger, G. Goldstein and H. D. Niall, Biochemistry 14 (1975) 2214.

(36) L. T. Hunt and M. O. Dayhoff, Biochem. Biophys. Res. Commun. 74 (1977) 650.

(37) N. Sugano, M. O. J. Olson, L. C. Yeoman, B. R. Johnson, C. W. Taylor, W. C. Starbuck and H. Busch, J. Biol. Chem. 247 (1972) 3589.

(38) T. H. Eickbush, D. K. Watson and E. N. Moudrianakis, Cell 9 (1976) 785.

(39) R. K. Busch, I. Daskal, W. H. Spohn, M. Kellermayer and H. Busch, Cancer Res. 34 (1974) 2362.

(40) L. C. Yeoman, J. J. Jordan, R. K. Busch, C. W. Taylor, H. Savage and H. Busch, Proc. Nat. Acad. Sci. USA 73 (1976) 3258.

(41) M. S. Rao, L. I. Rothblum and H. Busch, Cell Biology International Rpts., in press.

These studies were supported by the USPHS Cancer Programs Award 10893, awarded by the National Cancer Institute, DHEW, and generous gifts from Mrs. Jack Hutchins, the Kroc Foundation, the Pauline Sterne Wolff Foundation and the Davidson Fund.

DISCUSSION

W. CIEPLINSKI: Do you have any idea whether the synthesis of histone 2A-ubiquitin complex proceeds from a single message or whether ubiquitin is attached following histone 2A synthesis?

H. BUSCH: The question you raise is very important. I wish that I could give you a satisfactory answer.

W. CIEPLINSKI: The synthesis of ubiquitin and histone 2A perhaps proceeds separately?

H. BUSCH: I think it is very likely.

W. CIEPLINSKI: How does the extraction of nucleolar chromatin with 0.15 and 0.35 M NaCl etc. affect nucleosome structure?

H. BUSCH: We have published that in _Experimental Cell Research_ last year. The whole thing begins to unfold as you extract it with salt.

W. CIEPLINSKI: What happens if you add histones back?

H. BUSCH: When we add histones back, the system dies. I am confident that the histones are not the controlling element in the nucleolar read-out. But, let me say, that confidence could be shaken by one good experiment.

T. BORUN: There are fairly sensitive methods for looking at the nucleolar structure. For instance you can do CD studies during thermal denaturation and from plots of dΘ-dT you get all kinds of inflections. Recently it has been shown that subtle changes in prereplicative chromatin can be perceived using low concentrations of ethidium bromide and flow microfluorometry. Both of these are subtle indices of super structural organization.

I was concerned about the extraction. It is conceivable that if the superstructure were to be maintained during extraction, though a control protein had been removed, then the idea that you could add the protein back for reconstitution would be valid. But, if during extraction you caused a complete rearrangement by some kind of high salt induced annealing of superstructures then such a strategy would fail. Do you have any evidence bearing on the maintenance of nucleolar superstructure by sensitive techniques during extraction?

H. BUSCH: We have followed the changes with an electron microscope very carefully. We know that with high salt we do cause a marked unfolding of the nucleolar structure. Starting with naked DNA and adding histones we have not had any success in reconstitution. But, when we start with the 0.6 M extract which has characteristics that I have pointed out and now reconstitute, that is the first time we have seen any positive results.

J. WATSON: Is there evidence from your transcription experiments whether they involve initiation of new chains or are you just elongating pre-existing ones?

H. BUSCH: We have just found what appears to be an initiated fragment with a pppGp terminus. We are getting that from liver nucleoli and it looks as if it is possible to reinitiate with that system. But, in our system at the moment, we feel we are mainly elongating pre-existing chains.

K. McCARTY: Have you any quantitative information to suggest the number of nucleosomes with the A24 protein representing a modification of H2A as a covalent link with ubiquitin?

H. BUSCH: Well, it is either one in every 5th or two in every 10th nucleosome. But, we have not yet been able to show whether it is one in five or two in ten nucleosomes.

K. McCARTY: Have you made any attempts to use DNase II as a technique for enrichment of the modified nucleosomes?

H. BUSCH: We are doing studies with DNAse II. At the moment these studies have been held back because of the need to complete the structural studies.

V.G. ALLFREY: We have some observations that deal with the structure of ribosomal genes at the nucleosomal level. This is in *Physarum* where there is a well-defined rDNA satellite which has two ribosomal genes each situated near an end of a linear DNA molecule in a palindrome-like array. So the two genes are transcribed in opposite directions. You can purify this satellite DNA. It has all been mapped by restriction nuclease digestion. Dr. Edward Johnson, in my laboratory has made some studies of the structure of the ribosomal gene when it is active and when it is inactive. If you cleave with nucleases, such as staphylococcal nuclease, you cut the ribosomal DNA into fragments of the usual length. The core particle contains the 140 base pairs of DNA. However, when you subject those particles to sucrose gradients, the monomer shows two peaks. The slow sedimenting peak (which is only 5S) is greatly enriched in the ribosomal sequences. This was shown by hybridizing ^{32}P-labelled ribosomal RNAs to pieces of DNA in the various nucleosome peaks. So the DNA that was being transcribed behaves as though it is arranged in nucleosomes, in the sense that it is cut by staphylococcal nuclease in the regular way and you get 140 base pair core particles. But those cores, as judged by their sedimentation coefficients are fully extended, not compact. Now, with Physarum you can starve the organism so that it no longer makes ribosomal RNA. When "starved" chromatin is treated with staphylococcal nuclease there is no enrichment of the ribosomal genes in the more extended nucleosomal configuration. The rDNA sequences are found in the higher order nucleosomes. So when the ribosomal genes are active the conformation of the nucleosome has been altered. As a consequence of that change the DNA exists in a more extended configuration. That agrees with the electron microscopic studies indicating that the transcription complex at least for ribosomal RNA, has the DNA in nearly a fully-extended state, rather than tightly wrapped around nucleosomal beads.

H. BUSCH: I would like to add that these observations agree very well with observations relating to A24 and the loss of the particular structure which would permit the kind of uncoiling needed for transcription. Further, these comments support Franke's suggestion that the nucleosomes are different in activated and nonactivated nucleoli.

TRANSMEMBRANE CONTROL AND CELL SURFACE RECOGNITION

U. RUTISHAUSER and G.M. EDELMAN
Developmental and Molecular Biology
The Rockefeller University
1230 York Avenue
New York, New York 10021 U.S.A.

Abstract: The transmembrane control of cell surface receptor mobility and the molecular mechanism of cell-cell adhesion have been studied to provide a background for examining the relationship between cellular events such as mitosis and differentiation as well as morphogenetic phenomena involving the cell surface. The experiments on receptor mobility suggest that the diffusion of many cell surface components in the plane of the membrane is controlled in part by a modulating assembly that includes microtubules and microfilaments. The analysis of adhesion among neural cells of the chick embryo depends upon an immunological assay for molecules associated with the initial formation of cell-cell bonds. The results suggest that adhesiveness is under a developmental control reflected by the presence of a 140,000 M.W. protein on the cell surface. Antibodies against this protein have been used to analyze the role of cell recognition in tissue formation.

INTRODUCTION

The formation of tissue patterns during embryogenesis requires the coordination of several fundamental cell processes, including mitosis, motility, changes in shape, membrane-membrane interactions, differentiation, and death. For several years we have been interested in examining the possibility that part of the developmental program occurs via relationships between these cellular events and a transmembrane control mechanism involving cell surface and cytoplasmic molecules. Two quite different lines of investigation have been pursued: an analysis of cell surface receptor mobility and its perturbation by cytoplasmic microtubules and microfilaments, and studies on the mechanism and control of cell-to-cell adhesion during the development of embryonic neural tissue. The purpose of this paper is to describe the rationale behind these approaches, summarize the major findings to date, and indicate the probable direction of subsequent studies.

ISBN 0-12-045450-5

TRANSMEMBRANE CONTROL OF RECEPTOR MOBILITY

Taylor, Raff and coworkers (1) have observed that cell surface receptors (e.g. immunoglobulin on B lymphocytes) are arranged in a diffuse distribution, but that in the presence of a crosslinking agent, such as antibodies to immunoglobulins (anti-Ig), these receptors redistribute into patches which then coalesce into caps. Patch formation depends upon the local diffusion of the receptors in the plane of the membrane. Cap formation results from the movement of these cross-linked receptor complexes toward one pole of the cell and depends upon cellular metabolism. When fluorescein-labeled anti-Ig is used, both the patches and caps induced by this reagent can be observed directly.

In 1972 Yahara and Edelman (2) reported that if lymphocytes are first preincubated with 100 μg/ml concanavalin A (Con A) at 21°C, patch and cap formation of Ig receptors cannot be induced by anti-Ig (Table 1). This effect could be reversed by removing the cell-bound Con A with a competitive inhibitor such as α-methyl-mannoside. The effect of Con A on Ig receptor capping appeared to depend on the valence of the lectin. Con A is a tetramer with four identical subunits, each having a single saccharide binding site (3). When this molecule is treated with succinic anhydride, it dissociates into dimers without affecting the binding sites on each subunit (4). Unlike native Con A, the succinyl-Con A derivative does not inhibit patches and capping of Ig receptors. If, however, the cell-bound succinyl-Con A is itself cross-linked by incubation of the cells with antibodies to Con A, inhibition of receptor mobility is again observed (Table 1). Monovalent Fab' antibody fragments do not have this effect. Together these results strongly suggest that the valence of native Con A is critical in the immobilization of Ig receptors, presumably because of its role in cross-linking receptors for Con A on the cell surface (4).

A second key feature of this phenomenon was revealed by the observation that the effect of Con A on Ig receptor mobility is reversed in the presence of colchicine, vinblastine, vincristine, or podophyllotoxin (Table 1) (5). By analogy with the known effect of these drugs on microtubules (6), this suggested that assemblies of microtubule-like structures might be directly or indirectly involved in the immobilization of receptors by Con A. Lumicolchicine, a photo-inactivated derivative of colchicine did not affect receptor mobility, suggesting that the action of colchicine does not result from nonspecific interactions with the cell membrane. It has also been shown that this drug does not interact directly with Con A molecules.

TABLE 1

Perturbation of cell surface receptor mobility

Treatment	% B lymphocytes forming caps with anti-Ig
Control (anti-Ig alone)	87
Con A (100 μg/ml)	4
Succinyl-Con A (100 μg/ml)	86
Succinyl-Con A + anti-Con A	53
Succinyl-Con A + Fab' anti-Con A	80
Con A + 10^{-4} M colchicine	30
Con A + 10^{-5} M vinblastine	51
Con A + 10^{-4} M lumicolchicine	4

(Date from references 2,4,5)

Based on these results and a variety of additional data, (1,2,4,7-10), Edelman, Yahara and Wang hypothesized that the mobility of cell surface receptors is modulated by a cytoplasmic network of microtubules and microfilaments (10,11). This model has four basic features. 1) The existence of microtubular and microfilament assemblies beneath the plasma membrane. 2) Receptors on the cell surface exist in two states that are in equilibrium with each other: the attached or A state in which the receptors are immobilized by an association with the microfilament-microtubular network, and the free or F state in which this linkage is dissociated and the receptor can move more freely. 3) Cross-linking of receptors at the cell surface causes them to shift to the A state, which increases the degree of polymerization of cytoplasmic microtubules. 4) Microtubule polymerization in turn causes a shift of other receptors (not cross-linked at the cell surface) to the A state. In this model, the effect of Con A is reflected by the cross-linkage of receptors with a subsequent polymerization of microtubules that can be reversed by treatment with colchicine. In all likelihood, the linkage of receptors to microtubules is indirect via the microfilaments which are associated with the inner lamella of the plasma membrane.

An alternative model for receptor immobilization by Con A is that all of these receptors are simply bound together at the cell surface by the lectin. To test this possibility, experiments were carried out on the mobility of Ig receptors on lymphocytes bound to fibers coated with either antigen or Con A (12) (Figure 1). Cells bound to antigen-coated fibers and then incubated with anti-Ig formed caps similar to those observed in solution. Ig receptors on cells bound to Con A-fibers, however, remained in a diffuse distribution in the presence of anti-Ig. As in solution, this effect was reversed by the addition of colchicine. Because no more than 10% of the cell surface is in contact with the lectin-coated fiber, these experiments indicate that a localized interaction with Con A is sufficient to cause the immobilization of receptors over the entire cell surface. The results therefore rule out an external cross-linkage model, and are consistent with the hypothesis that receptor mobility is modulated by a highly cooperative surface modulating assembly (10).

To provide a more direct and quantitative estimate of the receptor modulation effect, the lateral mobility of molecules in the plane of the surface membrane was also determined using the fluorescence photobleaching recovery (FPR) method (13). In these experiments, the surface receptors are first labeled with a fluorophore by incubating cells with Fab' directed against cell surface components and conjugated with rhodamine. A small area (3 μm^2) of the cell surface is then bleached with a focused laser beam, and the diffusion of unbleached fluorophore into this area is monitored by a photon counter. Diffusion constants can be directly calculated from this data. Using mouse fibroblasts, rabbit Fab' fragments raised against mouse lymphoma cells, and Con A immobilized on platelets (instead of nylon fibers), it was observed that coverage of 4% or more of the cell membrane with Con A-platelets causes a 6-fold decrease in receptor diffusion (from about $D=2.4 \times 10^{-10}$ cm^2/sec to $D=4 \times 10^{-11}$ cm^2/sec) over the entire cell surface (Figure 2). This decrease was partially (50%) reversed in the presence of colchicine.

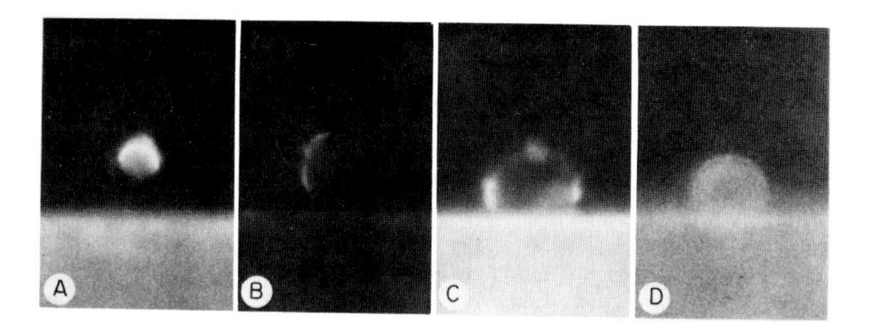

Fig. 1. Antibody-induced redistribution of immunoglobulin (Ig) receptors on B lymphocytes bound to nylon fibers: (A) cell bound to antigen-coated fiber, (B) cell bound to either an antigen or a Con A-coated fiber and incubated with 10^{-4} M, colchicine, (C) cell bound to antigen-coated fiber and incubated with sodium azide to prevent capping but not patching, (D) cell bound to Con A-coated fiber. (Adapted from reference 12).

Although the phenomenon of receptor modulation is clearly defined by these experiments, the molecular details of the modulating assembly remain to be established. In Figure 3 is shown an hypothetical model of the assembly, based on present information. In addition to microtubules and glycoprotein receptors for Con A, microfilaments are also included. Their presence is suggested by the work of Taylor et al. (1) on receptor-microfilament interactions during patching and capping (14) and by the observation that cytochalasin B, a drug that disrupts microfilaments, decreases the diffusion constants of membrane-bound molecules (15). Two alternatives to the structure of the modulating assembly are considered: that modulation induced by local crosslinkage of glycoprotein receptors represents alterations of submembranous components including a gelation of microfilaments or fibrils which results in restricted diffusion of other receptors; or that modulation results in enhanced binding of the cytoplasmic base of receptors to submembranous structures associated with microfilaments. In either case, the model assumes that microtubules are essential to the modulation process.

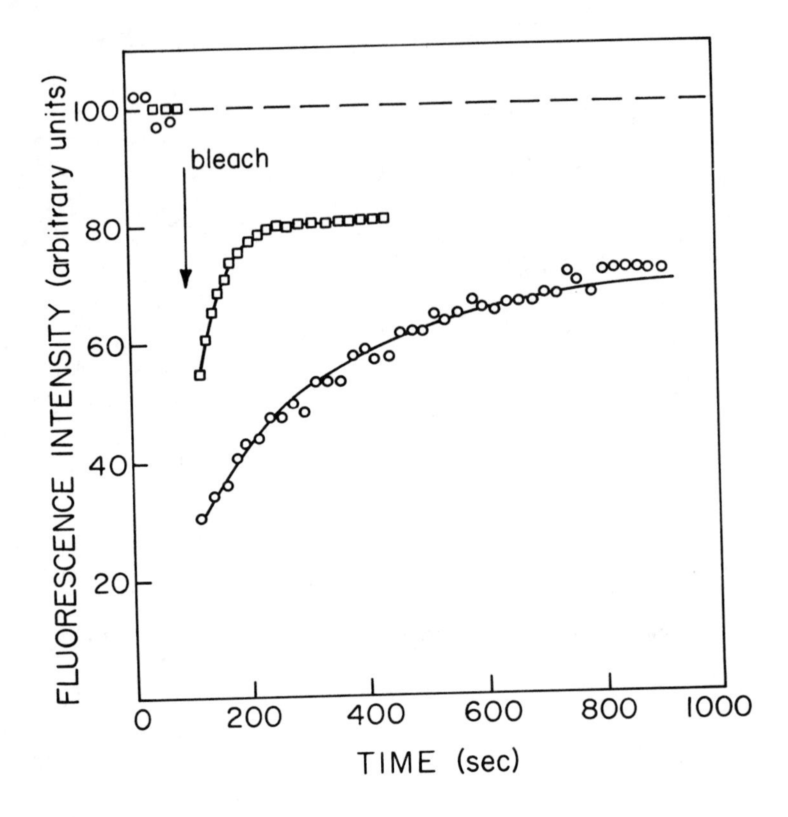

Fig. 2. Fluorescence photobleaching recovery curves of two cells labeled with rhodamine-labeled Fab. One cell had no Con A-platelets (□), the other had approximately 60 platelets covering 13% of its area (O). Both recoveries fit the theory for a single diffusion coefficient within experimental error (superimposed curves). D=1.25 x 10^{-10} cm^2/sec, and D=1.8 x 10^{-11} cm^2/sec, respectively (from reference 13).

In addition to elucidating details of its structure, the fundamental question remains as to the physiological function of the modulating assembly. One possibility is that some of the mechanisms that serve to coordinate cell functions during tissue formation operate by means of interactions in the assembly. Cell motility, for example, may well be affected by the state of the modulating assembly in that microfilaments appear to be associated with membrane movements. Other experiments from our laboratory suggest that the initial mitogenic stimulation of cells by lectins can be blocked by disruption of cytoplasmic microtubules (16, 17), and that transmission of this signal to the nucleus is prevented by modulating doses of Con A (18).

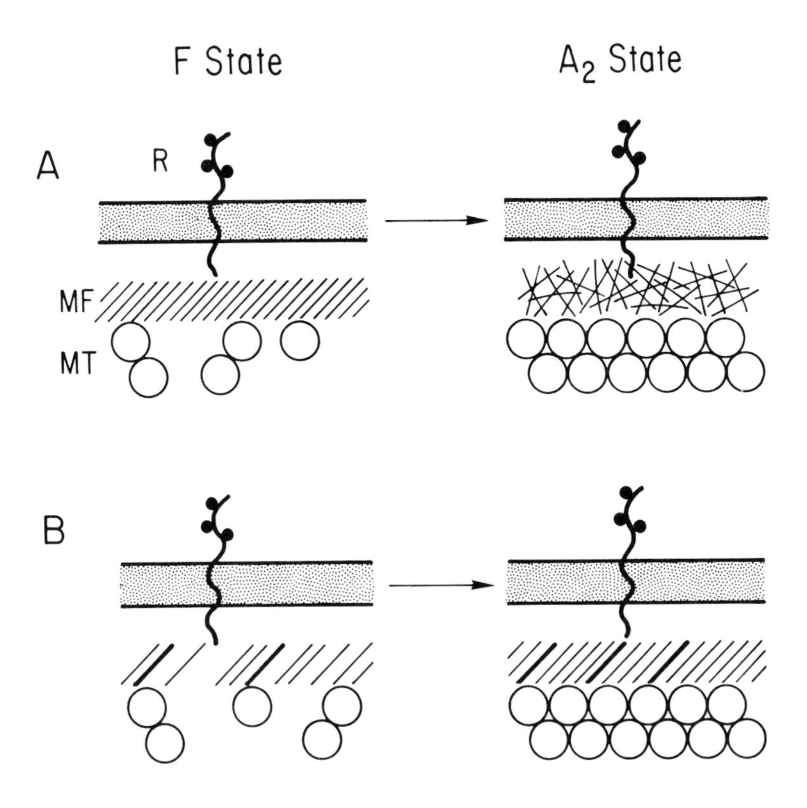

Fig. 3. Schematic representation of a surface modulating assembly showing two mechanisms by which a free or mobile (F state) receptor (R) might be influenced by cytoplasmic microfilaments (MF) and microtubules (MT) to become a less mobile (A state) receptor: (A) restricted diffusion of the receptor as a result of MF disorder and gelation; (B) anchorage of the receptor by interaction of its cytoplasmic base with a molecule associated with the modulating assembly (from reference 13).

As yet, however, the modulating assembly as a functional unit has been detected only through perturbation of cells with lectins. Part of the rationale for studying cell adhesion, in addition to its intrinsic interest for morphogenesis, is to investigate whether certain types of cell-cell contacts, in analogy to binding of cells to Con A-coated fibers or platelets, constitute a modulating event that serves to control other cell functions.

ADHESION AMONG NEURAL CELLS OF THE CHICK EMBRYO

Whereas cell surface modulation as a phenomenon is more clearly defined at the molecular than the physiological level, the converse is true for cell adhesion. Therefore, our present work has focused on achieving a better understanding of the adhesion process, in particular a description of molecules involved in initial binding between individual cells.

Almost 40 years ago, Holtfreter reported that cells obtained from different tissues would stick together randomly and then sort themselves out to form separate regions containing structures characteristic of the parent tissues (19). These observations indicated that there is selected adherence among cells, and suggested that this might be an important process in tissue formation.

Although Holtfreter's experiments provided a focus for later studies on cell recognition, the ensuing work led to quite divergent hypotheses on the mechanism of cell adhesion. One position has been that differences in adhesiveness reflect the presence of structurally distinct ligands on the cell surface (20,21). Alternatively, it has been suggested that aggregation of cells is not governed by particular molecules, but by differences in generalized interactions between large areas of the cell surface (22). A third model assumes that adhesion is mediated by a single mechanism that is subject to control at the cellular level (23,24). Whereas it might seem that tissue formation is too complex a phenomenon to involve nonspecific forces or a single molecular mechanism, it has been argued that histotypic segregation of cells can be accounted for by quantitative differences in adhesiveness rather than absolute binding specificities (25).

There are two basic questions at issue. First, how specific is cell adhesion, and second, what molecules cause the cells to bind to each other? To obtain information on the specificity of cell adhesion, we have examined the initial binding among neural cells from chick embryos of different ages.

To evaluate adhesion among different cell types, it is necessary to use an assay that detects individual binding events between pairs of cells with unambiguous identification of the cells involved. For our studies, we have used a modification (24) of the monolayer method (25) in which cells obtained from tissues by trypsinization are immobilized on culture dishes coated with waxbean agglutinin and incubated with cells in suspension that have been internally labeled with fluorescein. The number of cells that bind to the monolayer can be counted individually using a fluorescence microscope. To minimize the effect of the trypsinization, which damages cell surface proteins, all the cells are cultured in suspension for 12 hours prior to the assay.

Using this procedure, it was possible to test the specificity of binding among cells from different tissues (24). Retina and brain cells were chosen because this has been one of the classical systems for demonstrating tissue-specific adhesion. In contrast to earlier reports, however, we found that there is no absolute specificity in the binding among cells from the retina and brain, although the adhesiveness of these cells does change during development and at different ages in different tissues. As shown in Table 2, high levels of binding were obtained between pairs of cells from retinas of 8-day old chick embryos and from brains of 6-day old embryos. In both tissues, cells from older embryos were less adhesive, low levels of binding being obtained with 14-day retinal cells and 10-day brain cells. Although little binding occurred between an 8-day retinal cell and a 10-day brain cell, an 8-day retinal and a 6-day brain cell bound just as well to each other as to themselves. This observation implied that the mechanism of adhesion is the same for most cells from nervous tissues. As discussed below, the variation in binding with developmental age also suggests that adhesiveness may be important during a particular phase of nervous tissue development.

An Assay for Cell Adhesion Molecules - We have developed an assay for adhesion molecules based on the inhibition of cell aggregation by antibodies prepared against whole cells, and the identification of antigens recognized by these antibodies (27). The advantages of this procedure are that it does not require assumptions as to the number, composition, or mode of action of these molecules, nor does it require that they retain their biological activity.

As retinal cells in suspension bind to each other, the total number of particles decreases, and the rate of this decrease can be used as a measure of the aggregation process (28). To establish that decrease in particle number can be used to quantitate the adhesiveness of single cells, we compared results from this method with those obtained using the monolayer assay described above. These studies indicated that the rate of decrease in particle number is proportional to the rate of cell binding to a monolayer. In both assays the majority of cells in suspension aggregated or bound to the monolayer within 1 h, suggesting that the adhesiveness being measured is a property of neural cells in general.

TABLE 2

Binding among retinal and brain cells from embryos of different ages

Cell-cell binding between:*

Cell in monolayer	Cell in suspension	Cell-bound cells**
R_8	R_8	423
R_{14}	R_{14}	122
B_6	B_6	413
B_{10}	B_{10}	8
R_8	B_{10}	41
R_8	B_6	390

*R_8 and R_{14} are cells from retinas of 8-day and 14-day old chick embryos; B_6 and B_{10} are cells from brains of 6-day and 10-day old embryos.

**Expressed as the number of cells in suspension that bound to 1 mm^2 of cell monolayer (data from reference 24).

To obtain antibodies that can inhibit cell adhesion, rabbits were immunized with cells from retinas of 10-day old chick embryos (anti-R10), and Fab' fragments were prepared from the antibodies. The Fab' fragments, being monovalent, did not cause agglutination, but instead inhibited the aggregation of the retinal cells.

The approach we have adopted is based on the assumption that inhibition of aggregation by Fab' reflects the inactivation of particular cell surface molecules that mediate cell-to-cell adhesion. Alternatively, it is possible that anti-R10 Fab' inhibits simply by coating the cell surface. Therefore, two molecules that bind to the cell surface, a lectin derivative (succinyl-concanavalin A) that binds to glycoproteins but does not agglutinate cells (4) and Fab' fragments that bind a variety of carbohydrates (29) were also assayed for inhibitory activity. Although immunofluorescence techniques revealed that all three reagents bind in similar amounts to the surface of retinal cells, only anti-R10 Fab' decreased the rate of cell aggregation. This observation suggested that the effect of anti-R10 Fab' on adhesion is not the result of a nonspecific blockade of the cell surface.

The immunoassay for molecules involved in cell adhesion measures the ability of retinal cell molecules to neutralize the anti-R10 Fab', thus permitting aggregation to occur. A convenient source of such neutralizing substances was found to be supernatants from cultures of 10-day retinal tissue (TCS). Although these supernatants did not by themselves affect aggregation, they did reverse the inhibition by anti-R10 Fab' (Table 3). The amount of neutralization was found to be linear with respect to the logarithm of the supernatant volume added, and therefore the relative amount of Fab'-neutralizing antigen could be estimated. For purposes of quantitation, we defined one unit of neutralizing activity as the amount of antigen required to decrease the inhibition of adhesion by Fab' from 50% to 25%.

Isolation of a Cell Adhesion Molecule and Preparation of Specific Adhesion-Blocking Antibodies - Using the immunological assay described above, we identified and characterized the cell surface molecules recognized by the adhesion blocking antibodies in anti-R10 (30). Three steps were required: 1) purification of neutralizing antigens from TCS; 2) immunization with the purified activity to produce a specific adhesion-blocking antibody; and 3) identification of the cell surface molecules recognized by this specific antibody.

To purify the substances involved in the neutralization of anti-R10 Fab', the TCS mixture was fractionated by gel filtration and polyacrylamide gel electrophoresis and assayed for activity as described above. These procecures gave about a 500-fold increase in specific neutralizing activity with a total yield of 50%. The gel filtration of TCS

TABLE 3

Effect of tissue culture supernatant and anti-R10 Fab' on aggregation of cells from 10-day retinas

Assay	Anti-R10 Fab' (mg/assay)	Tissue culture supernatant (μl)	Aggregation (Δ%)*
Aggregation	0	0	41 ± 2
Effect of TCS	0	5-100	39 ± 2
Inhibition by Fab'**	1	0	18 ± 1
Neutralization of Fab' by TCS	1	50	33 ± 2

The standard deviation in triplicate assays is shown.

*Percent decrease in particle number after 20 min.

**Neutralizing activity is most reliably measured when the amount of Fab' added causes a 50% inhibition of cell aggregation (data from reference 27).

is illustrated in Figure 4a. Activity was found in a single region containing less than a tenth of the total protein. Further purification was achieved by polyacrylamide gel electrophoresis in Tris-glycine buffer (Figure 4b). Again, the activity was found in a single region.

To obtain specific adhesion-blocking antibodies, rabbits were immunized with the active fraction from polyacrylamide gel electrophoresis. After several intraperitoneal injections in complete Freund's adjuvant, antibodies were produced that inhibited aggregation of either retinal or brain cells by over 90%.

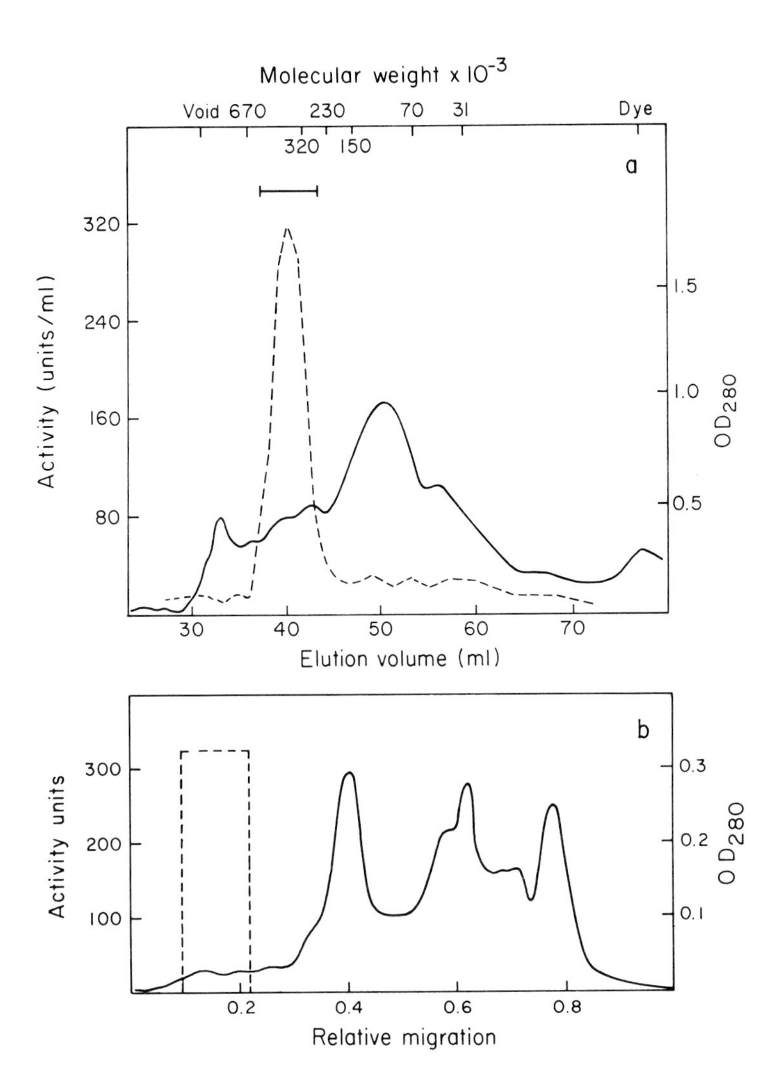

Fig. 4. Purification of Fab'-neutralizing activity from tissue culture supernatant. (A) Gel filtration. The solid line represents optical density at 280 nm and the dashed line indicates Fab' neutralizing activity. The active material was pooled for fractionation by gel electrophoresis. (B) Gel electrophoresis of the active material from (A). Active fractions were used to immunize rabbits (modified from reference 30).

To identify and characterize structures on the cell surface that are recognized by the antibodies that inhibit adhesion, cell membrane molecules were solubilized with a nonionic detergent and then mixed with the antibodies (31). In Figure 5a are shown membrane proteins, labeled with ^{3}H-leucine, that were immunoprecipitated by anti-R10. As would be expected with antibodies produced against whole retinal cells, a number of proteins were precipitated. In contrast, when antibodies to the purified activity were used (Figure 5b), the precipitate contained a polypeptide with a molecular weight in SDS of about 140,000 and another with an M_r of 40,000.

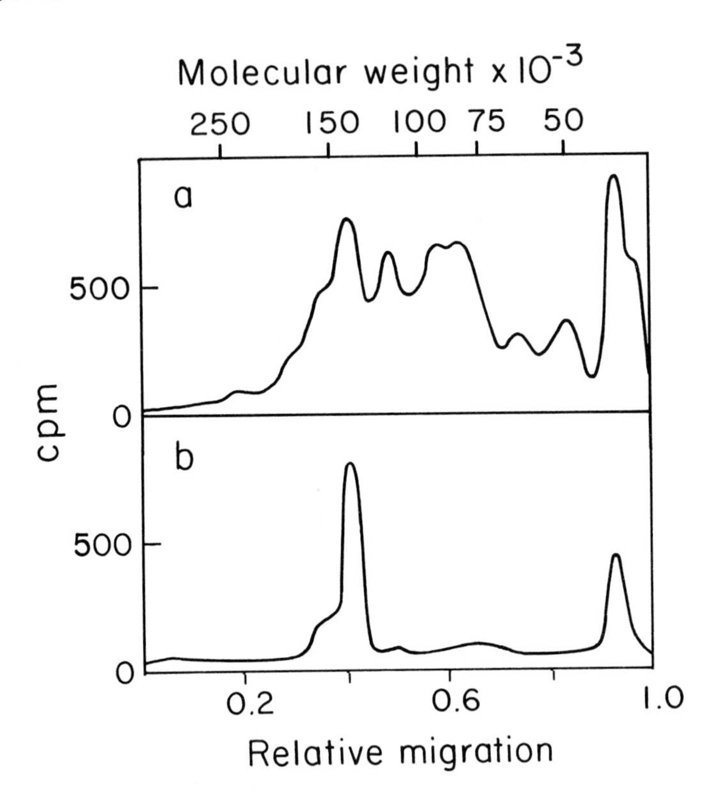

Fig. 5. SDS-polyacrylamide gel electrophoresis of immunoprecipitated ^{3}H-labeled proteins extracted from membranes of retinal cells from 10-day old chick embryos. (A) Immunoprecipitated with antibodies to whole retinal cells (anti-R10); (B) Immunoprecipitated with antibodies to the purified activity (redrawn from reference 30).

The 140,000 M_r component was specifically precipitated and has been named cell adhesion molecule or CAM. The 40,000 M_r component, however, does not appear to be involved in adhesion in that it was also precipitated by antibody from unimmunized rabbits. Other results suggest that this component represents actin that coprecipitates with antigen-antibody complexes. The major conclusion of these studies, therefore, is that inhibition of cell adhesion by either anti-R10 or by antibodies to the purified activity reflects their binding to a cell surface protein (CAM) having an apparent M_r in SDS of about 140,000.

The Function of CAM and the Role of Cell Adhesion in Development - If inhibition of cell adhesion by Fab' fragments involves inactivation of CAM, the simplest explanation would be that this molecule is a ligand in the formation of cell-cell bonds. Alternatively, CAM could be required to maintain some structural feature of the cell membrane that is necessary for binding. In either case, it might be expected that CAM would be found on the surface of cells in aggregates, particularly in areas of direct cell-cell contact. By using the peroxidase-anti-peroxidase procedure (32) with anti-CAM antibodies, CAM was localized in electron micrographs of sections through retinal cell aggregates formed after 20 min of incubation. The molecule was found to be uniformly distributed on the cell surface including areas in which the membranes of different cells were in close proximity.

A direct involvement of CAM in cell adhesion is suggested by experiments on the binding of retinal cells to nylon fibers coated with the antibodies against CAM. The adhesiveness of retinal cells varies with developmental age, and we have found (24) that this variation is closely correlated with the ability of the cells to bind to the fibers coated with anti-CAM (Figure 6). The results are consistent with the hypothesis that cell adhesiveness reflects the amount of CAM on the cell surface and indicates that this molecule may participate in the formation of bonds between a cell and another surface.

The relevance of cell aggregation *in vitro* to tissue formation has not yet been established. Nevertheless, certain correlations are consistent with the hypothesis that adhesiveness of cells represents an important parameter during embryonic development. Also shown in Figure 6 is the number of cells in the chick retina as a function of developmental age. Although the retina is not uniformly developed during these stages of embryogenesis, the results suggest that the period during which dissociated retinal cells are most adhesive immediately follows the time of maximal increase in cell number. It is coincident with the time during which the plexiform layers appear and precedes the formation of most synapses among retinal cells. Together, these observations suggest that CAM may play a strategic role in formation of nerve tissue. They do not, however, specify what this role is or how it is carried out.

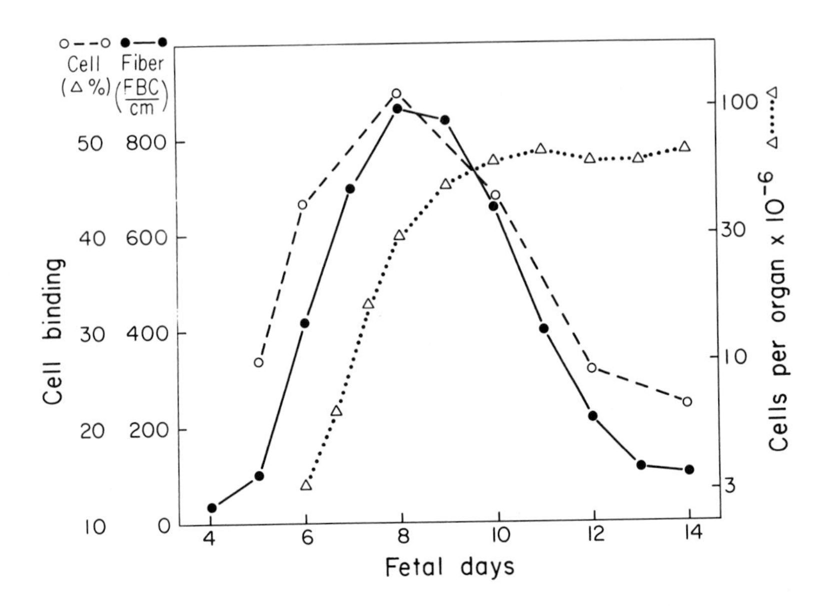

Fig. 6. Age dependence of retinal cell binding to nylon fibers coated with anti-CAM (o——o), retinal cell aggregation (o---o), and the number of cells obtained by dissociation of retinal tissue with trypsin (Δ...Δ). Cell-fiber binding is expressed as the number of cells bound to one edge of a one cm fiber segment; aggregation is expressed as the percent decrease in single cells (redrawn from reference 24).

The distribution of CAM in the retinas of 14- and 7-day chick embryos is shown in Figure 7. At day 14, most of the CAM is associated with regions almost entirely composed of neuronal processes, namely the two plexiform layers. At day 7, when these layers are not yet visible, CAM is uniformly distributed across the tissue. Further evidence that CAM is associated with the surface of neurites was obtained by growing retinal cells in culture to produce isolated bundles of nerve processes, and observing that these bundles were brightly stained by fluorescein-labeled anti-CAM.

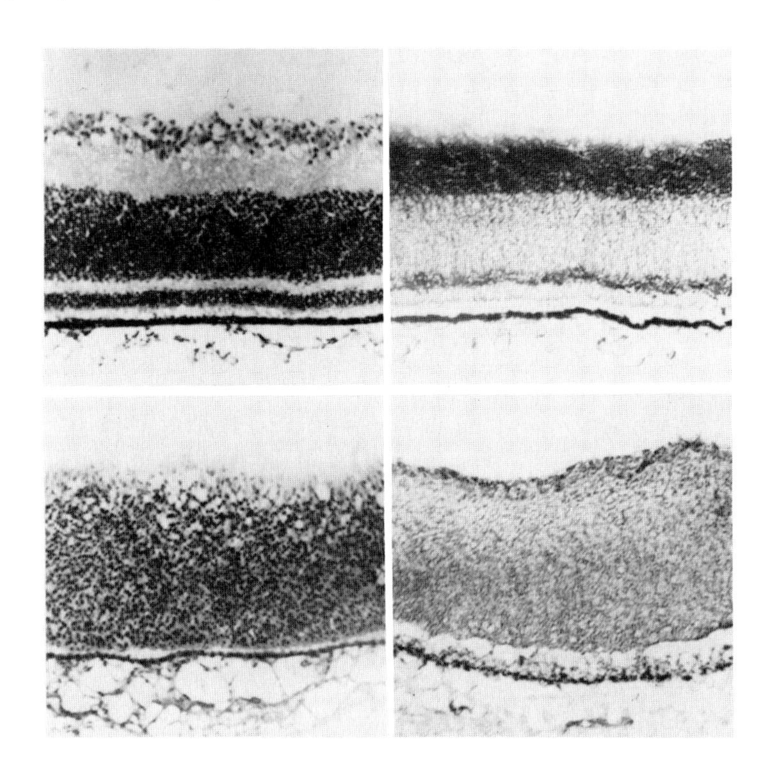

Fig. 7. Distribution of CAM in the retina of chick embryo. Left, top and bottom: hematoxylin-eosin staining of 14-day and 7-day tissues, respectively. Right, top and bottom: anti-CAM staining (darker areas) of 14-day and 7-day tissues, respectively.

The presence of a cell adhesion molecule on nerve cells and their processes during the time when the retinal cell and plexiform layers are being formed raises the possibility that it might function in the development of these layers. This possibility was tested by studying histotypic development in retinal cell aggregates. If aggregates of 8-day-old cells are maintained in culture for 7-8 days, large neurite regions containing synapses are formed (33,34). As in the retina, CAM was found to be preferentially distributed in the neurite regions. When aggregates that had been cultured for one day were transferred to a medium containing anti-CAM Fab' fragments, however, neurite regions did not appear.

Electron micrographs of these aggregates revealed that neurites were present, but had failed to sort out from the cell bodies to form distinct regions of tightly packed processes.

At this stage, therefore, it seems possible that a number of experiments can be carried out to provide information on at least one aspect of pattern formation in developing nerve tissue. The identification and purification of CAM, and the preparation of specific antibodies to this molecule represent our initial steps in this approach. It will be of primary importance to describe the chemical properties of CAM further, in particular to determine whether this molecule is a ligand in the formation of cell-cell bonds or whether it serves a controlling or supportive function in cell adhesion. These studies will include an analysis of the composition and structure of CAM, and of possible interactions with other molecules, membrane components or cytoplasmic structures that might affect or be affected by its function.

REFERENCES

(1) R. B. Taylor, P. H. Duffus, M. C. Raff and S. de Petris, Nature New Biol. 233 (1971) 225.

(2) I. Yahara and G.M. Edelman. Proc. Nat. Acad. Sci. USA 69 (1972) 608.

(3) G.M. Edelman, B.A. Cunningham, G.N. Reeke, Jr., J.W. Becker, M.J. Waxdal, and J.L. Wang. Proc. Nat. Acad. Sci. USA 69 (1972) 2580.

(4) G.R. Gunther, J.L. Wang, I. Yahara, B.A. Cunningham, and G.M. Edelman. Proc. Nat. Acad. Sci. USA 70 (1973) 1012.

(5) I. Yahara and G.M. Edelman. Nature 246 (1973) 152.

(6) L. Wilson, J.R. Bamburg, S.B. Mizel, L.M. Grisham, and K.M. Greswell. Fed. Proc. 33 (1974) 158.

(7) M.C. Raff and S. de Petris. Fed. Proc. 32 (1973) 48.

(8) I. Yahara and G.M. Edelman. Exp. Cell Res. 91 (1975) 125.

(9) S. de Petris and M.C. Raff. Ciba Foundation Symp. 14 (1973) 27.

(10) G.M. Edelman. Science 192 (1976) 218.

(11) G.M. Edelman, I. Yahara, and J.L. Wang. Proc. Nat. Acad. Sci. USA 70 (1973) 1442.

(12) U. Rutishauser, I. Yahara, and G.M. Edelman. Proc. Nat. Acad. Sci. USA 71 (1974) 1149.

(13) J. Schlessinger, E.L. Elson, W.W. Webb, I. Yahara, U. Rutishauser, and G.M. Edelman. Proc. Nat. Acad. Sci. USA 74 (1977) 1110.

(14) K.-G. Sundqvist and A. Ehrnst. Nature 264 (1976) 226.

(15) J. Schlessinger, D.E. Koppel, D. Axelrod, K. Jacobson, W.W. Webb, and E.L. Elson. Proc. Nat. Acad. Sci. USA 73 (1976) 2409.

(16) G.R. Gunther, J.L. Wang, and G.M. Edelman. Exp. Cell Res. 98 (1976) 15.

(17) D. McClain, P. D'Eustachio, and G.M. Edelman. Proc. Nat. Acad. Sci. USA 74 (1977) 666.

(18) D. McClain and G.M. Edelman. J. Exp. Med. 144 (1976) 1494.

(19) J. Holtfreter, in: Foundations in Experimental Embryology, eds. B. Willier and J. Oppenheimer (Prentice-Hall, Englewood Cliffs, New Jersey,1964)p.186.

(20) A.A. Moscona. J. Cellular Comp. Physiol. Suppl. 1 60 (1962) 65.

(21) J.E. Lilien, in: Current Topics in Developmental Biology, Vol. 4, eds. A.A. Moscona and A. Monroy (Academic Press, New York,1969)p.169.

(22) A.S.G. Curtis. The Cell Surface (Logos Academic Press, New York,1967)p.80.

(23) S. Roseman. Chem. Phys. Lipids 5 (1970) 270.

(24) U. Rutishauser, J.-P. Thiery, R. Brackenbury, B.-A. Sela, and G.M. Edelman. Proc. Nat. Acad. Sci. USA 73 (1976) 577.

(25) M. Steinberg. Science 141 (1963) 401.

(26) B.T. Walther, R. Ohman, and S. Roseman. Proc. Nat. Acad. Sci. USA 70 (1973) 1569.

(27) R. Brackenbury, J.-P. Thiery, U. Rutishauser, and G.M. Edelman. J. Biol. Chem. 252 (1977) 6835.

(28) C.W. Orr and S. Roseman. J. Molecular Biol. 1 (1969) 109.

(29) B.-A. Sela and G.M. Edelman. J. Exp. Med. 145 (1977) 443.

(30) J.-P. Thiery, R. Brackenbury, U. Rutishauser, and G.M. Edelman. J. Biol. Chem. 252 (1977) 6841.

(31) R. Henning, R.J. Milner, K. Reske, B.A. Cunningham, and G.M. Edelman. Proc. Nat. Acad. Sci. USA 73 (1976) 118.

(32) L.A. Sternberger, in: Electron Microscopy of Enzymes, ed. M.A. Hayat (Van Nostrand Reinhold, New York,1973) p.150.

(33) Z. Vogel, M.P. Daniels, and M. Nirenberg. Proc. Nat. Acad. Sci. USA 73 (1976) 2370.

(34) Z. Vogel and M. Nirenberg. Proc. Nat. Acad. Sci. USA 73 (1976) 1806.

This research was supported by grants AI-11378, AM-04256, and HD-09635 from the National Institutes of Health.

DISCUSSION

A. MARKS: How do you reconcile your apparent postulate that the biological signal generated by concanavalin A is due to receptor cross-linkage and yet when you score for a biological effect such as mitogenesis you get what is probably a better response with succinyl-Con A, a divalent lectin, or even monovalent Con A which apparently has no cross-linking potential? Is it fair to say that it is the binding of the lectin to the surface that initiates the biological response and that cross linkage may in fact block the transmission of the signal?

U. RUTISHAUSER: Yes. The crosslinkage of receptors at the cell surface is believed to interfere, via the modulating assembly,with the transmission of a mitogenic signal to the nucleus, but not in the generation of that signal. Thus tetravalent, divalent and monovalent Con A can all trigger the initial event, but only the tetravalent molecule is capable of blocking subsequent events such as DNA synthesis, blast transformation, and cell division. Furthermore, in cases where the mitogenic signal is generated by an agent other than Con A, the tetrameric lectin is still able to prevent the later steps.

A. MARKS: The only evidence that you presented for involvement of microtubules was the use of antimicrotubule drugs. The doses which you have shown were of the order of 10^{-4}M, and it is known that at those concentrations of antitubular drugs they in fact inhibit nucleoside transport in mammalian cells, probably by direct interaction with the component of a cell membrane. Have you considered the possibility that in fact the drug action in those experiments is also mediated by direct interaction with the cell membrane and does not involve microtubules at all. In other words, have you done these experiments at concentrations as low as 10^{-7}M and 10^{-6}M which is the dose response for 50% inhibition of tubulin depolymerization?

U. RUTISHAUSER: We have carried out the experiments at 10^{-6}M colchicine and obtained the same results as at 10^{-4}M. The reason for using the higher doses was simply to decrease the time required for the drug to penetrate the cell membrane and interact with cytoplasmic microtubules.

R.G. KALLEN: Does CAM interact with itself or with some other molecule on the cell surface? Do you have evidence for interaction between CAM molecules or have you alternatively immobilized CAM and looked for CAM - binding proteins that are cell surface molecules?

U. RUTISHAUSER: We are now making a considerable effort to answer this question, but at present we have no clear answer. There are a number of problems (the small amount of CAM available, the possible presence of inhibitors for binding, and possible requirements for a lipid environment) which make this a formidable task. Our approach requires a variety of experimental designs, including the use of CAM immobilized on surfaces and incorporated into vesicles.

I.B. SABRAN: In your measurements of the movements of the molecules on the surface of the cell, how long after your flash or your bleaching do you expect a biological impact to have been complete? Have you also considered the fact that bleaching may cause an effect on the microtubule structure itself?

U. RUTISHAUSER: The only purpose of the laser is to bleach the fluorophore. The modulation has been induced prior to the bleach and therefore we are able to observe the decrease in receptor mobility immediately after attenuation of the laser. As for damaging the cell with the laser, we cannot rule out such events but one does not observe any overt changes in the cell; repeated bleaches and measurements in the same area give the same diffusion constants, and if microtubules were destroyed one would expect receptor mobility to increase rather than decrease (as we observed when colchicine was added).

C. WEST: If you restrict the mobility of the group of molecules you identify as CAM by modulation, do you affect the adhesion between cells? Why don't you put CAM on a fiber and then add cells to the fiber and test for binding of other cells?

U. RUTISHAUSER: I can only say that addition of high doses of soluble Con A does not appear to affect cell adhesion. We have not as yet looked into the question with the FPR method. The multivalent lectins are often able to agglutinate cells themselves, so results in solution are somewhat difficult to interpret. As you indicated, more detailed studies should use immobilized Con A to eliminate this problem.

C. WEST: You added this group of molecules you identify as CAM to agglutinating cells, does it not affect the agglutination of those cells?

U. RUTISHAUSER: The CAM purified by the procedure I presented does not affect aggregation of retinal cells. As I mentioned, the demonstration of binding activity for a molecule solubilized from the cell surface membrane is a complicated problem, and it is probably too early to make much of this result.

I.B. FRITZ: Would you tell us more about the specificity of CAM? I couldn't tell from the EM's that you showed whether it was present also in the brain cells other than retinal cells. I presume that you obtained this antibody to the CAM from the retinal cells and culture. Is a similar molecule present on the brain cells?

U. RUTISHAUSER: Yes, the molecule is primarily found in retina, brain, spinal cord, and ganglia.

I.B. FRITZ: So your prediction then is that if you had cultured brain cells rather than retinal cells you would have gotten the same response?

U. RUTISHAUSER: Yes.

DNA FOLDING AND HISTONE ORGANIZATION IN CHROMATIN

G. Felsenfeld, R. D. Camerini-Otero, R. Simon,
B. Sollner-Webb, P. Williamson and M. Zasloff
Laboratory of Molecular Biology
National Institute of Arthritis, Metabolism
and Digestive Diseases
National Institutes of Health
Bethesda, Maryland 20014

Abstract: We have studied the organization of DNA within the nucleosome, the subunit of chromatin structure. The DNA within the nucleosome core is susceptible to attack by a variety of nucleases; the points of attack are restricted to sites that are separated by intervals of ten nucleotides on each strand. We compare the modes of action of three different nucleases. In each case, the enzyme has a characteristic cutting pattern in which cuts on opposite strands are offset, or staggered, relative to one another. Although the cutting sites are different for each enzyme, the centers of symmetry of their cuts all coincide. This suggests that there is a common periodic recognition site on the nucleosome for all of the enzymes.

We have been concerned with the role of various histones in interacting with the DNA of the nucleosome, and have shown previously that the arginine-rich histones, H3 and H4, play the central role in organizing nucleosome structure. We discuss this and other evidence concerning the effect of H3 and H4 on DNA folding and supercoiling. We present recent results concerning the interaction of the two H3 molecules within the nucleosome, and show that the sulfhydryl groups at position 110 are probably in sufficiently close contact to form a disulfide linkage without major perturbations of nucleosome structure.

We also consider the energetics of DNA bending

involved in nucleosome formation, and the sources of this energy in histone-histone interaction. The possible implications of these mechanisms for biologically important reactions such as transcription are discussed.

INTRODUCTION

Nucleosome structure involves a variety of specific interactions between the histones and DNA, and among the histones themselves. We have been interested in learning about the organization of DNA within the nucleosome, and about the role of the various histone components in forming the structure.

NUCLEASE ACTION

When nuclei are digested with any of a number of nucleases, the DNA that is digested first is the 'spacer' DNA (1,2,3) that connects the nucleosome 'cores' (Fig. 1). The spacer DNA varies in length from zero to about 100 base pairs, depending upon the source of the chromatin; the DNA of the core, however, is 140 base pairs long in all eukaryotic organisms so far examined (4,5).

Although the nucleosome is relatively resistant to nuclease attack, it eventually undergoes degradation also. In chromatin, attack within the nucleosome occurs at certain preferred sites, giving rise not to a continuous size distribution of small DNA fragments, but to a set of fragments of discrete size. For example, staphylococcal nuclease (Fig. 1) generates double-stranded fragments that are approximate integral multiples of ten base pairs in length (6). Pancreatic DNase (DNase I), spleen acid DNase (DNase II) and a number of other nucleases do not generate double-stranded fragment patterns of such well-defined size, but the single-strand gel patterns display a remarkable regularity (Fig. 2A), again corresponding to DNA molecules that are $10 \cdot n$ nucleotides long, with n an integer (7,8).

What is the origin of this regular array of fragments? The most naive interpretation of the single strand digest patterns would be that there are sites on each strand, at ten nucleotide intervals, that are accessible to attack by any nuclease. It will be noted at once that this does not specify the relationship between cuts on <u>opposite</u> strands. These may be offset, or staggered, with respect to one another by any number of base pairs between zero and nine. If the stagger is not zero, then when cuts on opposite strands are close together, the resulting partly digested DNA will fall apart, upon purification, into double stranded fragments with single stranded 'tails'. The length of the tails will be equal to the size of the stagger.

The presence of such structures is in fact revealed if partial digests of nuclei with DNase I are electrophoresed (9) as double-stranded, undenatured molecules (Fig. 2B). The blurred pattern obtained with this enzyme reflects some underlying regularity, but it is not as clear as that obtained with

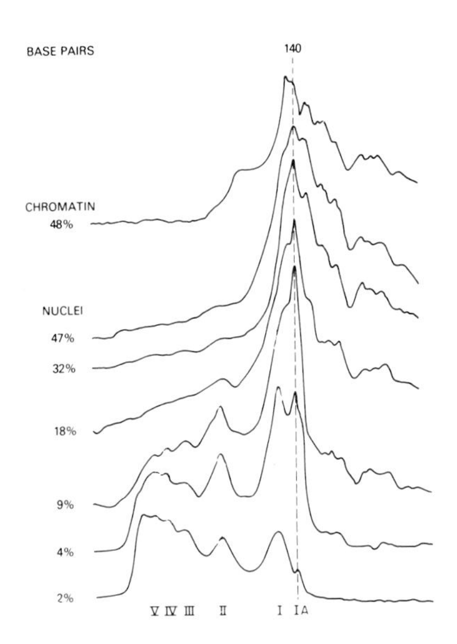

Fig. 1. Polyacrylamide gel electrophoresis of nuclear and chromatin digests. Purified DNA was run on 4% polyacrylamide slab gels, stained, photographed, and the negatives traced. Migration is from left to right. The percent of DNA made acid soluble is shown at the left. Roman numerals at the bottom refer to corresponding nucleosome oligomer size, IA is the DNA of the nucleosome core. (From Ref. 1).

staphylococcal nuclease. The reason is that the DNase I digest bands contain several components and also reflect the presence of rather long single stranded tails, while the staphylococcal nuclease digests have only rather short ones. This can be demonstrated qualitatively by treating the two digests with S1 nuclease, which digests single stranded DNA preferentially. If the DNA is subsequently denatured and then electrophoresed, the pattern of DNase I digest fragments is altered considerably, while that derived from partial staphylococcal nuclease digests change relatively little. This behavior suggests strongly that the cutting sites of the two enzymes are not identical (9).

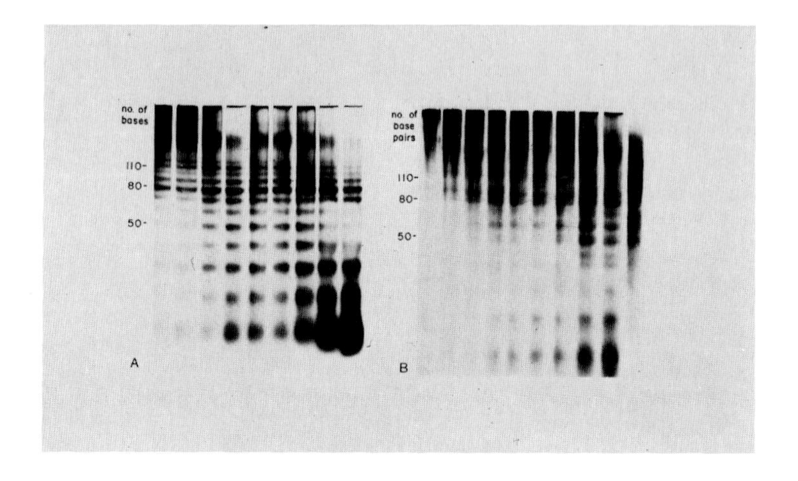

Fig. 2. DNase I digestion kinetics. Duck erythrocyte nuclei were digested with DNase I, and the isolated DNA electrophoresed on 10% polyacrylamide gels. (A) DNA denatured before electrophoresis. From left to right: 4, 8, 9, 12, 14, 16, 23, 34 and 53% of the DNA acid soluble. (b) Same samples, except that DNA was not denatured before electrophoresis. (From Ref. 9).

To measure the length of the stagger precisely, we have made use of DNA polymerase II (a gift of Dr. S. Wickner) to repair the single-stranded regions of the double-stranded digest fragments. The enzyme will add nucleotides to a recessed 3'-OH terminus; we have shown that repair is complete and faithful, using test templates generated by digestion of DNA with restriction endonucleases (9). If radioactive nucleoside triphosphates are used for the repair, the size of the recessed region can be determined by denaturation of the DNA product, and analysis of its size by electrophoresis and autoradiography.

When this procedure is applied to DNA from DNase I digests of nuclei, it is found that the recessed 3'-OH termini are elongated by eight nucleotides (9). This means that there must also exist 5'-P termini recessed by two nucleotides (9), though these of course are not extended by the action of the enzyme. We conclude that the stagger of cuts produced by DNase I is eight nucleotides (3'-OH recessed) and two nucleotides (5'-P recessed). Similar results have been obtained by Lutter (10).

Recently, we have carried out a similar analysis of the action of spleen acid DNase (DNase II). We find (11,12) that the double-stranded DNA digest electrophoretic pattern differs in detail from that generated by DNase I, though the dominant single-strand components of each double-stranded band in the DNase II pattern are identical to those appearing in the (roughly) corresponding DNase I band. The results are explained by the different stagger of cuts produced by DNase II. The difference is clearly revealed in the repair experiment using DNA polymerase II: 3'-OH termini are elongated by six nucleotides in this case (11,12). The cutting sites are staggered by six nucleotides (3'-OH recessed).

Though the results tell us that not all cuts made by the two enzymes can coincide, they do not determine the position of the two sets of cuts relative to one another. This can be accomplished, however, by studying the pattern produced by digestion of nucleosomes radioactively labelled at their 5' termini. The resulting single stranded DNA has lengths of $10n + 2$ nucleotides when DNase I is used, and $10n + 3$ nucleotides when DNase II is used in the digestion (11,12). In combination with the measurements of cutting stagger, this determines the relative location of all the cutting sites. The results are summarized in Fig. 3, together with the results obtained from a somewhat different analysis to determine the sites of action of staphylococcal nuclease (11,12). It is evident that all of the enzymes cut differently; however, all share a common local center of symmetry. It should be noted that such a center is a necessary (but not sufficient) condition for the presence of a true dyad axis in the nucleosome core itself.

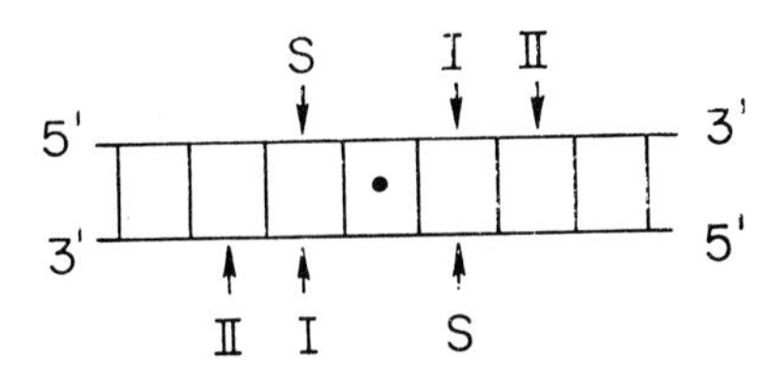

Fig. 3. Relative positions of cutting sites for DNase I, DNase II and staphylococcal nuclease (Marked I, II and S). The dot represents the dyad axis (From Ref. 11).

We conclude that a rather large portion of each ten base pair repeat is accessible to nucleases, but that each nuclease has a different characteristic cutting site occurring once every ten nucleotides on each strand. The separation or stagger of sites on opposite strands depends upon the particular enzyme used, but the center of symmetry of each pair of cuts is the same for all the enzymes studied. The results support a model proposed earlier (9) to account for the DNase I cutting pattern: The enzymes recognize a common site occurring once every ten nucleotides on each strand, but cut at a distance from this site and in a direction determined by the particular enzyme used. The recognition sites may, but need not, be coincident with the centers of symmetry. Unfortunately these results do not permit us to distinguish among various models for the folding of the DNA in the nucleosome. In particular, the results can be made consistent with either the 'kinked' structure or the continuously bent structure of DNA, both of which have been proposed (7,13) to account for the unusual digestion pattern we have been considering.

ROLE OF HISTONES

We have also been interested in the way in which the histones of the nucleosome core cause the compaction of nucleosome DNA. Although it might appear that much energy would be required to accomplish this deformation, the amount involved is probably not extraordinarily large. We can make an estimate (14) of this energy from the known average bending force constants of DNA. Landau and Lifshitz (15) have shown how to treat extended, semi-rigid molecules as flexible rods; their treatment relates the bending force constant to the persistence length, a parameter normally used to express the stiffness of a polymer coil. The persistence length can be deduced from the measurement of polymer properties (16). In the case of double-stranded DNA, the persistence length is about 600 Å (17), equivalent to a bending force constant of 260-360 Kcal-Å/rad^2-mole. We now assume some reasonable structure for the nucleosome roughly consistent with its known dimensions. We have estimated that a uniformly bent superhelix of diameter 110 Å containing 140 base pairs of DNA, and with a pitch of about 82 Å, would have a bending free energy of 20 to 28 Kcal per mole of nucleosome (14). Using more exact nucleosome core dimensions from recent crystallographic studies, the energy of bending for our hypothetical uniformly curved superhelix might be 24 to 34 Kcal per mole of nucleosomes.

The DNA is kept in this bent conformation by the histone octomer complex, which is probably attached to the DNA through the more basic regions of the histones. The attachment is quite strong at reasonably low ionic strength, so that the critical interactions required to keep the DNA bent are those between histones within the octomers. If these are too weak, they will not provide the force necessary to keep the DNA in the compact form. We have estimated the energy generated when histones, previously bound to unfolded DNA through their basic residues, are allowed to make additional contacts by folding of the DNA (14). We have used the known experimental values of the free energy of association of histones in solution, assuming that the histone contacts that stabilize dimer formation in solution are also involved in intranucleosomal contact. The solution free energies must be corrected to take into account the fact that, for our model folding reaction, the histones are already largely immobilized on DNA before folding occurs. The major correction requires that we take into account the rotational and translational entropies that are lost when histones first bind to unfolded DNA. When this is done, we find that one or two histone dimer interactions of average strength would be sufficient to overcome the resistance of DNA to bending. It is possible that there is some delicate equilibrium between folded and unfolded forms that is governed by relatively small changes in the histone interaction energy. It should be noted, however, that our calculation shows only that the folded form should ordinarily be stable; it does not show that nucleosomes are generated _in vivo_ using a pathway that involves folding of a previously formed, open nucleohistone complex.

Much work has been done to define the points of contact between the histones of the nucleosome. These studies involve the use of chemical crosslinking reagents, but the exact sites of attachment within the polypeptide chains have not been identified in most cases. We have made use of a cross-linking reaction involving a well-defined site on histone H3: The sulfhydryl group of the cysteine at position 110. This residue is conserved throughout evolution (with the exception that it is absent in yeast), and in all but a few species is the only sulfhydryl group present on any histone.

We have made use (18) of the natural tendency of H3 to form disulfide dimers to explore the effect of such dimer formation on nucleosome structure. We find that when all four core histones (H2A, H2B, H3 and H4) are reconstituted onto DNA, the substitution of a dimerized H3 molecule for the reduced form has no measureable effect upon the nuclease digest pattern of subnucleosomal fragments, nor on the ability of the histones to induce supercoil formation in closed circular DNA. When a mixture of the four histones is reconstituted onto DNA 140 base pairs long, a homogeneous particle with a sedimentation coefficient of about 11S is formed regardless of whether H3 monomers or disulfide dimers are used. We conclude that the two H3 molecules of the nucleosome are in close contact in the nucleosome in the region of the residue at position 110. Furthermore, if the nucleosome has a dyad axis, the disulfide linkage must lie on this axis. Our conclusions regarding the proximity of the two thiol residues are consistent with results obtained in energy transfer experiments, using fluorescent molecules coupled to the thiol groups of the cysteines (19,20).

Finally, we reiterate here that the arginine-rich histones H3 and H4 play the central role in nucleosome structure. We have shown that reconstitutes of DNA with H3 and H4 also have many of the properties of the nucleosome. Nuclease and protease digestion of such complexes gives rise to an array of discrete DNA and histone fragments similar to those obtained from chromatin (6,8). On a weight basis, a mixture of H3 and H4 is as effective as all four histones in the induction of supercoiling in closed circular DNA (14). These histones also induce folding in DNA of nucleosome core size (11,21). Combinations of histones lacking either H3 or H4 are unable to effect these changes in DNA. Results of

X-ray diffraction (22) and electron microscopic studies (21) also support these conclusions concerning the central role of the arginine-rich histones.

TRANSCRIPTION

We have also been interested in examining the structure of actively transcribed genes. Although *E. coli* RNA polymerase may be a useful probe for such regions, the difficulties associated with its use for *in vitro* transcription studies make it a cumbersome tool. Recently, hopes have been raised for a more effective method of detecting *de novo* synthesis of specific sequences corresponding to active genes by the introduction of mercury-substituted triphosphates (23) in the polymerization reaction. Newly made RNA should be mercurated, and therefore separable on a thiol-agarose column. We have recently found, however, that a second reaction of the polymerase totally confuses this assay (24,25). This reaction involves the use of endogenous messenger RNA as a template, to produce a mercurated anti-message strand. The endogenous message is now in a duplex with mercurated anti-message, co-purifies with it, and is ultimately detected by cDNA probes directed against the messenger RNA. The positive results obtained in this way depend upon the presence of the polymerase, but have nothing whatever to do with specific transcription of chromatin. Though the method can be modified to take this artifact into account, the procedures are tedious, and it is clear that great caution is necessary in interpreting results of all such experiments.

As in the case of bulk chromatin structure, the most useful probes of active genes appear to be the nucleases. The present view of an active structural gene derived from nuclease digestion studies is that such a gene, as judged by its response to staphylococcal nuclease, is covered with nucleosome-like structures (26,27,28); the structures are so modified, however, that they are unusually sensitive to pancreatic DNase (28).

If transcriptionally active genes are covered by histones, is it possible for an RNA polymerase molecule to transcribe the DNA? A partial answer to this question can be obtained using an artificial 'chromatin' template made by reconstituting the four core histones onto bacteriophage T7 DNA, and attempting to transcribe the template with *E. coli* RNA polymerase. We find (11,29) that when the salt concentration is raised to 0.45 M, chain propagation proceeds readily, producing long RNA molecules (60% of the mass with a number average size of 2000 nucleotides after 20 minutes). We conclude that under certain conditions the histone molecules need not be an insurmountable obstacle to passage of the polymerase. Our results tell us nothing, at this point, about the fate of the histones in the region being transcribed. Future experiments must determine whether they are momentarily displaced, slide, or dissociate into half-nucleosomes as the polymerase passes.

CONCLUSION

We have described briefly our studies of the internal architecture of the nucleosome. We hope to learn more about the points of contact between individual histones and histone segments, and specific regions of the DNA, so that we can explain in detail the observed regular arrangement of nuclease recognition sites on the nucleosome's surface. We also must know in detail the nature of those interactions between histones which stabilize the folded structure. With this information, it may be possible to understand some of the reactions that accompany the participation of nucleosomes in the biological activity of chromatin.

REFERENCES

(1) B. Sollner-Webb and G. Felsenfeld, Biochemistry, 14 (1974) 2915.

(2) R. Axel, Biochemistry, 14 (1975) 2921.

(3) B. Shaw, T. Herman, R. Kovack, G. Beaudreau and K. Van Holde, Proc. Natl. Acad. Sci. USA, 73 (1976) 505.

(4) N. R. Morris, Cell, 9 (1976) 627.

(5) D. Lohr, J. Corden, K. Tatchell, R. T. Kovacic and K. E. Van Holde, Proc. Natl. Acad. Sci. USA, 74 (1977) 79.

(6) R. D. Camerini-Otero, B. Sollner-Webb and G. Felsenfeld, Cell, 8 (1976) 333.

(7) M. Noll, Nucleic Acids Res., 1 (1974) 1573.

(8) B. Sollner-Webb, R. D. Camerini-Otero and G. Felsenfeld, Cell, 9 (1976) 179.

(9) B. Sollner-Webb and G. Felsenfeld, Cell, 10 (1977) 537.

(10) L. Lutter, J. Mol. Biol., in press.

(11) R. D. Camerini-Otero, B. Sollner-Webb, R. Simon, P. Williamson, M. Zasloff and G. Felsenfeld, Cold Spring Harbor Symp. Quant. Biol., 42 (1978) in press.

(12) B. Sollner-Webb, W. Melchior, Jr. and G. Felsenfeld, manuscript in preparation.

(13) F. H. C. Crick and A. Klug, Nature, 255 (1975) 530.

(14) R. D. Camerini-Otero and G. Felsenfeld, Nucleic Acids Res., 4 (1977) 1159.

(15) L. D. Landau and E. M. Lifshitz, Statistical Physics (Addison-Wesley, Reading, Mass. 1958) p. 478.

(16) H. B. Gray and J. E. Hearst, J. Mol. Biol., 35 (1968) 111.

(17) J. E. Godfrey and H. Eisenberg, Biophys. Chem., 5 (1976) 301.

(18) R. D. Camerini-Otero and G. Felsenfeld, Proc. Natl. Acad. Sci. USA, 74 (1977) in the press.

(19) M. Zama, P. Bryan, R. Harrington, A. Olins and D. Olins, Cold Spring Harbor Symp. Quant. Biol., 42 (1978) in press.

(20) A. Dieterich, R. Axel and C. R. Cantor, Cold Spring Harbor Symp. Quant. Biol., 42 (1978) in the press.

(21) M. Bina-Stein and R. T. Simpson, Cell, 11 (1977) 609.

(22) T. Moss, R. M. Stephens, L. Crane-Robinson and E. M. Bradbury, Nucleic Acids Res., 4 (1977) 2477.

(23) R. M. R. Dale and D. C. Ward, Biochemistry, 14 (1975) 2458.

(24) M. Zasloff and G. Felsenfeld, Biochem. Biophys. Res. Comm., 75 (1977) 598.

(25) M. Zasloff and G. Felsenfeld, Biochemistry, 16 (1977) 5135.

(26) R. Axel, H. Cedar and G. Felsenfeld, Biochemistry, 14 (1975) 2489.

(27) E. Lacy and R. Axel, Proc. Natl. Acad. Sci. USA, 72 (1975 3978.

(28) H. Weintraub and M. Groudine, Science, 193 (1976) 848.

(29) P. Williamson and G. Felsenfeld, Manuscript in preparation.

DISCUSSION

J. CARDENAS: Have you tried modifying the sulfhydryl groups, for instance with iodoacetamide, to see if this affects the reconstitution?

G. FELSENFELD: We have tried it, but we do not have any definitive results.

W. CIEPLINSKI: I have a question about the forces involved in the histone - histone association. I have the feeling that these forces would be obtained in a simple histone-histone association experiment and that they might not necessarily represent what is probably very likely a cooperative type of interaction that may happen in nucleosomes. So I was wondering whether you have any idea of the thermodynamics of a real cooperative interaction, otherwise I am not sure whether the model calculations are terribly relevant.

G. FELSENFELD: Of course there is always a problem with such model calculations. In what we have published we have been very careful to say that we are assuming that the interactions are similar to those in solution. In the opened-out nucleosome structure we can assume that there are still many histone-histone interactions. We have no way of estimating which they are or how they affect the subsequent histone-histone interactions that accompany folding.

W. CIEPLINSKI: When you are thinking of transcription that may involve some type of dynamic unfolding and folding back, this is really such a cooperative interaction, and therefore I am not sure that data from a simple reaction of the two histones in that medium would be relevant.

G. FELSENFELD: I did not say that only two histones were involved. I said that as few as one or two strong interactions would be enough to fold the DNA. Our imaginary unfolding reaction is obviously cooperative. This presents no problem. We simply add together all the relevant interaction free energies, if we know what interactions occur. Perhaps only one or two are involved. One has to be cautious about saying more than the results are suggestive. But I think you would agree that if there were an order of magnitude difference between the DNA folding and nucleosomal interaction energies you might be troubled.

BARRY L. MALCHY: Could you comment on Giorgiev's observation of nuclease digestion of E. coli DNA?

G. FELSENFELD: Are you asking me to comment on the observation that there are proteins that appear to protect bacterial DNA from digestion?

B. MALCHY: Could you comment on the regularity of the pattern that he observes when he does that digestion with nuclease?

G. FELSENFELD: Unfortunately I am not sure I remember the exact details of the results. But, as you know, there are basic proteins present in prokaryotes that are thought to be histone-like, so it is conceivable that one is dealing with a primitive system related to eukaryote chromatins. I think at this point it is too early to say.

B. MALCHY: Which could give you a similar 200-base paired karyotype cleavage?

G. FELSENFELD: Well, why not? The DNA of certain viruses yields a similar repeat pattern, though the virion contains no histones. The persistence length of DNA is of the order of 200 base pairs. One cannot take it literally, but it gives an order of magnitude limitation on the flexibility of DNA. Maybe if you are going to package DNA and try to make it compact without major discomfort, this is a lower size limit for continuous deformation, and many systems may try to do it that way.

B. MALCHY: Are you suggesting that because of the persistence lengths that you may get some regularity even if you attempt just to cleave DNA?

G. FELSENFELD: Well, no. You have the persistence length first, then maybe you select for the packaging system that has the shortest, energetically reasonable repeat length, and then the repetition of course is built-in because nature tends to make structures out of identical building blocks. So the fact that it is regular would simply be nature's economy; the fact that it is of the order of 200 base pairs might somehow be related to the bendability of DNA. That is just a speculation.

F. HAUROWITZ: You showed some of these DNA molecules, most as palindromes. Do you really believe that nucleosomes have to be palindromes?

G. FELSENFELD: No, they are not palindromes, but they do have a local symmetry of cutting sites conferred by the histones. The entire nucleosome core may possess a dyad axis. The histone octomer, in principle at least, could be so arranged as to maintain a two-fold axis, since it consists of two each of the four core histones. Such an axis would require symmetrically located cleavage sites measured from the 5' terminus of the antiparallel DNA strands.

H.S. SHAPIRO: You just answered the question that I was going to ask. Are you inferring, however, that if you partially reconstitute the nucleosome by adding H3 or H4 back you get the same size pattern of oligonucleotides when you digest with your nuclease?

G. FELSENFELD: When we use H3 and H4 alone in reconstitutes, the subnucleosome digestion pattern has the same bands as with whole chromatin, but the relative intensities are not the same. For example, with pancreatic DNAse you get the 10 nucleotide repeating single strand pattern, whereas with whole chromatin or nucleosomes, the 80 nucleotide long band is very prominent. Digests of reconstitutes with H3-H4 alone have all the bands of roughly equal intensity. So the relative accessibility of different sites in the repeat is different when only H3 and H4 are present.

H.S. SHAPIRO: How long is your digestion time? Is it short digestion with high concentration of enzymes?

G. FELSENFELD: We have examined a wide variety of conditions. Patterns may vary, but not significantly for the arguments I have presented.

A. KOOTSTRA: Since you are suggesting that some of the forces involved in the nucleosomal formation may be associated with the hydrophobic core, would you suggest then that first of all we have binding of the individual histones to the DNA, and then hydrophobic interaction to pull the system together?

G. FELSENFELD: No, that is why I wanted to make it absolutely clear that this is only a thought experiment. The folding reaction is contrived only to ask the question: would the folded structure be stable? Unfolding is a possible path, and we can in principle compute the free energy associated with this hypothetical reaction. It is definitely not obligatory that assembly proceed by this pathway. In fact Simpson and his collaborators have shown that you can chemically cross-link a histone core octomer, just the histones, and then add that to a solution containing DNA and get back a folded nucleosome structure. The experiment suggests that assembly does not proceed by a folding mechanism.

R.G. KALLEN: If your calculations are correct and your assumption is correct about the hydrophobic forces it seems to me that there might be a substantial temperature dependence which might alter the symmetry in the system.

G. FELSENFELD: Possibly. I do not know that anyone has looked at nucleosome folding as a function of temperature. However, Olins has shown that if you treat nucleosomes with urea, the histones remain attached to DNA, but the sedimentation coefficient drops dramatically suggesting that the nucleosome unfolds.

R. KALLEN: You have not done any digestion experiments at different temperatures?

G. FELSENFELD: No, not in a systematic way.

RELEASE OF 3 MICRON CIRCLES AFTER TREATMENT OF NUCLEI WITH TRYPSIN

Donald Riley and Harold Weintraub
Dept. Biochem., Moffett Labs, Princeton Univ., Princeton, N.J. 08540, U.S.A.

INTRODUCTION

It was previously shown that the actively transcribed globin genes in avian erythroblasts were preferentially sensitive to digestion by staphylococcal nuclease after, but not before, treatment of nuclei with trypsin (Weintraub & Groudine, 1976). This observation was interpreted to indicate that the active globin genes were associated with trypsin-sensitive proteins, presumably histones (Lacy & Axel, 1976). Since a great deal of work from this and other laboratories has led to the conclusion that trypsin treatment of bulk nucleosomes leads to only marginal disruption of structure, we thought that we might be able to visualize transcription units by virtue of the fact that their associated proteins would be preferentially attacked by trypsin and hopefully an altered morphology of active chromatin could be observed. Moreover, preliminary observations also indicated that trypsin is a very effective reagent for spreading out the entire contents of nuclei so that an analysis of the entire nuclear chromatin could be achieved.

Here we show that while 80% or so of trypsin treated chromatin is in the normal beaded configuration, from 1-10% of the nuclear content is released as large 3μ diameter circles. Since our detailed analysis of these circles is just beginning, many questions still remain unanswered; however, their morphology is so striking that we feel it appropriate to present a preliminary description of our findings in this Symposium.

RESULTS

Generation of Circles with Trypsin: Isolated nuclei from chick erythroblasts were treated with trypsin. The resulting viscous solution was then fixed with formaldehyde, touched to glowed, carbon-coated, electron microscope grids, washed with Photo-Flo, dried, stained with 5mM uranyl acetate and shadowed. Most of the chromatin in such a preparation is in the form of dense aggregates which are difficult to analyze; however, in such preparations we observe uniform circles of the type shown in Fig. 1. For a number of reasons, this type of preparation is inadequate since

ISBN 0-12-045450-5

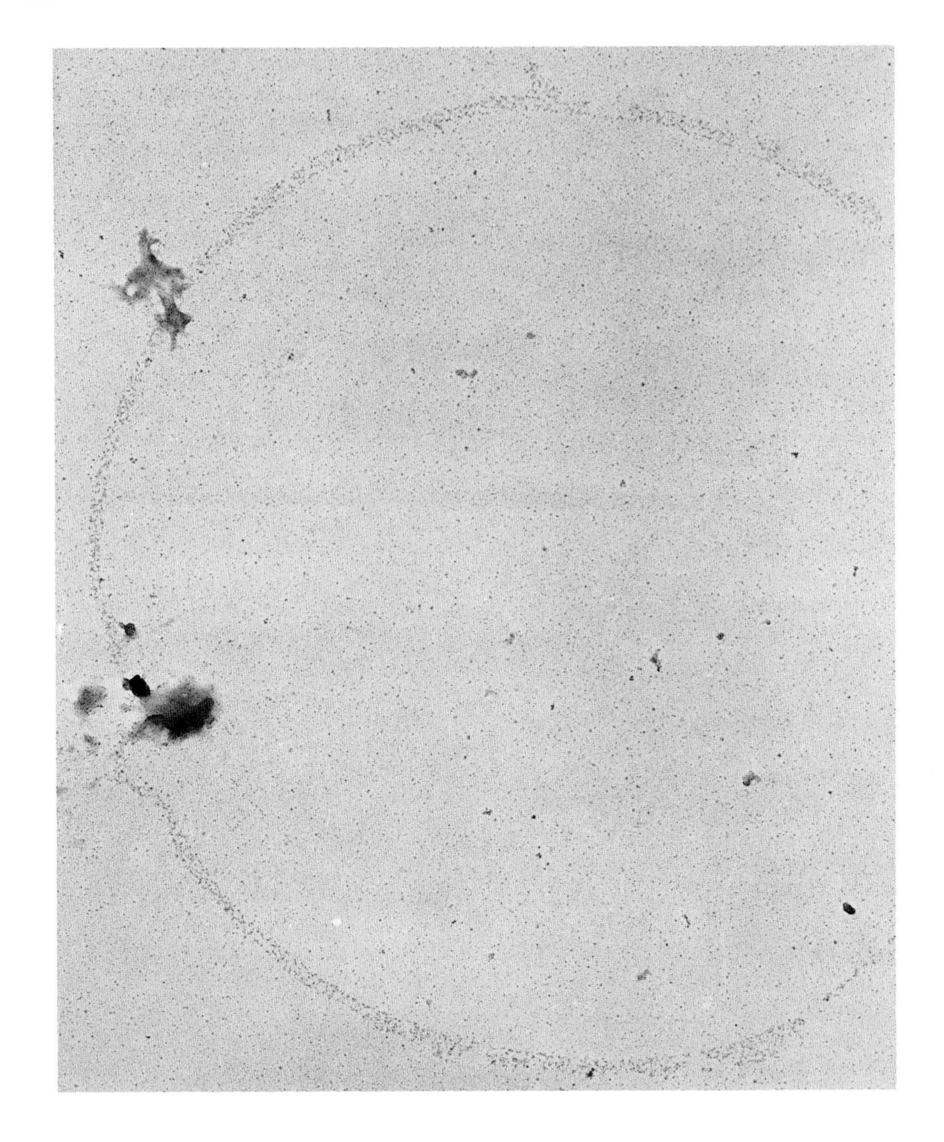

Fig. 1a.

Fig. 1 (a-c): Circles released after treatment of nuclei with trypsin. The circles have a uniform diameter of 2μ. An amorphous globule is also associated with about 30% of the circles.

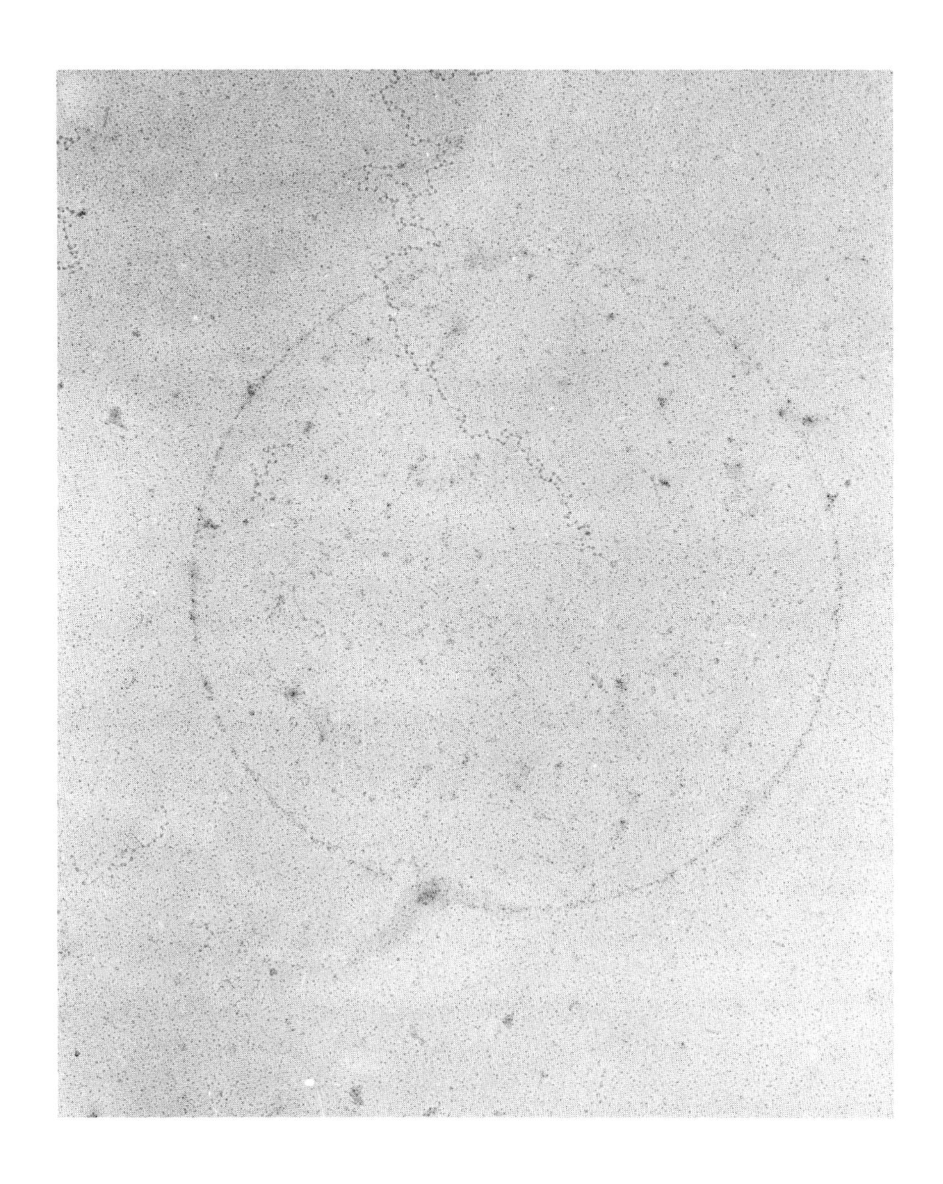

Fig. 1b.

Fig. 1c.

most of the nuclear contents cannot be analyzed. Consequently, we have modified this procedure in order to prevent chromatin aggregation. Our present procedures use a combination of very low levels of DNase I in conjunction with trypsin. The nuclease is used to decrease the size of the bulk chromatin (to about 50-1000 nucleosomes). This prevents aggregation, decreases the viscosity, and increases the frequency of circles. A second modification is that for convenience most of our experiments are now performed with MSB cells. These are a chicken lymphoid, leukemic line transformed by Mareks disease virus.

These circles have several properties which suggest that they may be artifactual as well as several properties not easily explained by artifact. Although the latter properties are interesting and should be considered seriously we cannot exclude an artifactual origin at present (see below). The circles are very uniform in diameter. In most preparations the circles represent 10-20% of the material on the grid, the rest of the material being in the typical beaded or nucleosome configuration (Fig. 2). Assuming the circles are DNA (see below) and that they have a compaction ratio of 1, then they are about 30kb in circumference. The distribution about this size is extremely uniform; however, a significant number of circles twice this size (60kb) is also observed. An amorphous globule is present on about 30% of the circles. Most often there is only one such globule per circle. The circles are extremely regular. We believe this regularity reflects an inherent rigidity since closed circular, relaxed PM2 DNA molecules added to these preparations and analyzed by the same techniques display a loose, floppy, morphology when visualized by the same methods. The circles do not stain uniformly like added free DNA or bulk chromatin in these preparations. Thus, there are sometimes gaps of about 100A° separating the densely stained regions that define the basic structure. Finally, 10-40% of the circles in any given preparation have "fuzzy" material emanating from the main backbone (Fig. 3). Very often, this material takes on a "gradient-like" character.

Factors Effecting the Generation of Circles: To test the possible artifactual origin of the circles, nuclei were left out of our procedures and the grids prepared in mock fashion. No circles were observed in these preparations. Although a small number of "arc-like" figures could be seen, these are consistantly an order of magnitude smaller than the types of circles shown in Figures 1 and 3. Moreover, grids prepared from nuclei not treated with trypsin (but treated with DNase) never show circles. Nuclei treated with trypsin in the presence of 2mM EDTA yielded a dramatically reduced

Fig. 2: Bulk chromatin released from nuclei after treatment with trypsin.

Fig. 3: Trypsin released circles associated with gradient-like material that is sensitive to RNase.

number of circles. This suggests divalent cations may be essential to the generation of circles by trypsin. Finally, circles are seen whether the preparation is fixed or not, whether it is stained or not, and whether it is shadowed or not although these factors affect contrast. Omission of the Photo-Flow rinse leads to greatly increased background levels so that both circles and beaded chromatin appear to be completely obscured.

Composition of Circles: The molecular composition of the circles was investigated using a variety of specific enzymes. As would be expected, the circles were resistant to high concentrations of trypsin. This does not imply that they lack protein since the histones, themselves, are extremely resistant to trypsin treatment as evidenced most dramatically by the retention of the beaded conformation in the nucleosomes present in these preparations.

Most importantly, pre-treatment of nuclei with levels of DNase I that reduce the average DNA size to 20-50 nucleosomes, results in no production of circles. Identical results are obtained with low levels of DNase II, micrococcal nuclease, and the restriction endonuclease Hpa II. The circles also appear sensitive to these nucleases after they are generated by trypsin treatment of nuclei. Thus, it is clear that DNA is at least one component responsible for circle integrity. (Our DNase I preparation contains no detectable protease activity as assayed by SDS gel electrophoresis of DNase I treated histones and no detectable RNase activity as measured by electrophoresis of DNase I treated globin mRNA on acrylamide gels).

Finally, we asked about the composition of the "fuzzy" material associated with circles. It was observed that incubation of nuclei with very low concentrations (5 μgm/ml) of RNase A (previously boiled for 30 minutes) completely eliminated the fuzz from the circles. These bare circles retained their characteristic "rigidity" and circularity. Identical results are obtained if the preparation of circles is heated to 65°C for 15 minutes.

Thus far, our results indicate that production of the circles is dependent on trypsin treatment of nuclei and is prevented by EDTA; the integrity of the circles depends on DNA and the fuzzy, often gradient-like material associated with the circles is sensitive to RNase A. The later observation suggests the possibility that the circles may be involved in transcription. To test this, cells were preincubated for 2 hours with actinomycin D or with α-amanitin. Circles were generated in the usual manner and examined.

While the normal frequency of circles was observed, none contained the fuzzy, gradient-like material.

Generality of Circles: Circles have been observed in preparation from HeLa cells, 16 day chicken erythrocytes, 4 day chicken erythroblasts, MSB cells, purified HeLa chromosomes, purified mini chromosomes from MSB cells, purified maxi chromosomes from MSB cells, and in Drosophila tissue culture cells. We have observed no differences (either in size, frequency, or associated RNase-sensitive material) in the circles obtained from any of these sources.

Frequency of Circles: It is extremely important to quantitate the percentage of chromatin present as circles. Of the material that adheres to the grid, about 10-20% is in the form of circles. (This calculation assumes a DNA compaction ratio of 3 for trypsin-treated nucleosomes and a compaction of 1 for circles). In addition it is important to estimate the relative affinities with which biological material adheres to these grids. To do this, equal amounts of pure PM2 DNA and chromatin were mixed and the relative adherence to the grid estimated. The PM2 DNA adhered about 4-6 times more efficiently than nucleosomes (assuming a compaction of 3) in our preparative conditions. Thus, if the circles stick to the grids in a manner that is more like DNA than nucleosomes then their lowest estimated frequency is 1-2% of the nuclear contents.

DISCUSSION

The minimal treatment of nuclei required to produce circles is incubation with trypsin. Additional steps in the procedure are designed to decrease trapping and increase the frequency at which circles are produced. The circles are sensitive to digestion by DNase I, DNase II, micrococcal nuclease, and the restriction nuclease, HPAII. We estimate that the circles represent from 1-5% of the total nuclear DNA and they are extremely uniform in diameter with a circumference of about 30kb; they are usually associated with a single globule and very often with fuzzy-RNase sensitive materials that tends to form a gradient along the circle axis and which disappears after pre-treatment of cells with actinomycin D or α-amanitin. Clearly, further information about these structures must await their biochemical isolation which is a main goal of our current research.

Possible Artifacts: The observations above must be interpreted with caution since, at this stage, we cannot eliminate the possibility that the circles are generated *in vitro* by a physical process. For example, a circular phase boundary (such as that of an air bubble) might orient RNA and DNA fragments liberated from nuclei, giving the impression of nuclease-sensitive circular images. As mentioned previously, there are two properties of these circles that are difficult to explain. The first is their apparent rigidity, while the second is their failure to stain continuously. Possibly, both features could be explained in terms of a poorly stained supporting substance. Alternatively, M. Gorovsky has argued very cogently that DNA that is preferentially sensitive to DNase I (as indeed these circles are) may be β-kinked. Such a structure would be expected to have the rigid properties of the circles we have observed (H. Sobell, personal communication).

How are the Circles Generated: We assume that these circles are normally attached to the main chromosome axis in much the same way as lampbrush loops. Biochemical evidence for such structures is now very abundant (Benyajati & Worcel, 1976; Paulson & Laemmli, 1977). How are they released? Clearly one possibility is that they are attached by means of a trypsin-sensitive linker. Alternatively, they may be generated by a trypsin activated nuclease. Precedent for the latter is plentiful in a number of biological systems (see for example, Clewell & Helsinki, 1969) and the finding that EDTA inhibits the release of circles certainly supports the nuclease explanation.

BIBLIOGRAPHY

(1) Benyajati, T. and Worcel, A. (1976). Cell 12, 83.

(2) Paulson, J. and Laemmli, U. (1977). Cell 12, 817.

(3) Clewell, D.B. and Helinski, D.R. (1969). Proc. Nat. Acad. Sci. USA 1159.

(4) Weintraub, H. and Groudine, M. (1976). Science 93, 848.

(5) Lacy, E. and Axel, R. (1976). Proc. Nat. Acad. Sci. USA 72, 3978.

CHANGES IN COMPOSITION AND METABOLISM OF NUCLEAR NON-HISTONE PROTEINS DURING CHEMICAL CARCINOGENESIS. ASSOCIATION OF TUMOR-SPECIFIC DNA-BINDING PROTEINS WITH DNAse I-SENSITIVE REGIONS OF CHROMATIN AND OBSERVATIONS ON THE EFFECTS OF HISTONE ACETYLATION ON CHROMATIN STRUCTURE.

V.G. ALLFREY, L.C. BOFFA and G. VIDALI

The Rockefeller University
New York, New York 10021

Abstract : The administration of 1,2-dimethylhydrazine to rodents leads to the induction of adenocarcinomas of the intestinal epithelium. Tumorigenesis is accompanied by characteristic changes in the composition and metabolism of non-histone nuclear proteins. Two protein classes of molecular weight 44,000 (TNP_1) and 62,000 (TNP_2) are synthesized at accelerated rates early in carcinogenesis and they eventually become the predominant non-histone nuclear proteins of the tumor. The increase in their rates of synthesis is detectable very early, long before any morphological indications of malignancy become evident. Similar proteins have been detected in human colonic adenocarcinomas, but not in non-malignant cells nor in embryonic intestinal epithelial nuclei. The two classes of tumor-associated proteins differ in their binding to homologous DNA covalently-linked to solid supports. TNP_1, which has a high affinity for DNA, also occurs in the dividing cell population of the tumor, whereas TNP_2 has little DNA affinity and is enriched in the non-dividing cell population. The complexity of both protein classes is being investigated by 2-dimensional gel electrophoresis, using new techniques of raster-scanning microdensitometry and computer-assisted graphics. When tumor chromatin is subjected to limited digestions with DNAse I, under conditions which selectively degrade transcriptionally active genes in other cell types, the TNP_1 proteins are preferentially released, suggesting that this subset of the nuclear proteins is associated with the transcribing regions of the genome in the tumor cell nuclei.

ISBN 0-12-045450-5

The structural basis for increased DNAse I sensitivity of transcriptionally active chromatin has been investigated by comparing the chromatins of tumor cell nuclei which differ in their contents of acetylated histones. Exposure of HeLa cells to 7 mM Na butyrate leads to an accumulation of the acetylated forms of histones H3 and H4. Although nucleosomes prepared from cells differing in their content of acetylated histones are similar to each other with regard to general physical properties (sedimentation coefficient, DNA-melting curves, and circular dichroism spectra), nucleosomes enriched in acetylated histones are much more susceptible to DNAse I attack. The view that acetylation of the histones alters chromatin configuration at the nucleosomal level is supported by electrophoretic analyses of the remaining histones at successive stages during DNAse I digestion. The most highly acetylated forms of histones H3 and H4 are preferentially released at early times, as would be expected if the acetylated forms were localized at regions of the chromatin that had been activated for potential transcription. A similar preferential release of acetylated histones after brief treatment with DNAse I has also been observed in nuclei from non-malignant cells, such as avian erythrocytes. Specific sets of non-histone proteins, including members of the high-mobility group (HMGs), are selectively released at the same time.

INTRODUCTION

It is a reasonable premise that malignant transformation, in molecular terms, involves a breakdown of the normal transcriptional control mechanisms that regulate cell growth and division. In an attempt to analyze the effects of chemical carcinogens on nuclear function, we have placed particular emphasis on the effects of the carcinogen on those chromosomal proteins which interact with DNA directly and regulate its structural organization and its template function. The experiments to be described deal with changes occurring in epithelial cell nuclei at successive stages in carcinogenesis induced by the administration of 1,2-dimethylhydrazine (DMH) to rodents (1-3). This system provides a model of chemical carcinogenesis with the advantages of organ specificity, high tumor incidence, and reproducible timing. The normal patterns of cell proliferation and differentiation in the crypts of the intestinal epithelium are modified, and aberrations in nucleic acid synthesis soon become evident (4,5). Accordingly, we have looked for changes in DNA-associated proteins at early times after exposure to the carcinogen. The results show that a characteristic subset of DNA-binding proteins is selectively synthesized before visible signs of malignancy appear. This subset of the total non-histone nuclear protein fraction is concentrated in the dividing cell population of the tumor. Some of the DNA-binding proteins are preferentially released during limited digestions with DNAse I, as would be expected if they were associated with transcriptionally active regions of the tumor chromatin. The reason for the selectivity of DNAse I attack on active or potentially active DNA sequences appears to be due to their association with highly acetylated forms of histones H3 and H4.

EXPERIMENTAL

Isolation of nuclei - Tumors were induced in CFN-Wistar male rats by weekly s.c. injections of DMH at a dosage of 20 mg/kg body weight. By the 21st week, more than 98% of the treated animals had multiple tumors of the colon (6). The tumor tissue was excised and homogenized, and a purified nuclear fraction was isolated by centrifugation through a 2.2 M sucrose density-barrier (7). In some experiments, the total nuclear fraction was further separated into classes differing in buoyant density and DNA-synthetic capacity as described previously (7). Similar fractionations were performed on normal mucosa and on epithelial tissue surrounding the tumors.

Extraction of nuclear proteins - The isolated nuclei were washed twice with 10 volumes of 0.14 M NaCl. Histones and other acid-soluble proteins were extracted in 5 volumes of 0.25 N HCl. The non-histone proteins remaining were then extracted in 0.1 M Na_2HPO_4 containing 6 M urea, 0.4 M guanidine-HCl and 0.1 % 2-mercaptoethanol (8). Yields of total non-histone proteins obtained by this procedure averaged 87 ± 2 % (6).

Electrophoretic analysis of non-histone nuclear proteins - In most of the studies to be described the non-histone nuclear proteins were complexed with sodium dodecylsulfate (SDS) and separated by electrophoresis in 9 - 14 % polyacrylamide gradient gels (9). For two-dimensional gel electrophoretic analysis, the proteins were first separated according to their isoelectric points by focusing in cylindrical polyacrylamide gels containing an Ampholine (LKB) pH gradient, modifying the O'Farrell procedure (10) by distributing the proteins throughout a riboflavin-containing acrylamide mixture prior to UV-induced polymerization (11, 12). After isoelectric focusing in the pH gradient, the proteins were separated according to size in 9 % polyacrylamide-SDS slab gels (9). The pI and molecular weight coordinates of individual protein spots were determined after staining with 0.1 % Coomassie Blue, or by autoradiography of 2-D gels containing radioactively-labeled nuclear proteins. The resulting photographic transparencies were analyzed by raster-scanning microdensitometry using a PHOTOSCAN System P-1000 (Optronics International, Inc., Chelmsford, MA). The absorbancy data, expressed in digital form, was plotted against pH and molecular weight coordinates as described elsewhere (12). Microdensitometric scans of optical density standards were used to establish the resolution and accuracy of this procedure for integration of spots of differing intensities, areas, contours and orientation (12). Computer-assisted graphics were employed to allow examination of protein peaks from different inclinations and directions, thus permitting a more incisive interpretation of heterogeneities in the surface-charge and size of individual proteins than has heretofore been possible (12).

DNA-affinity chromatography - Covalent linkages between the 5'-terminal phosphate groups of DNA and hydroxyl groups of Sephadex G-25 were formed by carbodiimide coupling according to the method of Allfrey and Inoue (13). The amount of covalently-bound DNA was determined by measuring the amount of DNA released from aliquots of the DNA-matrix by treatment with DNAse I (13). In order to avoid complications due to denaturation, the tumor nuclear proteins were extracted in 2 M NaCl, 20 mM Tris-HCl, pH 8.0, 0.1 mM phenylmethylsulfonyl fluoride (PMSF). After centrifugation at 100,000 x g for 24 hours, the extract was adjusted to a protein concentration of 0.2 mg/ml and mixed with a suspension of DNA-Sephadex G-25 containing 1 mg DNA per ml of the same high-salt buffer. The protein-DNA mixture was then subjected to successive dialysis steps against the buffer containing decreasing NaCl concentrations, in the order : 1 M; 0.6 M; 0.3 M; 0.14 M and 0.05 M. The suspension was then poured into a column for affinity chromatography. The proteins were eluted from the DNA column by stepwise increments in the salt concentration of the eluting buffer (10 mM Tris-HCl, pH 7.5, 10% glycerol, 0.1 mM PMSF, 1 mM EDTA, 1 mM 2-mercaptoethanol). The steps consisted of 0.05 M, 0.4 M, 0.6 M and 2 M NaCl. Fractions eluting from the DNA column at a given salt concentration were combined, dialyzed extensively against 10 mM Na phosphate buffer, pH 8.4, containing 0.1% SDS (sample buffer) and then concentrated by dialysis against 60% sucrose (w/v) in sample buffer. The samples were analyzed by SDS-polyacrylamide gel electrophoresis as described above.

Limited DNAse I digestions - The purified nuclei from DMH-induced colon tumors were washed extensively with 10 mM NaCl, 10 mM Tris-HCl, pH 7.0, containing 3 mM $MgCl_2$ and 0.1 mM PMSF. The nuclear pellet was resuspended in the buffer at a concentration of 1 mg DNA/ml and preincubated at 37° C. for 10 minutes. DNAse I (Sigma) was then added to a final concentration of 60 Kunitz units per ml. After 5 minutes at 37°, the reaction was halted by the addition of 0.1 volume of 0,1 M EDTA, pH 7.0. The incubation mixture was chilled in ice, sonicated briefly, and centrifuged at 2,000 x g for 10 minutes to separate the residual ("resistant") chromatin fraction from more completely degraded chromatin fragments in the supernate. Under these conditions, the supernatant fraction contains 10$\pm$ 2% of the total DNA of the original nuclear suspension. Histones and non-histone nuclear proteins were extracted from the residual chromatin as already described. The supernatant fraction containing the proteins released during DNAse I digestion was immediately frozen and lyophilized. Control nuclei were incubated in the absence of added DNAse I and processed under identical conditions. Nuclei from normal epithelial tissue were also subjected to the same treatment.

The selective release of acetylated histones and HMG

proteins from avian erythrocyte nuclei was carried out as described previously (14, 15). HeLa S-3 cells differing in their contents of acetylated histones and previously labeled with ^{14}C-thymidine and ^{3}H-thymidine were prepared as described elsewhere (16). The ^{14}C-labeled butyrate-treated cells were mixed with the ^{3}H-labeled control cells prior to isolation of the nuclei. The mixed nuclear suspension was treated with DNAse I, and the release of the isotopically-labeled DNAs was measured at different times, calculating the $^{14}C/^{3}H$ ratio for each time point (16). The proportions of the acetylated forms of histones H3 and H4 was determined by microdensitometry of the histone bands separated by electrophoresis in acid-urea gels (17).

RESULTS AND DISCUSSION

Selective Synthesis and Accumulation of Nuclear Non-Histone Proteins during DMH-Induced Carcinogenesis - The precise and orderly programming of cell growth and differentiation in normal tissues depends in large part on interactions between DNA and closely-associated chromosomal proteins that regulate chromatin structure and control the timing and extent of gene activity. The non-histone nuclear proteins, which have been strongly implicated in transcriptional control mechanisms, are known to differ in different normal tissues (e.g. 18); moreover, they are known to be altered in a variety of carcinogen-induced tumors (e.g. 6, 19-21) and transplantable tumor lines (21-25).

A major problem in interpreting most of the observed differences is the complexity of the cell populations examined. Many established tumor lines have different numbers of chromosomes; and nuclei from the different cell types in complex tissues such as the liver, brain, or intestinal epithelium have been shown to differ in their non-histone protein complement (7, 26, 27). Such differences in protein composition between different classes of nuclei in the normal tissue complicate interpretations of quantitative or qualitative differences between the normal and malignant states. The problem can be simplified by techniques which select more homogeneous populations of nuclei from normal and tumor tissues, e.g., by separation of nuclear classes differing in buoyant density, size, ploidy, or protein/DNA ratio (7,26,27). We have developed procedures for separating various classes of nuclei from the epithelial layers of the normal colon, and some of the changes that accompany normal differentiation in the crypts have been described (7). The approach has been extended to epithelial tumors induced by the carcinogen DMH (6,28) and characteristic changes in nuclear protein composition are observed during the course of carcinogenesis.

Previous autoradiographic studies using ^{3}H-thymidine have established that there is a gradient of DNA-synthetic activity at different levels of the intestinal crypts (29, 30). In normal epithelium, the cells proliferate in the lower and middle portions of the crypts, migrate toward the lumen of the intestine, and are gradually extruded from the surface. ^{3}H-thymidine incorporation into DNA is largely restricted to cells in the lower one-third of the crypts; DNA-synthesizing cells are much less common in the midregions of the crypts, and they are usually absent from the upper region. The distribution of DNA-synthesizing cells is significantly altered in colonic tumors. The gradient of DNA synthetic activity is much less pronounced in established adenocarcinomas and in pre-malignant growths, and many more dividing cells occur in the upper layers of the mucosa (31 - 33). In tumors induced by DMH, cells close to the luminal surface are found to incorporate ^{3}H-thymidine, in contrast to the inactivity of corresponding cells in normal epithelium (4).

The nuclei of both normal and malignant intestinal epithelial cells can be fractionated according to buoyant density to yield subsets of nuclei that differ in their protein composition and capacity for DNA synthesis (7, 28). It has been shown that the position of each nuclear subset within the density gradient corresponds to the histological localization of the cells of origin in the crypts, as judged by their capacity for DNA synthesis *in vivo*. This makes it possible to compare nuclear protein compositions as a function of cell type and stage in differentiation (7), and to characterize proteins associated with the dividing cell population of normal and malignant tissues.

Earlier electrophoretic comparisons of the non-histone proteins of normal epithelial nuclei and DMH-induced tumor nuclei indicated a characteristic shift in protein complement, with a striking increase in the concentrations of two classes of acidic proteins of molecular weights ca.44,000 (TNP_1) and ca.62,000 (TNP_2) which dominate the electrophoretic banding profiles of the non-histone proteins from the tumor cell nuclei (6). Fractionation of the tumor nuclei according to buoyant density indicated that the smaller tumor-associated protein class (TNP_1) was selectively localized in the dividing cell population and virtually absent from the non-dividing cells of the tumor mass (28). Similar proteins were detected in human adenocarcinomas and in derivative cell lines, such as HT-29 (28), but they were not observed in normal colonic epithelium nor in non-malignant hyperplastic growths (11).

Why do such proteins accumulate in tumor cell nuclei, and when does accumulation begin? This question has been approached by double-labeling experiments designed to compare the rates of synthesis of nuclear proteins in normal and DMH-treated epithelial cells at different times after administration of the carcinogen. Some results are illustrated in Fig. 1, which compares the incorporation of ^3H-leucine into the non-histone nuclear proteins of the colonic epithelium of DMH-treated animals with the incorporation of ^{14}C-leucine into the nuclear proteins of normal epithelial cells. In this test, carried out after 4 weekly injections of the carcinogen, the proteins were labeled *in vivo* for 2 hours. The epithelial layers of the colons of control and carcinogen-treated rats were homogenized together; nuclei were isolated from the combined homogenate, and the non-histone nuclear proteins were extracted and analyzed by SDS-polyacrylamide gel electrophoresis. Each band in the gel represents a mixture of ^{14}C-labeled protein from the control animals and the corresponding ^3H-labeled protein from the DMH-treated animals. Fig. 1 plots the distribution of each isotope as a function of protein mobility (molecular weight). The ratio of incorporation of ^3H-leucine to ^{14}C-leucine is plotted in the upper panel of the figure. This ratio serves to indicate the relative rates of synthesis of individual protein bands in the control and DMH-treated animals. The ratio is remarkably constant for most of the protein bands in the gel, except for two prominently-divergent bands in the 40-45,000 and 60-65,000 molecular weight regions. These proteins are synthesized much more rapidly in the DMH-treated animals. It is significant that the selective increase in synthetic rates of these two protein classes becomes evident before the appearance of pathological indications of malignancy (4).

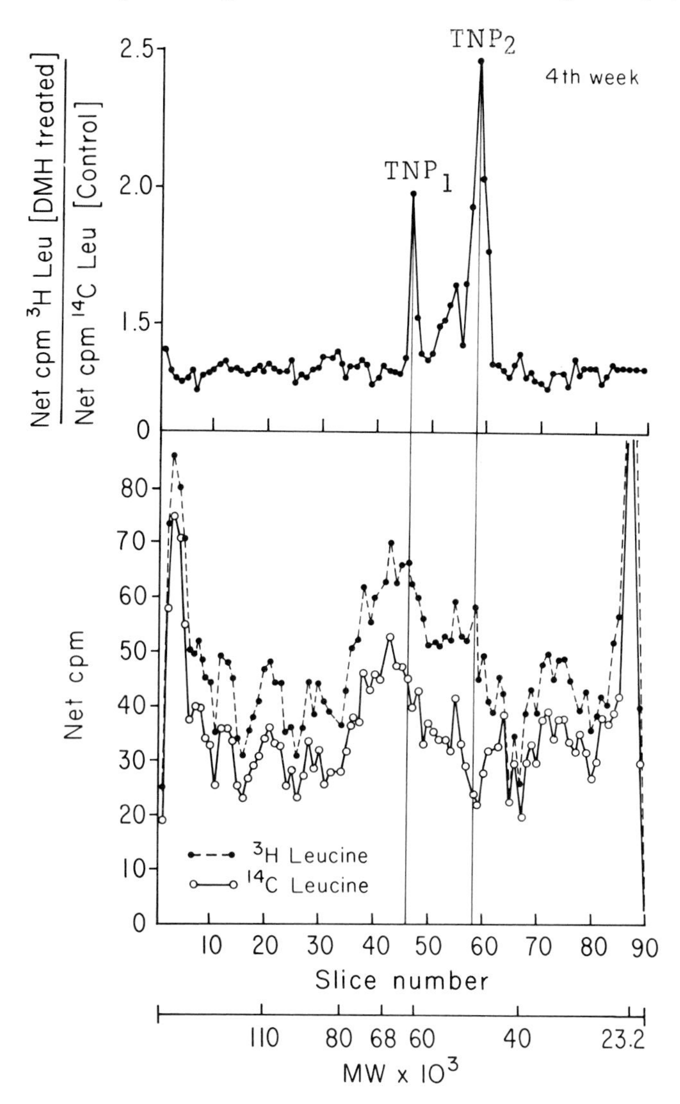

Fig. 1. Selective increases in synthetic rate of nuclear non-histone proteins TNP_1 and TNP_2 after 4 injections of DMH. The incorporation of leucine - ^3H into the nuclear proteins of DMH-treated animals is compared with leucine-^{14}C uptake into the nuclear proteins of control animals. In the upper panel, the $^3H/^{14}C$ ratio is plotted vs. protein Mol.Wt.

With more prolonged DMH treatment, the concentration of the tumor associated proteins increases, and the corresponding bands become more and more prominent in gel electrophoretic patterns. Quantitation of TNP_1 and TNP_2 at biweekly intervals has shown a progressive increase from the 7th to the 21st week, when over 98% of the animals have multiple tumor foci (6). TNP_1 and TNP_2 each comprise about 15% of the total non-histone protein in the tumor-derived nuclei, but, as pointed out earlier, they are not uniformly distributed throughout the dividing and non-dividing cell populations of the tumor mass.

Characterization of the Tumor-Associated Non-Histone Nuclear Proteins

Proteins banding in the regions of the gel corresponding to molecular weights ca. 44,000 (TNP_1) and ca. 62,000 (TNP_2) were isolated from the nuclei of DMH-induced tumors by preparative gel electrophoresis. The amino acid composition of both fractions was determined. TNP_1 and TNP_2 each show a predominance of glutamic and aspartic acids over the basic amino acids, lysine, arginine, and histidine (6). Both TNP_1 and TNP_2 are lacking in cysteine and methionine (a fact that eliminates the utility of ^{35}S-labeled amino acids as a probe for their synthesis). Proline is also absent.

The banding patterns of the nuclear non-histone proteins, as separated by unidirectional electrophoresis in SDS-polyacrylamide gels, do not adequately portray the complexity of the tumor non-histone protein complement. A better indication of the size and charge heterogeneity is provided by two-dimensional gel electrophoretic separations which combine isoelectric focusing in the 1st dimension with size separations in the 2nd dimension. We have modified the procedure of O'Farrell (10) to permit higher recoveries of proteins in the isoelectric focusing step (11, 12). Subsequent separations in SDS-polyacrylamide gels (at right angles to the original direction of charge-dependent migration) provide a 2-dimensional display of the nuclear proteins of DMH-induced adenocarcinomas. As described previously (11), the major bands in the 44,000 (TNP_1) and 62,000 (TNP_2) classes now appear to be much more complex. Both TNP_1 and TNP_2 have isoelectric points in the acidic range, in accord with the predominance of glutamic and aspartic acids in their overall amino acid compositions. TNP_2 shows considerable electrophoretic heterogeneity, with a major subgroup of proteins of molecular weight 61,000 with pI's ranging from 5.6 to 6.5, and another group of molecular weight ca. 63,000 with pI's between 6.2 and 6.8. The TNP_1 class appears simpler, and is more acidic; its major components focus over the pH range 4.85 to 5.25, but a single major component has a pI of 6.5. It is known that many of the tumor non-histone nuclear proteins are phosphoproteins (34) and it follows that some of the charge heterogeneity observed may reflect different levels of phosphorylation of the same polypeptide chains.

The analysis of the complex images provided by 2-dimensional gel electrophoresis is facilitated by raster-scanning microdensitometry of the stained gels or autoradiographs of gels containing isotopically-labeled nuclear proteins. The absorbancy of each spot is rapidly expressed in digital form, and a computer program has been developed for analysis and storage of the data (12). The absorbancy scans across 2-D gels can also be expressed graphically, as depicted in Fig.2 for the separation of marker proteins selected for their similarities to the tumor-associated proteins TNP_1 and TNP_2. The computer plot not only indicates the molecular weight and isoelectric point coordinates for each protein, but also reveals subtle heterogeneities in size and charge. When programmed to allow inspection of individual protein peaks from different inclinations or directions (Fig. 2), the system permits a more incisive interpretation of heterogeneities than simple visual inspection of the stained gel or autoradiograph could provide. For example, one can readily detect and measure alterations in electrophoretic mobility due to phosphorylation of nuclear proteins, or detect changes in molecular weight due to partial proteolysis. By incorporating appropriate internal standards and employing controlled photographic conditions, each gel can be analyzed to give contents, relative amounts, and specific activities of each of the separated protein spots. This approach can be particularly useful in the study of post-synthetic modifications of nuclear proteins which alter their charge (phosphorylation, acetylation, ADP-ribosylation, etc.), combining 2-D gel electrophoretic separations with autoradiographic procedures and computer-assisted graphics.

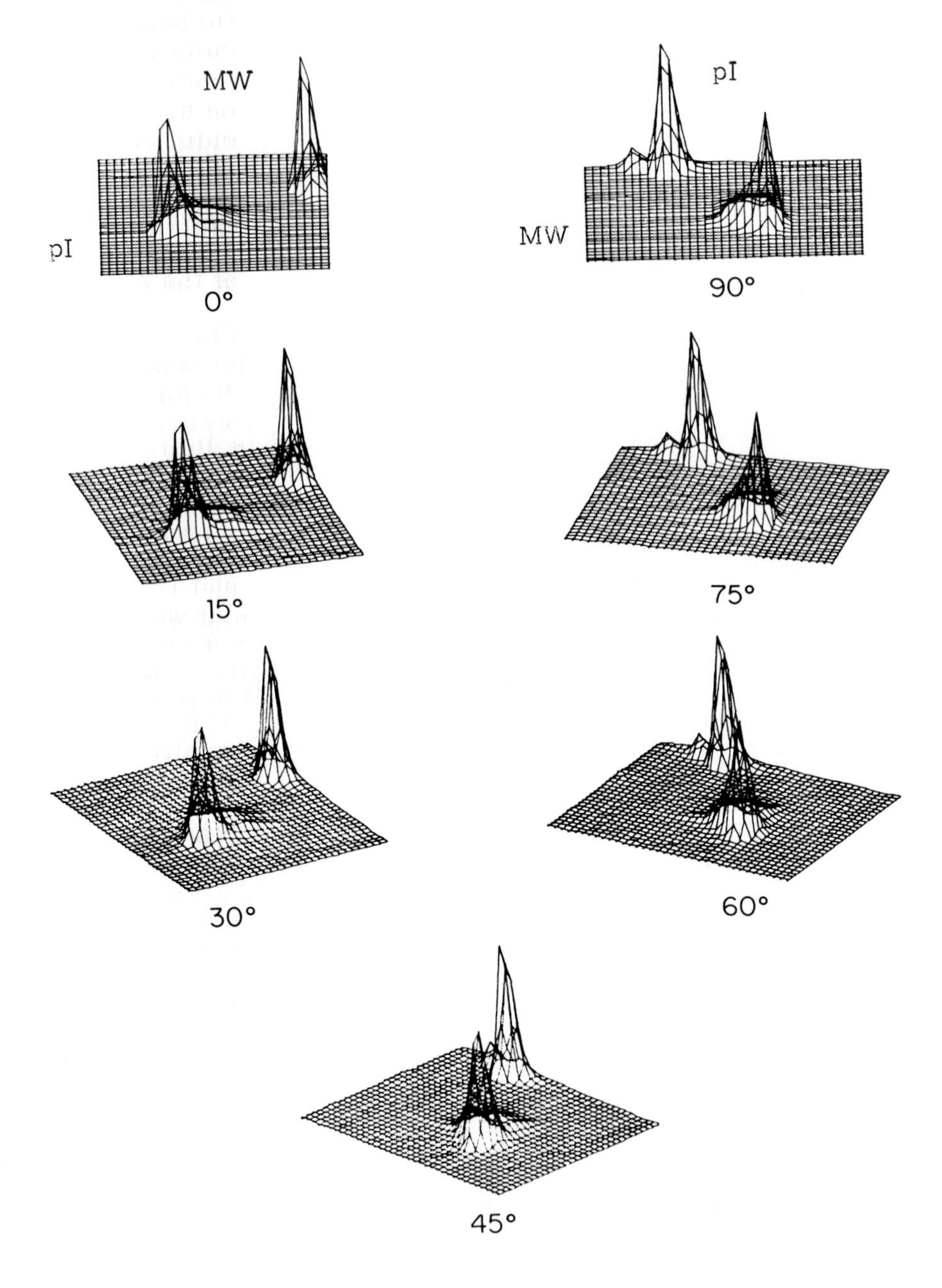
MW
pI
0°
pI
MW
90°
15°
75°
30°
60°
45°

Figure 2

Fig. 2. Computer-assisted raster scans of protein distribution in 2-dimensional gels. Protein standards — bovine serum albumin (MW 67,000; isoelectric point 5.4) and ovalbumin (MW 44,000; isoelectric point 4.8) — were first separated by isoelectric focusing in an Ampholine pH gradient. The gels were then sealed to slab gels containing SDS in 9% polyacrylamide, and the proteins are separated according to size by electrophoresis at right angles to the original direction of charge-dependent migration. After staining with Coomassie Blue, the absorbancy was measured by raster-scanning microdensitometry of a photographic transparency of the gel. The absorption data is plotted as a function of molecular weight and pI coordinates, with increasing molecular weight to the right and increasing pI to the left of the figures in the upper panel. Computer graphics permit viewing of the protein peaks from various directions, in this case from an inclination of 25° from the horizontal plane.

DNA-Binding by Tumor Nuclear Proteins

In the expectation that proteins involved in transcriptional control might interact directly with DNA sequences, we have studied the DNA-binding properties of the tumor nuclear non-histone proteins. The experimental approach employs DNA-affinity chromatography, as carried out on homologous double-stranded DNA covalently linked to Sephadex G-25 (13). This chromatographic procedure has been previously shown to reproducibly separate sets of nuclear proteins differing in DNA affinity, and to detect proteins which selectively interact with DNA sequences of different $C_{o}t$ values (35,36), or preferentially bind ribosomal-DNA sequences (37).

In order to investigate the DNA-binding properties of the tumor-associated proteins, the usual extraction procedure was modified to avoid denaturing agents. The nuclear proteins were extracted in 2 M NaCl buffered at pH 8.0, and the clarified extract was mixed with rat DNA bound to Sephadex G-25. After a series of dialysis steps to reduce the salt concentration to 50 mM NaCl, the DNA-Sephadex/protein mixture was poured into a chromatographic column, and the proteins were eluted by stepwise increments in the ionic strength of the eluting buffer. This chromatographic procedure provides subsets of the tumor nuclear proteins which were analyzed electrophoretically and compared with the original protein complement of the tumor nuclei. Fig. 3 shows the results of the fractionation.

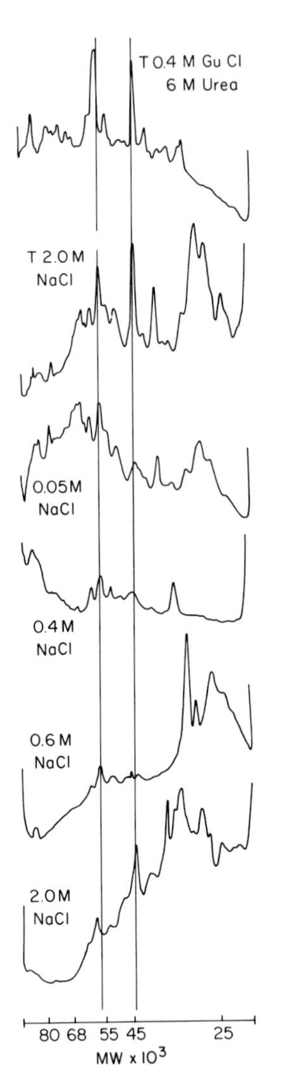

Fig. 3. DNA-affinity chromatography of tumor nuclear proteins. The upper panel shows the electrophoretic profile of the proteins extracted in 6 M urea- 0.4 M guanidine-HCl. Beneath it is the corresponding profile of the 2 M NaCl extract. The proteins of the 2 M NaCl extract were eluted from the DNA column by stepwise increments in the NaCl concentration of the eluting buffer, as indicated on the right of the corresponding electrophoretic profile. Note that the protein fraction of molecular weight 44,000 (TNP_1) is not displaced from DNA at low ionic strengths, but requires 2 M NaCl for its elution. Most of the TNP_2 peak is released from the column at low ionic strengths.

Proteins of the TNP_1 class (molecular weight 44,000) appear to be DNA-binding proteins, and they require high ionic strength for their elution from the column. In contrast, most of the TNP_2 class has little or no DNA affinity and emerges from the column at 0.05 M NaCl. (Control experiments using Sephadex G-25 columns that had been treated with the coupling reagent but not reacted with DNA (13) did not bind nuclear proteins, and both TNP_1 and TNP_2 were recovered in the 0.05 M NaCl "run-off" peak.)

Thus, the TNP_1 class of tumor nuclear proteins, although acidic in nature (as judged by overall amino acid composition (6) and low isoelectric point range (between 4.85 and 5.25)), binds strongly to DNA at physiological pH values. Whether the binding involves strong electrostatic interactions between regions of the polypeptide chains that are enriched in basic amino acids and DNA phosphate groups remains to be determined, but the dissociation of the complex by salt would suggest that TNP_1 binding to DNA probably has an electrostatic component. (This is also the case for well-defined and highly-specific gene regulatory proteins, such as the *lac* repressor in *E. coli* ; its interaction with the *lac* operon is known to require participation of the basic amino-terminal region of the otherwise acidic molecule (38).)

Evidence for an *in vivo* association of TNP_1 with tumor DNA sequences has been obtained by studies of proteins released during DNAse I digestion of tumor nuclei.

Selective Release of TNP_1 by Limited DNAse I Digestion of DMH-Induced Tumor Chromatin

Limited digestion of avian erythrocyte chromatin by DNAse I has been shown to result in a rapid and selective degradation of the globin genes, but these genes are not preferentially degraded in a non-erythroid cell, such as the fibroblast of the same organism (39). Similarly, the ovalbumin gene in oviduct chromatin is selectively degraded by DNAse I, but such sensitivity is not detected in liver nuclei (40). The differential sensitivity to DNAse I of transcriptionally-active DNA sequences, regardless of their rate of transcription (41), implies an altered conformation of the genes that are programmed for activity in a particular cell type. The reasons for the differential nuclease sensitivity of "active" DNA sequences are discussed in more detail later; the present discussion focuses on the use of DNAse I as a probe for the identification of proteins likely to be components of the transcription complex.

Nuclei were isolated from DMH-induced adenocarcinomas and incubated with DNAse I under conditions that released only 10$\pm$2 % of the total DNA. After this limited digestion, the nuclei were sonicated briefly and the suspension was centrifuged to separate a residual "resistant" chromatin fraction from more highly degraded chromatin fragments in the supernatant fraction. The proteins in both fractions were analyzed by SDS-polyacrylamide gel electrophoresis with the results shown in Fig. 4.

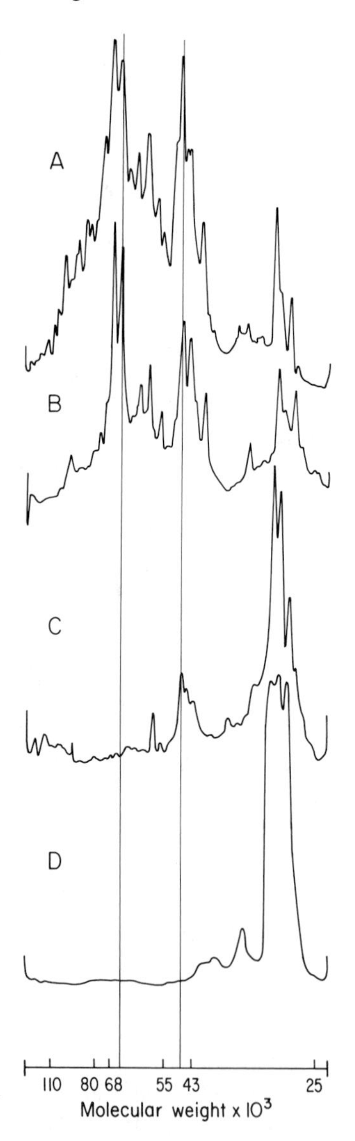

Fig. 4. Selective release of TNP_1 proteins during a limited digestion of adenocarcinoma nuclei by DNAse I. Aliquots of the nuclear suspension were incubated in the presence (B and C) or absence (A and D) of added DNAse I for 5 minutes. After centrifugation to separate the "resistant" chromatin fraction from more degraded chromatin fragments in the supernate, the proteins were extracted and analyzed by SDS-polyacrylamide gel electrophoresis. Densitometric tracings of the banding patterns are shown, with vertical lines indicating the positions of TNP_1 and TNP_2. Note that DNAse I digestion releases proteins of molecular weight similar to that of the TNP_1 class, but no loss of TNP_2 is observed. The "resistant" chromatin fraction after DNAse I digestion is depleted in TNP_1. Compare the profile in the DNAse I digested sample in (B) with the enzyme-free control in (A). Nuclei incubated in the absence of DNAse I lose very little protein; this pattern was obtained after a 10-fold concentration of the supernate.

A limited digestion of tumor chromatin by DNAse I releases some histones together with a class of nuclear proteins of molecular weight corresponding to that of TNP_1 (43-45,000 daltons). A smaller peak at 56,000 daltons is also observed (Fig. 4 C). It is significant that tumor nuclei incubated in the absence of DNAse I do not release proteins of the TNP_1 class (Fig. 4 D), nor is there any indication of a selective release of TNP_2 from tumor nuclei incubated in the presence or absence of DNAse I (Fig. 4 C and D). (It should be noted that control nuclei lose very little protein during the incubation period; the pattern in Fig. 4 D was obtained after a 10-fold concentration of the supernatant fraction.)

Examination of the electrophoretic profiles of the proteins remaining in the "resistant" chromatin fraction of control and DNAse I-treated tumor nuclei (shown in panels A and B of Fig. 4) indicates that limited DNAse I digestion has not diminished the proportion of TNP_2 remaining in the nuclei, but has released a considerable amount of TNP_1. This again indicates that the two major protein classes of the tumor nuclei are not alike in their nuclear localizations, and it confirms the chromatographic evidence in Fig. 3 that TNP_1 includes a class of DNA binding proteins. The presence of such proteins on actively transcribing genes is also indicated but not proven in these experiments.

Parallel experiments have been carried out on nuclei isolated from normal colonic epithelial cells. Limited digestion with DNAse I releases proteins of molecular weight 56,000 and 90,000 daltons (Fig. 5 C). The proportions of these proteins are reduced in the corresponding "resistant" chromatin fraction (Fig. 5 B), as compared with their proportions in the proteins remaining after incubation in the absence of added DNAse I (Fig. 5 A). Because all incubations were performed in the presence of the protease inhibitor, PMSF, and because control experiments have established that the added DNAse I is free of proteolytic activity, we conclude that the differential release of nuclear proteins by DNAse I digestion is not likely to be an artefact of proteolysis.

The reason for the selective release of proteins during limited DNAse I digestion was next investigated - with emphasis on the possibility that the selective degradation of transcriptionally active DNA sequences is due to post-synthetic modification of the associated histones.

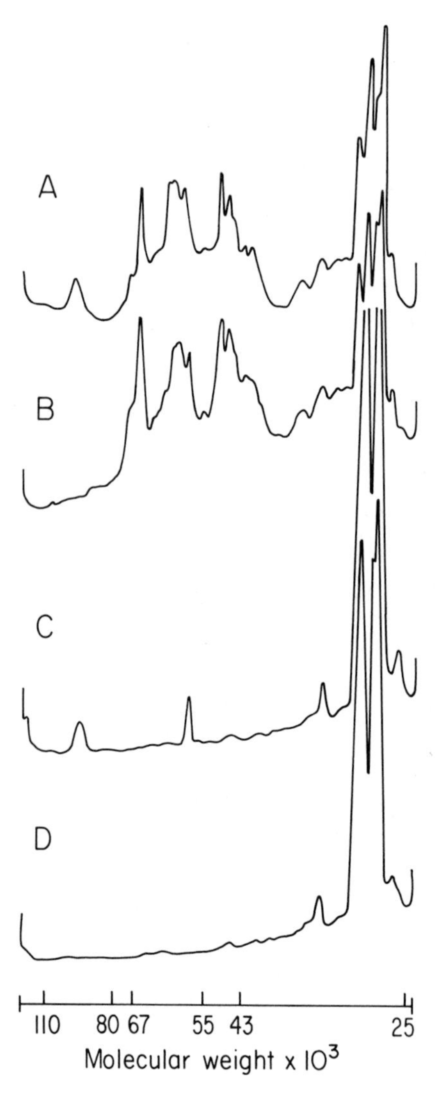

Fig. 5. Selective release of nuclear proteins during limited DNAse I digestion of normal colonic epithelial nuclei. Aliquots of the nuclear suspension were incubated in the presence (B and C) or absence (A and D) of added DNAse I for 5 minutes. After centrifugation to separate the "resistant" chromatin fractions from more degraded chromatin fragments in the supernate, the proteins were extracted and analyzed by SDS-polyacrylamide gel electrophoresis. Apart from the release of some histones, DNAse I treatment selectively releases proteins of molecular weight ca. 56,000 and 90,000. These proteins are not released from the control nuclei (D) incubated in the absence of DNAse I. The disappearance of the 90,000 MW fraction from the "resistant" chromatin fraction (B) is evident upon comparison of the electrophoretic profile with that of the control chromatin (A). The pattern shown in (D) represents a 10-fold concentration of the control supernatant fraction.

The Role of Histone Acetylation in the Differential DNAse I Sensitivity of Transcriptionally Active DNA Sequences

It has been repeatedly observed that DNA sequences with the potential for transcription in a particular cell type are selectively degraded during brief digestions of the nuclei with DNAse I. The corresponding genes are not selectively degraded during DNAse I-treatment of nuclei from cells in which they are not expressed (39-41). What is the reason for the increased susceptibility of 'active' genes to DNAse I attack?

On the premise that an altered conformation of the nucleosomes in transcription complexes may be due to post-synthetic modifications of the histones associated with the DNA-template, we have studied the effects of histone acetylation on the DNAse I sensitivity of tumor chromatin. There are many reasons why this premise warrants investigation. Since the discovery of histone acetylation in 1964 (42), there have been numerous correlations noted between increased acetylation of histones and gene activation, as induced by hormones, mitogens, and developmental stimuli, and decreased acetylation of histones when transcription is physiologically suppressed (or inhibited by carcinogens such as Aflatoxin B_1 (43)). Much of the evidence for temporal and spatial correlations between histone acetylation and gene activity has been reviewed recently (44,45). Of particular relevance to the present investigation of changes in nuclear proteins during carcinogenesis are observations that histones associated with the DNA of transforming viruses, such as SV 40 and polyoma virus, are much more acetylated than the corresponding histones of the host cells (46). The correlation between increased acetyl content of polyoma virus histones and malignant transformation is strengthened by the finding that non-transforming host-range mutants of polyoma virus fail to show a high level of histone acetylation (46). The transcriptionally active sequences of integrated adenovirus-5 genes in transformed hamster cells have recently been shown to be DNAse I-sensitive (47), and although the level of acetylated histones associated with the integrated viral genome has not yet been determined, the results on SV-40 and polyoma viruses suggest that viral gene activity may be generally associated with a high level of acetylation of the associated histones.

Our approach to this problem of the relationship between histone acetylation and DNAse I sensitivity of active genes is based on the recent observation by Riggs et al. (48) that exposure of HeLa cells to 5 mM Na butyrate leads to a progressive accumulation of the acetylated forms of histones H3 and H4. Butyrate is known to stimulate RNA synthesis in HeLa cells (49). We have confirmed the observation (48) that butyrate-treatment increases the level of histone acetylation in HeLa cells and further established that the modification occurs at the nucleosomal level. Mono-nucleosomes were prepared from control and butyrate-treated cells by digestion with staphylococcal nuclease and gradient centrifugation. The monosome fractions were compared with regard

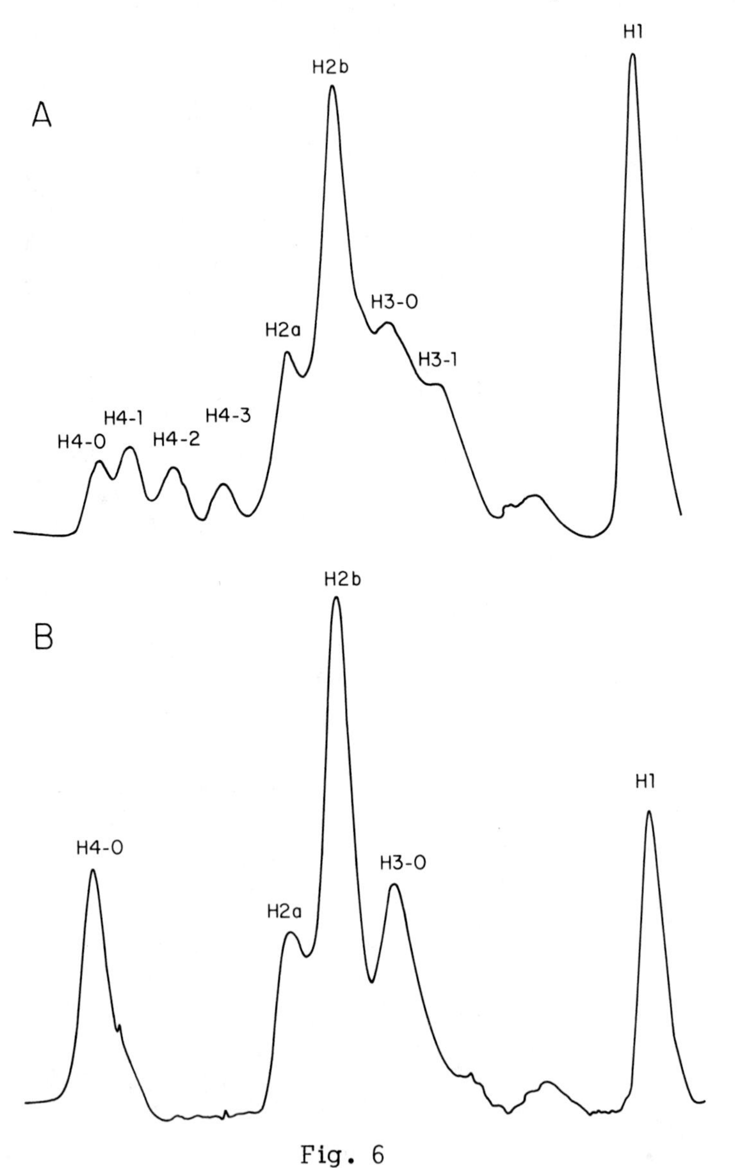

Fig. 6

Fig. 6. Changes in acetylation of nucleosomal histones H4 and H3 following a 21 hour exposure of HeLa S-3 cells to 7 mM Na butyrate in the culture medium. The nucleosomes of control and butyrate-treated cells were prepared by digestion of the isolated nuclei with staphylococcal nuclease, and the mono-nucleosome peaks were purified by sucrose density gradient centrifugation. The histones were extracted and analyzed by electrophoresis in 25 cm, 12 % polyacrylamide gels containing acetic acid and urea (17). Densitometric tracings of the stained histone bands are shown for the mono-nucleosomes of butyrate-treated cells in panel A, and for the control mono-nucleosomes in panel B. Note that increased substitution of the histone lysine residues by acetylation diminishes the electrophoretic mobility of the parent polypeptide chains. The mono- and multi-acetylated forms of histone H4 are present in high amounts after butyrate-treatment, as compared with their relatively low proportions in the control nucleosomes. Butyrate-treatment also increases the level of acetylation of histone H3 in the mono-nucleosomes.

---

to their contents of acetylated and non-acetylated forms of histones H4 and H3. (The modification of these histones by enzymatic acetylation involves a substitution on the lysine epsilon-amino group - with a corresponding reduction in net positive charge. The resulting charge differences account for the separation of the various modified and non-modified forms by electrophoresis in acid-urea-polyacrylamide gels (50).) Fig. 6 shows the densitometric tracings of the histone banding patterns in the chromatin monomer particles of control and butyrate-treated HeLa cells. It is clear that exposure to 7mM butyrate leads to a massive increase in the acetylated forms of these two key nucleosomal histones.

In order to determine why exposure to butyrate affects the proportions of the acetylated histone forms, we tested the effects of 7 mM Na butyrate on the uptake and turnover of radioactive acetate in the histones of HeLa S-3 cells in culture. For the uptake experiments, the control cells were incubated in the presence of ^{3}H-acetate, while the butyrate-treated cells were incubated with ^{14}C-acetate. The results (Table 1) show that 7 mM Na butyrate has little or no effect on the rate of incorporation of radioactive acetate into HeLa histones. The "turnover" experiments were more revealing.

In these experiments, cells were prepared containing histones labeled with either ^{14}C-acetate or ^{3}H-acetate. Cells containing ^{14}C-acetate-labeled histones were washed, resuspended in non-radioactive medium containing 7 mM Na butyrate, and incubated at 37° C. Aliquots were withdrawn at regular intervals for extraction of the histones and measurement of their residual ^{14}C-acetate content. In parallel experiments, cells containing ^{3}H-acetylated histones were incubated in the absence of butyrate, and samples were withdrawn at the same times for isolation of the histones and measurement of the remaining ^{3}H-acetate content. (Because the cells in both experiments were mixed at each time point during the "cold chase" before isolation of the nuclei and extraction of the histones, all radioactively-modified histones were exposed to the same conditions throughout their isolation, and differences in acetate retention must reflect differences in acetyl-group "turnover" in the living cells due to the presence of butyrate in the growth medium.) The results shown in Table 2 indicate that butyrate suppresses the turnover of acetyl groups in the histones. Since butyrate has no appreciable effect on the uptake of acetate groups, the net result is a progressive accumulation of the acetylated forms of histones H3 and H4, as observed in the isolated nucleosomes (Fig.6).

TABLE 1

Effects of Na butyrate on acetate uptake into HeLa histones

Time	Specific activities of histones labeled with ^{3}H-acetate *	^{14}C-acetate **	Ratio $^{14}C/^{3}H$
min	cpm/mg	cpm/mg	
5	3580	3610	1.01
10	4590	4410	0.96
15	7990	7360	0.92

* Control cells in butyrate-free MEM.

** Cells exposed to 7 mM Na butyrate in MEM.

TABLE 2

Effects of Na butyrate on acetate release from HeLa histones*

Time	Specific activities of histones labeled with ^{3}H-acetate **	^{14}C-acetate ***	Ratio $^{14}C/^{3}H$
min	cpm/mg	cpm/mg	
0	5710	6190	1.08
10	4280	5610	1.31
20	3460	5260	1.52
30	3060	5370	1.75
50	2420	5200	2.15

* HeLa cell histones were pre-labeled with either ^{3}H-acetate or ^{14}C-acetate. The cells were washed and suspended in non-radioactive media for the "cold-chase" experiments.

** Control cells in butyrate-free MEM.

*** Cells in MEM containing 7 mM Na butyrate.

What are the consequences of increased histone acetylation for chromatin structure ? We have compared the nucleosomes generated by staphylococcal nuclease digestion of control and butyrate-treated HeLa cell nuclei. The monomer peaks were purified on sucrose density gradients and compared with regard to sedimentation coefficient, circular dichroic spectra, and DNA melting curves. No significant differences were noted in any of these properties for nucleosomes differing as markedly in their content of acetylated histones as those shown in Fig. 6 (16). We then compared the chromatins of control and butyrate-treated HeLa cells for their sensitivity to DNAse I. In these experiments, the cells were labeled with radioactive thymidine in the following way: the control cells were incubated with ^{3}H-thymidine for 5 hours, washed, and then grown for 20 hours in non-radioactive medium. An equivalent cell suspension was labeled with ^{14}C-thymidine for 5 hours, washed, and also grown for 20 hours in non-radioactive medium; these cells were then placed in a medium containing 7 mM Na butyrate and incubated for an additional 21 hours, thus increasing the level of acetylation of their histones. The control cells and butyrate-treated cells were harvested and mixed prior to isolation of the nuclei. The mixed nuclear suspension was then treated with DNAse I, and the kinetics of release of ^{3}H-DNA fragments and ^{14}C-DNA fragments were compared. The results are summarized in Fig. 7, which plots the time course of degradation of the control (^{3}H) and highly-acetylated (^{14}C) chromatins.

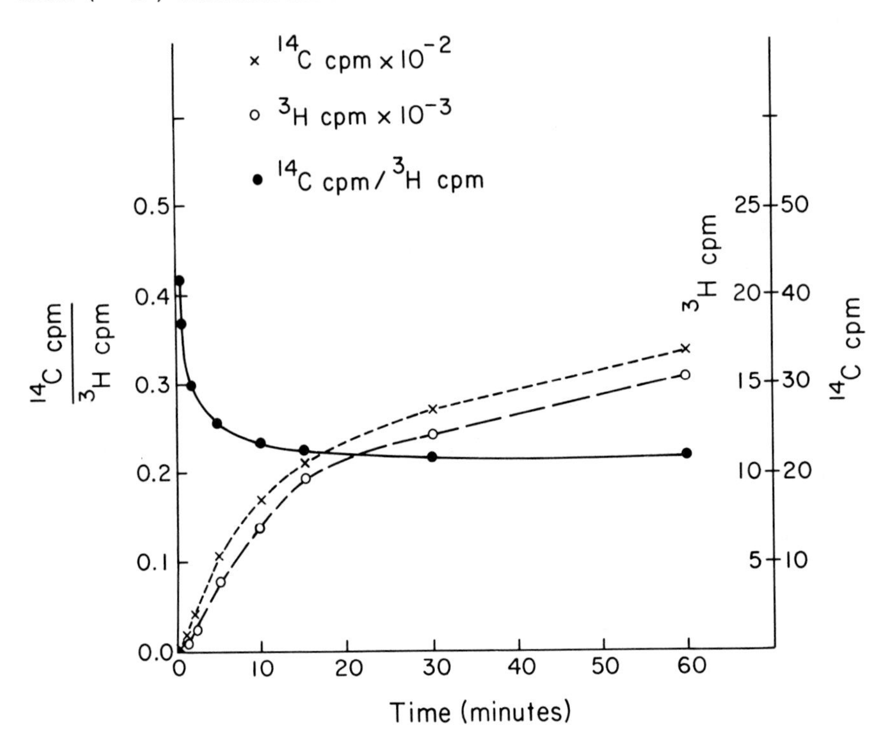

Fig. 7. Comparative kinetics of DNAse I digestion of HeLa chromatins differing in their content of acetylated histones. Cells were labeled with ^{3}H-thymidine and ^{14}C-thymidine. The ^{3}H-labeled cells were harvested, while the ^{14}C-labeled cells were incubated in 7 mM butyrate to increase the level of histone acetylation. Both cell populations were then mixed and nuclei were isolated. The mixed nuclear suspension was incubated with DNAse I and the kinetics of release of ^{3}H-DNA fragments and ^{14}C-DNA fragments were compared. The ratio $^{14}C/^{3}H$ was calculated for each time point. The early decrease in this ratio indicates that the highly-acetylated chromatin from butyrate-treated cells is more rapidly degraded than the chromatin of the control cells.

Fig. 7 also plots the ratio of $^{14}C/^{3}H$ for each time point during the DNAse I digestion. The change in the $^{14}C/^{3}H$ ratio at early times is particularly revealing, because it shows that chromatin containing the highly-acetylated histones is more susceptible to DNAse I attack.

Similar experiments using staphylococcal nuclease in place of DNAse I did not show this striking difference in the rates of degradation of chromatins from control and butyrate-treated HeLa cells. This result, taken together with the knowledge that staphylococcal nuclease selectively cleaves the DNA strands between the nucleosomes, whereas DNAse I can attack DNA enveloping the nucleosome core, strongly supports the view that acetylation of the nucleosomal histones H3 and H4, alters DNA interaction with the core histones and increases its susceptibility to DNAse I attack.

The previous conclusion is based on the assumption that the DNA sequences that are most rapidly degraded by DNAse I are associated with highly-acetylated forms of histones H3 and H4. To confirm this viewpoint, we have analyzed the histones remaining in the nuclei of butyrate-treated cells after a brief incubation in the presence or absence of DNAse I. The histone electrophoretic patterns are compared in Fig. 8, which shows that the proportions of the multi-acetylated forms of histone H4 (relative to the content of non-acetylated forms) are significantly lower after removal of only 10 % of the DNA. It follows that the most highly acetylated H4 molecules must have been released together with the most rapidly degraded DNA sequences.

Similar conclusions have been drawn from studies of histone release during limited DNAse I digestions of avian erythrocyte nuclei. The erythrocytes were incubated in the presence of ^{3}H-acetate in order to label the acetylated forms of the histones prior to isolation of the nuclei. Aliquots of the nuclear suspension were incubated with DNAse I under conditions which release only about 12 % of the DNA into the supernatant fraction (14). The histones remaining in the nuclei were then extracted, and histones H3 and H4 were purified by electrophoresis in SDS-polyacrylamide gels. The specific ^{3}H-activities of the residual histones from DNAse I-treated nuclei are compared with those of control nuclei (incubated without DNAse I) in Fig. 9.

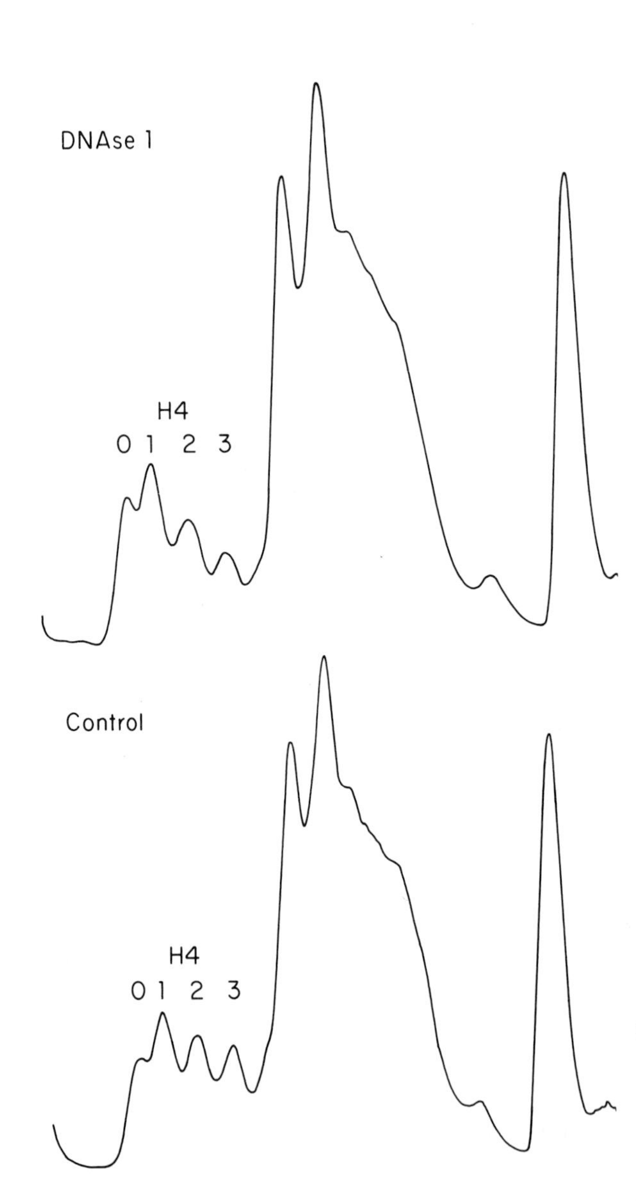

Fig.8. Selective release of multi-acetylated forms of histone H4 during a limited digestion of HeLa cell nuclei by DNAse I. The nuclei were prepared from cells grown in the presence of 7 mM Na butyrate to increase the proportions of highly acetylated H4 molecules. Aliquots of the nuclear suspension were incubated with DNAse I, or in its absence. After a brief digestion to solubilize 10 % of the nuclear DNA, the histones were prepared and analyzed in acid-urea-polyacrylamide gels. Note that the proportions of the di- and tri-acetylated forms of histone H4 (relative to the non-acetylated form) are much reduced after DNAse I treatment, as compared to the corresponding proportions in control nuclei that were incubated in the absence of DNAse I.

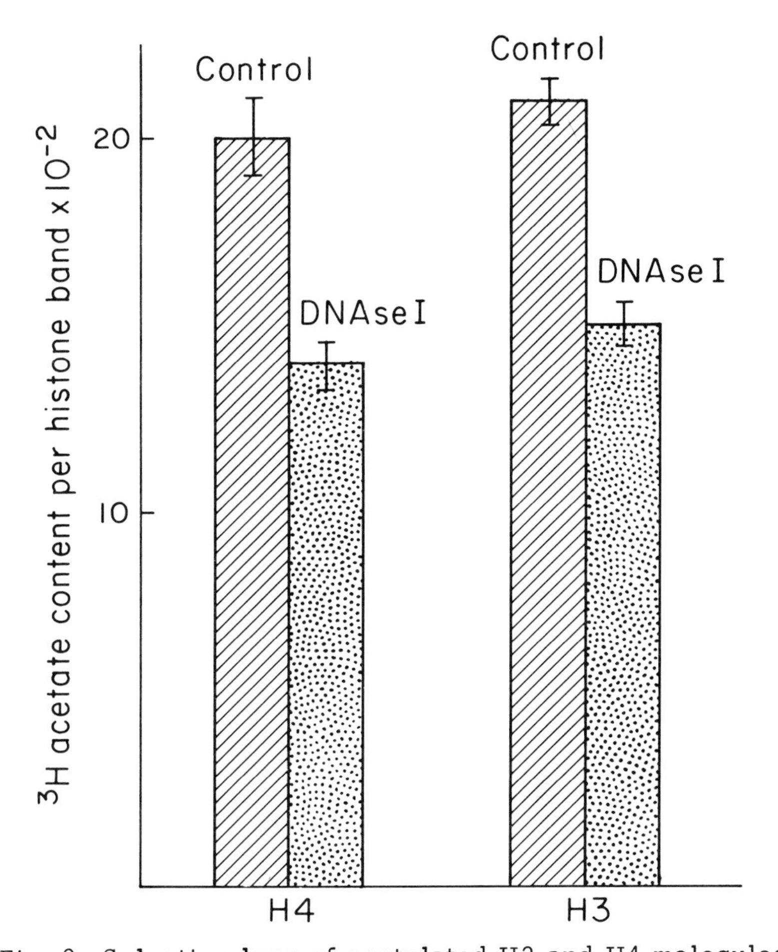

Fig. 9. Selective loss of acetylated H3 and H4 molecules from avian erythrocyte nuclei during limited digestion with DNAse I. The histones were labeled with ^{3}H-acetate in intact erythrocytes; the nuclei were isolated and treated with DNAse I to release 11-12 % of their total DNA content. The remaining histones were extracted and purified by electrophoresis in SDS-polyacrylamide gels. The specific ^{3}H-activities of histones H3 and H4 in the chromatin remaining after DNAse I-treatment are significantly lower than the corresponding ^{3}H-activities of control nuclei (incubated in the absence of DNAse I).

The specific ^{3}H-activities of both H3 and H4 remaining in the chromatin after removing only 11-12 % of the DNA are 30 % lower than the corresponding ^{3}H-activities of histones H3 and H4 in the control nuclei (incubated in the absence of DNAse I). It follows that the acetylated forms of the histones were preferentially released by DNAse I under conditions known to selectively degrade the transcriptionally active DNA sequences of the erythrocyte nucleus (39). The results are consistent with early suggestions that acetylation of the histones plays a role in gene activation (42), presumably by releasing structural constraints upon associated DNA sequences in the nucleosomes.

Selective release of non-histone proteins associated with "active" genes

The technique of limited DNAse I digestion has been employed for the identification and isolation of other proteins associated with transcriptionally active regions of the chromatin. The selective release of the TNP_1 protein class from DMH-induced adenocarcinomas has already been mentioned ; similar experiments have been carried out on avian erythrocyte nuclei (14). Duck erythrocyte nuclei were subjected to limited digestions with DNAse I under conditions which solubilize only 12 % of the total DNA. Proteins released into the supernatant fraction (prepared by low-speed centrifugation of the sheared nuclear suspension) were analyzed by electrophoresis in SDS-polyacrylamide gels. For purposes of comparison, the effects of staphylococcal nuclease were also tested, since the latter enzyme does not preferentially attack the globin genes in avian erythrocyte chromatin (39).

The supernatant fraction after DNAse I treatment shows the presence of a characteristic subset of nuclear proteins (Fig. 10). Among these are three prominent bands in the molecular weight region 28,200 - 29,500 daltons ; the corresponding bands in the chromatin remaining after DNAse I treatment are very faint (data not shown). These bands have been shown to correspond to the high-mobility-group (HMG) proteins originally described by Johns and coworkers (51, 52). When avian erythrocyte nuclei are treated with staphylococcal nuclease to release about 25 % of the total DNA into the supernatant, one does not observe a preferential release of the HMG proteins (Fig. 10). It follows that these proteins are not randomly distributed throughout the chromatin but are primarily associated with nucleosomes in a more DNAse I-accessible state. Their presence on "active" genes is consistent with an earlier report that HMG proteins stimulate transcription in vitro (52).

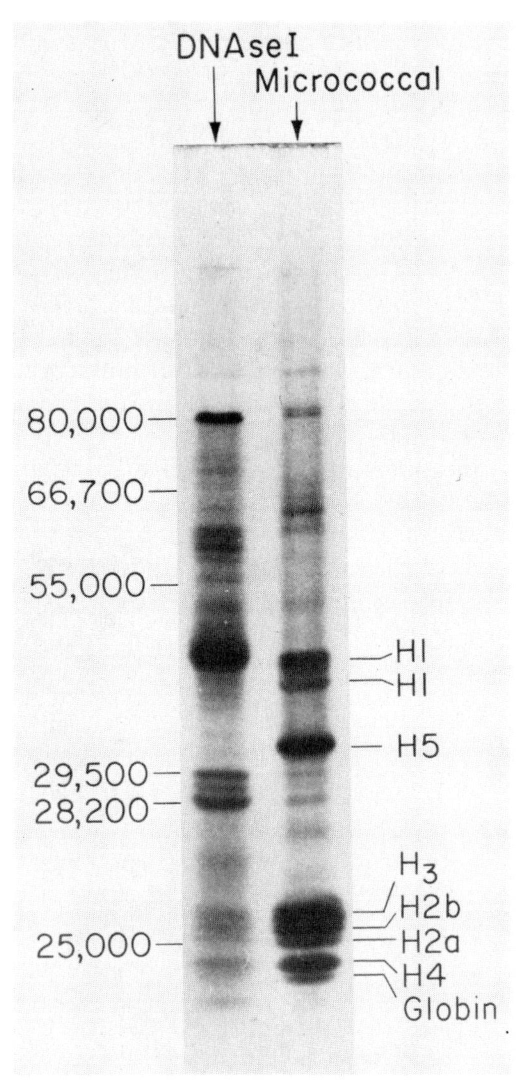

Fig. 10. Electrophoretic analysis of the proteins released during limited digestions of avian erythrocyte nuclei with (DNAse I (left panel) and staphylococcal nuclease (termed micrococcal nuclease in the figure)(right panel). Note that three prominent protein bands in the molecular weight range 28,200-29,500 (HMG proteins) are present in the supernatant after centrifugation of the DNAse I-treated nuclei. Only 11-12 % of the DNA was released under these conditions. In contrast, digestion of 24% of the nuclear DNA with staphylococcal nuclease did not result in an equivalent release of the HMG proteins, although a more extensive release of histones is evident, as expected for the greater degree of chromatin breakdown. DNAse I attack on the transcriptionally-active chromatin also releases a complex set of non-histone proteins.

Other non-histone proteins are also released during limited DNAse I digestions that are not released during comparable treatments with staphylococcal nuclease. Prominent among these are proteins of molecular weights 57-59,000 daltons and 80,000 daltons (Fig. 10). The functions of these proteins are not known, but in view of their likely association with transcriptionally active regions of the chromatin and their appreciable concentrations, it appears that such proteins must play a general, rather than a site-specific role in the synthesis or processing of erythrocyte nuclear RNAs.

A comparison of the electrophoretic banding patterns in Fig. 10 shows that a broad molecular-weight spectrum of non-histone proteins is released differentially, depending upon the enzyme used to disrupt the chromatin, but many more proteins are released with DNAse I. This is in agreement with much earlier evidence that template-active subfractions of chromatin from many tissues are enriched in their non-histone protein complement (53-58). Some of the proteins associated with the DNAse I-susceptible sequences are likely to be involved in transcriptional control mechanisms, and it would appear that this method of selective enzymatic release of a small fraction of the total nuclear proteins, combined with two-dimensional gel electrophoresis and high-resolution DNA affinity chromatography, will provide a powerful and simplifying approach to the identification and purification of regulatory proteins in the eukaryotic genome.

REFERENCES

(1) H.Druckrey, R.Preussmann, F.Matzkies and S.Ivanovic, Naturwissenshaften, 54 (1967) 285.

(2) H.Druckrey, in: Carcinoma of the Colon and Antecedent Epithelium, ed. W.J.Burdette (Charles C.Thomas, Springfield, Ill., 1970) p.267.

(3) M.J.Martin, F.Martin, R.Michiels, H.Bastein, E.Instreba, M.Bordes and B.Viry, Digestion, 8 (1973) 22.

(4) N.Thunherr, E.E.Deschner, E.H.Stonehill and M.Lipkin, Cancer Res., 33 (1973) 940.

(5) A.Hawks, R.M.Hicks, J.W.Holsman and P.N.Magee, Brit.J.Cancer, 30 (1974) 429.

(6) L.C.Boffa, G.Vidali and V.G.Allfrey, Cancer, 36 (1975) 2356.

(7) L.C.Boffa, G.Vidali and V.G.Allfrey, Exp.Cell Res., 98 (1976) 396.

(8) S.Levy, R.T.Simpson and H.Sober, Biochemistry, 11 (1972) 1547.

(9) U.Laemmli, Nature, 227 (1970) 680.

(10) P.H.O'Farrell, J.Biol.Chem., 250 (1975) 4007.

(11) L.C.Boffa and V.G.Allfrey, Cancer, 40 (1977) 180.

(12) L.C.Boffa, F.R.Kramer, B.A.Poole and V.G.Allfrey, manuscript in preparation.

(13) V.G.Allfrey and A.Inoue, Methods in Cell Biology, 17 (1977) 253.

(14) G.Vidali, L.C.Boffa and V.G.Allfrey, Cell, 12 (1977)409.

(15) V.G.Allfrey, G.Vidali, L.C.Boffa and E.M.Johnson, Proc. Vth Intern.Congr.Birth Defects (1977) in press.

(16) G.Vidali, L.C.Boffa and V.G.Allfrey, Proc.Nat.Acad.Sci. USA, 75 (1978) in press.

(17) S.Panyim and R.Chalkley, Arch.Biochem.Biophys., 130 (1969) 337.

(18) C.S.Teng, C.T.Teng and V.G.Allfrey, J.Biol.Chem., 246 (1971) 3597.

(19) M.Gronow and T.M.Thackrah, Eur.J.Cancer, 10 (1974)21.

(20) M.Gronow and T.M.Thackrah, Chem.Biol.Interact., 9 (1974) 225.

(21) L.C.Yeoman, C.W.Taylor and H.Busch, Cancer Res., 34 (1974) 424.

(22) C.B.Chae, M.C.Smith and H.P.Morris, Biochem.Biophys. Res.Commun., 60 (1974) 1468.

(23) L.R, Orrick, M.O.J.Olson and H.Busch, Proc.Nat.Acad. Sci.USA, 70 (1973) 1316.

(24) G.S.Stern, W.E.Criss and H.P.Morris, Life Sci., 14 (1974) 95.

(25) E.A.Arnold, M.M.Buksas and K.E.Young, Cancer Res., 33 (1973) 1169.

(26) F.Gonzales-Mujica and A.P.Mathias, Biochem.J., 133 (1973) 440.

(27) B.Wilson, M.A.Lea, G.Vidali and V.G.Allfrey, Cancer Res. 35 (1975) 2954.

(28) L.C.Boffa and V.G.Allfrey, Cancer Res., 36 (1976)2678.

(29) W.W.L.Chang and C.P.Leblond, in: Carcinoma of the Colon and Antecedent Epithelium, ed. W.J.Burdette (Charles C.Thomas, Springfield, Ill., 1970) p.197.

(30) M.Lipkin and H.Quastler, J.Clin.Invest., 41 (1962) 141.

(31) E.E.Deschner, C.M.Lewis and M.Lipkin, J.Clin.Invest., 42 (1963) 1922.

(32) E.E.Deschner, M.Lipkin and C.Solomon, J.Natl.Cancer Inst., 36 (1966) 849.

(33) M.Lipkin, Cancer, 34 (1974) 878.

(34) L.C.Boffa and V.G.Allfrey, unpublished experiments.

(35) V.G.Allfrey, A.Inoue, J.Karn, E.M.Johnson and G.Vidali, Cold Spring Harbor Symp.Quant.Biol., 38 (1974) 785.

(36) V.G.Allfrey, A.Inoue and E.M.Johnson, in: Chromosomal Proteins and their Role in the Regulation of Gene Expression, eds.G.S.Stein and L.J.Kleinsmith (Academic Press, New York, 1975) p.265.

(37) V.G.Allfrey, E.M.Johnson, I.Y-C. Sun, V.C.Littau, H.R. Matthews and E.M.Bradbury, in: Human Molecular Cyto-Genetics, ICN-UCLA Symp.Mol.Cell Biol., Vol.7, eds. R.S.Sparkes, D.Comings and C.F.Fox, 1977, in press.

(38) T.Platt, K.Weber, D.Ganem and J.H.Miller, Proc.Nat. Acad.Sci.USA, 69 (1972) 897.

(39) H.Weintraub and M.Groudine, Science, 193 (1976) 848.

(40) A.Garel and R.Axel, Proc.Nat.Acad.Sci.USA, 73 (1976) 3966.

(41) A.Garel, M.Zolan and R.Axel, Proc.Nat.Acad.Sci.USA, 74 (1977) 4867.

(42) V.G.Allfrey, R.Faulkner and A.E.Mirsky, Proc.Nat.Acad. Sci.USA, 51 (1964) 786.

(43) G.S.Edwards and V.G.Allfrey, Biochim.Biophys.Acta, 299 (1973) 354.

(44) V.G.Allfrey, in: Chromatin and Chromosome Structure, eds. H.J.Li and R.A.Eckhardt (Academic Press, New York, 1977) p.167.

(45) E.M.Johnson and V.G.Allfrey, in: Biochemical Actions of Hormones, Vol.5, ed. G.Litwack (Academic Press, New York, 1978) in press.

(46) B.S.Schaffhausen and T.L.Benjamin, Proc.Nat.Acad.Sci. USA, 73 (1976) 1092.

(47) S.J.Flint and H.M.Weintraub, Cell, 12 (1977) 783.

(48) M.G.Riggs, R.G.Whittaker, J.R.Neuman and V.M.Ingram, Nature, 268 (1977) 462.

(49) E.Ginsburg, O.Salomon, T.Sreevalson and E.Freese, Proc. Nat.Acad.Sci.USA, 70 (1973) 2457.

(50) L.J.Wangh, A.Ruiz-Carrillo and V.G.Allfrey, Arch.Biochem. Biophys., 150 (1972) 44.

(51) G.H.Goodwin and E.W.Johns, Eur.J.Biochem., 40 (1973) 215.

(52) E.W.Johns, G.H.Goodwin, J.M.Walker and C.Sanders, CIBA Foundation Symp., 28 (1975) 95.

(53) J.H.Frenster, V.G.Allfrey and A.E.Mirsky, Proc.Nat.Acad. Sci.USA, 50 (1963) 1026.

(54) K.Marushige and J.Bonner, Proc.Nat.Acad.Sci.USA, 68 (1971) 2941.

(55) E.C.Murphy, S.H.Hall, J.H.Sheperd and R.S.Weiser, Biochemistry, 12 (1973) 3843.

(56) R.T.Simpson and G.R.Reeck, Biochemistry, 12 (1973) 3853.

(57) D.Doenecke and B.J.McCarthy, Biochemistry 14 (1975) 1366.

(58) J.M.Gottesfeld, R.F.Murphy and J.Bonner, Proc.Nat. Acad.Sci. USA, 72 (1975) 4404.

This research was supported in part by grants from the USPHS (CA-14908 and GM-17383), the American Cancer Society (NP-228G) and the Rockefeller Foundation Program in Reproductive Biology.

DISCUSSION

K. McCARTY: In view of the fact that many of the nonhistone proteins undergo post synthetic modifications, sometimes exceeding that of histones, it is of interest to know if you have any evidence of acetylation and/or phosphorylation of TNP_1 (M_W44K) and or TNP_2 (M_W62K) which characterize colonic adenocarcinomas?

V.G. ALLFREY: We have not yet looked to see whether they are acetylated; we have looked at the phosphorylation of a lot of these nonhistone proteins in the tumor nuclei and many of them are highly phosphorylated; but I cannot answer the acetylation question.

K. McCARTY: If the TNP_2 is phosphorylated to a greater extent than TNP_1, it could afford an explanation to account for their difference in DNA affinity. What is known of their phosphorylation?

V.G. ALLFREY: The distribution of ^{32}P-labeled proteins has been studied only in one-dimensional gels, not in two-dimensional gels, and I cannot say that the DNA-binding proteins are phosphoproteins.

S. CRAIG: The redundant sequences of the genomes of eukaryotes are somewhat complex as suggested by a number of recent publications. I was wondering whether you have attempted to separate long from short redundant sequences in order to test whether your acidic proteins bind preferentially to one or the other of those components.

V.G. ALLFREY: No, we have not done experiments of that sort, largely because we felt that we would be dealing with enormously complex situations, just as you mentioned, involving different sequences and different lengths, different repeat numbers and large numbers of nuclear proteins, so instead we shifted our approach entirely to look at selective DNA-protein binding in a well-defined gene, the satellite DNA containing the ribosomal-coding sequences of Physarum, and I have mentioned some of that work earlier. The point is that there are proteins present in nucleoli that interact selectively with the rDNA satellite. In this system we can at least specify what gene we are talking about.

N. GHOSH: Is there any information as to whether other inhibitors of DNA synthesis, for example: hydroxyurea, methotrexate or ARA-C can induce HeLa cells to accumulate tetra acetyl H4 histones?

V.G. ALLFREY: I am not familiar with any evidence that hydroxyurea or other inhibitors of DNA synthesis influence histone acetylation. The butyrate works because it stops histone deacetylation and one can come up with reasonable enough mechanisms why that should be, but I do not know whether other agents which stop DNA replication or repair would influence levels of histone acetylation.

N. GHOSH: We have observed that sodium butyrate like hydroxyurea, methotrexate or ARA-C can inhibit DNA synthesis without interfering with RNA synthesis in HeLa cells. Therefore, is it possible that the tetra acetyl nuclear histones can have an indirect effect on some of the enzymes involved in DNA synthesis; e.g., thymidylate synthetase, ribonucleotide reductase or DNA polymerase?

V.G. ALLFREY: Well you are asking very interesting questions about how the structure of nucleosomes might be modified during DNA replication. In the interest of simplification we deliberately avoided that. What we did was label our cells with radioactive thymidine and then go through a "cold chase" for a full generation to be sure we were not dealing with newly replicated DNA sequences. So we think we are dealing with old DNA sequences and we are not dealing with this problem of structural alterations during replication. However, there is evidence from other cell systems, particularly in developing trout sperm, as studied by Dixon, that the modification of histones is also a very active event at those times, especially as the histones are stripped off the DNA and replaced by protamines. So there are a lot of reasons why histones might have to be acetylated when chromatin structure is altered. But what we are trying to do is focus in on the events that relate to transcription.

J. POUPKO: Do you have any evidence to indicate that these tumor specific acidic proteins are specifically associated with DNA in chromatin in vivo?

V.G. ALLFREY: The only evidence that we regard as direct evidence for in vivo association is the fact that they (at least one protein set) is released by DNAse 1.

P. SARIN: I wonder if you can comment on the expression of the 44,000 molecular weight protein (the NHP protein) -is it specific for the carcinogen-induced tumors or is it found in other tumors as well?

V.G. ALLFREY: No, it is largely specific for this tumor class (colonic adenocarcinomas) though we have not made a broad spectrum analysis of other types of tumors, - but we do not, for instance, see anything like it in the Morris hepatomas or in Novikoff hepatomas.

P. SARIN: Have you looked at any human leukemias or other solid tumors?

V.G. ALLFREY: Well, as I say, we have not investigated a lot of tumors, and no leukemic cell nuclei have been analyzed.

P. SARIN: Does the 44,000 dalton protein bind to DNA and change the melting temperature?

V.G. ALLFREY: I do not know its effect on DNA melting.

H.V. RICKENBERG: Are the acidic nuclear proteins which are formed in the tumor cells phosphorylated in ester or in acyl linkage ?

V.G. ALLFREY: Usually the serine and threonine residues are phosphorylated.

H.V. RICKENBERG: There is no evidence for any phosphorylation of glutamate or aspartate residues?

V.G. ALLFREY: I would have to say that it has never been looked at systematically.

J. KALLOS: Regarding the DNA binding proteins - do they bind selectively to double-stranded or single-stranded DNA?

V.G. ALLFREY: The test was against double-stranded DNA. I do not know if binding would be better for single-stranded DNA. On the other hand, if we thought we were dealing with something like unwinding proteins, we would expect to find them in the dividing cell population of the normal tissue, and we do not.

J. KALLOS: But you do not know whether or not there is any preference. Is modification by acetylation in any way related to trypsin treatment of histones?

V.G. ALLFREY: Dr. Weintraub had already shown that the treatment of histones on nucleosomes with trypsin leads to the cutting off of the amino terminal regions. And it has been shown by others, Whitlock and Simpson I believe, that the pattern of DNAse cutting is influenced if you remove those amino-terminal ends of the histones. So their results, I think, are fully compatible with our view that physiologically you do not cut off the histone termini - you simply neutralize the positive charges by acetylation, and thereby influence the architecture of the nucleosome.

R. PATNAIK: How do these acetylated histones perform in reconstitution with the DNA?

V.G. ALLFREY: We are doing reconstitution experiments in collaboration with Dr. Morton Bradbury. We intend to look at physical parameters such as neutron scatter, etc., but those experiments have not been completed yet.

L. HNILICA: Are those two proteins antigenic?

V.G. ALLFREY: We have not attempted to prepare antibodies for these proteins.

M. URBAN: In view of the action of sodium butyrate on the acetylation of HeLa histones and the protocol for analyzing RBC histone acetylation (i.e. ^{14}C-acetate labeling) I was wondering if instead of looking at overall acetylation you were instead looking at differences in the turnover of acetate on histones. Also, is it not possible that the regions of chromatin that were more susceptible to DNAse I digestion showing higher levels of histone acetylation were really the areas of chromatin where the acetyl modification groups turned over at a higher rate rather than a higher degree of acetylation?

V.G. ALLFREY: Well, though differential acetyl turnover might be a factor I think that the design, especially of the HeLa experiments rules it out, because we have higher degrees of acetylation in butyrate-treated HeLa cells, as you can tell from the gel banding patterns. And what we get upon DNAse I treatment is a selective release of the most-highly acetylated forms of H4. That implies that they are coming from those DNA sequences that are most susceptible to DNAse I attack.

M. URBAN: Do you have any idea what kind of DNA is being released from HeLa cell chromatin after the 10% DNAse I digestion?

V.G. ALLFREY: There I am really going on the basis of the work of other authors, notably Harold Weintraub and Richard Axel, that it is the active genes which are degraded under those conditions of limited DNAse I digestion.

M. URBAN: What about in chromatin from HeLa cells treated with sodium butyrate?

V.G. ALLFREY: Well - we have not made any attempt to assess the sequences which are degraded in HeLa cells during limited DNAse I digestion.

THE DEVELOPMENTALLY REGULATED MULTIGENE FAMILIES ENCODING CHORION PROTEINS IN SILKMOTHS

F.C. KAFATOS[‡], A. EFSTRATIADIS, M.R. GOLDSMITH[*], C.W. JONES, T. MANIATIS[+], J.C. REGIER, G. RODAKIS[‡], N. ROSENTHAL, SIM GEK KEE, G. THIREOS, AND L. VILLA-KOMAROFF

Cellular and Developmental Biology, Biological Laboratories
Harvard University, Cambridge, MA 02138

*Department of Developmental and Cell Biology
University of California, Irvine, CA 92717

+ Division of Biological Sciences
California Institute of Technology, Pasadena, CA 91125

‡also, Department of Biology, University of Athens
Athens 621, Greece

Abstract: The chorion (eggshell) of silkmoths consists of approximately 100 different proteins, which are produced by the follicular cells surrounding the oocyte, during a short period of differentiation. We summarize findings which reveal that the multiple chorion structural genes are members of evolutionarily related multigene families. We also discuss the approaches we are using in our studies on the developmental expression and evolution of these families.

INTRODUCTION

The temporal and quantitative expression of specific genes is precisely controlled during eukaryotic development. Molecular understanding of developmental controls on gene expression can be achieved by applying recently developed methods to the study of appropriate model systems, such as the chorion.

The chorion (eggshell) of silkmoths consists of approximately 100 different proteins, which are produced by the follicular cells surrounding the oocyte (1-6). The various proteins are synthesized during a two-day developmental period, after the oocyte is fully grown. As they are produced, they are secreted and deposited on the surface of the oocyte, eventually forming around it a strong proteinaceous protective layer, the eggshell. Although choriogenesis is a continuous process, individual proteins are not synthesized uniformly throughout, but during limited and characteristic time periods, both *in vivo* and in organ culture (3). In

ISBN 0-12-045450-5

other words, the chorion structural genes are expressed through protein synthesis in overlapping sequence, according to a strict developmental program.

Our earlier work, which defined the basic features of the chorion system in some detail, has been reviewed recently (2,7). Here, we summarize recent findings which reveal that chorion genes are members of evolutionarily related "multigene families." In addition, we review the methods which we are using in current studies on the evolution and developmental expression of these families.

THE CHORION PROTEINS

The nature of chorion genes as members of multigene families was revealed by detailed studies of the respective proteins (8-11). In the moth species used for most of our studies, *Antheraea polyphemus*, at least 4 molecular weight classes of chorion proteins can be distinguished: A (7000 to 11000 daltons), B (11000 to 15000 daltons), C (15000 to 20,000 daltons) and D (20000 to 30000 daltons). By far the most abundant classes are the As and Bs, which together account for 88% of the chorion mass. Resolution of the chorion proteins from individual animals by two-dimensional gel electrophoresis (2), reveals the existence of at least 18 different As and 25 different Bs. To date we have purified and partially sequenced 12 As and 6 Bs.

Although classification of the proteins into As, Bs etc. was originally based only on molecular weight (1), it soon became apparent that the members within each of the two major classes shared additional properties. Initial indications were the similar amino acid compositions, isoelectric points and solubility properties within each class (10). For example, although most As and Bs have acidic isoelectric points, the As tend to be somewhat more basic than the Bs; although both classes are very rich in non-polar amino acids (one-third of the total is glycine), individual As are consistently richer in cysteine (ca.8.4% of the residues) than individual Bs (ca. 5.7%). Definitive evidence came from sequencing studies.

The 12 A proteins examined to date have free N-termini. They have been sequenced from that end, in most cases for nearly half their total length. Two of the proteins, named A_4--c_1 and A_4--d_1 (9), have been sequenced almost completely. All As show extensive sequence similarities. Fig. 1 presents

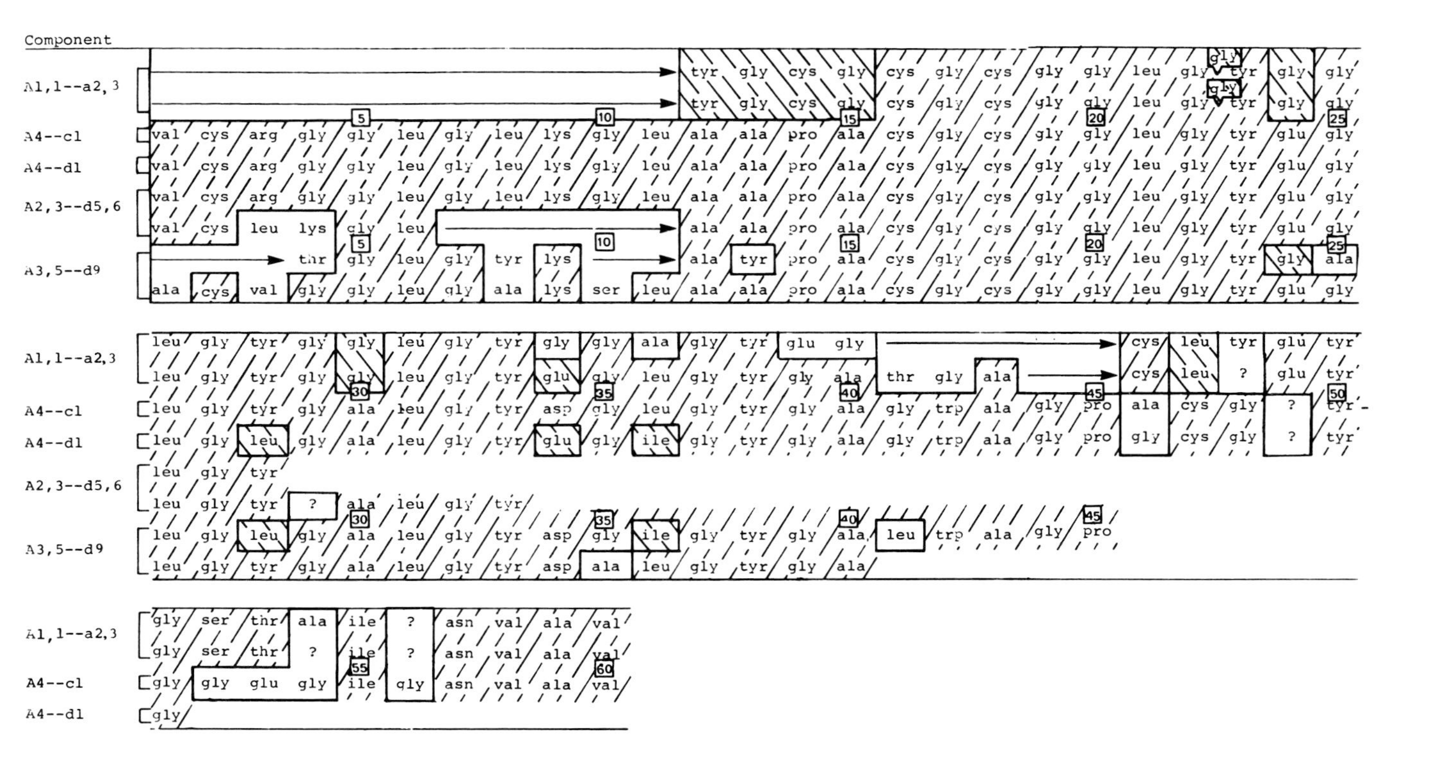

Fig. 1. Aligned sequences of A proteins from _Antheraea polyphemus_. For protein mixtures, multiple residues at individual positions were separated into 2 sequences. The sequences were aligned by introducing gaps ⟶ or insertions ⌴ . Shaded regions indicate sequence identities. From (8).

partial sequences of eight of these proteins, aligned so as to maximize sequence identity. The sequence similarities are so extensive, that it seems inescapable to conclude that the proteins are homologous (evolutionarily related), and not the products of convergent evolution. At the same time, it is clear that the proteins are encoded by different DNA sequences, since they differ by internal amino acid replacements, insertions or deletions. Similarily, the six Bs examined to date are both homologous and distinct (11). They could not be sequenced from their N-termini, because all are blocked. They have been sequenced to the extent of up to one-third of their length, beginning at an internal methionine, after cyanogen bromide cleavage. Comparison of the known B and A sequences revealed no lengthy (>3 residue) stretches of identical sequences. Thus A and B proteins are appreciably different and can be considered as two distinct classes.

CHORION GENES AS MULTIGENE FAMILIES

"Multigene families" have been defined (12) as groups of genes which are functionally similar, evolutionarily related and physically linked. The first criterion is clearly met by the genes encoding the structural proteins of the chorion. Evolutionary relatedness also seems established from the sequencing studies. The third criterion, physical linkage of the genes, also appears to be met, at least in the commercial silkworm, *Bombyx mori* (13). Genetic crosses were made in this species between inbred strains distinguished by a number of electrophoretic chorion protein variants. The chorion markers behaved as codominant Mendelian traits in F_1 crosses. Subsequent back-crosses of the F_1 to the parental strains revealed that a total of 15 out of 16 markers co-segregated, thus indicating that they are linked (Fig. 2). The linkage group was identified as that of the second chromosome (N=28). The possibility that the chorion markers represent the pleiotropic effect of a single modifier, rather than being encoded by multiple structural genes, appears unlikely (13): relevant evidence is the codominance of the chorion patterns in the F_1, and the demonstration in *A. polyphemus* that chorion proteins with different mobilities correspond to distinct primary sequences.

Recently it has become evident that various repeated gene families show two contrasting patterns of organization, dispersed and tandemly clustered (14,15). Although many chorion structural genes appear to be located on a single chromosome, it is not known how closely clustered they are.

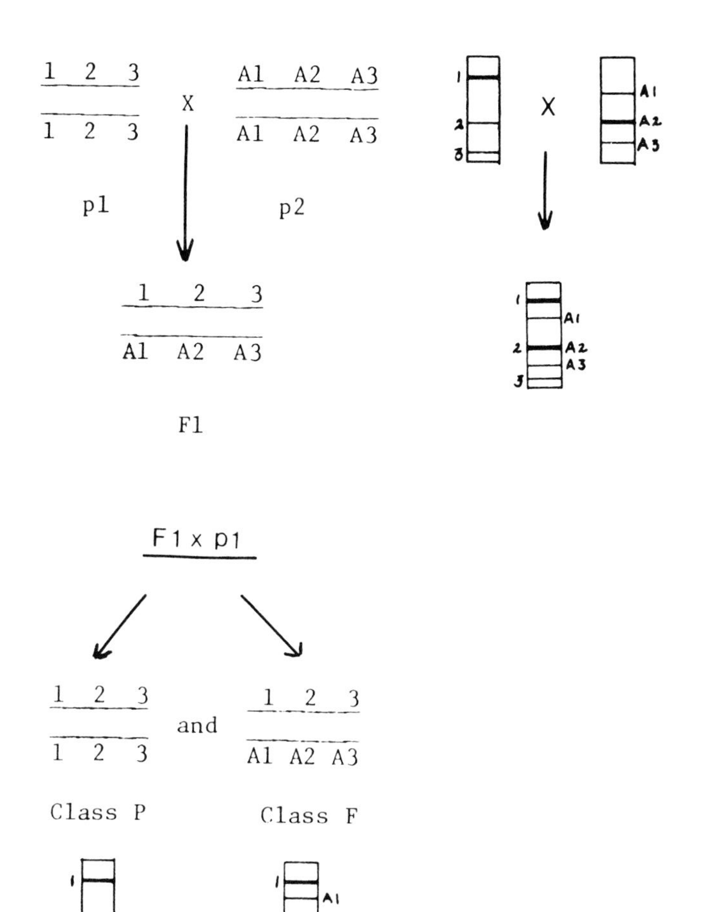

Fig. 2. Schematic representation of chorion proteins and their interpretation in terms of linkage of chorion genes in Bombyx mori (13). Each number (with or without prefix) corresponds to a structural gene whose product yields a protein at the indicated gel position; chromosome map positions of the linked genes are arbitrary. p_1 and p_2 are different homozygous inbred stocks. The diagrams indicate isoelectric focusing patterns of chorion proteins produced by parental, F_1 and testcross individuals.

Knowledge of the arrangment of these developmentally characterized genes will be important for understanding the mechanisms of their developmental regulation and evolution. We are now studying the organization of chorion genes both by genetic analysis and by recombinant DNA procedures.

CHORION cDNA CLONES

For studying the chorion system at the nucleic acid level we first isolated chorion mRNAs as a class. This can be accomplished easily using simple procedures (19), because of the biological specialization of the cells (> 95% chorion protein synthesis) and the characteristically small size of the proteins and mRNAs. When chorion mRNA preparations are translated in the wheat germ system, they yield mainly polypeptides precipitable by chorion-specific antibodies (Fig. 3).

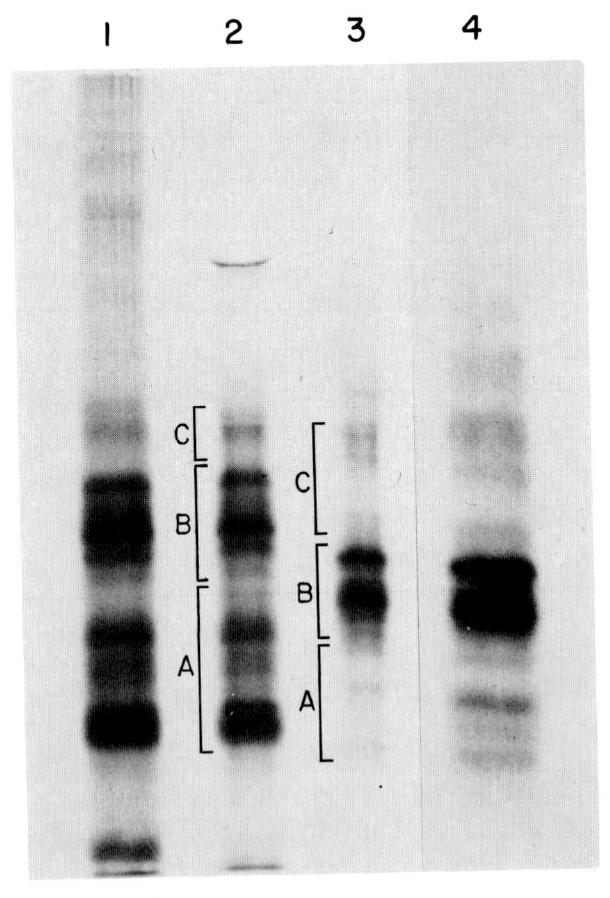

Fig. 3

However, the cell-free products appear to be higher in molecular weight than authentic chorion proteins. Like other secretory proteins, the chorion components appear to be synthesized as precursors bearing a "signal peptide" at their N-terminus (20,21). This hypothesis is supported by detailed comparison of specific molecular weight subclasses of chorion proteins with their putative *in vitro* precursors (unpublished results), using peptide analysis and incorporation of amino acids which are only found in certain protein subclasses (9-11). However, the high complexity of the chorion protein pattern makes rigorous characterization of individual cell-free products difficult. Thus, it is not possible to determine with certainty how much, if any, of the mRNA preparation corresponds to non-chorion mRNA sequences.

For analysis of the developmental controls on specific gene expression, it was clearly essential to obtain homogeneous probes for individual chorion genes. This was accomplished

---

Fig. 3. Cell-free translation of chorion mRNA in the wheat-germ system. Lane 1, total cell-free product. Lane 2, immunoprecipitated cell-free product (note the disappearance of most high M.W. products presumably encoded by endogenous wheat-germ messages, and of the lowest M.W. premature termination products). Lane 3, immunoprecipitated authentic chorion proteins. Lane 4, total authentic chorion proteins (not immunoprecipitated). Chorion mRNA was prepared and translated in the wheat germ system as described (17). The product was precipitated with ice-cold 10% TCA, resuspended in 6 *M* urea, 0.36 *M* Tris-HCl, pH 8.4, 0.05 *M* DTT, and carboxamidomethylated with one-half volume of 1.0 *M* iodoacetamide in 1.2 *M* Tris-HCl, pH 8.4. One aliquot was dissolved in SDS sample buffer for electrophoresis (Lane 1) and another was diluted 1:100 in 0.05 M phosphate, pH 7.5, 0.01 *M* NaCl for immunoprecipitation. To the latter sample 5 μg of unlabeled chorion was added as a carrier, together with a titrated amount of chorion specific antiserum (a gift of Dr. J. R. Hunsley). After an incubation of 2 hr at 25°C and overnight at 4°C, the sample was centrifuged at 13,000 X g for 15 min, and the pellet was washed 5X with immunoprecipitation buffer and dissolved in SDS sample buffer for electrophoresis (Lane 2). Approximately 80% of the radioactivity precipitated. In a parallel reaction, 5 μg of ^{14}C-carboxamidomethylated authentic proteins from mature chorion were immunoprecipitated to the extent of approximately 95% (Lane 3). A, B and C indicate the major size classes of chorion proteins (Lane 3) or their putative precursors observed in the wheat germ product (Lane 2).

by synthesizing double-stranded cDNAs (16) using as template an RNA preparation highly enriched in chorion mRNAs (17), inserting the cDNAs into linearized PML-21 plasmid molecules by the (dA)·(dT) procedure, and cloning the hybrid molecules in E. coli K 12. In this procedure, any one individual clone was expected to yield a probe representing in a highly faithful manner (18,19) only one of the many species of mRNA present in the initial mixture. This expectation was confirmed through detailed characterization of several clones by restriction endonuclease analysis and by cross-hybridization. Thus far, we have identified at least 20 clones as corresponding to distinct sequences (5,6).

Three of the distinct cDNA clones have been subjected to sequence analysis to date (unpublished results). All three clearly correspond to chorion proteins according to three criteria: amino acid content (e.g. very high Gly, high Cys), prevalence of certain repeating peptides (e.g. Gly-Leu-Gly, Cys-Gly) and longer stretches of sequence identity to known chorion proteins.

The cDNA insertion of one of these clones, 401, has been sequenced in its entirety. The coding strand was determined by previously used procedures (18), and the correct reading frame established by the observation that the other two frames result in nonsense codons. Apparently the insertion lacks the portion corresponding to the 5' untranslated region of the mRNA and the beginning of the coding region, including the initiator AUG. The remaining coding region corresponds to 171 amino acid residues (Fig. 4) and is followed by a UAA terminator and 49 additional nucleotides. At least part of the signal peptide appears to be represented: near its beginning the sequence codes for the hydrophobic cluster Ile-Leu-Ile-Leu, which is reminiscent of many signal peptides (22) but is not found in any known mature chorion protein.

The size of the coding region suggests that the 401 sequence codes for a high molecular weight B or a low molecular weight C protein. Hybridization analysis with chorion mRNA from all developmental stages of choriogenesis combined shows that clone 401 corresponds to a very abundant mRNA species, strongly supporting the first alternative (since the Bs are major and the Cs only minor components of the chorion). Direct evidence that 401 codes for a B protein came from comparison with the known chorion protein sequences (Fig. 4). The most extensive B protein sequence known (50 amino acids, from B_6--f_1) shows marked similarity to a portion near the middle of the 401 protein (75% sequence identity).

Our sequencing studies on mature B proteins had previously indicated the structure of only one-third of the molecule.

Fig. 4. Sequence of the protein encoded by chorion clone 401, and comparison to partial sequences of three different chorion proteins, one B and two As (italics). For B_6--f_1, the entire known sequence is shown (11). For $A_{1,1}$--$a_{2,3}$ only the N-terminal segment which shows homology to the 401 sequence is shown. Similarly, for A_1--c_1 only the C-terminal segment homologous to protein 401 is shown. Dots correspond to incomplete sequence, boxes indicate sequences which differ relative to the 401 protein, and overlining identifies repeating peptides.

The 401 sequence has now revealed the rest. Surprisingly, in addition to the segment which defines the 401 protein as a B, we recognize two shorter segments of marked similarity to the known A proteins (Fig. 4). One segment is 33 amino acids long and corresponds to the N-terminus of the As, while the other is 10 amino acids long and corresponds to the C-terminus of the As. These segments are found near each other in the B protein encoded by clone 401, within the C-terminal third of the molecule. By contrast, in the As they are found far apart, i.e. at the two extremes of the molecule separated by a ca. 50 amino acid long segment which bears no resemblance to the 401 sequence (or that of any other known B). Thus, if 401 is a typical B, it appears that As and Bs are homologous, although they are distinct enough to be classified into two different families.

The two segments which are common in As and Bs (at least the 401 protein) are notable for containing tandemly repeating peptides. The first (i.e. the one that in the As is found near the N-terminus), consists mostly of repeats of a pentapeptide, Gly-Leu-Gly-Tyr-Gly (2,8,9) and its variations. The other segment (i.e. the one found at the C-terminus) consists mostly of repeats of the dipeptide Cys-Gly. In this connection, it should be recalled that the chorion consists largely of fibers and that solubilization requires extensive reduction (2). A reasonable hypothesis is that the pentapeptide repeat segment forms a fibrous region, while the Cys-Gly repeat is involved in crosslinking of the fibers. These features of lepidopteran chorion proteins may have been fixed early in evolution, before divergence of the ancestral A and B genes. Needless to say, many specific models may be proposed concerning the evolution of chorion genes, and can be evaluated only after considerably more structural information becomes available. One possible model would be that the original ancestor of As and Bs consisted largely of the two common segments; this gene reduplicated and one copy became the A ancestor after receiving an internal insertion (between the two segments), whereas the other copy became the B ancestor after being spliced to a preceding DNA segment.

The extensive sequence similarities revealed by these studies raise the question whether the chorion cDNA clones will be of any use as pure probes, or whether cross-hybridization will vitiate our purpose in preparing them. We have performed extensive cross-hybridization studies which clearly established the usefulness of the probes. The studies defined the conditions (criterion) at which each sequence is hybridized specifically, i.e. only to itself and possibly to a very limited number of extremely similar sequences (allelic or otherwise).

For these studies, we modified the conventional filter-hybridization technique and developed a simpie but powerful "dot-hybridization" method. Quite simply, we spot on a single nitrocellulose filter disk equal amounts of denatured DNA from various clones, in dots of identical size. After baking and treating the filters with Denhardt's solution and bulk RNA and/or DNA, to prevent non-specific adsorption of nucleic acids, we hybridize the filters with the appropriate probe (usually in 50% formamide, 0.6 M salt, 50°C) and thoroughly wash them. Autoradiography reveals the intensity of hybridization of the various clones. If desired, the filters can be washed at higher temperatures to increase the specificity by melting-off mismatched hybrids (more stringent criterion). Alternatively, the hybridization may be performed at a higher temperature from the outset. The method is extremely useful for comparing simultaneously the extent of hybridization of many independent chorion clones with a single nucleic acid

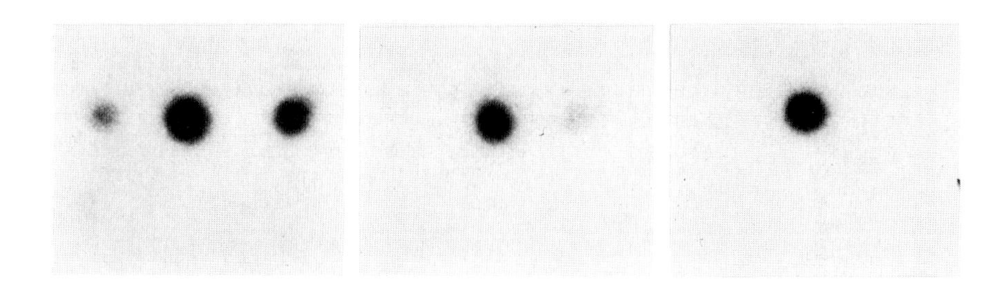

Fig. 5. Dot hybridization of DNAs at different criteria. Autoradiograms of three filters are presented. The radioactive probe was ^{32}P-cRNA made with the chorion insertion of clone 408 as the template. A filter hybridized to low criterion is presented at the left. It contains a row of 3 dots of DNA from chorion cDNA clones 401, 408 and 10 (left to right), plus a single dot of plasmid PML-21 DNA underneath the clone 401 DNA; hybridization was at 50°C in 50% formamide, 0.6 M Na^+. Autoradiograms of two replicate filters are also shown after melting-off mismatched hybrids by a 15-min wash with 50% formamide, 0.3 M Na^+ at 63°C (middle) and 71°C (right). Autoradiographic exposures have been adjusted to make the image of the self-hybrid (408 DNA-408 cRNA) roughly comparable in the three filters. At 63°C the clone 401 and PML hybrids have melted; at 71°C the clone 10 hybrid has melted as well.

probe. Moreover, the sensitivity is high because the autoradiographic signal originates from a very small area (dot).

Figure 5 documents the extent to which three independent chorion clones cross-react. These clones (401, 408 and 10) are the ones which were also subjected to sequence analysis. Clones 408 and 10 are extensively similar to 401 in sequence, so that we can safely classify them as coding for B proteins as well (unpublished results). DNA of the parent plasmid PML-21 was also spotted on the same filters, to serve as control. Filters spotted with these four types of DNA were hybridized with ^{32}P-cRNA, synthesized *in vitro* with *E. coli* RNA polymerase using as template the chorion DNA insertion of clone 408, which had previously been excised from plasmid DNA by the method of Hofstetter et al. (23) and purified by preparative polyacrylamide gel electrophoresis. Under the conditions used, the background hybridization with PML-21 DNA was insignificant even at moderately low criterion (left). The intensity of the self-hybrid, 408 cRNA-408 DNA, was perceptibly higher than that of the cross-hybrids, 408 cRNA-10 DNA and especially 408 cRNA-401 DNA. From the intensity of the spots we conclude that, relative to 408, 10 is a closer homologue than 401. At a more stringent criterion (63°C, middle), background and cross-hybridization with the distant homologue (401) were undetectable. At an even more stringent criterion (71°, right), cross-hybridization with the closer homologue (10) was also eliminated, and only the self-hybrid was detectable.

From follicles pooled according to developmental stage, we have obtained stage-specific crude mRNA preparations by Mg^{++} precipitation of polysomes and oligo(dT)-cellulose chromatography (17). These RNAs were fragmented and end-labeled in vitro with γ-^{32}P-ATP (24), and were used as probes in dot-hybridization experiments. The hybridized filters carried DNAs from 13 distinct clones, which are abundantly represented in total chorion mRNA. As a result of these experiments we now know at what stage of differentiation each of the genes corresponding to these 13 clones is represented in cytoplasmic polyadenylated RNA. Thus we can assert, for example, that 401 is a late gene, whereas 10 and 408 are early and 271 very early genes. Having this knowlege, we can now begin to ask at what level the timing of specific chorion gene expression is controlled (e.g. transcription, post-transcriptional RNA processing, etc.). As a first step, we are assessing the representation of specific gene sequences in pulse-labeled nuclear RNA.

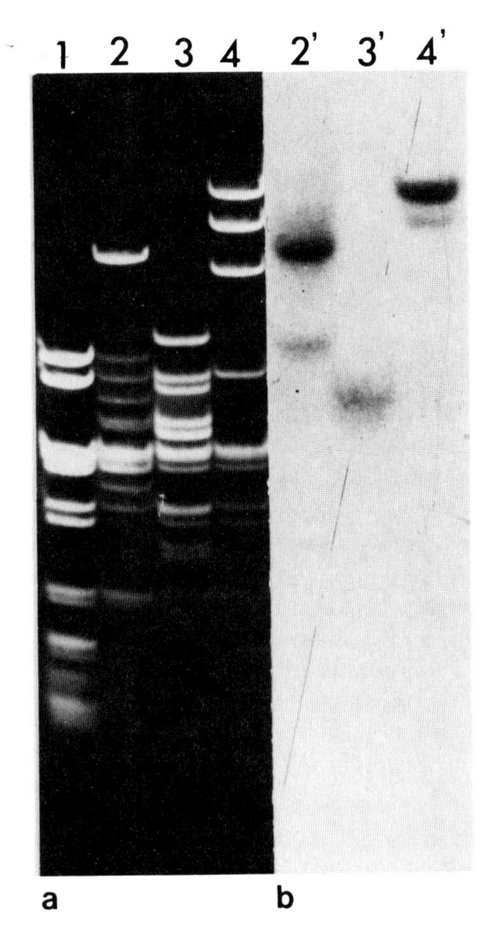

Fig. 6. Identification of fragments of cloned DNA containing sequences homologous to chorion mRNA.

a. Plasmid DNA was purified by equilibrium centrifugation of cleared lysates (27). DNA was digested with the restriction endonuclease Hha I (Bethesda Research Labs) and the products of digestion were electrophoresed on a 1.4% agarose gel. The DNA bands were stained with ethidium bromide and visualized under UV light. Lane 1 RSF-1030; Lane 2 RAP 30G6; Lane 3 RAP 9C5; Lane 4 RAP 30F4.

b. The DNA in the gel was denatured and transferred to a nitrocellulose filter (26). The DNA was fixed to the filter by baking under vacuum for 2 hrs at 80°C. The filter was soaked in a solution containing BSA, ficol and PVP (Denhardt's solution) and 4XSSC at 65° for 12 hrs. mRNA purified from total chorionating follicles (17) was fragmented to approximately 200 bases by brief treatment with alkali and labeled with γ-^{32}P-ATP (24). The labeled RNA was hybridized to the filter in a solution containing 4X SSC, 5X Denhardt's and 10 μg oligo-dT for 12 hrs at 65°. The filter was then washed extensively and used to expose X-ray film. The autoradiogram is shown. Lane 2', 30G6; Lane 3', 9C5; Lane 4', 304F. There was no hybridization to RSF-1030 (not shown).

CHORION CHROMOSOMAL DNA CLONES

We are also planning to use the cloned probes to study the organization of the respective chorion genes in chromosomal DNA (unpublished results). DNA was isolated from testes and very early ovarian follicles, and was inserted into the non-conjugative, ampicillin-resistant plasmid RSF-1030, using the poly(dA)·poly(dT) tailing method. RSF-1030 was chosen as the vector because it can be amplified and is not homologous to the cDNA vector, PML-21. Colonies containing chorion sequences were identified using end-labeled chorion mRNA as the probe in a Grunstein-Hogness screening assay (25). Of the 10,944 colonies screened thus far, 7 proved to have chorion sequences. Restriction analysis followed by transfer of the DNA to nitrocellulose paper and hybridization according to Southern (26), identified specific subfragments which bear chorion sequences (Fig. 6).

For greater completeness, we have also constructed an extensive library of λ phage Charon 9 clones carrying chromosomal DNA of *A. polyphemus* (10^6 recombinant plaques). Large numbers of phage plaques can be screened with ease (28), making it feasible to obtain clones of essentially all chorion genes. We are pursuing the characterization of these plasmid and phage clones, with a view towards understanding the linkage relationships of developmentally and evolutionarily characterized chorion genes and the nature of their associated DNA sequences.

ACKNOWLEDGMENTS

We wish to acknowledge the support of NSF, NIH and the Greek NSF (Hellenic National Research Foundation). We also thank M. J. Randell and L. DeLong for help in the preparation of the manuscript. L.V.K. was supported by the Helen Hay Whitney Foundation.

REFERENCES

(1) M. Paul, M. R. Goldsmith, J. R. Hunsley and F. C. Kafatos. J. Cell Biol. 55 (1972) 653.

(2) F. C. Kafatos, J. C. Regier, G. D. Mazur, M. R. Nadel, H. M. Blau, W. H. Petri, A. R. Wyman, R. E. Gelinas, P. B. Moore, M. Paul, A. Efstratiadis, J. N. Vournakis, M. R. Goldsmith, J. R. Hunsley, B. Baker, J. Nardi and M. Koehler in: Results and Problems in Cell Differentiation, Vol. 8, ed. W. Beermann (Springer-Verlag, Berlin 1977) p. 45.

(3) M. Paul and F. C. Kafatos. Devel. Biol.42 (1975) 141.

(4) F. C. Kafatos, T. Maniatis, A. Efstratiadis, Sim Gek Kee, J. R. Regier and M. Nadel, in: Organization and Expression of the Eukaryotic Genome, eds. E. M. Bradbury and K. Javaherian (Academic Press, Inc. London, 1977) p. 393.

(5) T. Maniatis, Sim Gek Kee, F. C. Kafatos, L. Villa-Komaroff and A. Efstratiadis, in: Molecular Cloning of Recombinant DNA, eds. W. A. Scott and R. Werner (Academic Press, New York, 1977) p. 173.

(6) Sim Gek Kee, A. Efstratiadis, W. C. Jones, F. C. Kafatos, H. M. Kronenberg, M. Koehler, T. Maniatis, J. C. Regier, B. F. Roberts and N. Rosenthal, Cold Spring Harbor Symp. Quant. Biol. 42 (1978) in press.

(7) F. C. Kafatos, in: Control Mechanisms in Development: Advances in Experimental Medicine and Biology, eds. R. H. Meints and E. Davies, Vol. 62 (Plenum Press, New York, 1975) p. 2959.

(8) J. C. Regier, F. C. Kafatos, R. Goodfliesh and L. Hood. Proc. Natl. Acad. Sci. USA 75 (1978) 390.

(9) J. C. Regier, F. C. Kafatos, K. J. Kramer, R. L. Heinrikson and S. Keim. J. Biol. Chem. 253 (1978) 1305.

(10) J. C. Regier, Ph.D. Dissertation, Harvard University (1975).

(11) G. C. Rodakis, Ph.D., Dissertation, University of Athens, Greece (1978).

(12) L. Hood, J. H. Campbell and S. C. R. Elgin. Ann. Rev. Gen., 9 (1975), 305.

(13) M. R. Goldsmith and G. Basehoar, submitted for publication (1978).

(14) L. H. Kedes, Cell 8 (1976) 321.

(15) D. J. Finnegan, G. M. Rubin, M. W. Young and D. S. Hogness. Cold Spring Harbor Symposia, 42 (1978) in press.

(16) A. Efstratiadis, F. C. Kafatos, A. M. Maxam and T. Maniatis. Cell 7 (1976) 279.

(17) A. Efstratiadis and F. C. Kafatos, in: Methods in Molecular Biology 8, ed. J. Last (Marcel Dekker, New York (1976) p. 1.

(18) T. Maniatis, Sim Gek Kee, A. Efstratiadis and F. C. Kafatos. Cell 8 (1976) 163.

(19) A. Efstratiadis, F. C. Kafatos and T. Maniatis. Cell 10 (1977) 571.

(20) G. Blobel and B. Dobberstein. J. Cell Biol. 67 (1975) 835.

(21) G. Blobel and B. Dobberstein. J. Cell Biol. 67 (1975) 852.

(22) A. Devillers-Thiery, T. Kindt, G. Scheele and G. Blobel. Proc. Nat. Acad. Sci. USA 70 (1975) 1154.

(23) H. Hofstetter, A. Schambock, J. Van Den Berg and C. Weissmann. Bioch. Biophys. Acta 454 (1976) 578.

(24) N. Maizels. Cell 9 (1976) 431.

(25) M. Grunstein and D. S. Hogness. Proc. Natl. Acad. Sci. USA 72 (1975) 3961.

(26) E. M. Southern. J. Mol. Biol. 98 (1975) 503.

(27) P. Guerry, D. J. LeBlanc and S. Falkow. J. Bact. 116 (1973) 1064.

(28) W. D. Benton and R. W. Davis, Science 196 (1977) 180.

DISCUSSION

A.C. STOOLMILLER: May I ask about the fidelity of this developmental sequence *in vitro*? You have described that it can be carried out in tissue culture at 25°. Are there temperature extremes beyond which development does not proceed? I would be interested in knowing what those high and low limits are. If one looks at the *in vitro* development near these temperature extremes is the fideltiy of the system still preserved?

F. KAFATOS: We have not done very much work along these lines. All we have shown is that at the physiological temperature of 25°C we have an essentially normal pattern of protein synthesis in terms of the species produced at the different stages and the time that it takes to go from one stage to the next. However, we have also observed that the absolute rate of protein synthesis decreases substantially in tissue culture, to about one-third, and yet the relative rates of protein synthesis and the timing of the program remain approximately normal. So we can say that whatever the process that stabilizes this program of gene expression, it is not very sensitive to the absolute rate of synthesis of specific chorion proteins.

D. ZOUZIAS: I would like to ask if you have done heteroduplex analysis of the chorion DNA clones derived from chromosomal DNA and of the clones derived from chorion mRNA?

F. KAFATOS: We are doing that now. We do not have any results we can report yet.

TRANSCRIPTION AT THE HEAT-SHOCK LOCI OF DROSOPHILA

B.J. McCARTHY, J.L. COMPTON, E.A. CRAIG & S.C. WADSWORTH
Department of Biochemistry & Biophysics
University of California, San Francisco
San Francisco, California 94143

The polytene chromosomes of Drosophila salivary glands respond to elevated temperature or interference with normal respiratory metabolism by inducing a specific set of puffs (1-3). These puffs are sites of intense synthesis of RNA (4-6). New proteins in numbers roughly equal to the number of heat-shock puffs appear in larval and adult tissues (7) and in cultured cells (4). In heat-shocked tissue culture cells pre-existing polyribosomes disappear while new ones, containing RNA complementary to the heat-shock loci, accumulate.

Many advantages are offered by this system as a vehicle for studies of the mechanism of gene regulation in eucaryotes. The response occurs both in polytene and diploid tissues making it possible to combine cytological observations with biochemical analysis. The response is rapid and apparently coordinated for several unlinked loci. Rapid progress has been made in isolating the new messenger RNAs, correlating them with the proteins and assigning them to genetic loci. In addition there exists the possibility of isolating each of the genes by molecular cloning techniques and thoroughly characterizing the mechanisms of transcriptional control. Unfortunately, there is, to date, no information as to the function of the induced proteins.

Induction of Puffing at Heat-Shock Loci of Isolated Nuclei In Vitro

The puffing response is a rather direct demonstration of the relationship between changes in chromosome structure and gene activation. For this reason we sought to obtain conditions in which the heat-shock response could be induced in vitro, thereby providing an assay for the inducing substances and processes. To this end it was necessary to define conditions for the isolation and preservation of polytene nuclei where they would respond to the signal and also retain chromosome morphology so that puffs could be subsequently scored. This can now be accomplished by disrup-

ISBN 0-12-045450-5

ting salivary gland cells in concentrated cytoplasm from cultured Kc cells (8).

Incubating D. melanogaster third instar larvae or their excised salivary glands at elevated temperatures induces nine puffs in the polytene nuclei therefore known as the heat-shock puffs (2). Four of these loci, 63BC, 67B, 87B and 93D, were selected to constitute the assay for in vitro puffing while the others were only informally monitored. The overall experimental design involves first the isolation of Kc tissue culture cell cytoplasm. Cells were either heat-shocked by incubating at 37°C for 30 min or maintained at room temperature as controls. After washing the cells were collected by pelleting in a glass Teflon homogenizer. Concentrated cytoplasm was prepared by homogenization without the addition of buffer and centrifugation at 4°C for 2 min at 8000g. The supernatant was designated control or heat-shocked Kc cell cytoplasm. Salivary glands were disrupted in this cytoplasm in a micro-homogenizer. Incubations were carried out for various periods at room temperature and puffing scored by standard techniques after nuclei were squashed. Sizes of the puffs at heat-shock loci were expressed relative to sizes of adjacent reference loci in order to correct for differences in polyteny among nuclei. The heat-shock loci scored and their corresponding reference loci are as follows: 63BC - 64A1; 67B - 66E1; 93D - 94A1; 87B - 87F1.

An example of the in vitro response obtained is illustrated in Figure 1. A nucleus incubated in heat-shocked cytoplasm shows a puff at 87B, the most prominent heat-shock locus. The same locus fails to puff when polytene nuclei are incubated in control cytoplasm. Similar observations were made for the other loci 63BC, 67B and 93D (8).

The results of such experiments may be expressed quantitatively in terms of the size distribution resulting from a large number of observations made of the four loci in question. The data are displayed graphically either as histograms or in terms of the percent of observations greater than a certain size (8). Alternatively, they may be summarized in tabular form simply as the median values of the set of observations made. The ratios of the median values for parallel experiments with heat-shocked and control cytoplasm thus provide a measure of the differential response to factors unique to the heat-shocked cytoplasm (Table 1).

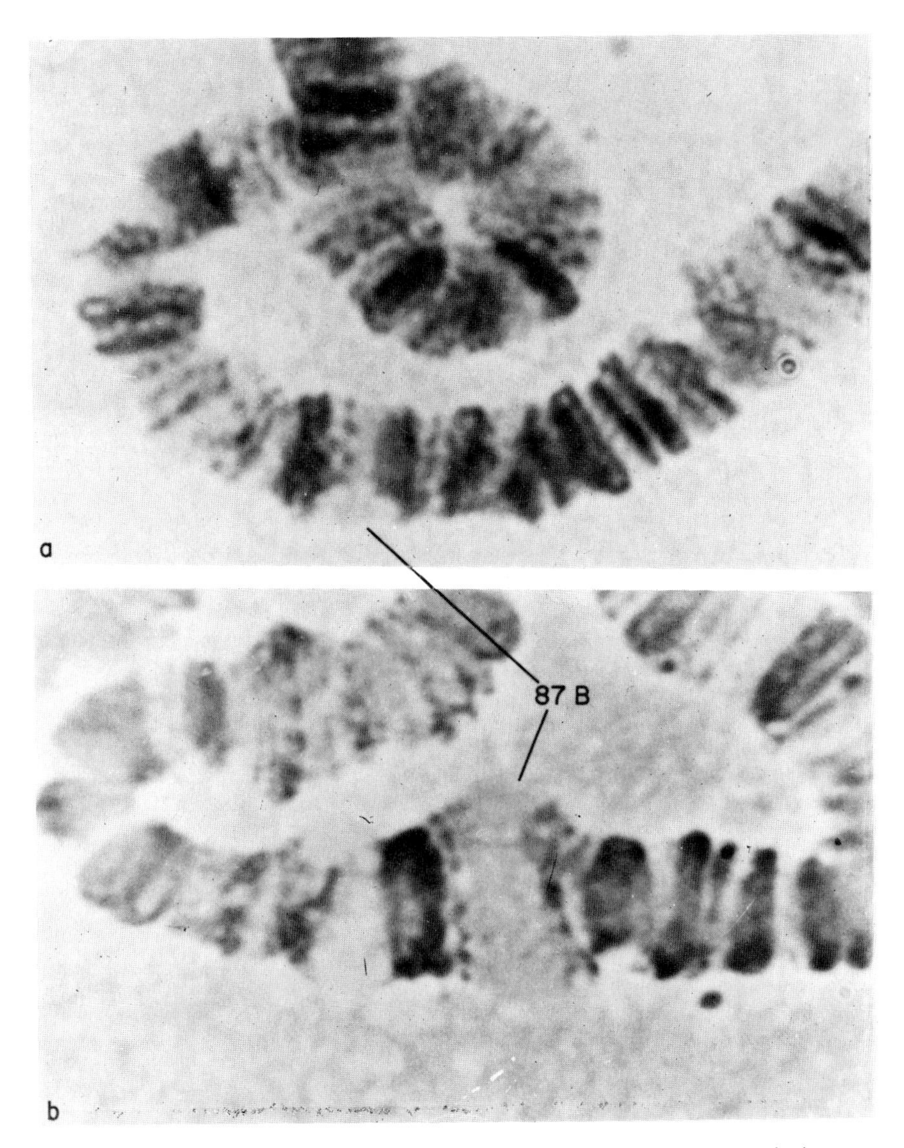

Fig. 1. A portion of chromosome IIIR from nuclei incubated in cytoplasm from (a) control or (b) heat-shocked Kc cells. The 87B heat-shock locus is indicated in each photograph.

TABLE 1

Ratio of the median values of puff sizes for nuclei incubated in heat-shock and control cytoplasm

Time of Incubation	63BC	67B	87B	93D
Zero	1.0	0.7	1.0	1.4
30 min	1.2	0.9	1.3	1.4
60 min	1.4	1.2	1.7	1.4
In vivo, full induction 15-min	1.7	1.4	2.0	1.9

Measurements were made on 30 or more nuclei incubated in heat-shock or control cytoplasm. Diameters of the four puffs were measured relative to their control proximal loci. Medians of the distributions were calculated and used to compute the ratio of heat-shock to control tabulated above.

The data may also be compared to those derived from measurements of nuclei prepared from intact glands, control or heat-shocked, which were dissected immediately prior to fixation. This provides a means of comparing the kinetics and magnitude of the in vitro response with the in vivo response. In general the distributions of sizes for 63BC, 67B and 87B in isolated nuclei after incubation in control cytoplasm are very similar to those in the nuclei prepared from intact glands. Incubating the nuclei for 30 or 60 minutes in cytoplasm from Kc cells heated to 37°C before disruption shifts the distributions to larger values intermediate between the in vivo control and heat-shock levels (Table 1). Figure 2 shows data for 87B from both an in vivo and an in vitro incubation. In contrast 93D does not respond differentially to the normal and heat-shock cytoplasm. Bonner and Pardue (6) have previously noted that 93D is induced to puff in Grace's medium while the other loci remained unpuffed.

A low level of induction is often elicited by control cytoplasm. The extent of induction and the proportion of

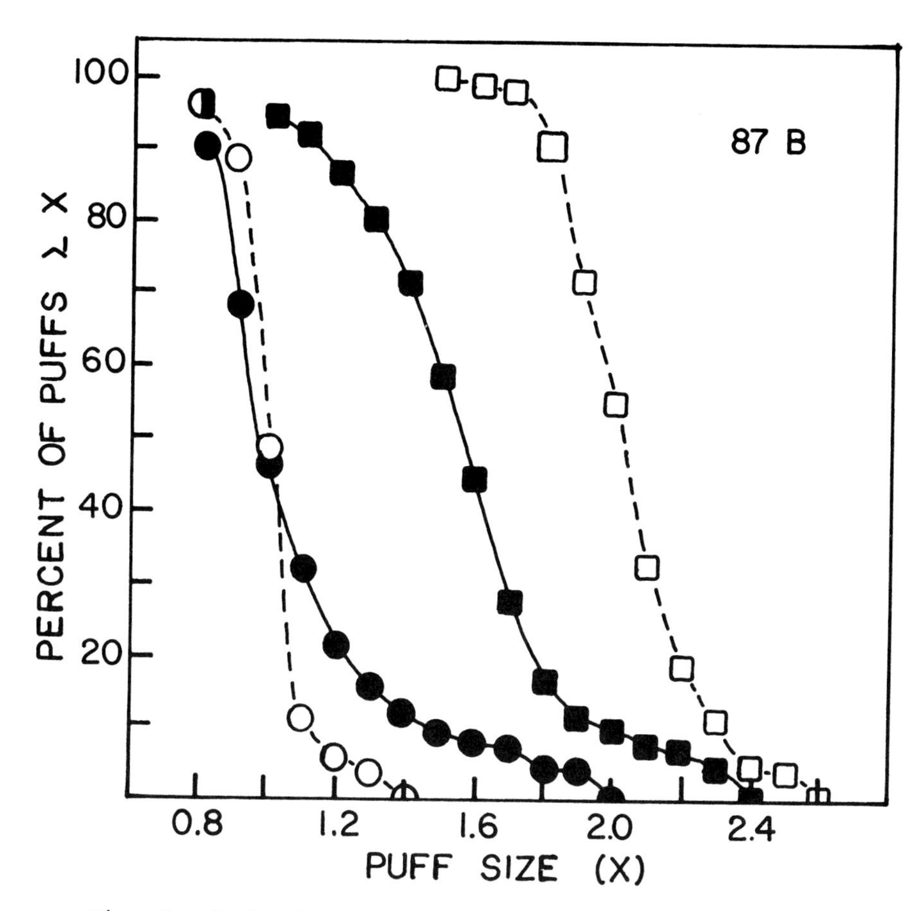

Fig. 2. Induction of the heat-shock puffs in vitro. The size of 87B is expressed relative to that of a reference locus, 87F1. Data are plotted as percent of observations greater than or equal to the sizes on the abscissa. The open symbols represent measurements on glands which were fixed immediately after being removed from larvae maintained at 25°C (o) or incubated at 37°C for 15 min (□). The solid symbols represent measurements of isolated nuclei incubated for 20 min in cytoplasm prepared from Kc cells which had been maintained at 25°C (•) or which had been incubated at 37°C for 30 min immediately before homogenization (■). Each curve represents at least 30 pairs of measurements.

puffed nuclei in control preparations is variable as is the extent to which nuclei puff in response to cytoplasm prepared from heated cells. Despite such variability there is a consistent difference in the response to the two different types of cytoplasm.

The kinetics of the in vitro response also differ from that in glands in vivo. A 15 minute period at 37°C is sufficient to maximally induce the heat-shock puffs in larvae. Furthermore those regress rapidly at 25°C. The sizes of 63BC, 67B and 87B increase progressively in the in vitro system and at about two hours approximate the maximal in vivo induction.

Transcription at Puffs Induced In Vitro

[^{3}H]CTP was added to the Kc cytoplasm at the time that the salivary glands were disrupted in order to test whether the puffs induced in the isolated nuclei were associated with a change in transcription. Autoradiograms of the two nuclei in Figure 3 demonstrate that incorporation of RNA precursors does occur in vitro and that different patterns result from incubation with cytoplasm from control and heat-shocked Kc cells. Preferential incorporation at the heat-shock loci is clearly obvious in the latter case (Fig. 3a). Transcription at specific loci was quantitated by expressing the number of grains over each region relative to the number over a reference region of approximately the same size after subtracting a background count from both. Since we assume that the number of grains appearing is a linear function of time this allows us to compare autoradiograms of different exposures.

Table 2 lists the means and standard deviations for grain counts made on nuclei labeled during incubation in control or heat-shocked cytoplasm and the differences between the means for the two preparations. It should be emphasized, however, that the procedures employed underestimate the difference between an induced locus and its reference region. For, if the autoradiograms were allowed to expose long enough to give more than a few grains in the reference regions the emulsion over the heat-shock loci was saturated. Therefore exposures which were still scorable in the heat-shock regions were used even though the reference regions were at background levels. In the extreme cases where correction for background gave no grains for the reference regions they were assigned the value of one, which automa-

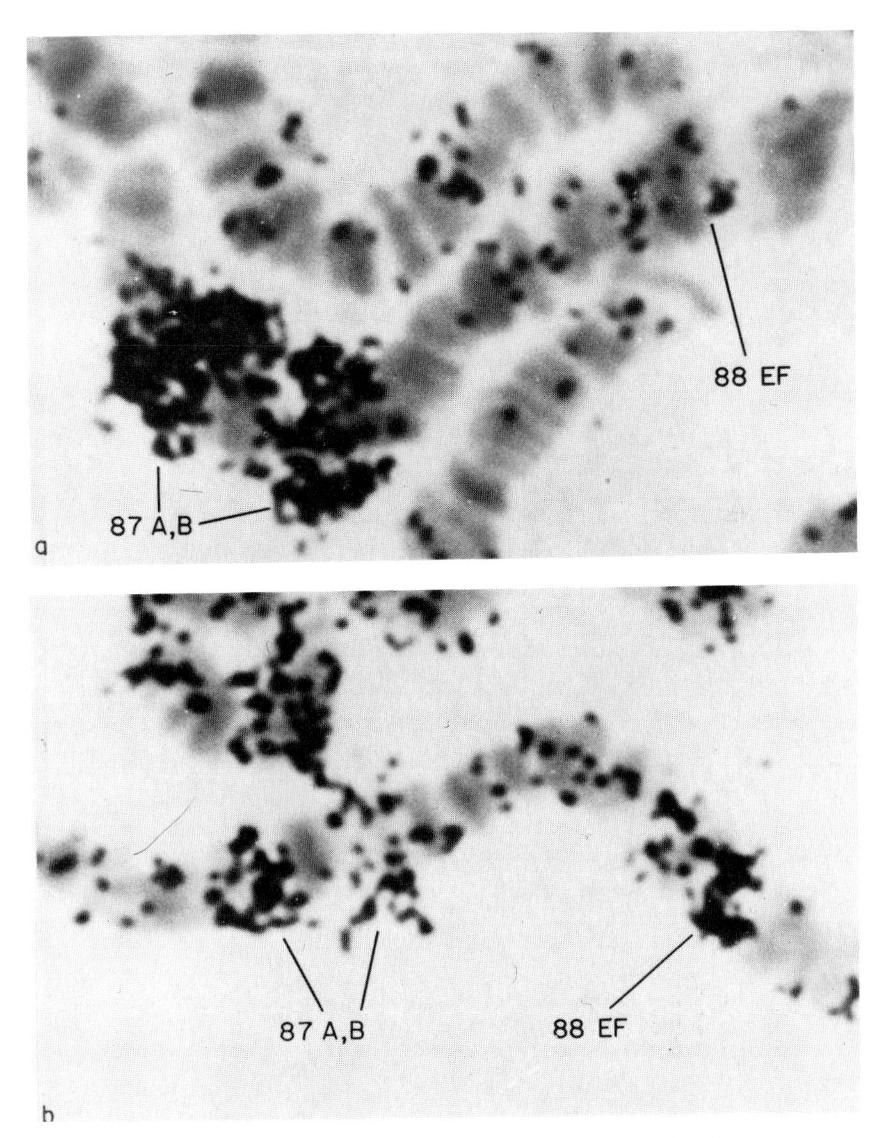

Fig. 3. Transcription of the heat-shock loci in vitro. Nuclei were incubated in vitro in ^{3}H-CTP containing cytoplasm from (a) heat-shocked or (b) control Kc cells. The heat-shock loci and the reference region are indicated in each photograph.

tically underestimates the relative transcription at the corresponding heat-shock locus. We thus feel that the increases in relative transcription between the two preparations shown in Table 2 are minimum estimates.

TABLE 2

Grain counts at specific loci in nuclei labeled during incubation in control or heat-shocked cytoplasm

	63BC	67B	93D	87B
Uninduced	3.8±1.0	3.3±0.9	5.8±1.1	4.3±1.0
Induced	15.0±2.2	23.9±2.5	21.3±2.1	30.7±2.8
Ratio	3.9	7.2	3.7	7.1

The means and standard deviations of the means of data from nuclei such as the two in Fig. 3. Uninduced and induced refer to nuclei incubated in cytoplasm from Kc cells maintained at 25°C and 37°C, respectively.

Some of the control nuclei analyzed by autoradiography appear to be partially induced. The level of background induction is similar to that observed in measurements of puff size. Even in the controls, unpuffed heat-shock loci incorporate some label as is the case for control intact glands. Thus induction at heat-shock loci probably involves a great increase in transcription rather than induction of quiescent genes.

Isolation of Recombinant Plasmids Containing Heat-Shock cDNA

As a first step in isolating and characterizing the heat-shock genes of Drosophila we inserted cDNA representing heat-shock induced messenger RNA into bacterial plasmids. Double-stranded cDNA was synthesized using polyadenylated RNA isolated from large or small heat-shock induced polysomes as template by methods similar to those used by others (9). Since prominent bands were visible on an acrylamide gel after restriction with endonuclease HaeIII, the total cDNA was cut with this enzyme, tailed with synthetic HindIII decamers and cloned in the HindIII site of pBR322. It is not yet clear how many of the heat-shock genes are present in this population of recombinant plasmids. Of the first twelve clones screened, four appear to represent the major loci 87A and 87BC. This conclusion was reached by preparing ^{3}H-cRNA from these plasmids and hybridizing it in situ to Drosophila chromosomes. Other recombinant plasmids are being examined in the same way.

It is also necessary to assign each recombinant plasmid to one of the heat-shock induced proteins. This is accomplished by hybridizing total polyadenylated heat-shock mRNA to the denatured plasmid DNA recovering the hybridized mRNA (10) and translating it in vitro (11). An example of this approach is given in Figure 4 which illustrates the translation of RNA hybridized to plasmid HS4 and HS12 DNA, two of the recombinants whose cRNA hybridized to the loci 87A,B. This experiment proves that the messenger RNA synthesized at these loci encodes the major 70,000 dalton heat-shock protein. The relationship of the 70,000 dalton protein to the 72,000 dalton protein is under investigation.

Isolation of Heat-Shock Genes

Although cloned cDNA is a very useful probe for many purposes, other questions can be approached only when the actual genes are available. We therefore employed the plasmids containing cDNA as probes to isolate homologous sequences from the genome. Drosophila DNA was cleaved with the restriction enzymes EcoRI, HindIII, PstI and BamHI. Aliquots of each were electrophoresed on a 1% agarose gel and fixed to a nitrocellulose filter by the method of Southern (12). Hybridization with ^{32}P nick translated probe made from plasmid HS4 was then used to determine the molecular weight distribution of homologous genome fragments (Fig. 5). A complex pattern was obtained since the genes at 87A,B appear to be some 5-6 fold repeated according to results of C_ot reduction experiments with HS4 DNA. For example two major high molecular bands exist in genome DNA cleaved with HindIII and EcoRI.

Having identified the genome fragments we began to isolate them. Drosophila DNA was cleaved with HindIII and fractionated on a sucrose gradient. The highest molecular weight portion, about 10% of the total was then ligated to HindIII cleaved and alkaline phosphatase treated pBR322. The clones of interest were selected by diluting the trans-

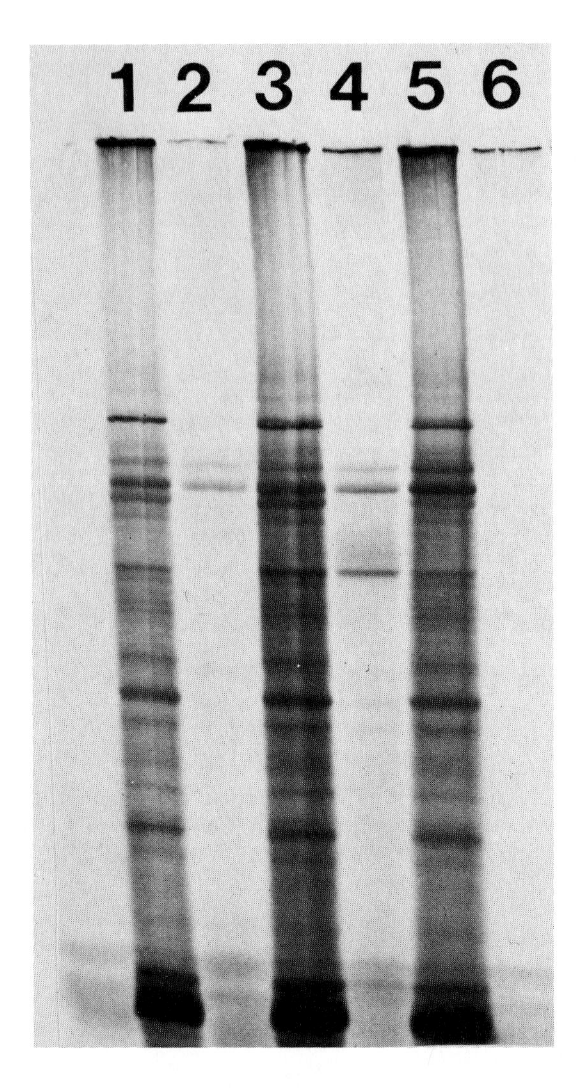

Fig. 4. Selection of heat-shock specific mRNA by cDNA clones. PolyA containing RNA isolated from polysomes of heat-shocked tissue culture cells was hybridized with either pHS4, pHS12 or pBR322 DNA and DNA-RNA hybrids separated from non-hybridized RNA on hydroxylapatite columns. The fraction-ateted RNA was translated in vitro using the cleaved rabbit reticulocyte system. The translation products were analyzed on a 10% SDS acrylamide gel.

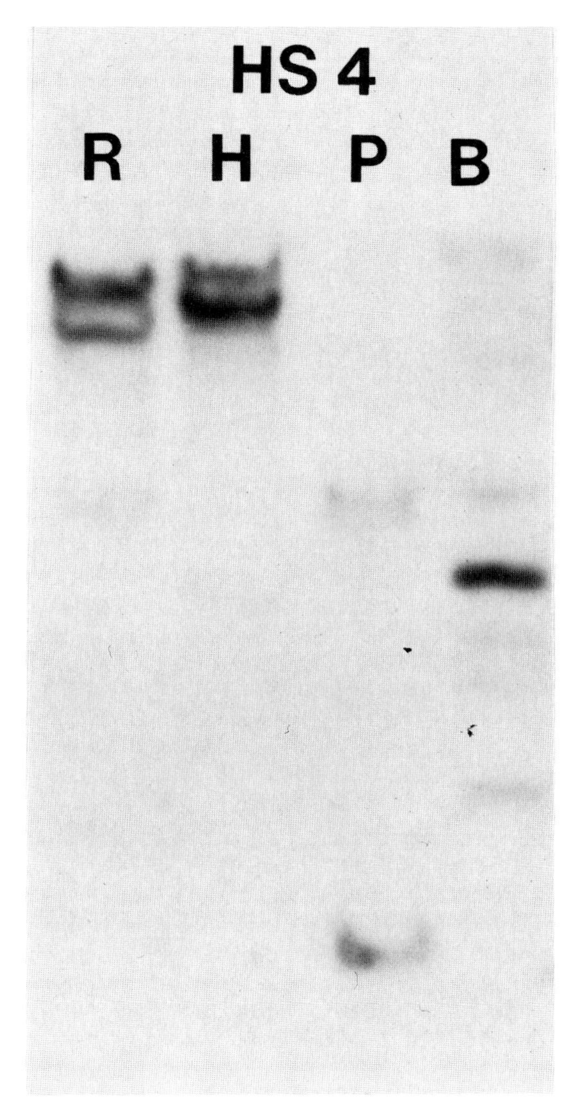

Fig. 5. Detection of fragments of Drosophila DNA containing sequences complementary to portions of the mRNA transcribed from loci 87AB. Drosophila DNA (10 ug) isolated from Kc tissue culture cells was cleaved with the following restriction enzymes: RI (a), HindIII (b), PstI (c), BamHI (d) and electrophoresed on a 1% agarose gel. The DNA was transferred and fixed to a nitrocellulose filter by the method of Southern and hybridized to the recombinant plasmid HS4 which contains a 210 base pair cDNA fragment complementary to DNA in loci 87AB labeled with ^{32}P by nick translation. Molecular

weights were determined by the rate of electrophoresis relative to RI cleaved lambda DNA and HindIII cleaved Ad2 DNA fragments.

---

formed culture and inoculating the 96 wells of each of 15 microtiter plates. The dilution was chosen to give an inoculum of about 20 cells per well. The cultures were grown up and transferred to sheets of Whatman filter paper according to Grunstein & Hogness (13). Colonies were lysed and screened with a nick translated probe made from the insert of plasmid HS4. Of the approximately 30,000 cells screened, two positives were obtained and subsequently isolated. The properties of one of those will be briefly described.

Analysis of Genome Clones

This recombinant plasmid, designated HS4G3, contained an DNA insert of approximately 9.5×10^6. To further confirm its homology with HS4 DNA, the cloned genome DNA was cleaved with HindIII, BamHI and PstI, electrophoresed and transferred to a nitrocellulose filter. The pattern of hybridization with a ^{32}P nick translated probe made from the HS4 cDNA fragment is shown in Figure 6. The pattern of bands is reminiscent of that in the total genome (Fig. 4) particularly the prominent PstI band at 0.7×10^6 and the two BamHI bands at 2.5×10^6 and 2.1×10^6.

Regions encoding messenger sequences were mapped in this DNA by R-looping (14). Total heat-shock mRNA was annealed with HS4G3 DNA under conditions where displacement loops are formed. When the preparation was examined in the electron microscope, three kinds of R-looped molecules were observed. Many contained one R-loop either at the end or in the middle of the DNA. Others contained two R-loops, either near the two ends or at one end and approximately in the middle (Fig. 7). Rare molecules contained three R-loops. It is apparent as well from the structure of the R-loops, that there are no insertions in these genes.

From a study of the electron micrographs and length measurements we conclude that this cloned DNA fragment contains three genes of approximately 2500 base pairs separated by spacers of approximately 1000 base pairs. Although it is homologous to 87A,B, we cannot state at present which locus it represents or whether it contains all the genes at that locus.

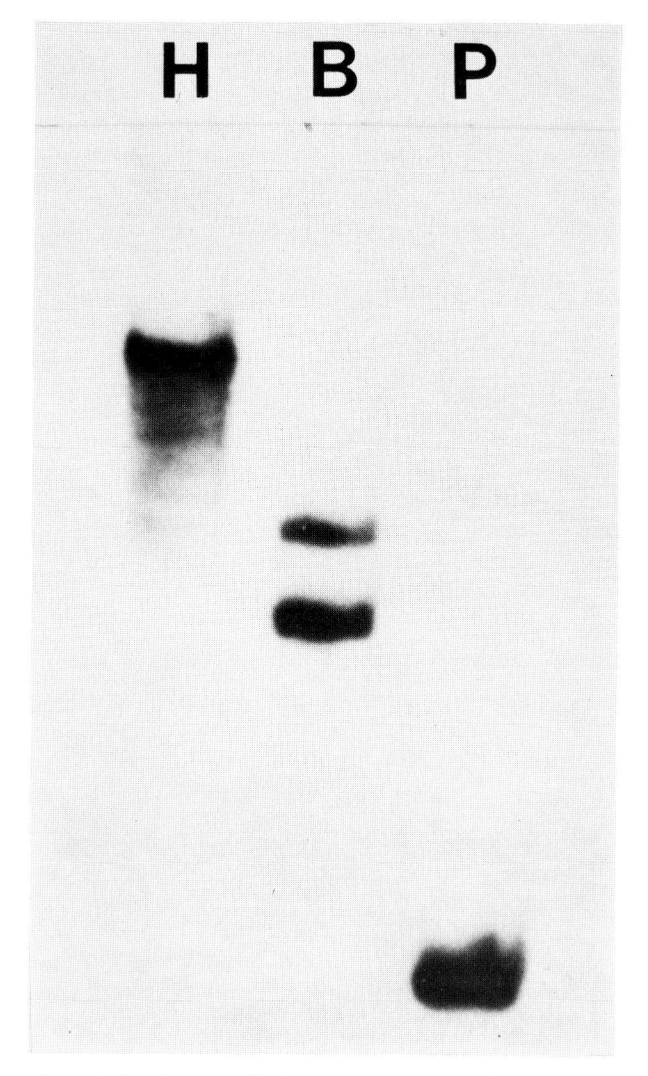

Fig. 6. Detection of fragments containing sequences complementary to portions of the mRNA transcribed from loci 87AB. HS4G1 DNA (100 ng) was digested with the following restriction endonucleases: HindIII (a), BamI (b) and PstI (c) and separated on a 1% agarose gel. The DNA was transferred and fixed to nitrocellulose filters by the method of Southern and hybridized to the isolated HindIII cDNA fragment (2×10^5 cpm) from HS4 labeled with ^{32}P by nick translation. Internal markers for molecular weights (RI cleaved lambda DNA; HindIII cleaved Ad2 DNA) were electrophoresed simultaneously.

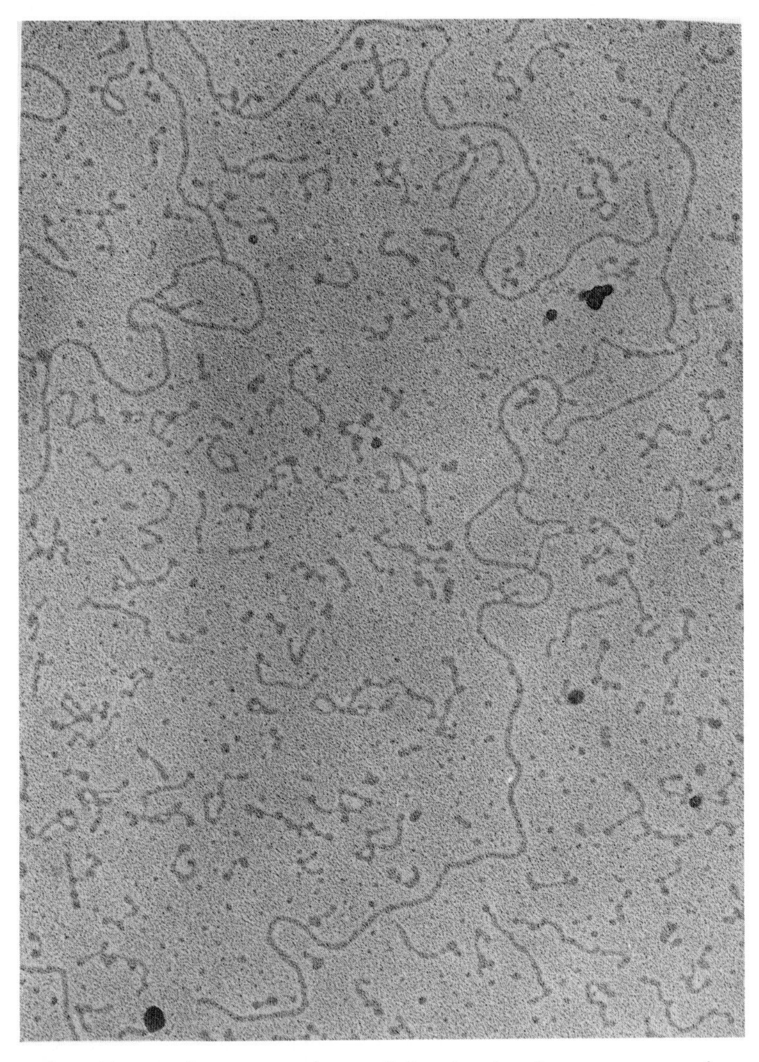

Fig. 7. R-loop mapping of heat-shock sequences in cloned DNA fragments. HindIII cleaved DNA from HS4G3 was hybridized with polyA containing heat-shock mRNA as described by Thomas et al. (15).

Conclusion

The elucidation of the mechanisms of selective control of gene expression is one of the fundamental problems of modern biology. A great deal is now known about several procaryotic systems, but very little is known about the types of control which function in eucaryotes. The complexity of the eucaryotic genome contributes to the difficulty of this work and makes the identification of appropriate model systems imperative. Because of the relatively small size of its genome and the extent of its genetic characterization, Drosophila is an attractive organism for this search. Specifically, the Drosophila heat-shock response seems to be a very useful eucaryotic model system. The results presented here illustrate many of its attractive features.

This research was supported by a grant from The National Science Foundation and postdoctoral fellowships awarded by the NIH (to JLC), the Damon Runyon-Walter Winchell Cancer Fund (to SCW) and the American Cancer Society (to EAC).

REFERENCES

(1) F. Ritossa. J. Cell Biol. 41 (1962) 876.

(2) M. Ashburner, in: Results and Problems in Cell Differentiation, Vol. 4, ed. W. Beerman, J. Reineif and H. Ursprung (Springer-Verlag, New York, 1972) p. 101.

(3) H.J. Leenders and H.D. Berendes. Chromosoma 37 (1972) 433.

(4) S.L. McKenzie, S. Henikoff and M. Meselson. Proc. Nat. Acad. Sci. USA 72 (1975) 1117.

(5) A. Spradling, S. Penman and M.L. Pardue. Cell 4 (1975) 395.

(6) J.J. Bonner and M.L. Pardue. Cell 8 (1976) 43.

(7) A. Tissieres, H.K. Mitchell and U.M. Tracy. J. Mol. Biol. 84 (1974) 389.

(8) J.L. Compton and J.J. Bonner. Cold Spring Harbor Symp. Quant. Biol. 42 (1977) in press.

(9) A. Ullrich, J. Shine, J. Chirgwin, R. Pictet, E. Tischer, W.J. Rutter and H.M. Goodman. Science 196 (1977) 1313.

(10) J. Lewis, J.F. Atkins, C.W. Anderson, P.R. Baum and R.F. Gesteland. Proc. Nat. Acad. Sci. USA 72 (1975) 1344.

(11) B.M. Paterson, B.E. Roberts and E.L. Kuff. Proc. Nat. Acad. Sci. USA 74 (1977) 4370.

(12) R.B. Pelham and J.J. Jackson. Eur. J. Biochem. 67 (1976) 247.

(13) E.M. Southern. J. Mol. Biol. 98 (1975) 503.

(14) M. Grunstein and D.S. Hogness. Proc. Nat. Acad. Sci. USA 72 (1975) 3961.

(15) M. Thomas, R.L. White and R.W. Davis. Proc. Nat. Acad. Sci. USA 73 (1976) 2294.

DISCUSSION

C. WEST: Can you heat shock the cytoplasm after you extract it from the cells, and does it have an effect?

B.J. McCARTHY: The problem is that if you isolate concentrated cytoplasm and then warm it up, it coagulates. So the experiment you suggest is basically impossible. You are dealing with a 25% solution of protein.

C. WEST: Have you been able to remove the mitochondria and observe an effect with cells that have been heat shocked before you have broken them?

B.J. McCARTHY: No, that is something which obviously we are going to try because of reports in the literature that a mitochondrial product is involved in the heat shock induction, but we haven't done the direct experiments.

R. OSHIMA: Do you know if the inducible substances consist of high or low molecular weight material - is it dialyzable from that cytoplasm?

B.J. McCARTHY: We don't know if its high or low molecular weight. We haven't done even the obvious dialysis exeriment.

T. MADEN: Are there any genetic data on the heat shock loci that might give information on the physiology of the response?

B.J. McCARTHY: There are no mutants which fail to respond to the heat shock. The only kind of genetic variants that exist are deficiencies which cover either 87A, 87B or both. That is part of the evidence that these loci code for the 70,000 dalton protein. But nobody to my knowledge has a mutant in responsiveness. In fact it's not obvious how one would select for it. If you have an idea I'd like to know about it.

T. MADEN: What is the phenotype of the mutants which are defective. You mentioned deletion mutants.

B.J. McCARTHY: If you delete one of the twin bands at 87, the mutant is viable. However the deletion of both of the bands leads to a homozygous lethal condition. Apparently, there are essential functions coded in the bands between 87A and 87B.

K. LATHAM: I was wondering if the relative oxidation state of the cytoplasm could be a physiologically relevant mechanism of regulation at the chromatin level, especially in light of the data that are known about the induction of enzymes just after birth in mammals when the oxidation state of the organism would change quite drastically.

B.J. McCARTHY: There are a lot of precedents in the literature suggesting that oxidation-reduction phenomena can affect enzyme activity. For example, there are cases where superoxide dismutase has been implicated as a controlling element. For example, superoxide dismutase can interact with guanylate cyclase and increase its activity. We also know that the cytoplasm is full of catalase because when we add peroxide it bubbles and the peroxide disappears.

J. MORRISSEY: Do you have any evidence that would rule out the possibility that there is really just a heat labile repressor protein that is being destroyed when you do the heat shock?

B.J. McCARTHY: No. Nothing direct at all.

STRUCTURE AND FUNCTION OF XENOPUS RIBOSOMAL GENES

R. H. REEDER, B. SOLLNER-WEBB, R. HIPSKIND,
H. L. WAHN AND P. BOTCHAN
Department of Embryology
Carnegie Institution of Washington
115 West University Parkway
Baltimore, Maryland 21210

Abstract: We review three lines of research which are aimed at understanding transcriptional regulation of the genes coding for 40S precursor rRNA in the frog *Xenopus*. First we describe a method for selectively labeling the 5' initiation terminus of the 40S precursor molecule. Using such terminally labeled precursors we have mapped the site of transcription initiation on the rDNA and can show that the initiation site is preceded by at least 200 bp of DNA whose sequence is conserved. We also find a limited heterogeneity at the 5' end of the precursor molecule.

In a second project we have isolated active rDNA chromatin from the amplified nucleoli of oocytes and have examined the histones on these transcriptionally active genes. Histones H_4, H_{2a}, H_{2b} and H_3 are present and so far cannot be distinguished from their counterparts on bulk somatic cell chromatin. The H_1 histones are not detected. We propose that removal of H_1 is necessary for gene transcription but secondary modifications of the four nucleosomal histones are not needed.

And finally, we describe a cell-free *in vitro* system in which *Xenopus* polymerase I will accurately reinitiate transcription on endogenous ribosomal genes. Use of this system should make it possible to precisely define all the molecular components needed for accurate transcription of the ribosomal genes.

INTRODUCTION

One of the major unsolved problems in present-day biology is the problem of how gene transcription is regulated in animal cells. We have chosen to attack this problem by focussing our research on a single type of gene -- the gene that codes for the large 40S precursor rRNA in the frog,

ISBN 0-12-045450-5

Xenopus. In this article we will review our current understanding of the structure of this gene and the proteins that interact with it during transcription.

BRIEF REVIEW OF rDNA STRUCTURE

Xenopus rDNA is an example of a multi-gene family (for review see ref. 1). In somatic cells the gene is tandemly duplicated about 450 times at a single locus (the nucleolar organizer locus) on one of the 18 haploid chromosomes. During oogenesis the rDNA also undergoes massive amplification and is packaged in thousands of extrachromosomal nucleoli which we will discuss later. For most of the work described in this article we have used various parts of the rDNA that have been spliced into bacterial plasmids and cloned in order to obtain large amounts of homogeneous material.

Figure 1 shows one repeating unit of rDNA. A repeating unit contains a transcribed gene region of 7.5 kb and a non-transcribed spacer region whose length may vary from 3 kb to more than 7 kb (2). The gene region is transcribed as a single precursor molecule that sediments at 40S and which is then processed into the mature 5.8S, 18S and 28S rRNA's of ribosomes. Transcription proceeds in the direction 18S → 5.8S → 28S (3,4). The non-transcribed spacer region is largely composed of repetitive simple sequence DNA (5). Spacers vary in length by virtue of containing more or fewer of these simple repetitive elements. Sequences in the gene region are under rigid selection pressure as shown by the fact that Xenopus rRNA shows some sequence homology with eukaryotic rDNAs as distantly related as spinach and pumpkin (6). Spacer regions, in contrast show detectable sequence variation even within the species Xenopus laevis (5).

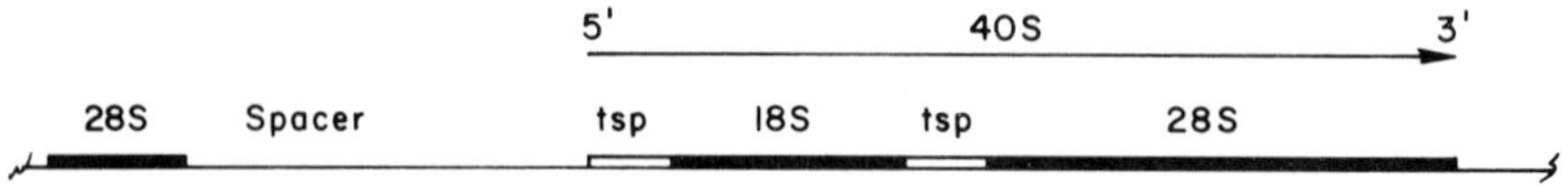

5' 40S 3'

28S Spacer tsp 18S tsp 28S

Fig. 1. Model of one repeating unit of X. laevis ribosomal DNA.

IDENTIFICATION OF THE PRIMARY TRANSCRIPT

One goal of our research has been to identify the promoter on rDNA and study the sequences which are necessary for RNA polymerase recognition. To do this we needed a method for identifying the 5' end of the primary transcript. It has been shown that when RNA polymerase initiates transcription *in vitro* it leaves a triphosphate terminus on the 5' end of the RNA chain (7). Therefore we expected that the primary transcript of *Xenopus* rDNA would have a triphosphate terminus. Moss (8) recently pointed out that the capping enzymes isolable from Vaccinia virions can be used to recognize such termini since they catalyze the reaction $pppG + SAM + (p)ppXpYpZp...... \rightarrow {}^{7m}GpppX^{m}_{p}YpZp......$. The vaccinia enzymes will cap any RNA that has a di or triphosphate terminus and thus offer a means to put a specific radioactive tag on the 5' end of any primary transcript. The Vaccinia capping enzymes will not cap a monophosphate or hydroxyl terminus and thus should ignore any ends due to secondary processing or random nuclease action.

Using the Vaccinia capping enzymes we have been able to show that a large fraction (up to 22% in some experiments) of the 40S precursor molecules have polyphosphate termini and can be capped *in vitro* (9). Digestion of the radioactively capped 40S RNA with base specific RNases shows that the majority of the cappable 40S molecules begin with the sequence (p)ppAAG.

From these results we conclude that the 40S rRNA represents the true primary transcript of *Xenopus* rDNA.

SEQUENCE ARRANGEMENT NEAR THE INITIATION SITE

In order to learn more about the DNA sequence near the initiation end of the gene we have mapped a number of restriction endonuclease sites in this region.

Figure 2a shows the result when SmaI sites were mapped on four different cloned fragments, each of which had a spacer of a different length. Starting at the EcoRI site within the 18S region and proceeding to the left, the SmaI sites are conserved in all four clones until a point about 0.9 kb from the Bam HI terminus in the spacer. At that point there is a conserved, non-conserved, boundary beyond which the SmaI sites are at variable locations. This conserved, non-conserved boundary coincides with the boundary

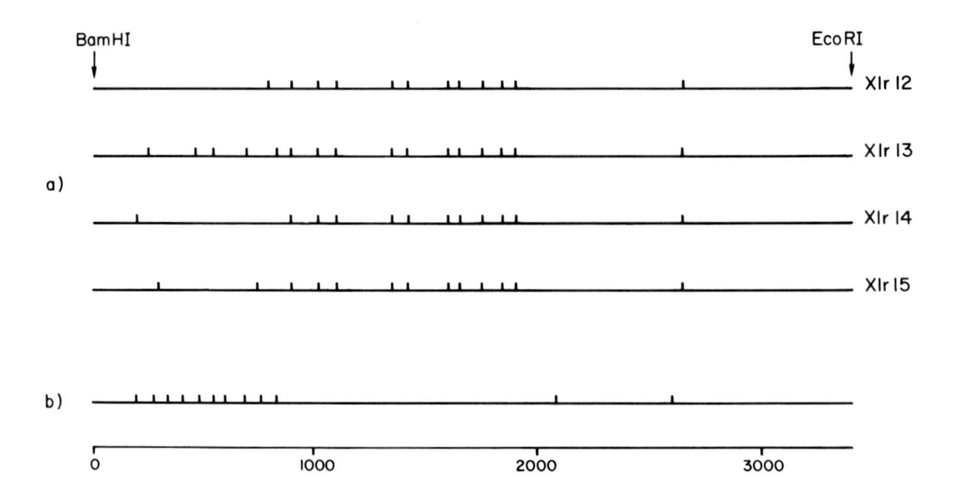

Fig. 2. Mapping of restriction sites near the 5' end of the 40S coding sequence.

a) Endonuclease SmaI sites mapped on four independently cloned fragments of rDNA. Going from the gene leftward into the spacer, cutting sites are conserved up to a point 0.9 kb from the Bam HI terminus and are not conserved beyond that point.

b) Mapping of AluI sites near the 5' end.

This endonuclease recognizes the repetitive element in the spacer.

---

between repetitive, simple sequence DNA and non-repetitive DNA. This is shown in Figure 2b where cutting sites for the endonuclease AluI are mapped. This enzyme recognizes a short repeating element of ∿60 bp in the region immediately to the left of the conserved, non-conserved boundary seen by SmaI.

LOCUS OF TRANSCRIPTION INITIATION ON rDNA

Previous mapping with the electron microscope has shown that 40S precursor transcription also initiates close to the repetitive non-repetitive and conserved, non-conserved boundary shown in Figure 2 (see ref. 5 for discussion). To define the site of chain initiation more closely we have tested the ability of radioactively capped 40S rRNA to hybridize to various restriction fragments spanning this region. A summary of several such experiments is shown in Figure 3. The 40S terminus hybridizes within a 200 bp SmaI

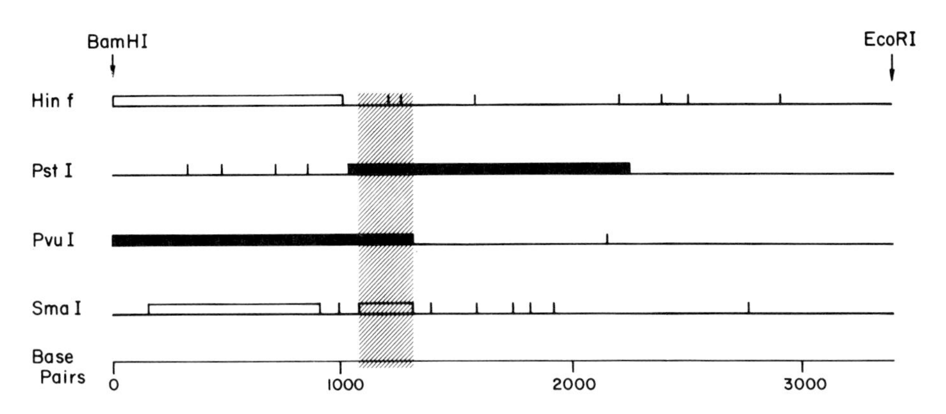

Fig. 3. Mapping the 5' end of the 40S sequence. The heavy lines indicate fragments which hybridize radioactively capped 40S RNA. Empty boxes indicate fragments that were tested but did not hybridize to the capped terminus.

fragment that is located at least 200 bp to the right of the conserved sequence and repetitive DNA boundary. Fragments produced by other enzymes will hybridize only if they overlap the 200 bp SmaI fragment. We conclude, therefore, that the site of chain initiation is preceded by at least 200 bp of a conserved, non-repetitive sequence whose function at present in unknown. A plausible speculation, however, is that this region might serve as a loading zone for RNA polymerase. Such a zone might be necessary to maintain the high polymerase density seen on Xenopus ribosomal genes (10).

Another unexpected result to emerge from these experiments is that the 5' end of the 40S rRNA is heterogeneous. Labeled termini hybridized to the 200 bp SmaI fragment were recovered, digested with RNase A (specific for C and U) and the fragments separated by charge on a DEAE column. As shown in Figure 4b, at least two different 5' terminal fragments were recovered, one about 9 nucleotides in length and the other about 10 in length. These two oligonucleotides are also two major species observed when capped 40S

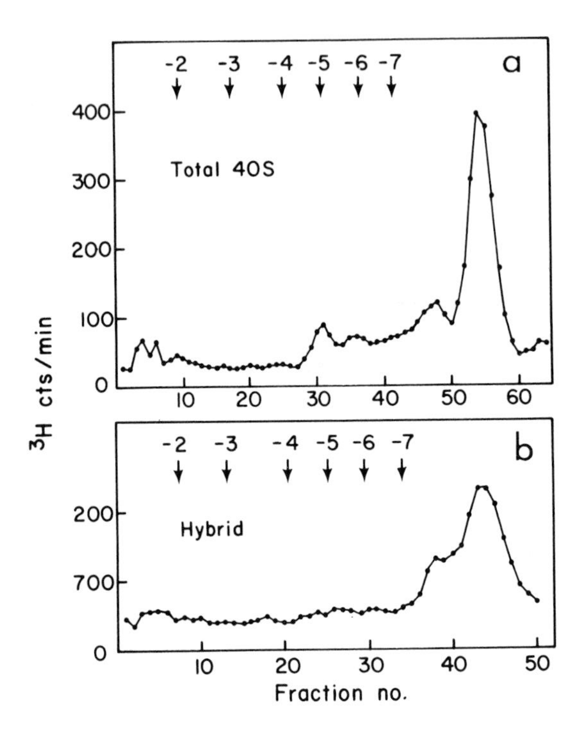

Fig. 4. RNase A oligonucleotides recovered from radioactively capped 40S RNA.

a) 40S RNA isolated from oocyte nucleoli by sucrose gradient centrifugation was radioactively capped, digested with RNA A, and the digest fractionated according to chain length on a DEAE column.

b) Capped 40S RNA was hybridized to the 200 bp SmaI fragment spanning the 5' end of the 40S sequence (see figure 3). The hybrid was recovered, digested with RNase A, and the digest fractionated as in a).

---

rRNA is digested with RNase A before hybridization (Figure 4a). Since the hybridization was to a cloned DNA fragment, it is likely that both termini are encoded by all 40S genes. It suggests that there are at least two alternate start sites for the RNA polymerase.

With the advent of rapid sequencing techniques it is now possible to sequence large stretches of DNA. It is still very difficult to know what significance to attach to most of these sequences. We believe that the above experiments lay a solid groundwork on which to interpret the nucleotide sequence studies of _Xenopus_ rDNA which are now in progress.

ISOLATION OF rDNA CHROMATIN

In the _Xenopus_ oocyte rDNA undergoes a massive and specific amplification to produce a nucleus containing over twice as much rDNA as it does bulk chromosomal DNA. This amplified rDNA is packaged in extrachromosomal nucleoli which can be isolated completely free of any bulk DNA contaminant (11). The isolated nucleoli have many of the characteristics expected of active ribosomal genes. They contain RNA polymerase I activity, 40S precursor rRNA, and in thin section show the morphology typical of active nucleoli. An unexpected result is that they also contain a nicking-closing enzyme activity that may play a role in transcription (11).

MEASUREMENT OF THE FRACTION OF ACTIVE rDNA.

Before we can use these isolated nucleoli to study active rDNA chromatin we need some quantitative measurement of the fraction of the genes that are transcribed. Our approach to this problem is based on the following considerations: 1) An active gene has 100-120 RNA polymerase molecules on it (10). Assuming each polymerase is about 500,000 daltons of protein, this means 50-60 x 10^6 daltons of polymerase protein per DNA repeat (gene plus spacer) of about 8 x 10^6 daltons. This suggests a very high protein to DNA ratio for an active ribosomal gene. 2) The detergent Sarkosyl will remove most proteins (including histones) from chromatin but will not remove RNA polymerase that is chain elongating (12; unpublished experiments of R. Reeder and H. Wahn). 3) In bouyant density gradients of Metrizamide (a non-ionic density gradient material) protein is dense while DNA is relatively light (13). Therefore, we predict that after treatment with Sarkosyl rDNA from amplified nucleoli will band at a high density in Metrizamide if it is loaded with RNA polymerase and will band at a lower density if it is inactive and not loaded. Figure 5 shows that this prediction is correct. After Sarkosyl treatment the rDNA bands in two fairly sharp peaks. The dense peak contains 75% of the rDNA and 100% of the RNA polymerase activity. The lighter peak contains 25% of the rDNA and no polymerase activity. Photographs of active amplified rDNA (10) show

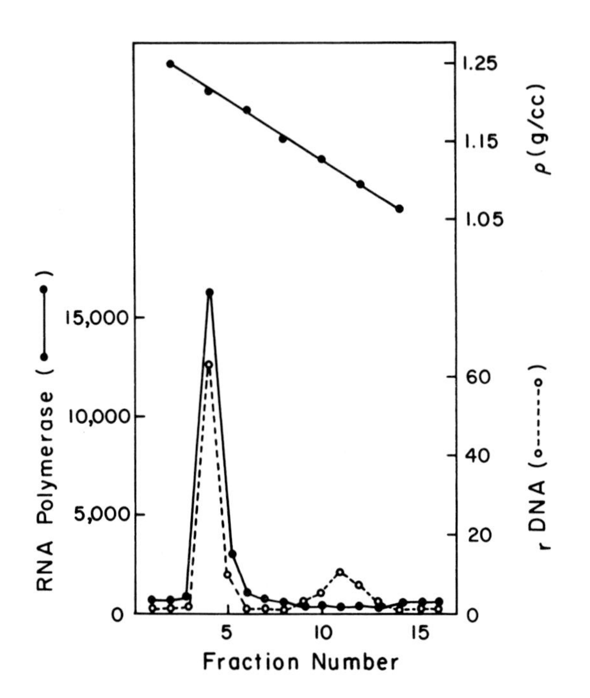

Fig. 5. Separation of active from inactive ribosomal DNA in Sarkosyl-Metrizamide gradients.

Isolated oocyte nucleoli were suspended in 0.5% Sarkosyl, layered over a preformed Metrizamide bouyant density gradient and centrifuged to equilibrium. After fractionation, aliquots of each fraction were assayed for rDNA content (O--O--O), RNA polymerase activity(●—●—●) and density (●—●—●). In this experiment 75% of the rDNA was in the active fraction.

---

that inactive genes are not interspersed among active genes. Therefore, we interpret this experiment to mean that in this preparation at least 75% of the genes were loaded with polymerase, i.e., active by our definition. We conclude that the large majority of the ribosomal genes in our preparations are active.

HISTONES ON ACTIVE rDNA

Active nucleoli contain proteins which we identify as the four nucleosomal histones (H_4, H_{2a}, H2b and H_3) by the following criteria: 1) They are acid soluble, 2) they comigrate with authentic histones from Xenopus cultured cells on SDS-acrylamide gels and 3) partial peptide maps of the nucleolar histones are identical with those of authentic histones (H. Wahn, unpublished). We have so far been unable to locate any protein that comigrates with cultured cell H_1 histones. We have begun to examine the nucleolar histones for secondary modifications.

Figure 6 shows a two dimensional acrylamide gel in which iodinated nucleolar proteins were mixed with unlabeled cultured cell histones. The normal H_1 histones are absent from the nucleolar proteins while H_4, H_{2a}, H_{2b}, and H_3 are all present and appear to co-migrate with the unlabeled markers. Differences in intensity are due to the fact that some histones iodinate better than others.

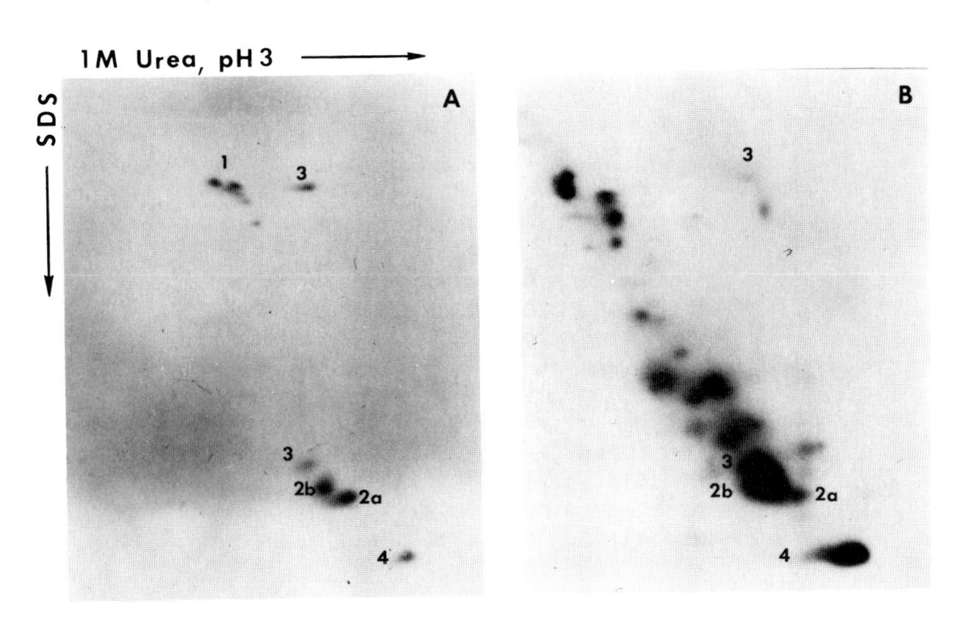

Fig. 6. Comparison of histones from nucleoli and cultured cells. An acid extract from nucleoli was labeled with ^{125}I and then mixed with unlabeled histones from cultured *X. laevis* cells. The mixture was then subjected to 2-D gel electrophoresis.

a) Cultured cell histones stained with Coomassie Blue.

b) Autoradiograph of ^{125}I labeled nucleolar histones.

In other experiments (data not shown) we have examined nucleolar H_4 on high resolution 2-D gels. The nucleolar H_4 migrates in the position of zero and mono-acetyl H_4 which is also true for H_4 from cultured cell chromatin. We find no evidence for hyper-acetylation of H_4 as is seen in cells treated with n-butyrate (14). This experiment is especially interesting in view of the fact that oocytes make and store large amounts of H_4. The H_4 in this stored pool all migrates as di-acetyl H_4 (15). This affords strong evidence that the histones we find in nucleoli belong there and are not artifactually picked up from the stored pool during the isolation procedure.

Our data lead us to believe that the major difference between active and inactive chromatin is the absence of histone H_1 on the former. The presence of secondary modifications on the four nucleosomal histones may not be obligatory for transcription.

TRANSCRIPTION OF rDNA IN ISOLATED GERMINAL VESICLES

Study of the regulation of eukaryotic transcription has been severely hampered by the lack of *in vitro* systems in which the RNA polymerases will initiate transcription with fidelity. One of the more promising systems for studying eukaryotic transcription appears to be the germinal vesicle (nucleus) of the *Xenopus* oocyte. Gurdon and collaborators have demonstrated transcription from various DNA's microinjected into the germinal vesicle (16). Although the results obtained are dramatic, the injection method seemed too laborious to use it to identify and study the molecular components of an accurate transcription system. To make the system more amenable to standard biochemistry we have tried manually isolating germinal vesicles from 50 to 100 oocytes at a time, homogenizing them with a glass rod, and running transcription reactions in a test tube.

We have studied the way in which the crude system transcribes endogenous rDNA already present in the germinal vesicle; we have also examined the effect of adding exogenous purified rDNA to the reaction. Figure 7a shows the time course of both the endogenous reaction and the reaction with added DNA. The endogenous synthesis continues for at least seven hours. At the end of that time the system had synthesized 115 RNA copies for each ribosomal gene present. (The calculation makes the worst assumption that the nuclei contain no significant triphosphate pools.) Such a large

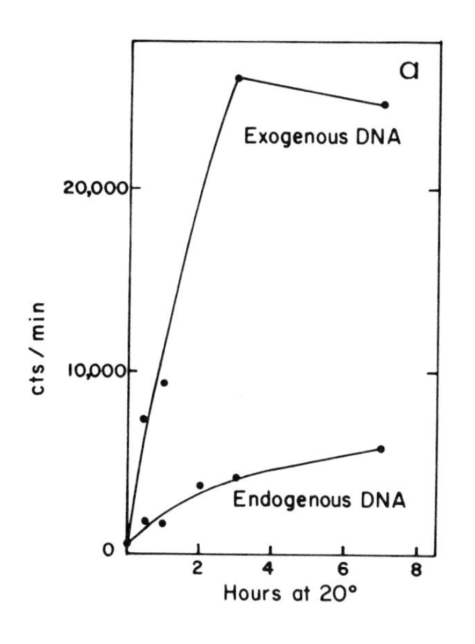

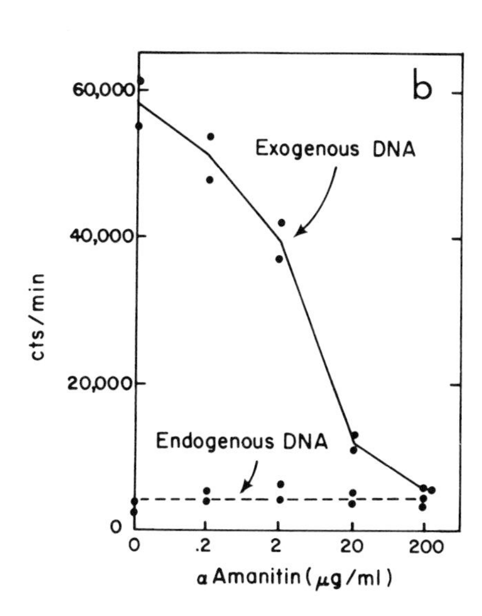

Fig. 7. RNA synthesis in oocyte germinal vesicle homogenates.

a) Time course of the reaction with and without addition of exogenous rDNA.

b) Effect of α-amanitin on the endogenous and exogenous reactions.

---

amount of RNA synthesis suggests that RNA chains are being re-initiated during the reaction. Addition of exogenous rDNA (about 125,000 pg per nuclear equivalent) resulted in a large stimulation of RNA synthesis. However, as we will show later, this stimulated synthesis is mostly aberrant.

To determine which RNA polymerase was responsible for the observed transcription we titrated both the endogenous and exogenous reactions with α-amanitin. Figure 7b shows that the endogenous reaction was completely resistant to α-amanitin which suggests that we are only detecting polymerase I transcription on amplified nucleoli. The exogenous reaction, however, seems to be due almost entirely to polymerase III since it is sensitive to α-amanitin in the range 0.2 to 200 μg/ml. Addition of 200 μg/ml α-amanitin brings the exogenous reaction down to the endogenous level again.

Both reactions were tested for their ability to select the coding (H) strand as opposed to the non-coding (L) strand for transcription. Figure 8 shows that the endogenous product hybridized exclusively to the H strand while the exogenous product hybridized symmetrically to both.

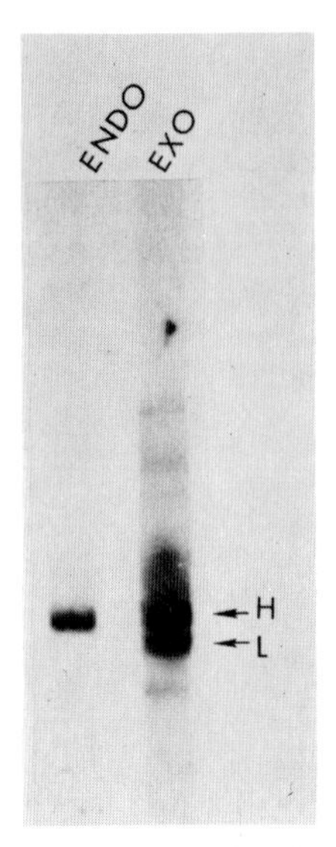

Fig. 8. Strand selection during *in vitro* transcription.

RNA made in the endogenous reaction or with exogenously added rDNA was hybridized to strand-separated rDNA. The endogenous product hybridized only to the coding (H) strand while the exogenous product hybridized to both strands.

We have also titrated the system with DNA to determine how much is needed for stimulation of synthesis. Figure 9 shows the interesting result that at lower levels of exogenous DNA (12.5 to 12,500 pg/nucleus) the system ignores the added DNA and does not transcribe it. Only when a large excess is added (125,000 pg/nucleus) does polymerase III begin to aberrantly transcribe.

In sum, these transcription experiments suggest that our crude system contains everything necessary for the endogenous system to re-initiate rRNA synthesis with fidelity. For reasons we do not yet understand polymerase I and II largely ignore exogenously added rDNA. At very high levels, however, polymerase III will transcribe the added DNA in an

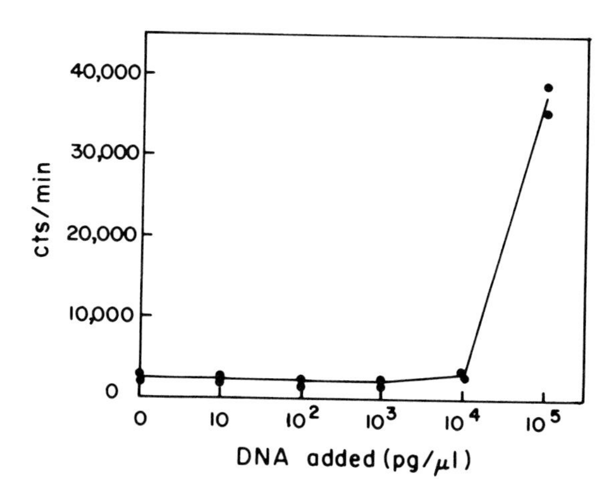

Fig. 9. Titration of germinal vesicle homogenate with exogenously added rDNA.

aberrant manner.

We felt it was important to obtain direct evidence for endogenous re-initiation of rRNA chains in this system. To do this we ran the endogenous reaction in the presence of unlabeled γ-thio-ATP and γ-thio GTP. The reaction also contained α-^{32}P-CTP to label RNA chains internally and unlabeled UTP. Others have shown that the γ-thio analogs of ATP and GTP are incorporated at normal rates into RNA chains and result in a reactive SH group at the 5' end of newly initiated RNA (17). After stopping the reaction the RNA was passed over a mercury agarose column to retain all molecules that had an exposed SH. The bound RNA was then eluted with dithioerythritol, an SH reducing agent. As shown in Figure 10a about 4.5% of the RNA made in the presence of the γ-thio-analogs stuck to the column. This was 25 fold more RNA than stuck when RNA was made in the absence of thio-triphosphates (Figure 10b). If the thio-groups are being incorporated at the 5' end we would expect that RNA bound to the column would hybridize preferentially to the 5' end of the gene as compared to RNA that did not bind to the column. Figure 11 shows that this result was obtained. At present we do not know the chain length of the RNA nor how stable the γ-thio-phosphate group is to hydrolysis. Therefore we cannot yet predict the

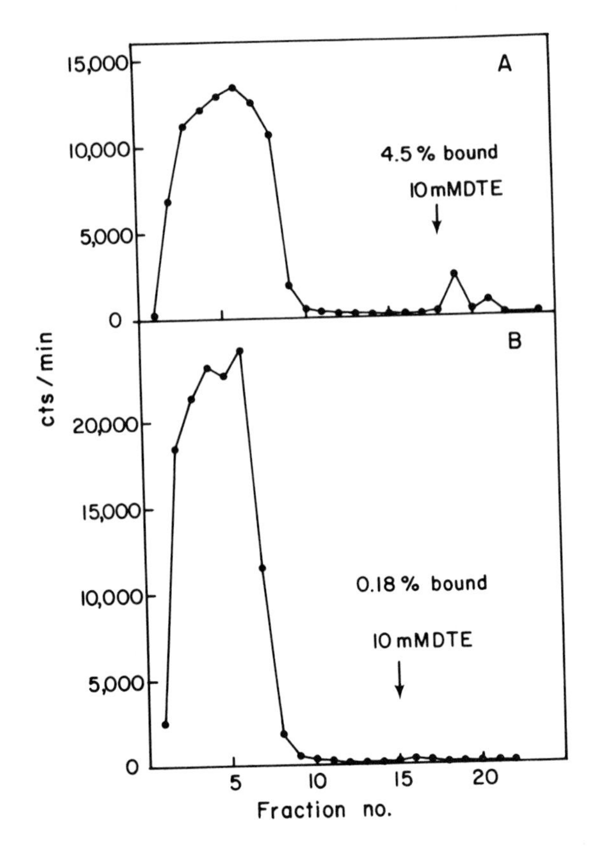

Fig. 10. Detection of RNA chain initiation using γ-thio-nucleoside triphosphates.

a) RNA made in the presence of γ-thio-ATP and γ-thio-GTP was passed through a mercury-sepharose affinity column. Material bound to the column was eluted with 10 mM dithioerythritol.

b) Same as a) except that γ-thio analogues were not present during RNA synthesis.

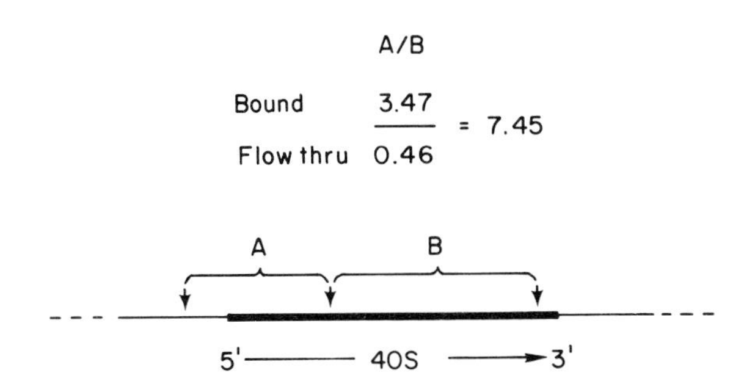

Fig. 11. Hybridization of RNA retained by mercury-sepharose to restriction fragments from the 5' end and the middle of the 40S coding sequence.

The relative positions of two restriction fragments (A and B) are indicated on the map. RNA bound to the column and RNA in the flow-through were hybridized separately to each fragment and the ratio of hybridization to A and to B was calculated.

---

theoretical amount of binding we should observe. However, these results, coupled with the extensive synthesis shown in Figure 7a, convince us that accurate chain initiation is occurring in the endogenous system.

Since the endogenous system has all the components needed for correct initiation, we should be able to dissect and identify those components and to persuade polymerase I to also initiate correctly on exogenously added rDNA.

ACKNOWLEDGEMENTS

This work was partially supported by an NIH grant to R. H. Reeder and by NIH fellowship grants to B. Sollner-Webb and P. Botchan. We thank Eileen Hogan for excellent technical assistance.

REFERENCES

(1) R. H. Reeder, IN: Ribosomes, eds. M. Nomura, A. Tissieres, and P. Lengyel (Cold Spring Harbor, New York, 1974) pp. 489-518.

(2) P. K. Wellauer, R. H. Reeder, D. Carroll, D. D. Brown, A. Deutch, T. Higashinakagawa and I. B. Dawid. Proc. Nat. Acad. Sci. USA 71 (1974) 2823-2827.

(3) R. H. Reeder, T. Higashinakagawa and O. L. Miller, Jr. Cell 8 (1976) 449-454.

(4) I. B. Dawid and P. K. Wellauer. Cell 8 (1974) 443-448.

(5) P. Botchan, R. H. Reeder and I. B. Dawid. Cell 11 (1977) 599-607.

(6) J. H. Sinclair and D. D. Brown. Biochemistry 10 (1971) 2761.

(7) U. Maitra and J. Hurwitz. Proc. Nat. Acad. Sci. USA 54 (1965) 815-822.

(8) B. Moss. Biochem. Biophys. Res. Comm. 74 (1977) 374-383.

(9) R. H. Reeder, B. Sollner-Webb and H. L. Wahn. Proc. Nat. Acad. Sci. USA 74 (1977) 5402-5406.

(10) O. L. Miller, Jr. and B. R. Beatty. Science 164 (1969) 955.

(11) T. Higashinakagawa, H. L. Wahn and R. H. Reeder. Dev. Biol. 55 (1977) 375-386.

(12) M. H. Green and T. L. Brooks. Virology 72 (1976) 110-120.

(13) G. D. Birnie, D. Rickwood and A. Hell. Biochim. Biophys. Acta. 331 (1973) 283-294.

(14) M. G. Riggs, R. G. Whittaker, J. R. Neuman and V. M. Ingram. Nature 268 (1977) 462.

(15) E. D. Adamson and H. R. Woodland. J. Mol. Biol. 88 (1974) 263-285.

(16) J. E. Mertz and J. B. Gurdon. Proc. Nat. Acad. Sci. USA 74 (1977) 1502-1506.

D. D. Brown and J. B. Gurdon. Proc. Nat. Acad. Sci. USA 74 (1977) 1064-2068.

(17) A. E. Reeve, M. M. Smith, V. Pigiet and R. C. C. Huang, Biochemistry 16 (1977) 4464-4469.

DISCUSSION

R.C. LEIF: At a density of 1.22 Metrizamide is not physiological, it is much better than sucrose, but it is by no means physiological.

R. REEDER: I agree that the Metrizamide density gradients are not exactly physiological. However, the salt concentration in them is close enough to physiological that RNA polymerase, nicking-closing enzyme and a lot of other fragile things come through the gradients and we still see what looks like normal structure in the isolated nucleoli.

R.C. LEIF: The other point is that DNA would be expected to be lighter than protein because hydrated DNA has a density of about 1.07. This was shown by Hearst and Vinograd in 1961 (Proc. Natl. Acad. Sci., 47, 1005).

R. REEDER: I agree. I do not find that surprising.

H. SMITH: It is sort of interesting that you do not find any histone I in your gels. I was interested in whether or not you have tried to resolve your gels to see if you have an A24 protein as Dr. Busch has shown in his system.

R. REEDER: There are several things we should look at in those gels and we have not done so yet. For instance we should look to see if HMG proteins are there and things like that. At the moment I cannot tell you since we have not done the experiments.

J.B. GURDON: May I refer to the first part of your talk concerning the initiation site for ribosomal RNA synthesis? Could you comment on the fact that you commonly see small transcripts in the supposed spacer region as viewed under the EM?

R. REEDER: Yes, there has been a little bit of controversy in the literature as to just where transcription does begin on these genes. As you referred to, people from Franke's laboratory have shown that you occasionally see transcription complexes out in the spacer. In fact, Marco Crippa has shown that if you add 5-fluorouridine to oocytes the spacer seems to disappear completely. People have argued that transcription actually begins out in the spacer, but that the RNA is rapidly processed until you get to the beginning of the 40S region and then it becomes stable. Well, we cannot rule out that there is an initiation site in the spacer. But if there is, it is so labile we cannot find it. My feeling is that both Franke's and Crippa's results actually result from a failure of termination. This would mean that the spacer chromatin is actually in an active configuration and that the only thing which keeps it from being transcribed normally is the fact that there is a stop signal upstream from it. That stop signal is not 100% effective. Or, if you mess it up with 5-fluorouridine, then the polymerase can bomb right on through the spacer. There is nothing in their experiments which says the things they see have to be due to initiation events as opposed to failure of termination.

J.B. GURDON: If I may just continue the point...I am not quite clear why or how incomplete termination would explain these short transcripts which is what I am specifically referring to.

R. REEDER: You have to make the additional assumption that RNA synthesized from spacer sequences is recognized as aberrant by the cell and is processed even as it is being transcribed.

J.B. GURDON: These short transcripts are the remaining bits which have not been cut off - is that right?

R. REEDER: Right, I am basing this scheme on the fact that we cannot find any initiation events out there in the spacer.

ANTIGENIC PROPERTIES OF NON-HISTONE PROTEINS IN CHROMATIN

L.S. HNILICA, J.F. CHIU, H. FUJITANI,
K. HARDY AND R.C. BRIGGS
Department of Biochemistry
Vanderbilt University School of Medicine
Nashville, Tennessee 37232

Abstract: Tissue and cell-specific antibodies were elicited by injecting rabbits with dehistonized chromatin preparations. The specificity of these nuclear antigens was shown to change with differentiation and carcinogenesis. While several experimental tumors in rats appeared to share common chromatin-associated nuclear antigen(s), spontaneous human malignancies differed in their immunological specificity from each other as well as from the tissues of their origin. Experiments on Novikoff hepatoma and regenerating rat liver revealed that the nuclear antigens may be physically and immunologically heterogeneous.

The specificity of chromatin-associated nuclear antigens was found to depend on the association(s) of chromosomal nonhistone proteins with homologous DNA. Treatment of chromatin with DNases as well as proteases virtually abolished its immunological activity. Immunocytochemistry with horse radish peroxidase conjugated antibody localized the described antigens in the cell nucleus. Selective extraction of chromosomal nonhistone proteins followed by chromatography on hydroxylapatite and Sephadex yielded a protein fraction which after association with homologous DNA contained essentially all the immunological activity assayable with antiserum to Novikoff hepatoma dehistonized chromatin. This still heterogeneous active fraction was represented in polyacrylamide gel electrophoresis by three polypeptide bands (m.w. 50-60,000).

INTRODUCTION

Chromatin can be regarded as a highly ordered structural and functional complex of interacting components, principally DNA, RNA, histones and nonhistone proteins. Obviously, the biological function of this giant macromolecular complex cannot be fully understood without detailed knowledge of the biochemical properties of its interacting components. While his-

ISBN 0-12-045450-5

tones in their molecular simplicity are extensively characterized and the full understanding of their biological functions may be near, detailed studies of the chromosomal nonhistone proteins have hardly begun. The formidable complexity and heterogeneity of this protein class prevented the systematic, one by one isolation and analysis applied so successfully to histones. Our better understanding of the chromosomal nonhistone proteins will depend on the detection and application of suitable markers which could be linked to specific properties and functions of selected protein molecules. The exceptional sensitivity and selectivity of antigen-antibody interactions has the potential of becoming a major tool for studies on the function and structure of chromatin and its components.

More than a decade ago, Henning et al. (1) and Messineo (2) reported that nuclear proteins, especially nucleohistones are immunogenic and can elicit specific antisera. These early observations did not stimulate much interest in the immunochemistry of chromosomal nonhistone proteins, perhaps because of their extreme heterogeneity and poor solubility in physiological solvents. Essentially all the attention was focused on DNA and other nuclear antigens associated with autoimmune diseases, especially systemic lupus erythematosus.

Discoveries that chromosomal nonhistone proteins may play an essential role in transcriptional control of selected genes during cytodifferentiation (Reviewed in 3-5) rekindled interests in the antigenic properties of chromosomal nonhistone proteins. In an elegant study, Chytil and Spelsberg (6) showed that dehistonized chromatin can elicit tissue-specific antibodies in rabbits. Their observations were quickly confirmed and extended (7-9), and increasing experimental evidence indicates that the immunological properties of chromosomal nonhistone proteins can be of considerable advantage in investigations of their biological functions (10-15).

EXPERIMENTAL

Isolation of nuclei and chromatin.

Adult male Leghorn chickens were made anemic by daily injections of 1% neutralized phenylhydrazine (10 mg/kg). The polychromatic primitive erythrocyte count was determined daily and after reaching 60-70%, daily bleeding was administered (12 ml/day) until these cells represented 95-97% of the total red cell population. Usually 4-5 days on phenylhydrazine followed by 2-3 days of bleeding were sufficient. Animals with less than 95% of polychromatic primitive erythrocytes (reticulocytes) were not used. Mature erythrocytes were obtained from the wing vein of untreated animals. Unless specified, all preparative work was performed at 4°C or in an ice bath. Blood was collected in 0.15 M saline/0.015 M sodium citrate solution containing 0.01% heparin. The method of Evans and Lingrel (16) with the addition of 0.5% Triton X-100 wash was used to remove membrane ghosts. The washed nuclei were extracted extensively with 50 mM Tris-HCl, pH 7.4 - 24 mM KCl - 5 mM $MgCl_2$ - 0.2 mM $CaCl_2$ (TKMC) until no visible hemoglobin remained. Finally, the nuclei were purified by centrifugation through 1.8 M sucrose in TKMC for 30 min. at 17,000 x g. Pelleted nuclei were either used immediately or suspended in 2.3 M sucrose in TKMC and stored at -20°C. In several experiments the reticulocyte population was further purified by gradient centrifugation as described by Brasch et al. (17).

Transplantable tumors were maintained in Sprague Dawley (Novikoff and 30D hepatomas and Walker carcinosarcoma) or Buffalo (Morris hepatomas) rats. Experimental hepatomas were produced by feeding Fisher rats with a diet containing 10% corn oil and 0.06% N,N-dimethyl-p-(m-tolylazo) aniline (3'-MDAB) or α-naphthylisothiocyanate (noncarcinogenic liver stimulant). Human tissues were obtained from the Department of Pathology and the Vanderbilt University Hospital. HeLa cells, S_3 strain, were the gift of Dr. D. Friedman (Department of Molecular Biology, Vanderbilt University).

Where applicable, the method of Chauveau et al. (18) as modified by Blobel and Potter (19) was employed for the isolation of nuclei. Novikoff hepatoma and other tumor nuclei were isolated by the osmotic shock and Chaikoff press procedure (20). Chromatin was prepared from the isolated nuclei using the method described by Spelsberg et al. (21).

Fractionation of Chromatin.

Isolated chicken erythrocyte or reticulocyte nuclei were suspended in TKMC - 1 mM phenylmethylsulfonylfluoride solution, stirred for 5 min in an ice bath and centrifuged at 1200 x g for 10 min. Nuclear pellets were resuspended in 0.25 M sucrose - 5 mM Tris-HCl, pH 8.0 to a concentration of 0.5 - 1.0 mg/ml (with respect to DNA content) by gentle homogenization in a hand operated glass homogenizer fitted with a loose Teflon pestle. The nuclei were allowed to swell for

30 min and then 40 ml aliquots were sheared in 20 sec. intervals at 4°C and maximum power for a total of 120 seconds using a Branson Model W185 sonifier equipped with the large control horn. After a brief centrifugation at 200 x g to remove debris and unbroken nuclei, the final sonicates were pooled and centrifuged at 12,000 x g for 30 min giving a chromatin pellet CP_1 and a soluble supernatant. A 20 fold concentrate of TKMC was added dropwise to the stirred supernatant to bring the final concentration to 1.0 x TKMC. After stirring for 5 min in an ice bath the suspension was centrifuged at 3,000 x g for 20 min yielding a chromatin pellet CP_2 and a soluble chromatin supernatant CAS. All three fractions as well as a sample of the original untreated chromatin were made 6 mM EDTA, 1 mM phenylmethylsulfonylfluoride and dialyzed against 200 vols of 1 mM EDTA and then distilled water overnight. The soluble active chromatin fractions (CAS) were concentrated in dialysis bags with Sephadex G-200 to a final concentration of approximately 2-5 OD_{260}/ml.

Chromatin from Novikoff hepatoma and other tissues was fractionated following the scheme developed in our laboratory (22). Briefly, this method consists of three sequential extractions with 5 M urea. The first extraction (5 M urea in 50 mM Na-phosphate buffer, pH 7.6) removes the majority of chromosomal nonhistone proteins. This extraction is followed by the removal of histones in 5 M urea, 2.5 M NaCl, buffered with 50 mM Na-succinate buffer to pH 5.0. Finally, immunologically specific fraction of chromosomal nonhistone proteins (along with a variety of other DNA-binding proteins) can be obtained by dissociation of the Na-succinate pellet with 5 M urea, 2.5 M NaCl, buffered with 50 mM Tris-HCl buffer to pH 8.0. The three principal chromosomal protein groups resulting from this fractionation scheme are the UP, HP and NP proteins respectively.

Preparation of antigens and immunization

Isolated chromatins were dehistonized by homogenization in 5.0 M urea, 2.5 M NaCl, 50 mM Na-phosphate buffer, pH 6.0 (modification of a method developed in our laboratory, ref. 21) and centrifugation at 100,000 x g for 30 hrs. The pellets contained DNA and nonhistone protein-DNA complexes. After rinsing the walls of the tubes with distilled water, we resuspended the pellets in 2 mM Tris-HCl (pH 7.5), using a glass homogenizer fitted with a Teflon pestle. The suspension was stirred at 4°C overnight and centrifuged at 1,000 x g for 5 minutes to sediment undissolved material. The supernatant containing the nonhistone protein-DNA complexes was used for immunization.

Immunization

Equal volumes of complete Freund's adjuvant and dehistonized chromatin (0.5-1.0 mg DNA/ml) were mixed and homogenized in a Teflon pestle-glass homogenizer immediately before immunization. About 100-200 μg (as DNA) of dehistonized chromatin was injected into the toe pads (hind feet) of male New Zealand rabbits weekly for 2 weeks and then intramuscularly for 4-6 additional weeks. Finally, 1 week later, a booster of a fourth of the original dose was given intravenously. The rabbits were bled 7 days after the last injection, and the sera were used in immunoassays.

Immunoassays

The complement fixation technique of Wasserman and Levine (23) was used to test immunospecificity of antibodies. The antisera were decomplemented by heating at 56°C for 30 minutes. Lyophilized guinea pig serum complement, washed sheep erythrocytes, and rabbit anti-sheep erythrocyte serum were purchased from Cappel Laboratories (Downingtown,Pa.). Isotonic Tris-HCl buffer, pH 7.4 (0.1 M NaCl - 10 mM Tris 0.5 mM $MgSO_4$ - 0.1 mM $CaCl_2$ - 0.1% bovine serum albumin) was used for all dilutions. The reaction mixtures containing, in a total volume of 0.8 ml, 0-10 μg of DNA as chromatin, antiserum (0.1 ml rabbit antiserum diluted 50-100 times), and complement (0.2 ml of 200 x diluted restored guinea pig serum, Cappel Laboratories) were incubated at 4°C for 17 hours. To each sample 0.2 ml of sensitized sheep erythrocytes was added and the mixture was incubated at 37°C for 30 minutes. We determined the extent of hemolysis by an absorbance at 413 nm. All assays were corrected for anticomplementarity of antisera.

Localization of Antigens

The localization of antigens in the cells was accomplished by horseradish peroxidase bridge or fluorescence technique (24,25). Air-dried tissue sections were fixed in acetone for 10 minutes, and incubated for 30 minutes in a humidified chamber with antiserum diluted 10-20 times. Thereafter, cells were washed well in phosphate-buffered saline (pH 7.2) for 15 minutes and reincubated under similar conditions with sheep anti-rabbit γ-globulin conjugated with horseradish peroxidase. After a final wash in phosphate-buffered saline, slides were stained with saturated aqueous solution of 3,3'-diaminobenzidine and 0.005% H_2O_2.

Analytical methods

A modified Lowry et al. (26) procedure was used for protein analysis with calf thymus histone or bovine serum albumin fraction V as standards. DNA was assayed according to Burton (27). Polyacrylamide gel electrophoresis was performed as described by Laemmli (28).

RESULTS AND DISCUSSION

Biochemical and electrophoretic studies on chromosomal nonhistone proteins indicate their tissue and species specificity. These observations are more recently supported by immunological experiments. Although the experimental evidence is still quite limited, it appears that some nuclear protein antigens reflect the phenotypical specificity of cells and perhaps may play an important role in regulating the process of cellular differentiation.

Tissue and cell specificity of nuclear antigens.

As was already mentioned, dehistonized (6-8) or whole chromatin (9) is antigenic and can elicit the formation of specific antibodies. In Chytil and Spelsberg's experiments (6) an antiserum to dehistonized chromatin from the oviducts of immature chicks stimulated with diethylstilbestrol for 15 days was assayed in the presence of dehistonized chromatin preparations from stimulated chick oviduct, liver, heart and spleen. Clearly significant differences in complement fixation indicated the presence of tissue specific antigenic molecules in the assayed chromatin preparations. The extensive tissue specificity of dehistonized chromatin preparations was confirmed in our laboratory (7,8). Antisera to dehistonized chromatins from calf thymus, rat liver and Novikoff hepatoma fixed the complement in the presence of its corresponding antigen and only to a much lesser extent in the presence of chromatin from another tissue. These results (reproduced in Fig. 1) established that the immunological tissue specificity of chromosomal nonhistone proteins can be demonstrated in higher vertebrates is not limited to chicken tissues. Additionally, these experiments showed clear immunological differences between the chromatin of normal liver and malignant tumor, i.e. Novikoff hepatoma.

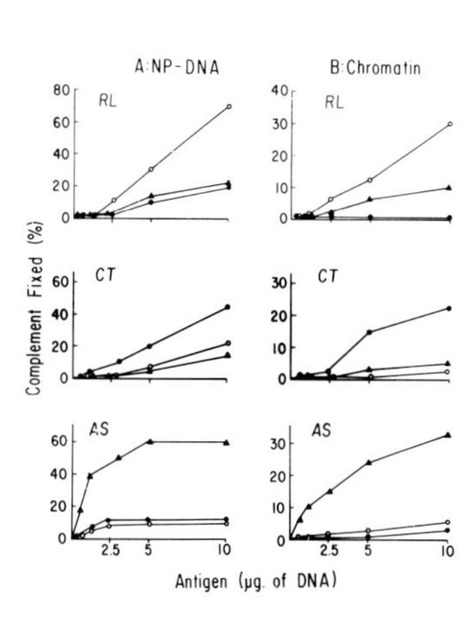

Fig. 1. Complement fixation of nonhistone protein-DNA complexes (A) and chromatin (B) from rat liver (O - O), calf thymus (● - ●), and Novikoff hepatoma (▲ - ▲) in the presence of antiserum to the dehistonized chromatins of the individual tissues.

Cell differentiation can be described as carefully programmed expression and repression of great numbers of genes comprising the cellular genome. In addition to the phenotypically distinct and often morphologically identifiable changes associated with differentiation, more subtle modifications of genetic activity take place in response to various stimuli (hormonal, environmental, dietetic, etc.). Since chromosomal nonhistone proteins are believed to play essential roles in regulating transcriptional expression of specific genes, the immunological properties of chromatin were investigated in differentiating cells and tissues.

Using antibody to dehistonized chromatin from adult rat liver Chytil et al. (29) found that liver chromatin from 19 day old rat fetuses essentially did not fix the complement in the presence of this antiserum. The complement fixation increased gradually _post partum_ and reached maximum in mature rats. Similar changes in the immunological specificity of chromosomal nonhistone proteins were seen in chromatins of immature chick oviducts undergoing estrogen-induced differentiation (30). In the presence of antiserum to dehistonized chromatin from fully differentiated oviducts (15 days on diethylstilbestrol), chromatin from untreated, immature chicks did not fix the complement. However, the complement fixation increased with the time of treatment and reached full extent in the presence of fully stimulated (15 days on diethylstilbestrol) oviduct chromatin. Thus the degree of immunological reactivity of the dehistonized chromatins showed a gradual estrogen-dependent transition which coincided with the morphological and biochemical development of oviduct.

A more dramatic display of differentiation associated changes in immunological specificity of chromatin was observed in maturing chicken reticulocytes. Daily administration of phenylhydrazine and blood letting to adult chickens resulted in an almost complete replacement of mature erythrocytes by polychromatic primitive erythrocytes (reticulocytes). Antiserum to dehistonized reticulocyte chromatin fixed the complement only in the presence of reticulocyte chromatin but not in the presence of chromatins from mature erythrocytes or other tissues (Fig. 2). Discontinuation of treatment in the pre-

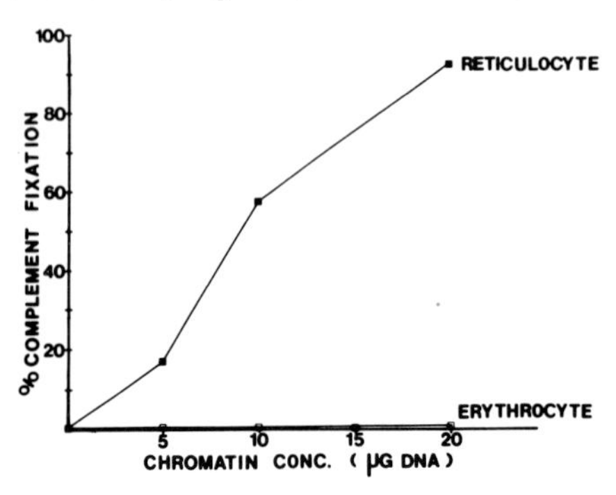

Fig. 2. Complement fixation of chicken reticulocyte (■ - ■) and erythrocyte (□ - □) chromatins in the presence of antiserum to dehistonized reticulocyte chromatin.

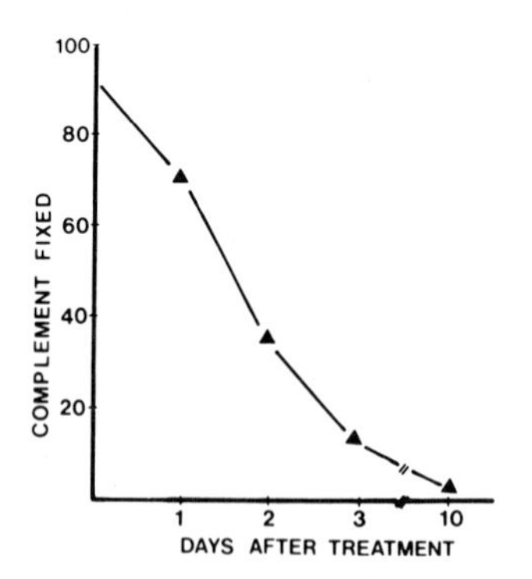

Fig. 3. Loss of reticulocyte-specific nuclear antigen(s) during red blood cell maturation. Chickens were made anemic by a daily schedule of phenylhydrazine administration and blood letting. Chromatin was prepared from their red blood cells on days 0,1,3 and 10 after the discontinuation of treatment and assayed by complement fixation.

sence of antiserum to chicken reticulocyte chromatin resulted in rapid gradual maturation of reticulocytes with a concomitant disappearance of the reticulocyte-specific immunological reactivity (Fig. 3).

The several examples of immunological tissue and cell specificity of chromatin shown and discussed in this section point to extensive changes of qualitative composition of chromosomal proteins. These changes occur during differentiation and can be detected immunologically. When better characterized, these antigens may serve as useful markers for studies on cell differentiation.

Nuclear antigens in cancer and development.

Assuming that cells may become transformed to malignancy by errors in transcriptional control mechanisms (differentiation) which result in hereditary changes characteristic for the cancerous phenotype, we have initiated investigations into immunological properties of chromatin during neoplasia. Initial studies on immunological tissue-specificity of chromatin revealed dramatic differences in complement fixation between normal rat liver and Novikoff hepatoma. As illustrated in Fig. 4, only rat liver chromatin fixed the complement in

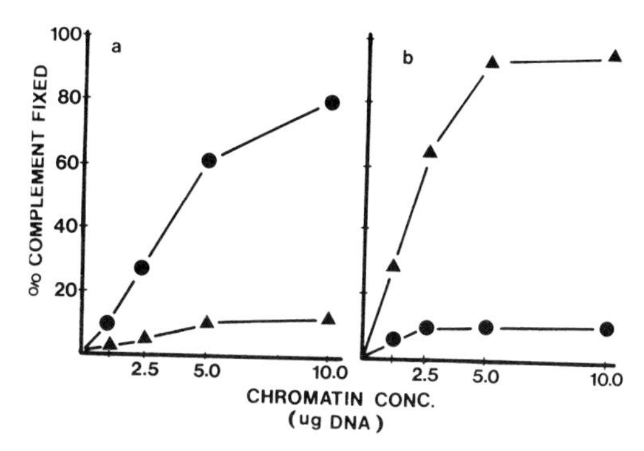

Fig. 4. Immunological specificity of normal rat liver (● - ●) and Novikoff hepatoma (▲ - ▲) chromatins. The assays were performed in the presence of antisera to dehistonized chromatin from rat liver (a) or Novikoff hepatoma (b).

the presence of antiserum to dehistonized rat liver chromatin. Novikoff hepatoma chromatin was inactive. The opposite was true when the complement fixation was performed in the presence of antiserum to Novikoff hepatoma dehistonized chromatin. Chromatins from three transplantable rat tumors did not differ from each other by the complement fixation assay (Fig. 5). It appears as if the neoplastic process changed the specificity of the original tissue into an immunologically new type, common to at least the tumors compared in Fig. 5.

The antiserum to Novikoff hepatoma chromatin also fixed the complement in the presence of chromatins from Morris hepatomas. These experiments were performed to study the possibility that the immunological differences between normal tissues and malignant neoplasms may reflect the extremely rapid

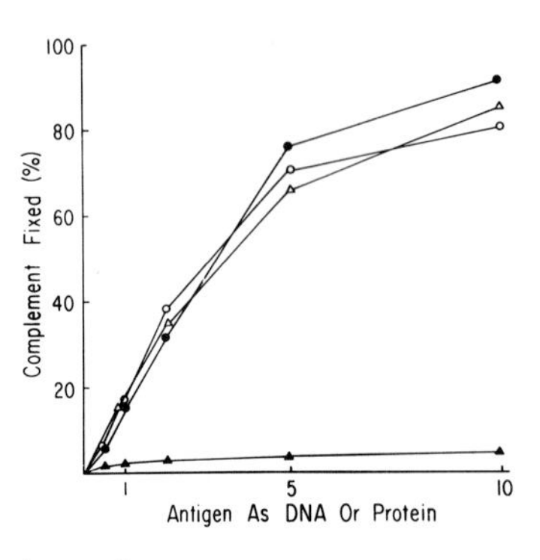

Fig. 5. Complement fixation of chromatins isolated from three transplantable rat tumors. The assays were performed in the presence of antiserum to dehistonized Novikoff hepatoma chromatin. Novikoff hepatoma (O - O), transplantable 30D ascites hepatoma (● - ●) and Walker 256 carcinosarcoma (Δ - Δ). Normal rat liver (▲ - ▲).

growth of the latter. Morris hepatomas represent a suitable model for testing this possibility. The complement fixation of chromatins obtained from four Morris hepatomas differing in their growth rates was assayed in the presence of antiserum to dehistonized Novikoff hepatoma chromatin. As can be seen in Fig. 6, the poorly differentiated, fast growing 7777 and 3924A hepatomas compared well with the rapidly growing Novikoff tumor.

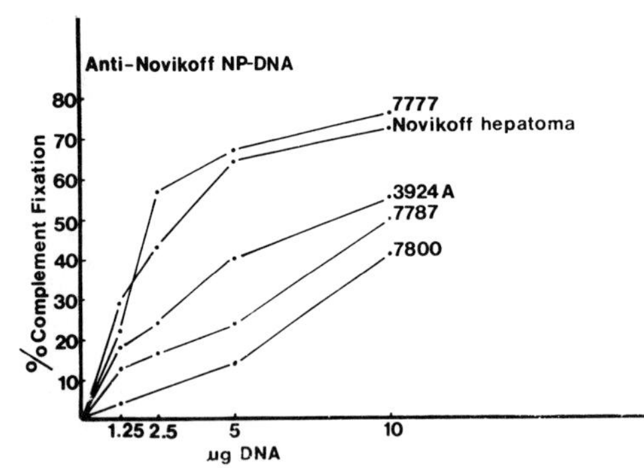

Fig. 6. Complement fixation of dehistonized chromatins from Novikoff hepatoma and Morris hepatomas 7777, 7787, 7800 and 3924A in the presence of antiserum to Novikoff hepatoma dehistonized chromatin.

The slowly growing and better differentiated 7800 and 7787 hepatomas fixed the complement to a lesser extent than the fast growing tumors. However, the differences in complement fixation did not warrant a conclusion that the observed immunological tumor-specificity of DNA-nonhistone complexes in chromatin reflected the expression of growth-associated antigens. This conclusion obtained further support from our experiments on hepatectomized rats. Although chromatins isolated from livers of rats 24 and 48 hrs after hepatectomy fixed the complement in the presence of antiserum to dehistonized Novikoff hepatoma chromatin (the 6, 12 and 72 hr samples were essentially nonreactive) immunoabsorption of the antiserum with 48 hrs regenerating rat liver chromatin abolished the immunoactivity to regenerating rat liver chromatins while only diminishing the Novikoff hepatoma activity (31). These experiments suggest the presence of at least two kinds of antibodies in the antiserum to Novikoff hepatoma dehistonized chromatin. One type is oriented against antigen(s) common to rapidly proliferating tissues while the other appears specific for Novikoff hepatoma (and perhaps other transplantable rat tumors).

The heterogeneity of anti-nuclear antibodies is further emphasized by nuclear antigens reported to be specific for various tissues or disease. Akizuki et al. (32) recently purified a soluble nuclear protein (Ha antigen) which reacted with antibodies circulating in blood of patients with sicca syndrome (Sjögren's syndrome) or patients with sicca syndrome associated with systemic lupus erythematosus. The antibodies against this protein were present only infrequently in patients with rheumatoid arthritis or systemic lupus erythematosus without sicca syndrome and absent in patients with other connective tissue disease. This relatively small protein (m.w. approx. 43,000) migrated as a single band in polyacrylamide electrophoresis.

Various nuclear antigens were reported to be associated with virus transformed cells (33-36), or tumors of nonvirus etiology (2, 37, 38). The best characterized nuclear antigen was isolated from Novikoff hepatoma nuclei by Yeoman et al. (11). This nuclear antigen 1 or NAg-1 is a moderately acidic glycoprotein (acidic/basic amino acids 1.4) with lysine as NH_2 terminal and molecular weight of approx. 26,000. Antibodies to this protein formed precipitin bands in the double-diffusion immunoprecipitation assay with chromosomal proteins of Novikoff hepatoma, Walker 256 carcinosarcoma, and 18-day fetal rat liver. Chromosomal proteins from normal or regenerating rat liver, heart and kidney were not reactive.

Antiserum to dehistonized Novikoff hepatoma chromatin was found, in our laboratory, to also react with chromatin preparations from fetal rat liver. As can be seen in Fig. 7, chromatin from fetal (18 day) liver fixed the complement extensively. The complement fixation decreased rapidly after birth and liver chromatins of 3 week old rats were essentially not reactive (similar to adult rat liver chromatin). In this behavior, the nuclear antigens described in this chapter resembled the Novikoff hepatoma NAg-1 protein isolated by Yeoman et al. (11). However, as will be shown later, the nuclear antigens present in dehistonized Novikoff hepatoma chromatin differ from the NAg-1 protein by their extractability from chromatin, molecular weight and, most of all, dependency for immunological specificity on their association with homologous (rat) DNA. Apparently antibodies to dehistonized Novikoff hepatoma chromatin are quite heterogeneous indicating the need for extensive fractionation and purification of their antigens.

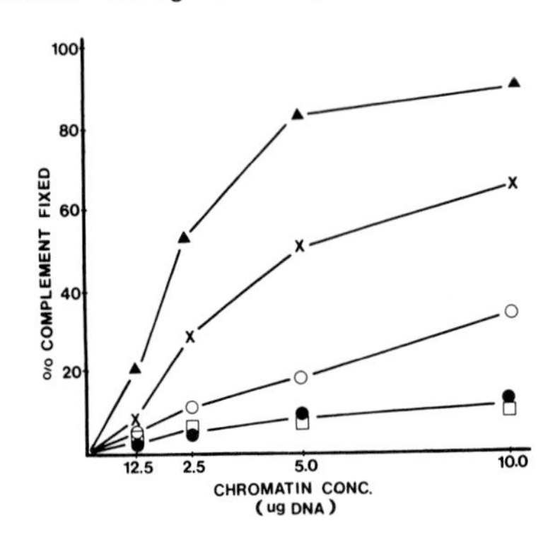

Fig. 7. Complement fixation of chromatins isolated from developing rat liver and Novikoff hepatoma assayed in the presence of antiserum to dehistonized Novikoff hepatoma chromatin. Fetal - 18 day (X - X), one week (O - O), three weeks (□ - □) and adult (● - ●) rat livers. Novikoff hepatoma (▲ - ▲).

Nuclear antigens in hepatocarcinogenesis

Immunological changes of chromosomal nonhistone proteins associated with neoplasia pose some practical applications. If the malignant tumor associated antigens are indeed specific for the cancerous phenotype, their appearance can become a marker for malignant transformation. To test this possibility, Fisher rats were fed a diet containing N,N-dimethyl-p-(tolylazo) aniline (3'-MDAB) or a noncarcinogenic stimulant α-naphtylisothiocyanate. The animals were sacrificed in weekly intervals and their liver chromatins assayed in the presence of antiserum to dehistonized Novikoff hepatoma or 3'-MDAB produced hepatoma chromatins. As can be seen in Fig. 8 as soon

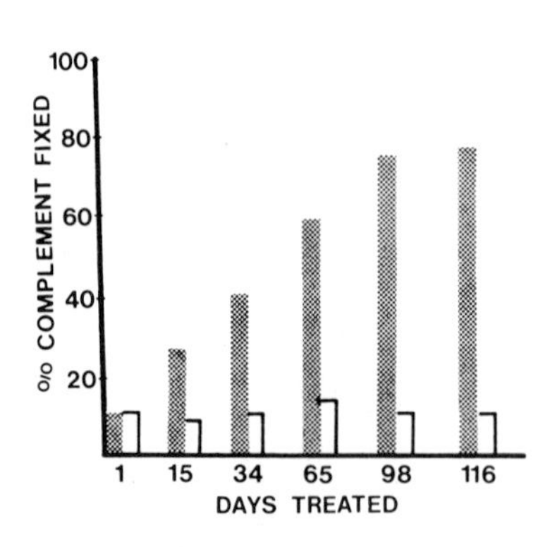

Fig. 8. Complement fixation of chromatins isolated from livers of rats maintained on a diet containing 3'-methyl-4-dimethylaminoazobenzene (shaded bars) or α-naphthylisothiocyanate (open bars). The assays were performed in the presence of antiserum to dehistonized Novikoff hepatoma chromatin.

as two weeks after the initiation of carcinogenic diet, the immunological specificity of hepatic chromosomal nonhistone protein-DNA complexes began to change from a type characteristic for liver to a new type found in Novikoff hepatoma. While liver chromatins of rats on the α-naphtylisothiocyanate (a noncarcinogenic compound producing extensive bile duct cell proliferation similar to that produced by 3'MDAB) diet remained negative, the immunological activity of liver chromatins of rats maintained on 5'MDAB diet converted almost entirely to the hepatoma type in nine weeks.

The described experiments indicate that carcinogenesis is indeed accompanied by profound changes in the immunological character of chromosomal nonhistone proteins in the affected cells. Rat liver and other organs, however, represent a relatively poor experimental model. Various cells populating most organs may respond differently to the carcinogenic diet. More dramatic changes in the immunological specificity of chromosomal nonhistone proteins after a relatively short exposure to carcinogens in tissue cultures were observed in our laboratory. These observations are in a good agreement with the experiments of Zardi et al. (9) who reported significant immunological changes in chromosomal nonhistone proteins between WI-38 cells and their SV40 transformed 2RA counterpart.

Nuclear antigens in human cancer

All the experiments described here were performed on experimental animals. Since carcinogen or virus induced tumors differ immunologically from spontaneous malignancies (39), it was necessary to investigate the immunological specificity of chromosomal nonhistone protein-DNA complexes in spontaneous neoplasms. Antisera to dehistonized chromatins of human breast or lung carcinomas were found to be highly specific when assayed in the presence of corresponding chromatin preparations. As can be seen in Fig. 9, the antiserum to human

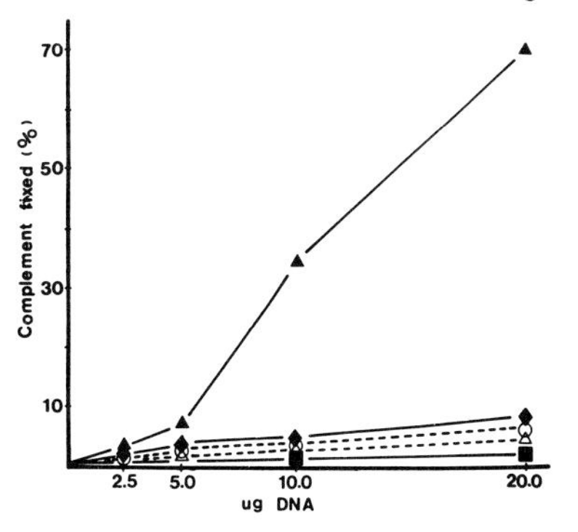

Fig. 9. Complement fixation of chromatins from human breast carcinoma (▲ - ▲), normal human breast (△ - △), human breast benign tumor (◆ - ◆) normal human lung (O - O) and human placenta (■ — ■). The assays were performed in the presence of antiserum to dehistonized human breast carcinoma chromatin.

breast carcinoma dehistonized chromatin fixed the complement extensively only in the presence of breast carcinoma chromatin. The complement fixation was negative or marginal in the presence of chromatins from normal human lung, breast, placenta or a benign tumor of the breast. Similar results were obtained when antiserum to human lung carcinoma dehistonized chromatin was assayed in the presence of lung carcinoma chromatin (positive) or chromatins of normal lung or breast, breast carcinoma, human placenta, HeLa cells and Novikoff hepatoma (all negative). This strict immunological tissue specificity (Fig. 10) of

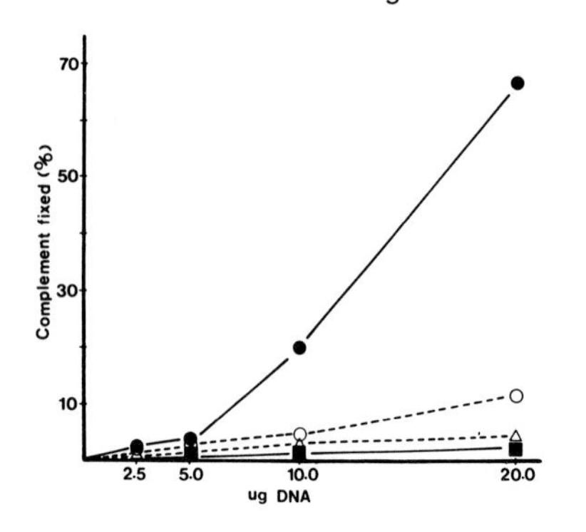

Fig. 10. Complement fixation of chromatin preparations from human lung carcinoma (● - ●), normal human lung tissue (O - O), normal human breast, breast carcinoma, human placenta and HeLa cells (Δ - Δ) or rat Novikoff hepatoma (■ - ■).

chromosomal nonhistone protein-DNA complexes in spontaneous human malignancies differs significantly from our findings on experimental neoplasms. As is shown in Figures 5 and 6, chromatins from various experimental tumors fixed the complement in the presence of antiserum to Novikoff hepatoma dehistonized chromatin and did not exhibit measurable specificity of their nuclear protein antigens. However, when chromatins isolated from human lung and breast carcinomas were assayed in the presence of antiserum to dehistonized Novikoff hepatoma chromatin, both human tumor chromatins fixed the complement significantly, regardless of their species heterology (Fig. 11). Interestingly, chromatins from normal human tissues

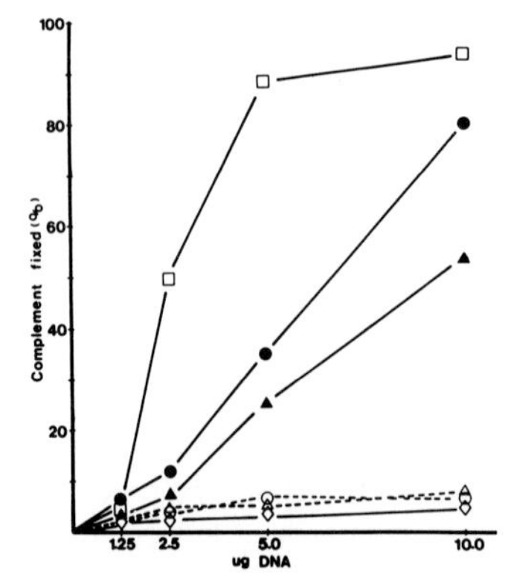

Fig. 11. Complement fixation of chromatins isolated from rat Novikoff hepatoma (□ - □), human lung cancer (● - ●), human breast carcinoma (▲ - ▲), normal human breast tissue (Δ - Δ), normal human lung (O - O) and normal rat liver (◇ - ◇). The assays were performed in the presence of antiserum to dehistonized Novikoff hepatoma chromatin.

did not fix the complement in the presence of Novikoff hepatoma antiserum.

The lack of species specificity exhibited by antiserum to Novikoff hepatoma chromatin in assays with human neoplasms confirms our earlier observations on Ehrlich ascites (mouse) and canine transmissible sarcoma (40). Chromatins of these two tumors fixed the complement significantly (although to a much lesser extent than Novikoff hepatoma chromatin) in the presence of antiserum to dehistonized Novikoff hepatoma chromatin. These observations further support our notion that the nuclear antigens of Novikoff hepatoma chromatin are heterogeneous.

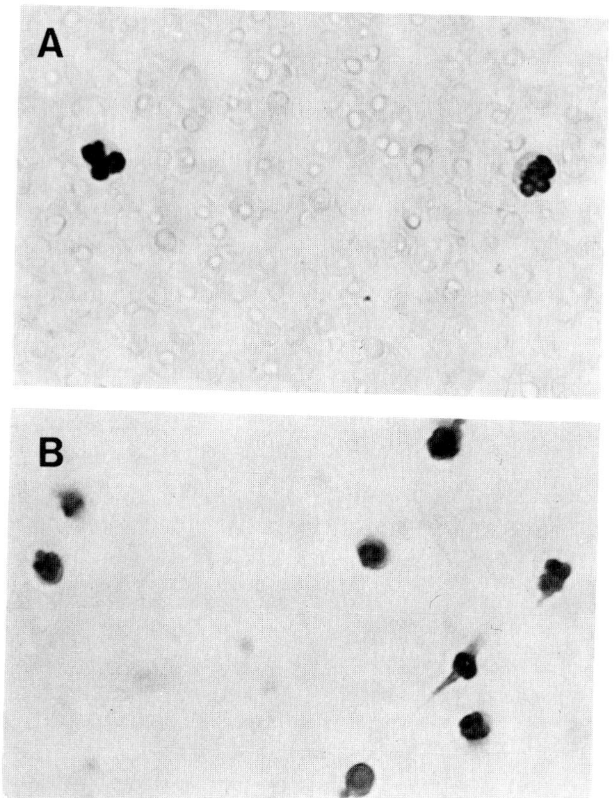

Fig. 12. Indirect localization of antigen(s) specific for human granulocyte nuclei using a horseradish peroxidase conjugated second antibody.
A: Peripheral blood smear from normal human adult (250X).
B: Peripheral blood smear from human adult diagnosed as acute granulocytic leukemia (250X).

The immunological specificity of chromatin proteins in human tissue is further illustrated in Figures 12a and b. Antiserum raised to dehistonized chromatin of cells tentatively identified as polymorphonuclear leukocytes stained (fluorescence or horse radish peroxidase method) only cells of this lineage in normal human blood smears (Fig. 12a). The antiserum also localized in the nuclei of leukemic cells derived from the granulocytic series (Fig. 12b). Apparently, the chromatin associated nuclear antigens consist of a variety of proteins which immunogenic properties present a complex picture of multiple antigenic determinants. There may be several separate antigens in Novikoff hepatoma chromatin or the principal immunogenic molecule or complex may have multiple antigenic determinants. Distribution of the immunogen as well as its conformation in chromatin may determine its specificity.

Characterization of nuclear antigens.

Since all the experiments discussed here were performed with isolated native or dehistonized chromatin, it is reasonable to expect that such chromatin may be contaminated by fragments of cellular membranes and other particulate materials of cytoplasmic origin. If strongly immunogenic, these contaminants may be responsible for the observed immunological properties of isolated chromatin. Immunohistochemical localization experiments using either fluorescein or horse radish peroxidase conjugated antibodies showed clearly the nuclear localization of the antigens described in this chapter (Fig. 12,13). Chromatin associated antigens, specifically interacting with antibodies elicited by injections of dehistonized or native chromatin have been localized in the nuclei of rat liver (ref. 25, 31, and Fig. 13), human lung carcinoma (41), polymorphonuclear leukocytes (Fig. 12), cultured cells and lymphocytes (42,43) Syrian hamster tsAF8 and human LNSV cells (44) and 3T6 or WI-38 fibroblasts (45). In all instances the localization of antigens was clearly nuclear and reflected the tissue and species specificity detectable by complement fixation assays.

The possibility that nuclear membrane associated antigens react in the complement fixation assays of isolated chromatins in the presence of appropriate antisera (nuclear membrane contamination of isolated chromatin) was investigated by Wilson and Chytil (46). These authors immunized rabbits with nuclear membrane isolations and assayed the resulting antisera by complement fixation. Rat liver chromatin, native or dehistonized, did not react with the antiserum to liver nuclear membranes.

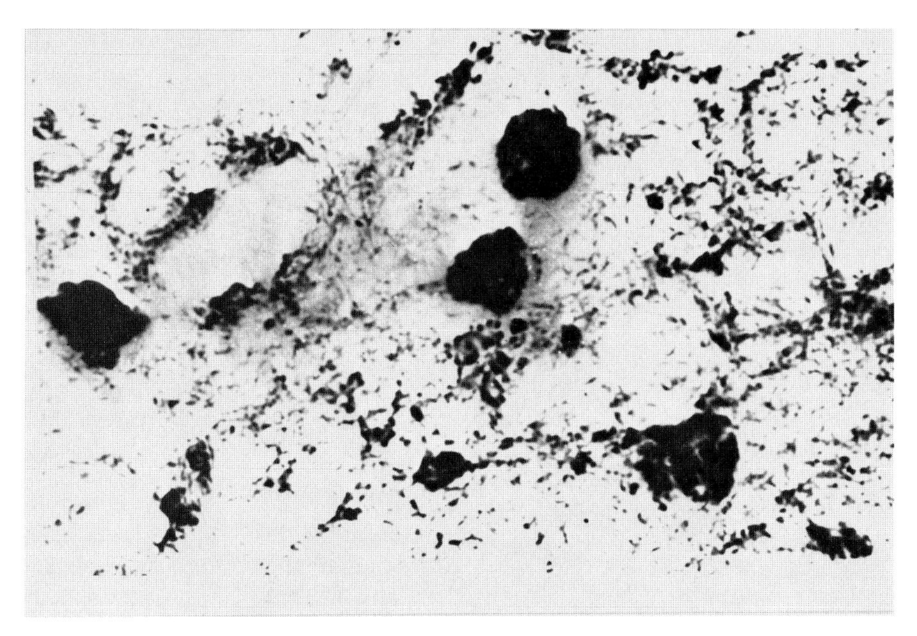

Fig. 13. Localization of rat liver nuclear antigens by the horseradish peroxidase bridge technique. The localization was performed in the presence of rabbit antiserum to dehistonized rat liver chromatin.

This shows that the immunogenicity of dehistonized chromatin preparations was not caused by contaminating membrane fragments.

The presence of DNA in chromatin preparations used in our immunological experiments raises the possibility that the antibodies may be directed to specific complexes between chromosomal proteins and DNA. As documented by the dehistonization experiments, exposure of chromatin to 5 M urea and 2 M NaCl did not destroy its immunogenicity. This permitted reconstitution experiments designated to identify the particular fraction of chromatin responsible for its immunological tissue specificity. It was established that although the presence of DNA was essential for the formation of immunologically active complexes, the nonhistone proteins were donors of the observed tissue specificity (7, 8, 22, 40, 47-49). A characteristic protein-DNA exchange experiment is illustrated in Fig. 14. The dependence of tissue-specific complement fixation on the source of chromosomal nonhistone proteins is quite clear. Some of the nonhistone protein-DNA complexes are also immunologically species specific since in reconstitution

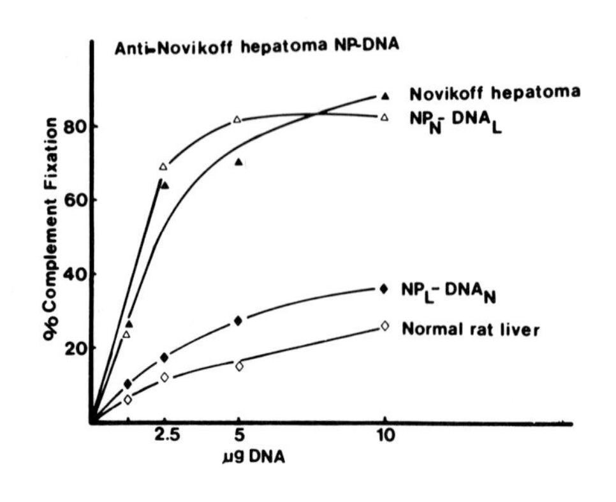

Fig. 14. Complement fixation of chromatins isolated from Novikoff hepatoma (▲ - ▲) and normal rat liver (◇ - ◇) as well as complexes reconstituted from Novikoff hepatoma (N) or rat liver (L) DNA and DNA-associated chromosomal nonhistone proteins NP. The assays were performed in the presence of antiserum to dehistonized Novikoff hepatoma chromatin.

experiments, DNA from other species or synthetic polyanions could not substitute for the homologous DNA (8, 22, 48-51). The species specificity of nuclear protein antigens was also reported by Zardi et al. (42) and Tsutsui et al. (43, 44).

Our search for chemical identity of the nuclear antigens present in various tissues is organized along two experimental avenues. In one, we are attempting to fractionate chromatin by physical methods with immunological activity of the fractions being the screening procedure. In the other, fractionation of chromosomal proteins by selective extraction and chromatographic procedures is the principal tool for seeking out the specific antigens. Hopefully, combination of these two approaches should provide us with information on the relationships of nuclear antigens to chromatin structure and conformation as well as on their biochemical and physico-chemical characters.

By employing a fractionation scheme based on limited shearing of chromatin and its subsequent precipitation by divalent cations we have obtained a transcriptionally active fraction of chicken reticulocyte chromatin (52). This fraction contained essentially all the chromatin RNA and was quite rich in chromosomal nonhistone proteins. Although it represented only about 1% of the chromatin DNA, it contained most of the DNA-associated, reticulocyte-specific nuclear antigens. As can be seen in Fig. 15, the CP_1 and CP_2 fractions of reticulocyte chromatin which represented 4.5 and 95% of total chromatin DNA were nearly immunologically inactive while the activity of the CAS fraction (less than 1% of total DNA) was extremely high. Fractionation of mature erythrocyte nuclei

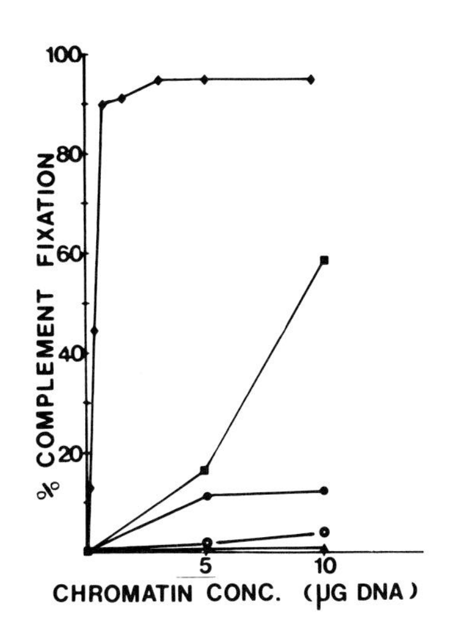

Fig. 15. Complement fixation of chicken reticulocyte and erythrocyte chromatin fractions isolated by shearing and precipitation with divalent cations (52). The assays were performed in the presence of antiserum to dehistonized chicken reticulocyte chromatin. Unfractionated reticulocyte chromatin (■ - ■), reticulocyte fraction CP_1 resistant to shearing (● - ●), reticulocyte fraction CP_2 isolated by precipitation of sheared chromatin with divalent cations (▲ - ▲), fraction of sheared reticulocyte chromatin soluble in the presence of divalent cations (♦ - ♦). This transcriptionally active fraction CAS contained less than 1% of the total reticulocyte chromatin DNA. Erythrocyte chromatin fractions CP_1 and CP_2 did not fix the complement (▲ - ▲) while the fixation of the active erythrocyte fraction CAS (0 - 0) was only marginal.

by the same technique did not produce any fractions containing antigenic material present in the reticulocytes. Dehistonization of erythrocyte chromatin did not change its immunological behavior, i.e., the dehistonized erythrocyte chromatin was inactive. Similar to rat liver or Novikoff hepatoma, the complement fixation of reticulocyte chromatin or its fractions could be abolished by digestion with pronase or DNase II. However, digestion with DNase I reduced the complement fixation of reticulocyte chromatin to about 50% of the controls. This differential inactivation of nuclear antigenic protein binding sites on reticulocyte DNA points to a possible nonrandom distribution of DNA-protein antigens in chromatin.

The other technique applied to the fractionation of nuclear antigens was based on selective extraction of chromosomal proteins by buffered solutions of 5 M urea. According to the method developed in our laboratory (22), the bulk of chromosomal nonhistone proteins can be extracted from chromatin by 5 M urea dissolved in 50 mM Na-phosphate buffer (UP protein fraction). The remaining chromatin (UC) can be then dehistonized by extraction with 2.5 M NaCl, 5 M urea in 50 mM Na-succinate buffer (HP proteins). Finally, the nonhistone protein fraction which remains attached to DNA in buffered 5 M urea (the first step in our procedure) can be obtained by extraction of the dehistonized pellet (HC) with 2.5 M NaCl, 5 M urea in 50 mM Tris-HCl buffer, pH 8.0 (NP protein fraction). The identity of proteins present in these three fractions of chromatin (i.e., nonhistone proteins UP, histones HP, and DNA-binding nonhistone proteins NP) has been established by electrophoretic, amino acid and DNA-interaction analysis (8, 22, 40, 48-51).

For a more rapid extraction of the antigenic DNA-binding chromosomal nonhistone proteins, the pellet after the first extraction with buffered urea (UC pellet) can be solubilized in buffered 5.0 M urea, 2.0 M NaCl or KCl and subjected to chromatography on hydroxylapatite. The immunologically active chromosomal nonhistone proteins elute with 50 mM phosphate buffer (Fig. 16).

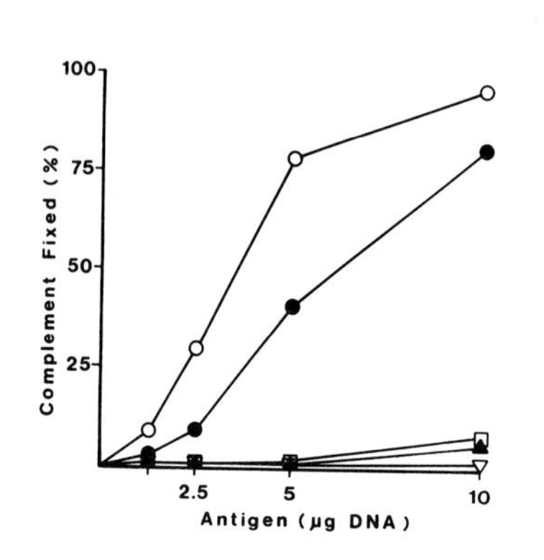

Fig. 16. Complement fixation of chromosomal protein fractions isolated by chromatography on hydroxylapatite. The concentrated fractions were reconstituted with rat spleen DNA and assayed in the presence of antiserum to dehistonized Novikoff hepatoma chromatin. Fractions eluting with 200 and 500 mM potassium phosphate contained mostly DNA and were assayed without the addition of exogenous DNA. Unfractionated control Novikoff hepatoma chromatin (● - ●), Novikoff hepatoma fractions eluting with 50 mM (O - O) and 100 mM (▲-▲) potassium phosphate, rat liver fraction eluting with 50 mM potassium phosphate (□ - □) and all other Novikoff hepatoma or rat liver fractions (▽ - ▽).

The 50 mM phosphate buffer fraction as well as the NP protein fraction are still heterogeneous. The low molecular weight portion of this mixture (three major protein bands of m.w. approximately 12-16,000 daltons) was not antigenic when reconstituted with homologous DNA and assayed by complement fixation. The immunologically active high molecular weight portion could be resolved by preparative polyacrylamide gel electrophoresis into several major bands (Fig. 17). When separated and cut according to the pattern in this Figure, only proteins recovered from gel segment 4 were antigenic (Fig. 18).

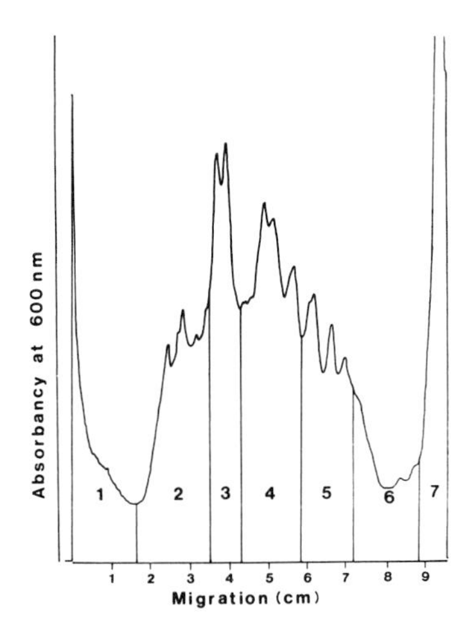

Fig. 17. Polyacrylamide gel electrophoresis of Novikoff hepatoma fraction of chromosomal nonhistone proteins eluting with 50 mM potassium phosphate. The electrophoresis was performed in the presence of 1% sodium dodecyl sulfate and the direction of migration was from left to right. The numbers indicate sections into which the slab was cut to facilitate the elution of the separated proteins.

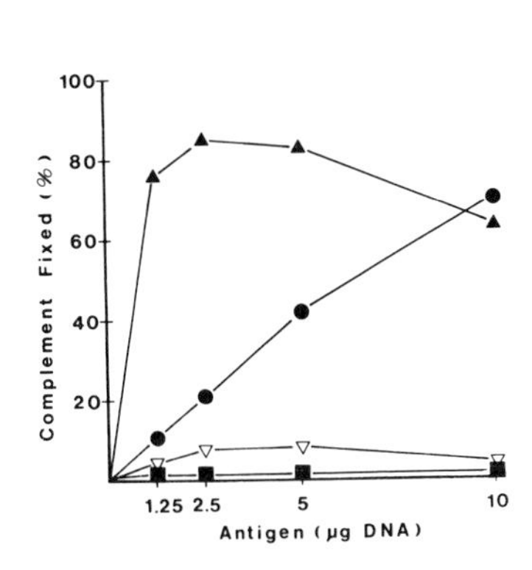

Fig. 18. Complement fixation of fractions obtained by preparative electrophoresis illustrated in Fig. 17. All samples were eluted from crushed gels and sodium dodecyl sulfate removed by ion-exchange chromatography. The recovered protein fractions were reconstituted with rat spleen DNA and assayed in the presence of antiserum to dehistonized Novikoff hepatoma chromatin. Unfractionated control chromatin from Novikoff hepatoma (● - ●), electrophoretic fractions 4 (▲ - ▲) and 7 (▽ - ▽), all other electrophoretic fractions (■ - ■).

These experiments narrow down our search for the DNA-associated nuclear antigens in Novikoff hepatoma cells to three major protein bands of molecular weight between 50-60,000 daltons.

CONCLUSIONS

Although not yet fully characterized, the antigenic complexes described here are of considerable interest. Their immunological specificity has been shown to change with cell differentiation and organogenesis. Malignant transformation resulted in profound changes of the nuclear antigens and their complexes with DNA. The DNA-associated nuclear antigenic protein(s) of Novikoff hepatoma which partial purification was described in this article differs from other nuclear antigens reported in the literature. It does not resemble the Ha protein antigen (m.w. approx. 43,000) recently purified from calf thymus nuclei by Akizuki et al. (32). This antigen reacts with antibodies in sera of patients with sicca (Sjörgren's) syndrome and is not tissue or species specific. The Novikoff hepatoma DNA-associated protein antigen also differs from the NAg-1 nuclear antigen characterized recently by Yeoman et al. (11). Although also present in Novikoff hepatoma nuclei, the NAg-1 is extractable at physiological salt concentrations and its molecular weight is only about 28,000. However, the DNA-associated Novikoff hepatoma antigen may be present in the 0.15 M NaCl insoluble immunogenic precipitate described by Busch et al. (10). Apparently, many nuclear proteins are antigenic. It is reasonable to anticipate that after their appropriate characterization, these antigens and their antibodies can be used for probing the structural and functional properties of chromatin and chromosomes.

REFERENCES

(1) N. Henning, W. Frenger, F. Scheiffarth and A. Assaf, Z. Rheumaforsch, 21 (1962) 13.

(2) L. Messineo, Nature (London), 190 (1961) 1122.

(3) J.F. Chiu and L.S. Hnilica, in: Chromatin and Chromosome Structure, ed. H.J. Li (Academic Press, New York, 1977) p. 193.

(4) S.C.R. Elgin and H. Weintraub, Ann. Rev. Biochem., 44 (1975) 725.

(5) G.S. Stein, T.C. Spelsberg and L.J. Kleinsmith, Science, 183 (1974) 817.

(6) F. Chytil and T.C. Spelsberg, Nature New Biol., 233 (1971) 215.

(7) K. Wakabayashi and L.S. Hnilica, J. Cell Biol., 55 (1972) 271.

(8) K. Wakabayashi and L.S. Hnilica, Nature New Biol., 242 (1973) 153.

(9) L. Zardi, J.C. Lin and R. Baserga, Nature New Biol., 245 (1973) 211.

(10) H. Busch, L.C. Yeoman, R.K. Busch, J.J. Jordan, M.S. Rao, C.W. Taylor and B.C. Wu, Cancer Res., 36 (1976) 3399.

(11) L.C. Yeoman, J.J. Jordan, R.K. Busch, C.W. Taylor, H.E. Savage and H. Busch, Proc. Natl. Acad. Sci. USA, 73 (1976) 3258.

(12) L.M. Silver and S.C.R. Elgin, Proc. Natl. Acad. Sci. USA 73 (1976) 423.

(13) L.M. Silver and S.C.R. Elgin, Cell, 11 (1977) 971.

(14) L.M. Bustin, FEBS Lett., 70 (1976) 1.

(15) L.S. Hnilica, J.F. Chiu, K. Hardy and H. Fujitani, in: Cell Differentiation and Neoplasia, ed. G.F. Saunders (Raven Press, New York, 1978) in press.

(16) M.J. Evans and J.B. Lingrel, Biochemistry, 8 (1969) 3000.

(17) K. Brasch, G.H.M. Adams and J.M. Neelin, J. Cell Sci., 15 (1974) 659.

(18) J. Chauveau, Y. Moule and C. Rouiller, Exptl. Cell Res., 11 (1956) 317.

(19) G. Blobel and V.R. Potter, Science, 154 (1966) 1662.

(20) J.A. Wilhelm, A.T. Ansevin, A.W. Johnson and L.S. Hnilica, Biochim. Biophys. Acta 272 (1972) 220.

(21) T.C. Spelsberg, L.S. Hnilica and A.T. Ansevin, Biochim. Biophys. Acta, 228 (1971) 550.

(22) J.F. Chiu, M. Hunt and L.S. Hnilica, Cancer Res., 35 (1975) 913.

(23) E. Wasserman and L. Levine, J. Immunol., 87 (1961) 290.

(24) P.K. Nakane, J. Histochem. Cytochem., 18 (1970) 9.

(25) F. Chytil, in: Methods in Enzymology, Vol. 40, eds. B.W. O'Malley and J.G. Hardman (Academic Press, New York, 1975) p. 191.

(26) O.H. Lowry, N.J. Rosebrough, A.L. Farr and R.J. Randall, J. Biol. Chem., 193 (1951) 265.

(27) K. Burton, Biochem. J., 62 (1956) 315.

(28) U.K. Laemmli, Nature (London), 227 (1970) 680.

(29) F. Chytil, S.R. Glasser and T.C. Spelsberg, Develop. Biol. 37 (1974) 295.

(30) T.C. Spelsberg, W.M. Mitchell, F. Chytil, E.M. Wilson and B.W. O'Malley, Biochim. Biophys. Acta, 312 (1973) 765.

(31) J.F. Chiu, F. Chytil and L.S. Hnilica, in: Onco-Developmental Gene Expression, eds. W.H. Fishman and S. Sell, (Academic Press, New York, 1976) p. 271.

(32) M. Akizuki, M.J. Boehm-Truitt, S.S. Kassan, A.D. Steinberg and T.M. Chused, J. Immunol., 119 (1977) 932.

(33) G.J. Todaro, K. Habel and H. Green, Virology, 27 (1965) 179.

(34) M. Suzuki and Y. Hinuma, Int. J. Cancer, 14 (1974) 753.

(35) K.O. Fresen and H. ZurHausen, Proc. Natl. Acad. Sci. USA 74 (1977) 363.

(36) G. Klein, B. Giovanella, A. Westman, J.S. Stehlin and D. Mumford, Intervirology, 5 (1975) 319.

(37) S. Perez-Cuadrado, S. Haberman and G.T. Race, Cancer, 18 (1965) 193.

(38) D.J. Carlo, N.J. Bigley and Q. Van Winkle, Immunology, 19 (1970) 879.

(39) D.W. Weiss, Cancer Immunol. Immunother., 2 (1977) 11.

(40) J.F. Chiu, C. Craddock, H.P. Morris and L.S. Hnilica, FEBS Lett., 42 (1974) 94.

(41) J.F. Chiu, L.S. Hnilica, F. Chytil, J.T. Orrahood and L.W. Rogers, J. Natl. Cancer Inst. 59 (1977) 151.

(42) L. Zardi, J.C. Lin, R.O. Petersen and R. Baserga, in: Control of Proliferation in Animal Cells, eds. B. Clarkson and R. Baserga (Cold Spring Harbor Lab. Press, 1974)

p. 729.

(43) Y. Tsutsui, I. Suzuki and K. Iwai, Exptl. Cell Res. 101 (1976) 202.

(44) Y. Tsutsui, H.L. Chang and R. Baserga, Cell Biol. Int. Reports 1 (1977) 301.

(45) K. Okita and L. Zardi, Exptl. Cell Res., 86 (1974) 59.

(46) E.M. Wilson and F. Chytil, Biochim. Biophys. Acta, 426 (1976) 88.

(47) T.C. Spelsberg, A.W. Steggles, F. Chytil and B.W. O'Malley, J. Biol. Chem., 247 (1972) 1368.

(48) K. Wakabayashi, S. Wang, G. Hord and L.S. Hnilica, FEBS Lett., 32 (1973) 46.

(49) K. Wakabayashi, S. Wang and L.S. Hnilica, Biochemistry, 13 (1974) 1027.

(50) J.F. Chiu, S. Wang, H. Fujitani and L.S. Hnilica, Biochemistry, 14 (1975) 4552.

(51) S. Wang, J.F. Chiu, L. Klyszejko-Stefanowicz, H. Fujitani, L.S. Hnilica and A.T. Ansevin, J. Biol. Chem. 251 (1976) 1471.

(52) K. Hardy, J.F. Chiu and L.S. Hnilica, Biophys. J., 17 (1977) 218a.

Original research described in this communication was supported by National Cancer Institute Grant CA-18389 and Contracts N01-CP-55730 and N01-CB-53896.

DISCUSSION

R.A. HICKIE: Did you look at the complement fixation of the preparations from regenerating liver?

L. HNILICA: I think that Dr. Chiu did look at regenerating rat liver and there was cross reactivity during the peak of DNA synthetic activity which was at 18-24 hours. However, when he did immunoabsorption, he could remove this component, and he still retained the Novikoff specific complement fixation. That again suggests that the Novikoff hepatoma nuclear antigens are either heterogenous or their antigenic sites are heterogenous. The immunoabsorption did not remove the component specific for the Novikoff hepatoma chromatin. I know what you are probably aiming at. There was a protein antigen described by Dr. Busch's laboratory which is also present in Novikoff hepatoma, it is the NAg-1 antigen. The NAg-1 is present in fetal liver and also in regenerating rat liver. I believe it is also found in other tumors, but it is not present in normal rat liver. However, its molecular weight is about 28,000, which is about half of what we see for the DNA-binding antigens in Novikoff hepatomas. Since the electrophoresis has been done in SDS, there is essentially little chance for aggregation and our molecular weight range of 50-60,000 is correct.

M. URBAN: You have discussed differences in the antigenicity by complement fixation between dehistonized chromatin of normal and malignant tissues, i.e. most of the studies were conducted on normal liver, Novikoff hepatoma or induced tumors. I wonder if you also could detect differences in the antigenicity between dehistonized chromatin of liver and dehistonized chromatin of spleen. In other words can you detect non-malignant tissue specific differences in chromatin by your assay.

L. HNILICA: Yes. We have done experiments with rat thymus, rat liver and rat kidney and the immunological differences hold. Also, Chytil did some similar experiments with normal chicken tissues. He could produce antibodies to normal tissues which did not react with chromatin from another normal tissue.

M. URBAN: Don't you think it is surprising that you saw antigenic differences between reticulocyte chromatin and erythrocyte chromatin when in essence, at the reticulocyte stage, differentiation has already ceased, the chromatin is condensed, the avian erythrocyte-specific histone 5 is already present, there is very little transcription of the chromatin, and the cells have ceased dividing. In other words reticulocytes are very similar to mature erythrocytes.

L. HNILICA: I do not know if the reticulocyte maturation and differentiation can be equated. I mean, you are right that the difference between reticulocyte and erythrocyte is definitely smaller than, let us say, between hepatocyte and erythrocyte. All I can say is that the results we have obtained are real.

T. MADEN: Would you care to comment on whether there may be any relationship between your findings and those described by Dr. Allfrey on nonhistone proteins this morning and possibly also those described by Dr. Busch earlier?

L. HNILICA: I do not know if I can speak for Dr. Allfrey. However, he did tell me that he has not yet looked at the antigenicity of his two proteins. The molecular weight of his second protein is very similar to the Novikoff group. Of course Dr. Allfrey looked at tumor of colon and we looked at Novikoff hepatoma. It is likely that each tissue or tumor has its own specific nuclear protein antigen. Regarding Dr. Busch's work, there are two or three major differences between the NAg-1 and the Novikoff hepatoma antigens described here. First is the molecular weight. For the NAg-1 it is 28,000 as determined by polyacrylamide gel electrophoresis in SDS. In our experiments, although we do not know which one of the three protein bands is the antigen, the molecular weights of the three bands approximate between 50 to 60,000 daltons, about twice the molecular weight of the NAg-1. The second difference is in extractibility - the NAg-1 is extracted with 0.35 M NaCl. The antigen we are looking at survives this extraction. Actually, most of our chromatins, especially during our earlier work were extracted with 0.3 or 0.35 M NaCl because John Johns published a paper about 5 or 6 years ago where he showed that isolated chromatin is heavily contaminated by cytoplasmic particles or protein which can be removed by extracting chromatin with 0.35 M NaCl. Since we wanted to make sure that our chromatin does not contain any cytoplasmic contamination we extracted most of our chromatin preparations before dehistonization with 0.35 M NaCl. This treatment will remove the NAg-1. The last difference is the requirement of our antigen for DNA. The NAg-1 can be assayed by immunodiffusion and DNA is not needed in these assays.

Dr. Busch published a paper in *Cancer Research* which reviewed work from his laboratory. Here he described antibodies to nucleolar proteins and also some antibodies to other nonhistone proteins which were not solubilized by extraction with 0.3 M NaCl. However, I do not know whether in Dr. Busch's experiments the proteins that remain associated with DNA in chromatin extracted with 0.3M NaCl contain our type of antigens.

PHYSICAL ASPECTS OF THE GENERATION OF MORPHOGENETIC FIELDS AND TISSUE FORMS

A. GIERER
Max-Planck-Institut für Virusforschung
Spemannstr.35, 74 Tübingen, West Germany

Abstract: Prepatterns (morphogenetic fields) arising in initially near-uniform tissues play a major role in morphogenesis. A set of biological features occurring frequently in development can be explained on the basis of molecular kinetics with short-range autocatalytic activation in conjunction with longer range inhibiting effects (lateral inhibition).

Evidence will be discussed suggesting that combinatorial schemes are involved in specifying areas and subareas in morphogenesis.

Real form of tissues often arises by evagination or invagination of initially nearly flat cell sheets. Self-regulatory features suggest that the complex mechanisms determining cell form and interaction can often be subsumed under the minimizing of potential functions of the geometrical characteristics of the cell. Stability of the free cell sheet requires non-linear relations between potential and boundary areas corresponding, for example, to contraction of intracellular fibers or to capping of membrane components. Prepatterns interfering with potentials at defined locations can affect mechanical properties, especially bending moments, leading to tissue evagination and the formation of defined structures. Shell theory is suitable for computer calculations. The emphasis on bending moments distinguishes the generation of such biological from most technical and architectural forms.

PATTERN FORMATION: BIOLOGICAL TYPES AND PHYSICAL BASIS

Development is a complex process combining many different elementary mechanisms, such as cell proliferation, determination, differentiation, movement and death. Spatial order within multicellular tissues and organisms could be envisaged to result from (a) self-assembly of various preexisting differentiated cell types giving rise to defined structures (1,2),

ISBN 0-12-045450-5

(b) time-controlled processes: if a structure grows out sequentially in time, with a sequence of determining steps occurring in the course of growth, order in time may be directly converted into order in space, (c) generation of structure within originally near-homogeneous cells or tissues, be it with or without growth.

The third "morpholactic" type of pattern formation plays a major role in developmental biology. For its understanding it is generally assumed that first an invisible spatial distribution of physical features called prepattern or morphogenetic field (for instance, a substance gradient specifying "positional information" (3))is formed, which in turn elicits local responses of the cells to differentiate and/or generate form. The actual existence of such prepatterns can be inferred from certain transplantation experiments (4). The chemical, structural basis of prepattern formation is not yet known. It seems appropriate, however, to introduce the same basic "demystifying" assumption that underlies the explanation of all other aspects of cellular biology studied thus far, namely that it is eventually based on the interaction and movement of molecular compounds in plasma and on membranes. One then expects pattern formation to be accountable for by reaction kinetics of a type where the concentration c_i of any compound changes in time as function f_i of the concentrations of various compounds (to deal with interaction) and a spreading term to account for movement (e.g. by diffusion):

$$\frac{\partial c_i}{\partial t} = f_i(c_1 \ldots c_n) + D_i \frac{\partial^2 c_i}{\partial x^2} \qquad i=1\ldots n$$

This type of equations, while rather non-committal with respect to mechanisms, imposes stringent constraints on the construction of theories and models; so stringent, indeed, that one may ask whether spatial patterns can be generated on this basis at all.

That this is possible was discovered twenty-five years ago by Turing (5). He showed that patterns can be formed if there are two substances reacting auto- and crosscatalytically, and spreading by diffusion.

Linear reaction kinetics do not generally lead to patterns with biological features, whereas general non-linear kinetics are too varied to be of much explanatory value. As a starting point, we may specify a set of basic biological features to be explained, and ask for the kinetic requirements of molecular interactions accounting for these properties.

The following set of features occur in many systems and stages of development:

1.) Striking structures (e.g. a head in a regenerating section of hydra) are formed starting from near-uniform tissues.
2.) A prepattern precedes and directs activation of structures (e.g. of a hydra head).
3.) There is polarity, a rather unspecific, possibly slight tissue anisotropy that may direct the orientation of an asymmetry without affecting its final form. In some cases, like hydra, polarity is due to a gradient of a scalar property, e.g. of a concentration gradient of some compound (6).
4.) Many morphogenetic fields are self-regulating. In particular, a part of a tissue may produce all structures at reduced size.
5.) There is short-range activation: Certain small transplants, and other stimuli, can induce a secondary organizing center (forming, for instance, a secondary head in hydra).
6.) Further, there are inhibitory effects of wider range: pre-existing structures, such as a head in hydra, can inhibit the development of like or similar structures nearby (4). Such inhibitory effects can account for the spacing of repeating structures, for instance, of leaf rudiments in plant development.

PATTERN FORMATION BASED ON AUTOCATALYSIS AND LATERAL INHIBITION

We may now ask which are the (minimal) requirements for reaction systems that explain the basic features of biological development just listed. For this problem we have proposed the following solution (7):

1.) There have to be at least two compounds or compound groups interacting auto- and crosscatalytically; one activating, the other inhibiting. Inhibition can also be mediated by depletion of a compound required for activation.

2.) The range of the inhibitor must be considerably larger than the range of the activator (lateral inhibition). Range is defined as the mean distance between production and decay or removal of a compound.

3.) Total area must considerably exceed the range of activation.

4.) To generate patterns, local activation within a near-uniform initial distribution must be self-enhancing due to the autocatalytic features of the system.

5.) The inhibitor effect must ensure that average activation is limited to prevent an overall autocatalytic explosion, and

6.) the inhibitory reaction must be relatively fast as compared to the activating one, to ensure stability of the pattern produced.

These conditions can be put in mathematical terms to assess equations as to whether they do or do not lead to such patterns, and to generate theories and models for pattern formation (7, 8). Many different molecular models involving no features unusual in molecular biology can produce such patterns and only biochemistry could decide between them. Initiation of pattern formation may occur by preexisting asymmetries (polarity), random fluctuations, or external induction.

Figure 1a shows a computer simulation based on such a model. Formation of a gradient is initiated by a small peak in the left half (which may be due to random fluctuations, or external induction). This initial advantage is self-enhancing according to autocatalytic activation; inhibitor produced in the activated region spreads into a wider area so that activation (left) can proceed only at the expense of deactivation elsewhere. Autocatalytic activation proceeds until a stable distribution is reached. The final form is essentially determined by activator range and other system parameters, and is rather indifferent to details of initial distributions. Only the cue to orientation is taken from them and a shallow initiating gradient leads essentially to the same pattern (Fig. 1b).

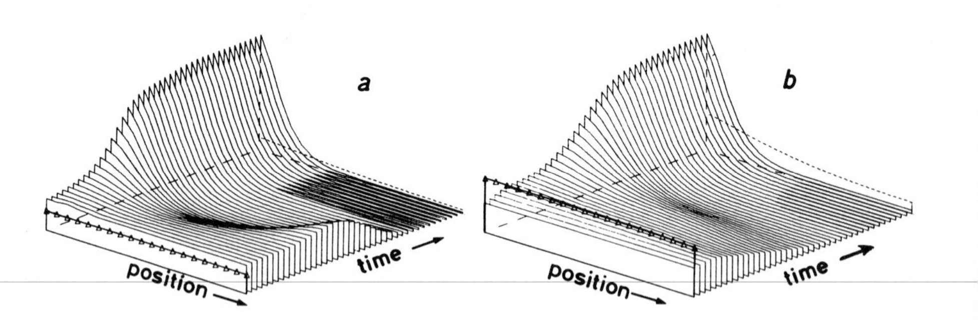

Fig. 1. Pattern formation by lateral inhibition. An activator pattern may be initiated by a small random fluctuation or initial induction (left within fig. a) or by an asymmetric distribution of sources -Δ-Δ-Δ-, such as a shallow gradient of activator producing enzymes (fig. b).

The example shows that morphogenetic gradients to specify positional information can easily be generated by such mechanisms. In fact, if a closed field is somewhat, but not much, larger than the range of activator, the pattern formed is always asymmetric, forming a gradient (9). In particular a gradient is formed if a pattern-forming system operates while a field grows beyond the minimal size required for pattern formation. If initiated by random fluctuations, the orientation of the gradient is also random. However, for most embryological systems and their substructures orientation is determined by polarity, that is, by some (possibly slight and unspecific) initial asymmetry of the developing or surrounding tissue.

In wider fields a symmetric pattern may be formed; secondary structures can be initiated in areas sufficiently far out of the range of inhibition of a primary peak; and periodic patterns can be generated. Some types of patterns in two dimensions (cell sheets) are exemplified in fig. 2. Properties

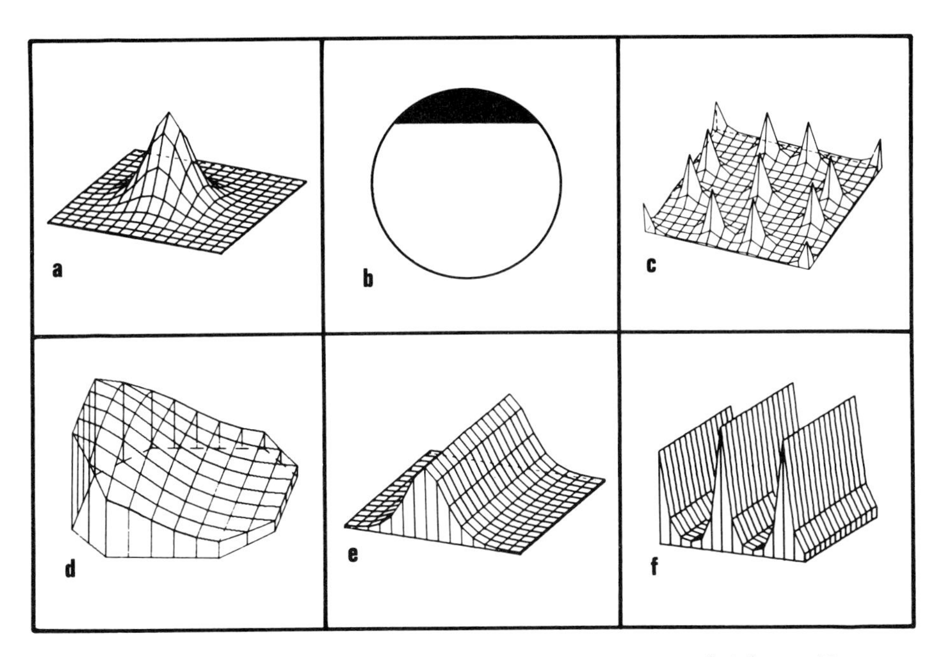

Fig. 2. Two-dimensional patterns, e.g. within cell sheets. Depending on parameter choices, one may obtain a single peak on a plane (a) or a sphere (b), regularly spaced peaks (c), or a gradient (d). In anisotropic tissues a ridge of activation (e) or a periodic pattern in one of the two dimensions (f) may be generated (11).

of various systems, including hydra, the developing retina and insect embryos, have been simulated on the basis of the theory (7, 9, 10, 11).

The theory is applicable to intracellular as well as intercellular patterns (9) to account for anisotropic features of the cell, such as form, extension of processes and chemotactic movement. One can estimate that a fraction of a minute is sufficient to produce a focus of activity in a small area of the cell membrane oriented by a shallow external gradient.

The principle of lateral inhibition - short-range activation in conjunction with longer range inhibition - is the simplest way to generate patterns by physical molecular kinetics. There may be cases with many interacting components in which a distinction of activatory and inhibitory effects is not unambiguously possible. However, the occurrence of short-range activation (induction) and long-range inhibition (spacing of like or similar structures) is well documented experimentally in embryology and regeneration. Very complex patterns can be produced by combination of simple pattern-forming systems of the lateral inhibition type.

EFFECTS OF PREPATTERNS AND MORPHOGENETIC FIELDS; RECURSIVE MECHANISMS

Prepatterns are expected to determine cell responses, such as cell determination, differentiation, proliferation, movement and death. We may consider cell determination as transition between multiple stable states of a cell (12,13). For instance, there may be products (e.g. of a gene) feeding back on the rate of production in such a way that activity is bistable, with distinct "on" and "off" states. If a prepattern affects the probability of transition, this leads to a continuous distribution of "on" cells in response to a continuous prepattern (Fig. 3a); alternatively, the response of the cell may occur above a certain threshold; this would divide an area, in response to a continuous prepattern, into strikingly different segments (Fig. 3b). More complex cell responses would lead to more complex subdivisions of an area. Thus it is possible that there are several different control circuits turned on by a prepattern above certain thresholds or in certain concentration ranges, either in an additive (Fig. 3c) or in a mutually exclusive manner (Fig. 3d). In the latter two cases a cell response may depend not only on the local value of the prepattern, but also, in a recursive manner, on the stage of determination already reached in or near a given positon. An extreme model for recursion would be to postulate that the prepattern determines only the terminal section of the pattern

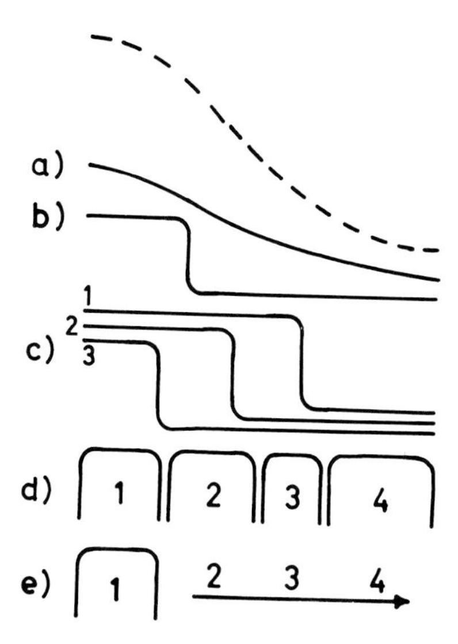

Fig. 3. Possible cell responses to morphogenetic gradients (---) include (a) graded and (b) discontinuous distribution of cell types, (c) several thresholds of determination, and (d) a sequence of mutually exclusive states. In cases (c) and (d) activation may depend on history and neighbouring areas; e.g. there may be primary activation at a terminal, followed by sequential determination (e).

which then initiates a sequence of inductions across the tissue (Fig. 3e).

Some recursive features are found experimentally, e.g. in the regeneration of imaginal discs and their products in insects. Excised proximal parts regenerate distal parts, but distal parts often duplicate to produce another distal part. Perhaps some structures are determined sequentially in the course of outgrowth in a proximo-distal direction, but there is experimental evidence against generalizing this simple explanation: Rudiments of structures to be regenerated have been found before any growth occurred (14), and in intercalary regeneration relatively more distal cells participate in the production of relatively more proximal structures (15). Thus cell communication between adjacent areas codetermines the fate of the regenerate and suggests that prepattern formation, possibly on the basis of lateral inhibition, is involved. However, additional features are necessary to limit regeneration to those parts that are missing, perhaps by sensing discontinuities in the sequence of structures (11, 15). Further studies are required for a better understanding of such recursive mechanisms.

COMBINATORIAL MECHANISMS: IS THERE AN "AREA CODE" IN MORPHOGENESIS?

In the course of development of an organism areas and subareas become sequentially determined to develop into different structures; eventually thousands of different areas arise, some distinguishable by cell type and composition, but many by more refined features of form and other quantitative parameters.

It appears likely that there are only few prepattern-forming systems and these become activated repeatedly in different stages and locations of development to subdivide areas. If activation can cause the same control circuit to be switched on in different locations, a limited set of biochemical circuits would suffice to define in a combinatorial manner a very large number of different states of the cells. Effects of prepatterns to specify areas and subareas could then lead to a combinatorial subdivision in analogy to an area code: Different regions would be primarily defined by different combinations of control circuits turned "on" or "off".

There is some indirect evidence suggesting the existence of combinatorial specifications. The scheme of transitions occurring in transdetermination of imaginal discs in insects has been interpreted in terms of a combinatorial scheme (12). In the course of cell differentiation there is a sequence of determining steps from the fertilized egg to the fully differentiated cell types, each step involving a decision between few (perhaps only two) pathways. This system of sequential branching calls for a specific explanation. Combinatorial schemes in which the switching on of some circuits is a prerequisite for switching on others, seem adequate to account for it (13). More direct evidence for a combinatorial specification has been discovered by Bohn (16). If different segments of cockroach legs are grafted onto each other, intercalary growth occurs in most cases where structures are missing; however, if different segments are grafted onto each other at corresponding intrasegmental positions (say, one third from the distal end) no intercalary regeneration is initiated. Thus, for the discontinuity signal initiating regeneration, intrasegmental specification seems to be similar in different segments. This indicates that specification is combinatorial, with different specifiers referring to segment type (such as femur or tibia) and to intrasegmental position. Similar experiments have demonstrated independent specification of the angular position.

Thus, in the insect there appears to be at least combinatorial specification of body area (which determines, for instance, leg formation), segment, intrasegmental position and angular position within the leg. One may estimate that this combination of specifiers is sufficient to distinguish between some 10^3-10^4 subareas in the insect. Twice the number of specifiers would lead to an order of 10^6-10^8 distinguishable subareas, an order of magnitude that seems sufficient even for area specification of the mammalian nervous systems. While these considerations suggest that some combinatorial area specification is involved in morphogenesis, it is not proposed that the scheme is simple, say digital, nor that it applies for all tissues.

CAUSES OF TISSUE EVAGINATION

One of the most interesting features of higher organisms is their specific three-dimensional form. Its development is again complex, involving cell movement as well as growth patterns, but often structures can be traced back to rudiments which in turn arise by evagination or invagination of initially nearly flat cell sheets. For instance, the blastula invaginates to form the gastrula which folds to produce the neurula, and buds often arise by evagination.

To account for evagination of cell sheets, one might postulate locally enhanced cell proliferation, or lateral forces extending over the sheet to cause folding. However, embryological evidence suggests that often evagination occurs in areas that do not show unusual growth patterns, and evagination is not suppressed if the evaginating area is excised. This indicates that evagination is a local process in which bending moments are produced to generate curvature of a cell sheet.

What determines this curvature? Different mechanisms probably combine to affect cell form and interaction in the sheet. Membrane components affect adhesion between cells (2). This is a dynamic process in which cell surfaces may protrude and retract, membrane components are inserted and removed, and may move and cap within the membrane itself. Cell communication across intercellular junctions (17), and interaction of cell surfaces with the medium probably affect cell metabolism and form. The formation of and interaction with extracellular matrices may contribute. Further, intracellular fibers, such as microtubules and microfilaments are expected to participate in defining cell form (18). Contractile mechanisms may, but need not, be involved in these cases. If they were, they could explain changes of cell form only in conjunction with other mechanisms, such as anisotropic anchorage or extensions of fibers. Probably the effects mentioned interact dynamically to specify cell form and interaction in cell sheets.

MINIMAL POTENTIAL, STABILITY AND FORM OF CELL SHEETS

One cannot expect to relate tissue form directly to such a complex combination of molecular events; rather a phenomenological theory is required with parameters that are interpretable by molecular processes and at the same time permit the derivation of the real form of tissues. Further, a comprehensible physical theory requires introducing approximations and simplifying assumptions which are consistent with biological evidence. The main assumption in the theory to be described (19) is the notion that tissue form corresponds to a state of minimal potential as generalization of minimal energy.

The notion of minimal energies of adhesion has been successfully introduced by D'Arcy Thompson (20) in discussing cell form and by Steinberg (2) in the generation of spatial order by sorting out. Ljapunov has shown that the notion of minimizing potentials can be applied to analysing general stability characteristics of dynamic systems. If there are parameters which change in time as function of various parameters, stable steady states may result. In this case they correspond to a minimum of a potential function, at least in the environment of the stable state. The physical consequence is that the system has regulatory features to correct distortions.

Such regulatory features are experimentally found in developing systems to an impressive extent. Often tissues can be desaggregated into cells to reform tissue-like structures by sorting out; intracellular cytoarchitecture can often be broken down reversibly; morphogenetic processes can be interrupted by inhibitors and resumed after their removal; in some cases tissue evagination can even be reverted and repeated (21). All these regulatory effects suggest that the form generated corresponds to a steady state rather than to a sequence of time-determined all or none processes and may be describable by minimizing potentials, even in states far from the stable state.

The principle of minimal potential to describe tissue form does not imply that regulatory capacities persist indefinitely; after a rudiment is laid down, irreversible processes of fixation, growth and differentiation may occur, followed by further morphogenesis in subsections.

We will consider the evagination of a free cell sheet as the prototype for tissue evagination, since cell sheets coming off a surface or carrying underlying tissue with them can be treated in a formally similar manner. For each cell in the sheet we distinguish external, internal and intercellular surfaces or boundary areas (f_a, f_b, f_c). Inside-out asymmetry of cell sheets is often directly visible and can be explained as being induced by the difference between internal and external medium. As discussed above, we may further take it for granted that there are mechanisms generating prepatterns to distinguish an area of the cell sheet from the rest, initiated either internally or by external induction. All effects of cell communication across the sheet are subsumed under such prepatterns. Cell response is taken as minimizing potentials, the potential of the sheet being additive with respect to the potential of the cells. Potential in general is a function of geometric as well as internal parameters of the cell, but if we are interested in geometry only, we may take the internal variables as eliminated, and express potential as function of geometry only (This approximation need not always be valid). Using external,

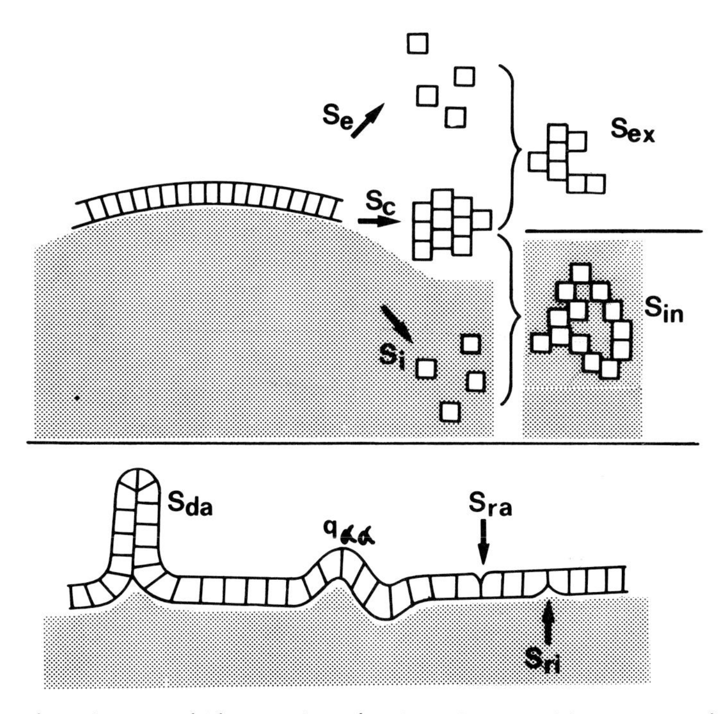

Fig. 4. Stability criteria for free cell sheets with different inside and outside media. Stability must hold against decay into outside medium (S_e) and inside medium (S_i); clumping (S_c); and any combination of intercellular with external (S_{ex}) or internal (S_{in}) surface areas. $q_{\alpha\alpha}$ is stability against bending ("undulation"), S_{ra} and S_{ri} against cleavage between intercellular boundary areas.

internal and intercellular boundary areas as geometric variables does not imply that the molecular effects reside exclusively in these areas, since other geometric features, such as edges, corners or internal cross-sections, can be expressed in terms of boundary areas and vice versa.

Requirements for a theory of evagination are the stability of the cell sheet per se, and the localized generation of bending moments and curvature by prepatterns. For cell forms accounting for sheets of variable thickness and curvature there is no complete mathematical solution to determine stability against all conceivable alternative states. However, an approximative stability assessment has been proposed, based on the following conditions (Fig. 4): The cell sheet must be stable against decay into inside and outside medium and against clumping; a more refined version is stability against <u>any</u> combination of external and intercellular surfaces, and <u>any</u> combination of internal and intercellular surfaces, corresponding to spongy structures, lines or sheets extending into the media. Further requirements are stability against bending (positive curvature in one area at the expense of negative curvature in another leading to undulation) and against cleavage through the sheet between intercellular surfaces.

It can be shown that linear relations between potential and boundary areas

$$P = c_a f_a + c_b f_b + c_c f_c$$

cannot meet the stability criteria for cell sheets against clumping and decay. It follows that no stable isotropic distribution of molecules within the surface, however complex in composition, could account for a stable cell sheet. Some non-linear feature is required. For instance, inclusion of saturation terms

$$P = c_a f_a - \frac{g_a \kappa_a f_a}{1+\kappa_a f_a} + c_b f_b - \frac{g_b \kappa_b f_b}{1+\kappa_b f_b} + c_c f_c$$

suffices to guarantee stability for wide ranges of parameters with respect to all stability criteria mentioned. Such non-linear terms can be interpreted in many ways, for instance as saturation or capping of membrane components, or as the insertion or contraction of fibrous structures close to a boundary area.

The stability assessment has been extended to laterally anisotropic tissues with different (e.g. proximo-distal and dorso-ventral) axes, to cells of variable volume (resulting for instance,from vacuolization), and to multiple cell layers.

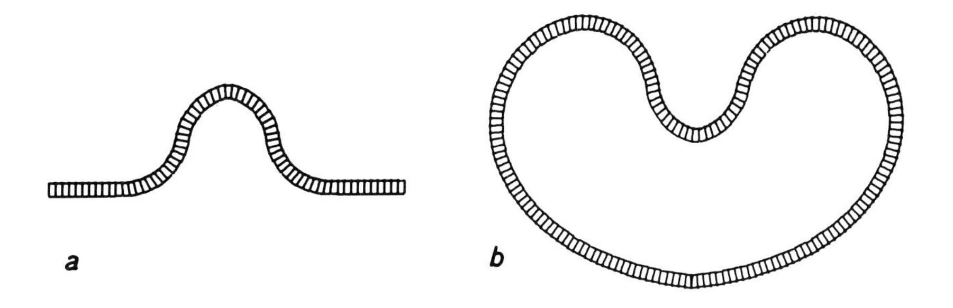

Fig. 5. Curvature in one dimension within a two-dimensional cell sheet caused by a ridge of activation. (a) Evagination of a layer supported by a surface, (b) invagination of a cylinder parallel to the axis.

TISSUE MECHANICS AND THE GENERATION OF FORM

Since potential of the cell sheet is taken as additive with respect to the cells, minimal potential for the sheet is given by mechanical laws in which forces and moments are derivatives of cell potentials with respect to intercellular distances and angles. These can easily be calculated for potential functions. Once potentials consistent with the stability of a cell sheet are assigned to cells, local changes of parameters by prepatterns lead to bending moments, curvature and form.Generally, the form assumed by a cell sheet will be a complex function of prepatterns, boundary conditions etc. A few simple prototypes will be discussed. If a prepattern forms a stripe of activation in a cell sheet supported by a surface, this may cause a ridge of evagination of the cell sheet (Fig. 5a). A stripe of activation parallel to the axis of a cylinder will cause evagination or invagination, the latter resembling the process of neurulation (Fig. 5b). Of particular interest are structures with curvatures in two dimensions. Simple prototypes are given by rotationally symmetric forms. Shell theory is suitable for computer calculations of such structures.

Fig. 6a shows an example with strong bending moments generated in an activated cap area and weaker bending moment in the surrounding areas. The complex structure produced is due to the interplay of curvatures in the two dimensions. Sheets are "reflected" if they come too close to the axis. Thus, tissue mechanics in two dimensions may lead to complex and sometimes anti-intuitive features.

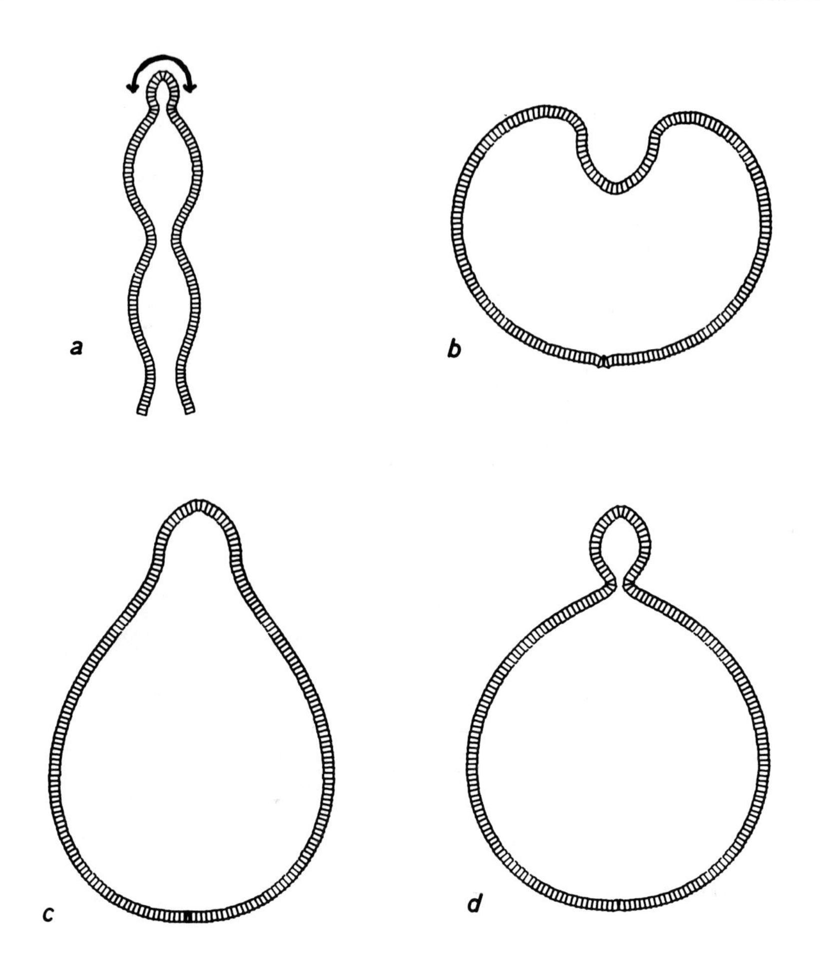

Fig. 6. Curvature in two dimensions (rotational symmetry, axis of rotation vertical). (a) In a cap area (<—>), bending moment is induced to be higher than in the remaining tissue. The complex structure results from the interplay of curvatures in two dimensions. (b-d) Focus of activity (see fig. 2b) is assumed to generate localized negative or positive contributions to bending moments in a closed sphere, leading to invagination (b), or to evagination with bending moments increasing from (c) to (d).

Figure 6b-d shows invagination and evagination of a closed rotationally symmetric cell sheet. A prepattern with a focus of activation (see fig. 2b) is assumed to lead to negative or positive contributions to bending moments. No ad hoc assumptions on boundary conditions are required; internal physical processes within the closed structure suffice to account for its three-dimensional form.

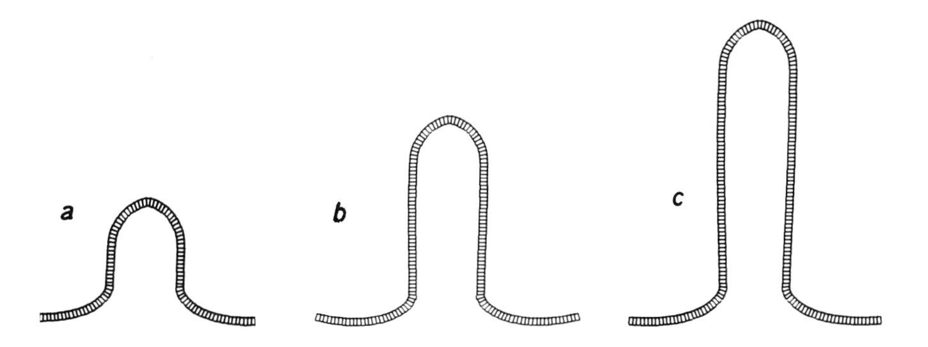

Fig. 7. Computer model for the generation of elongated structures upon budding. Activation is assumed to induce strong bending moments in a constant cap area, and weaker moments in the body area increasing from a-c.

Many structures formed in biology are elongated. Elongation can result if a cap area is activated to generate a strong and a surrounding area to produce a weaker bending moment; if resistance to curvature is weaker in a cap as compared to the surrounding area; if effects of coordination favour near-cylindrical structures; or if a cap area generates tissue anisotropy with bending moments in the surrounding area that induce curvature mainly in the angular dimension leading to near-cylindrical configurations. Elongation can also result from complex combinations of such effects. Fig. 7 shows a computer simulation of the generation of an elongated structure starting from a nearly flat cell sheet (such as occurs during budding of hydra).

The few examples are to show how prepatterns may induce in initially nearly flat cell sheets the formation of defined structures. For a more detailed account the reader is referred to the original paper (19).

Shell theory is also suitable for dealing with more complex cases and less symmetric structures, and additional features, such as growth, differentiation and mechanisms of fixation, may be introduced into the formalism.

The theory is applicable not only to cell sheets but also to membranes or complexes of membranes with extracellular matrices, to account for the form of individual cells. Thus, fig. 6b-d may be taken as a model of invagination or evagination of a cell membrane; an intracellular prepattern activating a small membrane area may lead, for instance, to the insertion of molecules into the inner part of the membrane to generate bending moments causing invagination.

It is not claimed that all tissue forms can be explained on the basis of the theory with the principle of minimizing potentials. On the contrary, there will be cases dominated by effects which are neglected in our approach, such as friction, shear and mechanisms that are primarily time-controlled. On the other hand, the self-regulatory features often observed in morphogenesis suggest that the principles of minimal potentials and theories based on them have a range of applications relating molecular mechanisms to real form.

The mechanics employed to calculate tissue forms are conventional, but within the framework of shell theory emphasis is placed on local bending moments generated within cell sheets (and not, for instance, on tangential forces, or forces exerted externally). This distinguishes such biological morphogenesis from the generation of most inorganic forms in nature, be they technical, architectural, or geological.

The crucial quantitative role of boundary areas - internal, external and intercellular - of cells in a tissue in defining the stability and form of cell sheets suggests that the same quantities may also be sensitively involved in other regulatory processes, especially cell proliferation.

ACKNOWLEDGEMENTS

I am much indebted to Lynn Graf and Dr. Hans Meinhardt for the critical reading of the manuscript. Figures 4-7 are reproduced with permission (19).

REFERENCES

(1) A. Moscona. Exp. Cell Res. 22 (1961) 455

(2) M.S. Steinberg, Science 141 (1963) 401

(3) L. Wolpert. Curr. Topics in Devl. Biol. 6 (1971) 183

(4) G. Webster and L. Wolpert. J.Embryol.exp.Morphol. 16 (1966) 91

(5) A. Turing. Phil.Trans.Roy.Soc. 237 (1952) 32

(6) A. Gierer, S. Berking, H. Bode, C.N. David, K. Flick, G. Hansmann, H. Schaller and E. Trenkner. Nature 239 (1972) 98

(7) A. Gierer and H. Meinhardt. Kybernetik 12 (1972) 30

(8) A. Gierer and H. Meinhardt. Lectures on Mathematics in the Life Sciences 7 (1974) 163

(9) H. Meinhardt and A. Gierer. J.Cell Sci. 15 (1974) 321

(10) H. Meinhardt. J. Cell Sci. 23 (1977) 117

(11) A. Gierer. Curr.Topics in Devl.Biol. 11 (1977) 17

(12) S. Kauffmann. Science 181 (1973) 310

(13) A. Gierer. Cold Spring Harbor Symp.Quant.Biol. 38 (1973) 951

(14) D. Bulliere. Am.Embryol.Morphog.5 (1972) 61

(15) H. Bohn. Wilhelm Roux' Arch.Entwicklungsmech.Org. 167 (1971) 209

(16) H. Bohn. Wilhelm Roux' Arch.Entwicklungsmech.Org. 165 (1970) 303

(17) W.E. Loewenstein and Y. Kanno. J.Cell Biol. 22 (1964)

(18) E. Lazarides and K. Weber. Proc.Nat.Acad.Sci.(USA) 71 (1974) 2268

(19) A. Gierer. Quart.Rev.Biophys. 10 (1977)

(20) W. Thompson d'Arcy. On growth and form. Cambridge University Press, 1952

(21) B.S. Spooner and N.K. Wessels. Proc.Nat.Acad.Sci. (USA) 66 (1970) 360

DISCUSSION

R.C. LEIF: Going back to your single cell aggregation with the hydra, with the field theory is it possible to grow those cells up in spinner culture for lets say a week, in which case any field forming substances which determine polarity should be diffused or absent.

A. GIERER: Isolated cells from hydra lose their capacity to sort out properly if they are kept for more than a few hours, unfortunately, so that this experiment cannot be done up to now.

T. MADEN: What does Morphogenetic Field Theory have to say about model properties such as the development of multiple vertebrae, in other words about the formation of recurrent features in an animal as exemplified by the somites of higher organisms.

A. GIERER: Well, periodic structures are easily accounted for on the basis of this theory. The question, however, is whether those cases in which one has a defined but large number are due to pre-patterns of this type. The theory can explain, in principle, a fairly defined number of multiple peaks if one initiates them in sequence with patterns and subpatterns and sub-subpatterns, but it is not clear whether this is the real explanation for somites. One could also imagine that other mechanisms more reflecting the cell response to fields by determination and differentiation, possibly in conjunction with mechanisms analogous to counting, account for these exact but large numbers. This is an open but very interesting question.

J.D. WATSON: Alfred, can I ask a question? You never mention microfilaments. When do you think microfilaments in your form of morphogenesis might come together?

A. GIERER: I think I briefly mentioned microfilaments as one of those elements making up the cytoarchitecture and, indeed, I think that in the response of cells to generate form of tissues microfilaments are very important. Whether active contraction really matters is not yet known. In any case, contraction in itself would not constitute an acceptable explanation, it looks more as if a steady state is approached where anisotropic anchorage, extension, removal and perhaps contraction of fibers together with the other effects combine to co-determine the form of a cell.

THE CELL-TO-CELL MEMBRANE CHANNEL IN DEVELOPMENT AND GROWTH

Werner R. Loewenstein
Department of Physiology and Biophysics
University of Miami School of Medicine
P.O. Box 520875
Miami, Florida 33152 U.S.A.

Abstract: The hypothesis that the channels in permeable cell junctions are conduits for the cell-to-cell transmission of growth controlling molecules predicts that tissue cells of normal growth (growth-control competent) be capable of cell-cell transmission (channel-competent), and that channel-incompetent dividing tissue cells be growth-control incompetent. A survey of cells of normally growing organs and tissues and of deviant incompetent cells bears this prediction out. Fusion of channel-competent human cells with channel-incompetent mouse cells produces channel-competent and growth-control competent hybrids so long as they have the sufficient chromosome complement of the human parent cell. Upon loss of human chromosomes, in particular chromosome No. 11, the hybrids switch to channel incompetence and growth control incompetence. Channel competence and growth control competence behave as genetically insegregable traits, as does the pair of the opposite phenotype.

Another approach explores the hypothesis that the channels are selective conduits for the cell-to-cell transmission of differentiation-controlling molecules, i.e., conduits with a variable, developmental-stage dependent bore. Simultaneous probing of the cell junction of *Chironomus* salivary gland with fluorescent-labelled molecules of different size shows that the effective channel bore undergoes reduction at a critical stage of larval development; the molecular weight limit for permeation reduces from $\geq$ 1600 to < 700 daltons.

ISBN 0-12-045450-5

INTRODUCTION

Cells of organized tissues have channels in their junctions that provide a path for direct flow of hydrophilic molecules between cell interiors (1,2). These channels appeared early in phylogenetic development; they are present in primitive sponges and throughout the phylogenetic scale up to man. As a general rule, a given cell in a tissue interconnects via such channels with many neighbors (1). Thus, all cells in a salivary gland (3,4) or in a thyroid acynus (5) are interconnected, or cells in liver (6) and skin (7,8) are widely interconnected forming continuous systems from within.

The cell-to-cell channel has a bore of at least 14 Å, wide enough to permit the passage of peptide molecules up to 1600 daltons. One may expect, of course, electrostatic interactions between charged permeant molecules and the polar groups lining the channel, but for molecules up to about 500 daltons such interactions evidently do not greatly hinder their passage. Given these properties, the cell-to-cell channel is expected to permit transmission of a broad range of cellular molecules: inorganic ions, metabolites, small hormones, high energy phosphates, vitamins, nucleotides, cyclic nucleotides, etc. It would not be too surprising therefore if such an ubiquitous and ancient hole between cells has adapted to many cellular functions. I discuss here the possibility of functions in differentiation and growth control.

THE CELL-TO-CELL CHANNEL AS A CONDUIT FOR MORPHOGENS

In regard to the first possibility, our general hypothesis is that the cell-cell channels are conduits for differentiation-inducing molecules (9). In dealing with this possibility, I lean on the facts that the cell-cell channels are present in the embryo (9-16), where early on they interconnect all cells (17); and that the effective bore size of the channel (in the adult animal) is variable in a graded manner, at least under experimental conditions (18,19). Our working hypothesis is that the channel bore varies at different stages of development, permitting different morphogenetic molecules to be sieved out by their size or charge (9). A selective intercellular transmission mechanism of this kind, with regional programs in the embryo, would be genetically economical, indeed.

An early hint in this direction was provided by our work on *Chironomus* salivary gland. The space constant of voltage attenuation along a chain of cells was found to fall one order of magnitude between the fourth-instar and prepupal stages of larval development, while input resistance and cell length remained relatively constant (20). This suggested a marked fall in junctional conductance during development (by transmission line analysis (21), a fall by a factor of 10 to 40). Recently, Caveney (22) showed in an interesting series of experiments that the cell-cell conductance in *Tenebrio* epidermis changes during larval development and that one of the changes can be mimicked by the hormone ecdysterone. Cell-cell conductance depends on the effective bore size of each channel and on the number of channels in a junction, and so these results and the earlier ones either meant that the channel bore changes during development or, less interestingly, that the number of channels changes.

R. Wagner and I have now made permeability measurements on *Chironomus* salivary gland that permit a distinction between these alternatives. The channels were probed simultaneously with molecules of different size (and different color label). In each permeability test, a mixture of two or three molecular species was microinjected into a cell, and the transit velocities across junction were determined. Here if, and only if, the channel bore changes, should the junctional transit velocities of differently sized molecules be changed by different factors. Indeed, at a critical stage of the fourth instar--under the experimental conditions between the tenth and eleventh day of this stage--the junctional transit velocities of the larger molecules of the test series were found to fall more than those of the smaller ones. In fact, in most cases, at the critical stage, junctional transit was entirely blocked for the larger molecules (e.g., $LRB(Leu)_3(Glu)_2OH$, 1157 daltons), while that for the smallest molecule continued (DANS SO_3H, 251 daltons) (R. Wagner & W.R. Loewenstein, unpublished work).

These results are encouraging. They reveal that the molecular size limit for junctional permeation changes in this developing system, and similar analyses of other systems should soon tell us how general this phenomenon is.

THE CELL-TO-CELL CHANNEL AS A CONDUIT FOR GROWTH-CONTROLLING MOLECULES

I have given elsewhere the *a priori* arguments for a role of the cell-cell channels in the regulation of cellular growth and I have illustrated these with a simple self-regulating model *sui generis* for junction-connected cell systems (23-25).

Here I restate only the basic hypothesis and discuss the experimental results so far obtained in testing it.

The general hypothesis is that the cell-to-cell channels are conduits for the cell-cell transmission of molecules necessary for growth control (1,23). Even in this general form, without specifying the control loop to which these molecules pertain, the hypothesis is of immediate experimental interest. Its corollary is that absence or obstruction of the channels results in disturbance of cellular growth (23). And so the hypothesis makes the testable predictions: (i) a dividing cell population competent in regulating its growth is channel-competent; and (ii) a dividing cell population which is channel-incompetent is growth-regulation incompetent.

Our strategy for testing these predictions was (a) to examine competence of a broad range of normally growing cell types; (b) to search for deviant cell types that are channel-incompetent and to see whether these are also growth-control incompetent; and (c) to try to trace the two incompetencies to a common genetic defect. A straightforward genetic approach would have been to use one-step mutant cells, but we were unable to produce them. The next best strategy was to use segregants for step *c*, as produced by the spontaneous loss of chromosomes in somatic cell hybrids between channel-competent and channel-incompetent cells. This approach, based on loss of whole chromosomes, falls short, of course, of demonstrating a correlation between the channel and growth control at the single gene level; but it permits, at least in principle, to demonstrate such a correlation at the single chromosome level.

The first prediction was born out by the work of the past ten years. All cells of organized tissues examined, capable of dividing, were found to be channel-competent. The examination covered a wide variety of tissues (1,25), including tissues in conditions where cell division is fast and generalized, such as during regeneration (liver) or wound healing (skin) (7). The only normal tissue cells known to be channel-incompetent are (adult) skeletal muscle fibers and nerve fibers. But these cells--and nicely fitting the hypothesis--are no longer capable of dividing; while they are capable of dividing, in the embryonic state, they are also channel-competent (26,11).

The testing of the second prediction was more difficult. The procurement of channel-incompetent cell strains alone has been a major task and the demonstrations of the incompetencies of cell-cell transmission and of growth control are technically more involved than the demonstrations of the respective competencies. Moreover, a meaningful test of the hypothesis requires the establishment of a strict correlation between the incompetencies, which again is complex. But sufficient results are now on hand for an evaluation of the hypothesis.

Channel-incompetent cell types were obtained by X-irradiation of epithelioid embryonic cells (27), or by searching for random mutants among various cell lines. A total of 17 channel-incompetent cell strains were thus isolated (28,29). In all cases, the channel incompetent cells were also growth control incompetent by *in vitro* and *in vivo* criteria (25,27-29).

In the case of these 17 cell strains, the primary search was for channel incompetence; and only after the junctional transmission properties of the original clones had been tested and their channel defect had been established, were the growth properties of these clones tested. This experimental sequence is the most satisfactory one for testing the hypothesis. The simpler procedure, with the sequence in reverse, namely, a survey of channel competence among cell lines with known growth abnormalities (cancerous), came up with 7 cancer cell types (hepatoma, ref. 27,30,31; neuroblastoma, ref. 32,33; breast carcinoma, ref. 34,35) which appear fully channel incompetent and 8 types in which the bore or the number of the channels is abnormally small (36,37) or the rate of channel formation abnormally slow (36).

The result that the occurrence of (full) channel incompetence was invariably paralleled by growth control incompetence is naturally very satisfying in terms of the hypothesis; but the crucial question is whether the two incompetencies are genetically linked. In the absence of knowledge of the identity of the growth controlling molecules, the demonstration of such a linkage offers the only feasible way to cut short the risk of an epiphenomenon. For an analysis of this question, Azarnia and I made somatic cell hybrids between channel-competent and channel-incompetent cells. In the various hybrid cells we produced, the channel competence behaved like a dominant trait: the early hybrid generations, with the nearly complete chromosome complement of both parent cells, were channel-competent, and this went hand in hand with their growth control competence (38,29). Upon chromosome loss, in unstable hybrids, many phenotypic traits segregated from each other, but channel competence never segregated from growth control competence, nor did the opposite traits segregate from each other (29).

I shall give, as an example, the results obtained with a human cell / mouse cell hybrid. Here the channel-competent (and growth-control competent) parent cell was a human fibroblast and the incompetent parent counterpart, a derivative (Cl-1D) of a mouse L cell. The experimental criteria for channel competence were the presence of cell-cell transmission of electrolytes (electrical coupling) and fluorescent tracer molecules (about 300 daltons), and the presence of gap junctional intramembranous particles as seen in freeze-fracture electronmicroscopy. The criteria for growth control competence were density dependence of cellular growth *in vitro* (for the parent and all hybrid cells) and tumorigenicity in immunosuppressed mice (parent cells and relatively stable hybrids). Both parent cells were genetically marked by enzyme defects, facilitating hybrid selection; and the rate of human chromosome loss was adequate for segregant analysis (29).

The hybrid cells were channel-competent and growth-control competent so long as they had the sufficient chromosome complement of the human parent. As the hybrid generations lost human chromosomes, some of the hybrid clones switched to channel incompetence, and this was invariably paralleled by a switch to growth control incompetence. A total of 41 hybrid clones were analyzed (26 clones, see refs. 29,39; plus 15 clones in more recent work, R. Azarnia & W.R. Loewenstein, unpublished). There was segregation of several biochemical and morphological traits. Yet in no instance did channel competence segregate from controlled growth or, in the revertants, channel incompetence from uncontrolled growth. In agreement with the hypothesis, the competencies of cell-cell transmission and normal growth behaved like genetically linked characters.

We have now also made an attempt at identifying the chromosome(s) whose loss is responsible for the switch to channel incompetence. Chromosome analysis (by quinacrine and Giemsa banding) of 32 hybrid clones indicates that the loss of the human chromosome pair No. 11 is associated with simultaneous loss of the capacity for cell-to-cell transmission of electrolytes and fluorescent tracers and of gap-junctional membrane particle aggregates (R. Azarnia & W.R. Loewenstein, unpublished work). Chromosome 11 thus seems to carry genes necessary for the formation of the cell-to-cell channel. However, it is not excluded that other chromosomes carry such genes in addition.

REFERENCES

(1) W.R. Loewenstein, Ann. N.Y. Acad. Sci., 137 (1966) 441

(2) W.R. Loewenstein, Cold Spring Harbor Symp. Quant. Biol., 40 (1975) 49

(3) W.R. Loewenstein and Y. Kanno, J. Cell Biol., 22 (1964) 565

(4) B. Rose, J. Membrane Biol., 5 (1971) 1

(5) A. Jamakosmanovic and W.R. Loewenstein, J. Cell Biol. 38 (1968) 556

(6) R.D. Penn, J. Cell Biol., 29 (1966) 171

(7) W.R. Loewenstein and R.D. Penn, J. Cell Biol. 33 (1967) 235

(8) W. Nagel, Nature, 264 (1976) 469

(9) W.R. Loewenstein, Devel. Biol., 19, Sup. 2 (1968) 151

(10) E.J. Furshpan and D.D. Potter, Curr. Topics Devel. Biol. 3 (1968) 95

(11) J.D. Sheridan, J. Cell Biol., 37 (1968) 650

(12) M.V.L. Bennett and J.P. Trinkaus, J. Cell Biol., 44 (1970) 592

(13) J.T. Tupper and J.W. Saunders, Devel. Biol., 27 (1972) 546

(14) C. Slack and A.E. Warner, J. Physiol., 232 (1973) 313

(15) A. Warner, J. Physiol., 235 (1973) 267

(16) R.A. DiCaprio, A.S. French and E.J. Saunders, Biophys. J. 14 (1974) 387

(17) S. Ito and N. Hori, J. Gen. Physiol., 49 (1966) 1019

(18) B. Rose, I. Simpson and W.R. Loewenstein, Nature, 267 (1977) 625

(19) W.R. Loewenstein, Y. Kanno and S.J. Socolar, Nature, (1978) (in press)

(20) W.R. Loewenstein, S.J. Socolar, S. Higashino, Y. Kanno and N. Davidson, Science, 149 (1965) 295

(21) W.R. Loewenstein, M. Nakas and S.J. Socolar, J. Gen. Physiol., 50 (1967) 1865

(22) S. Caveney, Science, 199 (1978) 192

(23) W.R. Loewenstein, Perspectives in Biol. & Med., 11 (1968) 260

(24) W.R. Loewenstein, in: Eighth Canadian Cancer Conference, Honey Harbour, ed. J.F. Morgan (Pergamon Press, Toronto 1969) p. 162

(25) W.R. Loewenstein, Biochim. Biophys. Acta Rev. on Cancer, (1978) (in press)

(26) D.D. Potter, E.J. Furshpan and E.S. Lennox, Proc. Nat. Acad. Sci., 55 (1966) 328

(27) C. Borek, S. Higashino and W.R. Loewenstein, J. Membrane Biol., 1 (1969) 274

(28) R. Azarnia, W. Larsen and W.R. Loewenstein, Proc. Nat. Acad. Sci., 71 (1974) 880

(29) R. Azarnia and W.R. Loewenstein, J. Membrane Biol., 34 (1977) 1

(30) R. Azarnia and W.R. Loewenstein, J. Membrane Biol., 6 (1971) 368

(31) R. Azarnia, W. Michalke and W.R. Loewenstein, J. Membrane Biol., 10 (1972) 247

(32) P.G. Nelson, J.H. Peacock and T. Amano, J. Cell Physiol., 77 (1971) 353

(33) R.P. Cox, M.J. Krauss, M.E. Balis and J. Dancis, J. Cell Physiol., 84 (1974) 237

(34) I.S. Fentiman, J. Taylor-Papadimitriou and M. Stoker, Nature, 264 (1976) 760

(35) I.S. Fentiman, and J. Taylor-Papadimitriou, Nature, 269 (1977) 156

(36) R. Azarnia and W.R. Loewenstein, J. Membrane Biol., 30 (1976) 175

(37) C.M. Corsaro and B.R. Migeon, Proc. Nat. Acad. Sci., 74 (1977) 4476

(38) R. Azarnia and W.R. Loewenstein, Nature, 241 (1973) 455

(39) W. Larsen, R. Azarnia and W.R. Loewenstein, J. Membrane Biol. 34 (1977) 39

The work in the author's laboratory was supported by research grants from the National Cancer Institute, U.S. National Institutes of Health and the U.S. National Science Foundation.

DISCUSSION

C. GRIMMELIKHUJZEN: I just want to make a remark. Dr. Alfred Gierer, the previous speaker, showed that the activators in hydra should have short range action, and that inhibitors should have low range action. Our group found that the activators have a molecular weight of about 900 to 1,000, and that inhibitors have a molecular weight of about 500. You show that the junction channels can discriminate between molecules of these sizes so that an inhibitor could easily pass and an activator might not. This could be an explanation why activators have a short range and why inhibitors have a long range.

E. HERTZBERG: I have two questions that bear on calcium regulation and junction function. One comes as a result of some work done by Miles Epstein and Bernie Gilula. They suggested that while calcium regulation may be a very real phenomenon, as you showed originally in Chironomus, the calcium control does not seem to work when you move out of arthropods and insects into mammalian cells. Have you extended your observations using aequorin and calcium to mammalian systems?

W.R. LOEWENSTEIN: Yes, the observations have been extended in the meantime. The junctional channel regulation by Ca works in mammalian cells, namely sheep and dog heart cells, as reported by Délèze and by DeMello. It works also in cells of the embryos of the vertebrates Triturus and Xenopus, as shown by my colleagues Ito and Kanno. Channel closure is not as simple to demonstrate in cultured mammalian cells because their junctions seem especially well protected by internal Ca sequestering mechanisms. But by proper experimental design, Flagg-Newton in our laboratory was able to show the Ca regulatory mechanism in a variety of cultured mammalian cells too. So we may be quite confident now that cell-cell channel regulation by Ca^{2+} is a general mechanism.

E. HERTZBERG: Also, could you comment on Anne Warner's observations, which came out in Nature recently, indicating that an increase in cellular acidity, which some have suggested comes about as a result of calcium micro-injection, might be the actual control point for communication via gap junctions?

W.R. LOEWENSTEIN: Yes, this is an important point. Thomas and Meech discovered that elevation of cytoplasmic Ca^{2+} concentration in snail nerve cells causes fall in intracellular pH, and Warner, taking up this lead, found that a fall in internal pH produced by high external CO_2 concentrations leads to junctional uncoupling. The question now is whether the pH- and Ca effects on junctions are independent. My colleagues Rick and Rose are investigating this point. They have shown so far, by use of aequorin, that a lowering of internal pH causes release of intracellular Ca. So the pH effect may conceivably be via the Ca mechanism. But this last point still remains to be shown.

D.G. KERLEY: I would like to ask two questions. First, when you hypertonically treat these cells in order to separate the junctions, do they re-form on return to isotonic medium?

W.R. LOEWENSTEIN: Well, yes, and it doesn't seem to matter how you separate the cells. You can, in fact, separate them manually as Ito and I have done some years ago, and then put them together again in another spot on the membrane. They will then make viable channels within a matter of one to two minutes.

D.G. KERLEY: Is anything known about the mechanism of this re-formation?

W.R. LOEWENSTEIN: Not much yet, except that apparently immediate protein synthesis is not required. This was shown by J. Sheridan. The only other thing that I could tell you is that when channels form slowly enough one can, by means of high-resolution methods, detect their opening, the opening of the single individual channels as quantal steps of cell-cell conductance. This, Kanno, Socolar, and I have just done. Based on these experiments, one may envision the channel formation process as a stable interaction of subunits from the two apposing membranes, an interlocking of protochannels as it were. Most simply, perhaps, the formation process may be imagined as the sequence: (1) arrival of the channel subunits at the region of cell contact - and here the time required for channel formation seems long enough for the subunits to arrive even by random lateral diffusion in the two membranes - and (2) interlocking and opening of the protochannels. A striking feature of the cell-cell channel revealed by these experiments, is its stability. Once formed it persists in the open state for at least several minutes, and closures are not seen unless cytoplasmic Ca^{2+} concentration is elevated.

THE INTERCELLULAR TRANSFER OF MOLECULES IN TISSUE CULTURE CELLS: A KINETIC STUDY BY MULTICHANNEL MICROFLUOROMETRY

ELLI KOHEN AND CAHIDE KOHEN
Papanicolaou Cancer Research Institute,
POB 236188, Miami, Florida

ABSTRACT: *In situ* kinetic studies on the intercellular transfer of molecules can be achieved by multisite microfluorometry of tracer or intracellular fluorochromes, in combination with microinjection of tracers (e.g. fluorescein, 6:carboxy fluorescein, fluorescein isothiocyanate glutamic) or metabolites (e.g. glucose-6-P). Other fluorescence photographic or fluorescence scanning techniques are available to study the intercellular transfer of fluorochromes, but the sensitivity and temporal resolution of multichannel microfluorometry are required to determine the kinetics of intercellular flux and the intercellular transfer of metabolites (or their catabolites) detected via NAD(P) $\rightleftarrows$ NAD(P)H transients. The intercellular transit time τ has been evaluated in a variety of cell cultures (liver, glia, cultivated pancreatic islet cells, fibroblasts) where intercellular transfer of tracer is confirmed by visual observation. In these cultures τ varied from 0.4 sec to 1 sec. No intercellular transfer is detected in highly malignant ascites (EL2) and melanoma (HPM) cultures. "Communicating territories" are defined in clusters of liver and pancreatic islet cells, as the tracers are seen to spread preferentially into some of the secondary or tertiary cells adjacent to the cell injected with tracer, while there is no spread into other adjacent cells. The intercellular transfer of metabolites (e.g. glucose-6-P or one of its catabolites) is observable in the NCTC 8739 spontaneously transformed cells and L cells grown with the mitochondrial inhibitor atractylate. In preparations of cultivated pancreatic islets, which show extensive spread of tracers within a cluster (e.g. 10-15 cells), reverse NAD(P)H transients (i.e. NAD(P)H $\rightleftarrows$ NAD(P) instead of NAD(P) $\rightleftarrows$ NAD(P)H) are observed. Using suitable biological preparations (e.g. cultivated pancreatic islet cells with α, β and Δ cells exhibiting different endocrine activities) intra and intercellular metabolic interactions may be studied within a whole functional system.

ISBN 0-12-045450-5

INTRODUCTION

It has been postulated that the intercellular transfer of putative regulatory molecules (1-4) can influence differentiation and development during embryogenesis. Indeed Kanno and Loewenstein (5) found in 1964 that relatively large molecules can diffuse from cell to cell via permeable regions in the cell junctions (6). There are also the facts of extensive communication by junctional channels in embryonic development (7-9) as well as impaired communication in some cancerous cell strains (10-17). The size range of molecules permeating the junctional channels (e.g. up to 1200 daltons) (18,19) could include growth-controlling molecules (3,20). The identity of the growth-controlling molecules is unknown at this time. Due to this difficulty, as a first step, studies have to be initiated on simple cell systems (models) with tracer molecules. It is therefore important to develop the methods which could ultimately make possible the *in situ* detection of presumptive regulatory molecules and their intercellular transfer, within the multicellular assembly represented by differentiating cells.

The *in situ* detection of molecules in living cells requires the use of optical techniques such as microspectrophotometry (21) or microfluorometry (22). While such methods have been used successfully by Chance et al. (23) as well as others (24-26) for *in vivo* studies on the exposed and perfused organs of whole animals, the resolution of the method for studies on individual contiguous cells is considerably enhanced if tissue culture material is used. In reference to *in vitro* differentiation, cultures of differentiating metanephrogenic embryonal tissue in the induced state have been described by Saxen et al. (27-31). Similar cultures of differentiating muscle tissue have been described by G. Yagil (personal communication). Therefore, what is required is the development of an optical method for studies on the intercellular transfer of molecules in cultured cells (32,33).

In situ kinetic studies on the intercellular transfer of molecules can be achieved at high temporal resolution and sensitivity by optical methods such as multisite microfluorometric monitoring (33) of tracer fluorochromes in combination with microinjection of tracers or metabolites (33). Using a recently described method, multichannel microfluorometry (33), two categories of fluorochromes are available for such studies:

1) exogenous fluorescent tracers, such as fluorescein (5,6) or other fluorescence-labelled molecules (18,19) (e.g.

fluorescein isothiocyanate, 6-carboxyfluorescein, cf. I. Simpson, J. Flagg-Newton, personal communication).

2)intracellular fluorochromes, e.g. NAD(P)H (34,35) which may via changes in their oxidation-reduction state (and thereby fluorescence), help to detect the intercellular movement of non-fluorescent metabolic intermediates (e.g. glycolytic phosphate esters), when the latter are microinjected into a selected cell.

Undoubtedly, the second category of tracers is more suited to evaluate functionally and in a dynamic context intra- and intercellular metabolic interactions (36,37). However, the results of such studies will be better understood and interpreted if they can be correlated with what is already known in terms of intercellular communication. On this basis, exogenous fluorescent tracers (3,6,18,19) have been used to establish a system of equivalence between microfluorometric findings on cell-to-cell transfer of chemicals and electrical measurements (38,39) on cell-to-cell coupling (cell-to-cell junctional passage of ions), using cell systems or lines known to be electrically coupled (12,38,39) or uncoupled (10-17). The preliminary results of microfluorometric studies (33) show that in cells which are coupled electrically, the intercellular transfer of fluorescent tracers can be followed kinetically; conversely there is no such transfer in electrically uncoupled cells.

MATERIALS AND METHODS

I. Apparatus

The multichannel microspectrofluorometer, Fig. 1 (40) used for such studies consists in principle of an inverted microscope with darkfield illumination, a slit in the image plane to define the cell region viewed, intermediate optics and a silicon intensified tube (multichannel detector) operated in conjunction with a multiscaling computer. The output of the multiscaling computer is recorded on magnetic tape and processed from tape in a Univac computer.

The microspectrofluorometer can be operated in two modes:

1) in the topographic mode, adjacent cell regions (e.g. each region 0.4μ x 4-20μ) are imaged on a unilinear array of 100-150 channels on the detector target. The fluorescence is scanned every 64 msec for the whole array of channels.

2) in the spectral mode the fluorescence emission of a single cell region is dispersed via a prism or grating and analyzed wavelength by wavelength; thus a multichannel display of the fluorescence spectrum is obtained. This mode

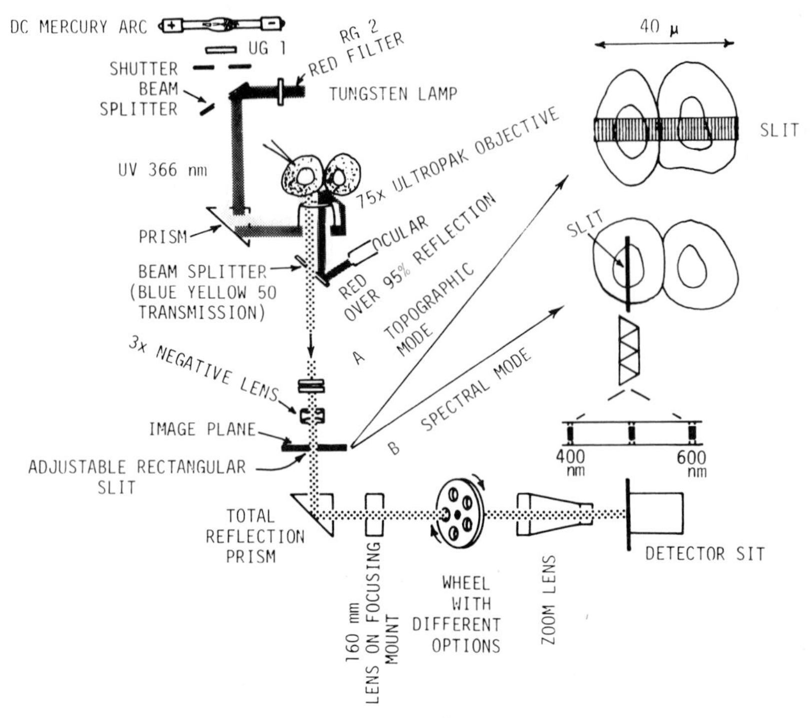

FIGURE 1

Instrumental arrangement which makes it possible to follow fluorescence changes visually in a whole cluster of adjacent cells and via a multichannel detector in at least two adjacent cells. The beam splitter under the Ultropak objective reflects towards the ocular the near totality of red light for cell and microinstrument visualization. 50% of blue-yellow fluorescence is also reflected for visual observations, while the other 50% is transmitted toward the detector. An adjustable rectangular slit in the image plane delimits the field of view. At least two adjacent cells are included in the field of view, one of which (see micropipette) is microinjected with tracers or metabolites using a microelectrophoretic or a piezoelectric-pressure technique. Each cell is scanned diametrically 16 times a sec., about 50 channels viewing each cell. The optical arrangement before the detector includes a wheel with different prism options or an empty turret. If the empty turret is placed on the light path, topographic dispersion is obtained, with different regions of the two adjacent cells viewed by different channels (see topographic mode). If a prism is placed on the light path, a fluorescence spectrum is obtained from a single cell region delimited by the slit (see spectral mode). The topographic or spectral dispersion is adjusted by a zoom lens. The detector is a silicon intensified tube (SIT). The drawing is not to scale.

---

has no direct application for the study of intercellular communication, which requires a topographic arrangement. However, the spectral mode is helpful in determining the identity of the fluorochrome observed in a particular cell. Such an example will be shown.

Two types of observations are possible in the topographic mode:

1) the fluorescence levels at specific times can be plotted against topography, i.e. topographic plots of fluorescence in selected scans (Fig. 2).

2) the fluorescence changes at specific intracellular sites can be plotted against time, i.e. time plots of fluorescence changes at specific sites (Fig. 3).

The optical arrangement includes a special dichroic filter (designed and built by Ing. W.Olsen, Optisk Laboratorium, Lingby, Denmark) which reflects toward the ocular nearly 100% of red light, but only 50% of the blue-yellow fluorescence, while the rest of the fluorescence is transmitted towards the detector (Fig. 1). Thus, changes in the fluorescence of injected cell and neighbors can be followed both visually and by multichannel microfluorometry. Red light is used for visualization of the cell and microinstruments during microinjection of tracers or metabolites.

II. Cells and Culture Material

Cells known to be electrically coupled or uncoupled were used. Among those known to be electrically coupled: Chironomus salivary gland cells (5,6,18-20,38), rat liver (12), glia culture (J. Barrett, personal communication) (41), fibroblast culture (B cells) (G. Flagg-Newton, personal communication), pancreatic islet cells in monolayer culture (42-47).

Among those known to be electrically uncoupled: L cell (48-51).

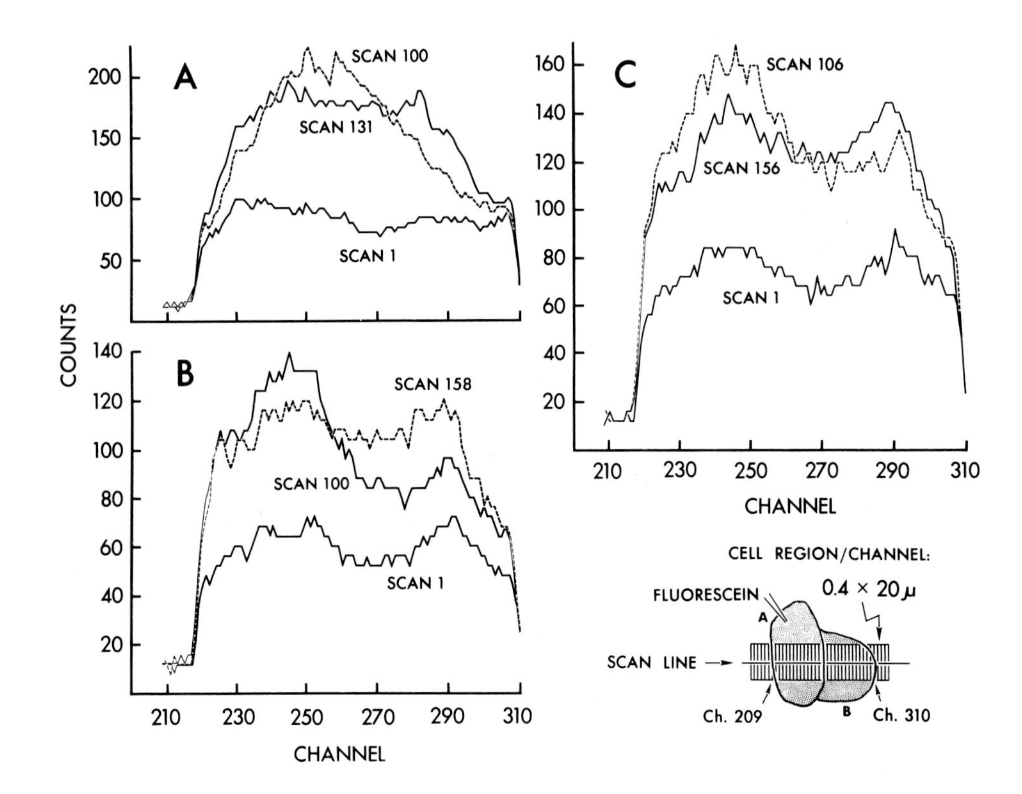

FIGURE 2

Topographic observations of intercellular transfer. In this example, the observations are carried in Chang liver cells. As the fluorescence tends to return to initial level 15-60 sec after injection of fluorescein, the results of three consecutive injections (A,B,C) in cell A of the same couple are shown. The injection time was 10 msec for first injection, 50 msec for second and third injections. The channels viewing the cell couple are seen in the right lower corner. Each cell is imaged on about 50 channels. In the fluorescence scan of the two cells before injection of tracer, the intercellular region appears as a valley between two prominences (see scan 1). Following injection of fluorescein there is first a rise predominantly in the region of the injected cell (scans 100,100 and 106) respectively for A,B,C and subsequently equilibration between injected and neighbor (scans 131, 158 and 156). Thus the kinetics of the fluorescence changes are different over the channels corresponding to neighbor. Decrease of fluorescence intensity throughout the whole injected cell is accompanied by rise in the neighbor, which excludes an artefact due to scatter. The crossover point between the earlier scans (100,100,106) and later scans (131,158,156) corresponds to the intercellular region (around Ch. 260). Experiment M16/F3-7. 1 scan = 64 msec.

Since microfluorometric findings on the intercellular transfer or non-transfer of molecules generally correlated well to electrical coupling or uncoupling in these cells, a basis was provided to extend such studies to other cell lines for which there is no prior knowledge on electrical coupling. Among such lines: EL2 ascites cancer cells (35), (Harding Passey) (52) HPM melanoma cells, NCTC 8739 cells (spontaneously transformed clone derived from mouse embryo mince) (53,54). Chang liver cells were also used (55,56).

EL2 cells and melanoma cells failed to show intercellular transfer of molecules, while the results with NCTC 8739 and Chang liver were variable.

For observation of NAD(P) $\rightleftarrows$ NAD(P)H transients with glucose-6-P, in the case of L cells it was necessary to minimize the Pasteur effect (57) by partial anaerobiosis, or growth in presence of mitochondrial inhibitors (atractylate (58), oligomycin (59)).

Monolayer cultures of pancreatic islet cells (42,60-63 were maintained in different concentrations of glucose (5.6 , 16.7mM), for 24 h. prior to the experiment. Electrophysiological evidence for the coupling of β cells in microdissected islets of mouse pancreas was obtained by H.P. Meissner (47). Electrical coupling in the type of monolayer cultures used for microfluorometric studies was demonstrated by C. Pace, Dept. Phys. Univ. Miami.

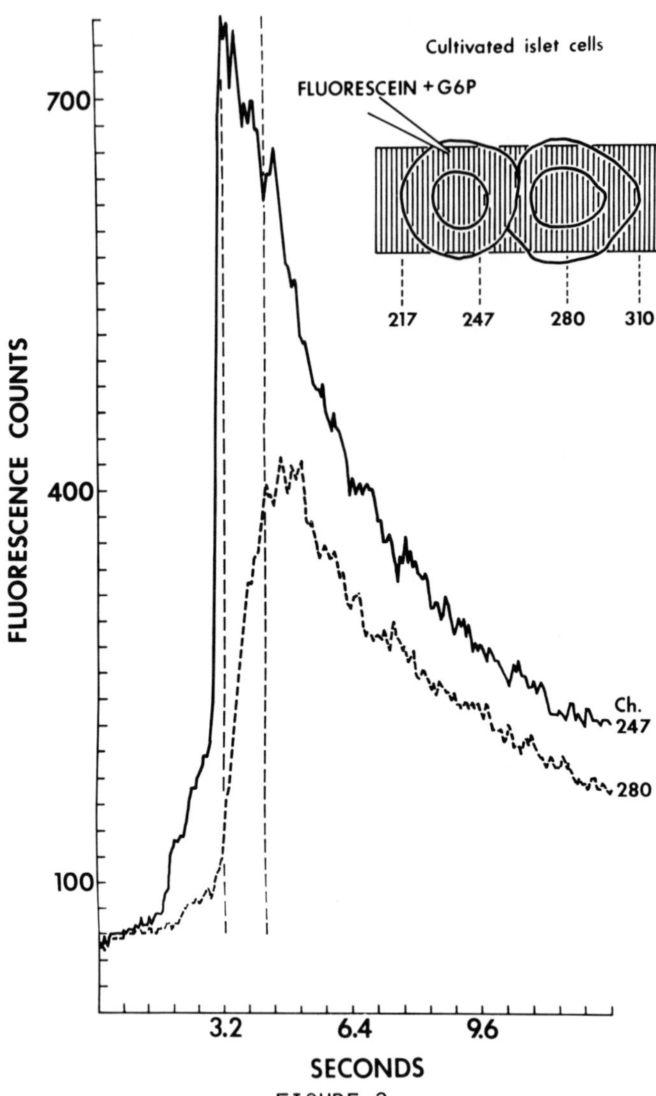

FIGURE 3

Curves used in the evaluation of intercellular transfer. Generally two columns of the matrix representing the channel by channel (site by site) fluorescence changes in the two cells (see right upper corner) are plotted. The solid curve shows the fluorescence change at the center of the cell injected with tracer and the curve in broken line shows the fluorescence change at the center of the neighbor. It is noteworthy that as indicated between the two vertical lines, the fluorescence at the center of the neighbor cell continues to rise while it has already started to drop in the injected cell. Intercellular transit time (τ) can be evaluated from such plots, as the time interval between the maximum intensity in injected cell and the maximum in neighbor. The injection time was 250 msec. Experiment M161/F8.

III. Addition of Tracers

The introduction of dyes or metabolites is achieved by microelectrophoresis (64,65) or a piezoelectric microinjection technique (66). The fluorescent tracers used were fluorescein, fluorescein isothiocyanate (FITC) glutamic and: 6-carboxyfluorescein. The last two tracers were obtained from Drs. I. Simpson, B. Rose, and J. Flagg-Newton of the Department of Physiology and Biophysics, University of Miami School of Medicine, the first to show the usefulness of these compounds in their studies on cell-to-cell communication. For microfluorometric observations FITC glutamic and 6-carboxyfluorescein were injected into mammalian cells. Glucose-6-P was injected as metabolite.

To minimize the occurrence of microclots in the micropipette tip, fluorescein was diluted and the microelectrophoretic drive maintained by addition of a glucose-6-P solution. In such combination injections,the spectral mode (40) was used to demonstrate that the fluorescence observed was largely due to fluorescein and not to reduction of NAD(P) which is quite understandable in view of the much higher quantum yield of fluorescein as compared to NAD(P)H.

In mammalian culture cells currents up to 10^{-7} amp, lasting from 25-500 msec are used for microinjection of metabolites. In the case of dyes, the fluorescence due to the dyes is seen to decay within 30 sec. to a few minutes. The possible mechanisms for such decay are discussed in a preceding paper (33). Even if the decay would be due to extrusion of tracer from injected cell, extrusion cannot account for intercellular transfer to neighbor cells, since:

1) decay kinetics are slower than intercellular transfer kinetics;
2) in the case of non-communicating cells decay occurs with no transfer to neighbors.

IV. Recognition of Intercellular Transfer and Technical Safeguards.

Intercellular transfer can be recognized in topographic plots (Fig. 2) and time plots (Fig. 3).

Fluorescence changes are followed in at least two adjacent cells, one of which is injected with tracer or metabolites (Fig. 2). A matrix is obtained with e.g. rows corresponding to scans and columns to different sites in injected cell or neighbor. Cell-to-cell junctions are assumed to occur where there is a topographic discontinuity in the intensity during the time course of fluorescence.

Due to the speed of the intracellular transfer, the kinetics of the fluorescence change are practically comparable for all the channels viewing the injected cell. In the same way the fluorescence in the channels viewing the adjacent neighbor seems to change together. On this basis two topographic regions are identified, i.e. injected cell and neighbor. The intermediary region channels should correspond to the intercellular junctional region.

In the optical arrangement used, scatter of fluorescence emission from a cell injected with tracer over an adjacent neighbor is minimal. This is amply demonstrated by the numerous negative trials in which there is no transfer of fluorochrome from injected cell into neighbor (Fig. 4), i.e. there is no change at all or a 5% change over the channels viewing the neighbor.

Intercellular transfer is further confirmed by the time course of the fluorescence changes in injected cell and neighbor, since a decrease in injected cell is accompanied by a rise in the neighbor. However, if diffusion within the injected cell is relatively slow, fluorescence may decrease at the injection site and increase at surrounding sites within injected cell, with diffusion-related scatter over the neighbor. This could simulate intercellular transfer, but with an important difference: in diffusion-related scatter the increase of fluorescence over neighbor is simultaneous with an increase in the injected cell at sites surrounding the injection site. In genuine intercellular transfer the increase at all sites within the neighbor continues, after the fluorescence has started to decrease throughout all

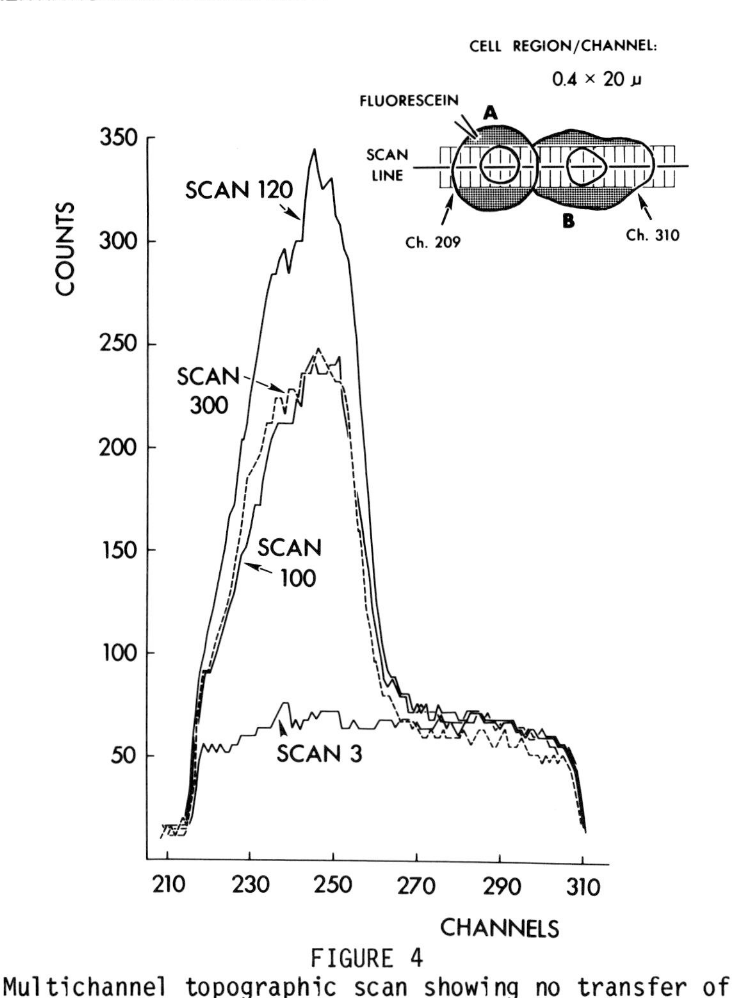

FIGURE 4

Multichannel topographic scan showing no transfer of fluorescein in another couple of Chang cells. The cells and the multichannel array are seen in the right upper corner. Cell A (injected) extends from about channel 215 to channel 260, cell B (neighbor) from about channel 260 to channel 310. The topographic fluorescence scan prior to injection of fluorescein does not show a valley between two eminences, as could be expected, e.g. around channel 260, cf. Fig. 2. This is probably because cells A and B were quite flat. The results of scans about 6 sec (scan 100) and 7.2 sec (scan 120) after injection of fluorescein are seen. The fluorescence rise time within injected cell is ∿1.2 sec. This long rise is due to the fact that the injection was performed by a piezoelectric pressure technique instead of microelectrophoresis.

While the physical phenomenon of piezoelectricity is quite rapid, in piezoelectric-pressure microinjections there is a delay due to the time it takes to break the resistance at the tip. The fluorescence increase is limited to cell A with practically no change in cell B. Obviously, the contribution of scatter from cell A to the fluorescence measured over the channels viewing cell B is totally negligible. The optical arrangement is such that every channel will indeed view essentially only a corresponding cell region with practically no scatter contribution from other regions. Furthermore, the crossover phenomenon so clearly observed in Fig. 2 (i.e. first major rise in cell A, then decrease in A with rise in B) is totally absent here. Experiment M20/F11.

---

of the injected cell (crossover or "connected vessel" phenomeon). Thus, two territories (injected cell and neighbor) are clearly defined with different time courses of fluorescence, a situation which cannot be duplicated by any kind of scatter.

Delays due to intracellular transfer could obscure the observation of the intercellular transfer time if the latter were to proceed nearly at the same speed as the former. Earlier two-site microfluorometric studies have revealed (67) that in mammalian cells the intracellular transfer of fluorochromes proceeds at a velocity of a μ/few msec, i.e. ∿ 0.1 sec for crossing the average mammalian cell (shorter than the lower limit of intercellular transit time). In a typical example calculated from a couple of pancreatic islet cells (Expt. M161/8) intracellular transfer was completed in ∿ 0.1 sec, while the intercellular transit time was 1.2 sec.

Upon injection of fluorescent tracer in the surrounding medium at a few μ from one of the observed cells there is rapid dilution of the dye into the medium, within about 0.1 sec.

Visual observations indicate that measured fluorescence changes are genuinely due to the presence of fluorescent tracer within a cell (via injection or cell-to-cell transfer) as the cells filled with dye are sharply identified, among their neighbors or cells from other clusters.

The observed phenomenon cannot be reproduced by movements of the micropipette loaded with fluorescein, as at the high concentration of fluorescein used its fluorescence is self-quenched, the pipette is withdrawn from the field of observation following a rapid injection (e.g. $\sim$ 25-250 msec), the same phenomena are observed when a cell is injected outside the field viewed by channels, but there is transfer of fluorescein into cells viewed by channels.

VI. Transfer of Metabolic Intermediates

Metabolic intermediates, which are not fluorescent themselves, elicit upon microinjection, associated changes in the redox state of endogenous fluorochromes, i.e. NAD(P) $\rightleftarrows$ NAD(P)H transients (35). The transients are followed by corresponding changes in the blue fluorescence emission of the observed cells (i.e. NAD(P)H emission with maximum at 465-475 nm). The cell-to-cell transfer of metabolites can be followed by associated NADP $\rightleftarrows$ NAD(P)H transients in neighbors.

VII. The Definition of Parameters for the Determination of Intercellular Transit

From topographic (Fig. 2) and time plots (Fig. 3) the intercellular transit times τ or cell-to-cell equilibration times can be evaluated for injected dyes (or metabolites), see Table I.

It is also possible to attempt some computation of intercellular flow (F) from the kinetics of fluorescence changes in injected cell and neighbor.

To reduce the complexity of the evaluation the following assumptions are made:

1) There are n cells in contact with cell <u>a</u> each with an intercellular junctional contact area of $d_{ai} \cong A/n$ in reference to cell <u>a</u>

2) The intensity at the center of each cell in contact with cell a, is Ib(t) at instant t.

Thus, as derived in a previous paper (33) the intercellular flow

$$F = \frac{\pi\, d_a l_a\ (dI_a(t)/dt)}{6\ \ I_a(t)/ha\ -\ I_b(t)/h_b}$$

for cells assimilated to flattened ellipsoids (d_a, l_a, h_a = cell dimensions, $dI_a(t)/dt$ = derivative of fluorescence intensity in injected cell). Cell thickness

h_a cannot be measured but may be estimated from a flattened ellipsoid model and assumed equal for neighbors ($h_a = h_b$).

The error introduced by the assumption can be computed from the partial derivatives of F in reference to the variables d_{ai}, $I_b(t)$ and h_b. It is concluded that the higher the intensity I_a is at the center of the primary cell in reference to the intensities I_b in neighboring cells, the smaller will be the error introduced by the assumptions.

RESULTS

Topographic observations of intercellular transfer in a variety of mammalian cell lines in culture

Prior to injection of fluorescent tracers, the observed fluorescence consists of emission due to intracellular fluorochromes, e.g. NAD(P)H and flavoproteins (mainly the former under experimental conditions used). The topographic fluorescence pattern (Fig. 2) can help to identify cell-to-cell boundaries, e.g. as a valley between two prominences if the two cells are more or less of comparable thickness and shape. In the case of very flat cells (e.g. monolayer culture of cultivated pancreatic islet cells) the prominence and valleys may become imperceptible.

Upon microinjection of fluorescein or another tracer, in consecutive topographic scans a "connected vessels" phenomenon is observed (see Methods) with fluorescence ultimately equilibrating between injected cell and neighbor. The cross-over region between early topograhic curves showing prominent fluorescence in the injected cell and later curves showing decrease in injected vs. rise in neighbor, coincides with the region defined as the cell-to-cell boundary.

Cell-to-cell equilibration of tracer is completed within from a few hundred msec. to a few sec (see Table I). Depending upon the cell type, conditions or spread to several secondary and tertiary cells, different patterns of complete or incomplete equlibration are observed.

Table I

Average transit times (τ) of molecules*

F = intercellular flux per unit concentration difference (μ^3/min) between adjacent cell and neighbours is also indicated, as approximated.

n = number of observations

CELL TYPE	SPECIAL GROWTH CONDITION	ADDITIONS TO THE EXPERIMENTAL MEDIUM OR SPECIAL CONDITIONS	MICROINJECTION (MICROELECTROPHORESIS)	n	τ (sec)	F ($10^4 \times \mu^3$/min)
CHIRONOMUS	--	10^{-3}M Ca^{++}	FLUORESCEIN	2	6.5	(148)
CHANG LIVER	--	10^{-3}M Ca^{++}	FLUORESCEIN	1	1.8	(18)
RAT LIVER	--	10^{-3}M Ca^{++}	FLUORESCEIN	4	0.6	(24)
GLIA CULTURE	--	10^{-5}M Ca^{++}	FLUORESCEIN ISOTHIOCYANATE GLUTAMIC + GLUCOSE-6-P	4	0.4	(54)
GLIA CULTURE	--	10^{-5}M Ca^{++}	FLUORESCEIN (0.5M) + GLUCOSE-6-P (1M)	2	0.4	(44)
FIBROBLAST CULTURE (B CELLS)	--	10^{-5}M Ca^{++}	FLUORESCEIN ISOTHIOCYANATE GLUTAMIC + GLUCOSE-6-P	2	0.4	(66)
FIBROBLAST CULTURE (B CELLS)	--	10^{-5}M Ca^{++}	FLUORESCEIN (0.5M) + GLUCOSE-6-P (1M)	3	0.4	(396)

* In rare instances there was intercellular transfer of tracer in crowded preparations of L cells, with τ varying from 0.2 to 4.7. In L cells grown for 21-24 days in 3×10^{-6}M atractylate there was some transfer of glucose-6-P or a catabolite leading to reduction of NAD; τ was about 0.1 sec.

Kinetic analysis of intercellular transfer

Liver (12), glia (J. Barrett, personal communication) and B cells (G. Flagg-Newton, personal communication) are known to be coupled and communicate by other techniques. L cells are known to show no communication (48-51) or maybe at best minimal communication (49), substantiated in electron microscopic examination by lack of gap junctions (48).

In these microfluorometric studies rat liver, glia cells and fibroblasts (B cells) showed intercellular transit (equilibration) times τ from 0.4 to 0.6 second, i.e. a maximum fluorescence in a contiguous neighbor was reached 0.4 to 0.6 sec after the maximum due to injection was detected in injected cell (Table I). The Chang liver cultures showed irregular communication and the intercellular transfer time was longer, i.e. 1.8 sec. (Table I, Figs. 2,3).

In these studies most of the L cells examined failed to reveal intercellular transfer of tracers except for rare instances. (See Table I footnote).

When F, the intercellular flux, is computed tentatively (on the basis of assumptions described), for quite identical transit times there are larger variations of fluxes (see Table I).

For a given value of transit time τ, total outward fluxes from injected cell to neighbors will be larger, if the intercellular junctional area is larger.

Also, to complete cell-to-cell equilibration within a time τ, the flux (mole per unit time) will have to be larger if the cells are larger. In the case of chironomus with a flux 6 times greater than in liver cells, the transfer time can be 6 times longer as there is a larger volume to fill.

Intercellular transfer of tracer molecules in cultivated pancreatic islet cells

Monolayer cultures of pancreatic islets present a suitable material for studies on intercellular transfer of tracer molecules, as demonstrated by passage of injected tracers into several adjacent cells forming a cluster (Fig. 5). Also detailed morphometric analysis of intercellular gap junctions is being made of these cells in another laboratory (Histology, Univ. Geneva). In these cultures clusters of

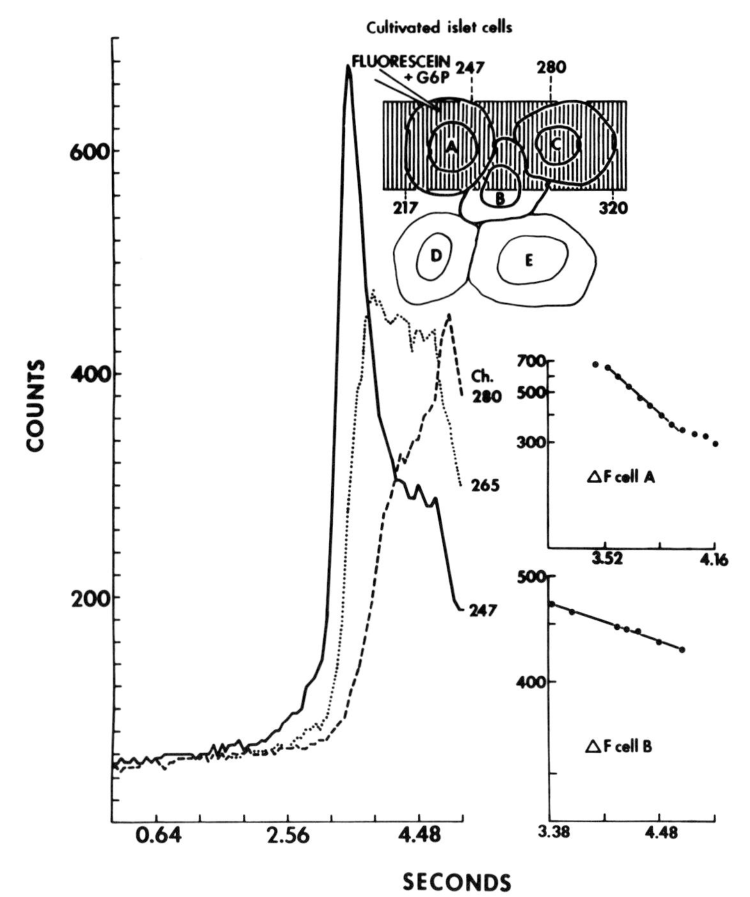

FIGURE 5

Intercellular transfer of fluorescein between three adjacent cultivated pancreatic islet cells. Primary cell (injected with fluorescein) = cell A; secondary cell = cell B; tertiary cell = cell C. The time course of fluorescence changes ($I_A(t)$, $I_B(t)$, $I_C(t)$ near the centers of cells A,B and C is shown simultaneously. Cell A: ch. 247, cell B: ch. 265, cell C: ch. 280. $I_B(t)$: ch. 265 continues to rise after $I_A(t)$: ch. 247 has started to drop. Similarly $I_C(t)$: ch. 280 continues to rise after I_B has started to drop. The intercellular transit time between cells A and B can be evaluated from the difference of the scan number at which maximum intensity is reached in cells B and A respectively; similarly for the transit time from B to C.

The intercellular fluxes F out of cell A and B can be evaluated from the slopes of decay in cells A and B (see text). Since the sink for cell B (i.e. cells C,D,E) is ∿ three times larger than the sink for cell A(i.e. cell B), the decay slope of cell B should be steeper than the decay slope of cell A, but the real slope of cell B cannot be seen here. The difference between cell A and B could be due to the fact that the introduction of tracer into cell A is by microinjection, while the introduction into cell B is via intercellular transfer. Thus when the injection time is terminated within 250 msec the fluorescence in cell A will have only a decrease component due to intercellular transfer, while the intercellular transfer into cell B will continue for about 0.4 sec more until a maximum is reached in cell B and cell B equilibrates with cell A. During that time the fluorescence in cell B will have both a rise and a decrease component and when a maximum is reached in cell B, fluorescence has already increased significantly in cell C, which then accounts for a decrease in B to C flux. This is perhaps an oversimplified description, as there could be other explanations, such as extrusion from cell A or loss from the site of injection, smaller junctional areas or lower junctional permeability between cell B and C,D,E. The injection time was 250 msec. Experiment M161/F7.

---

epitheloid endocrine cells are surrounded by fibroblastoid cells. Microinjection of fluorescein revealed:

1. Transfer of tracer from epitheloid cell into neighboring epitheloid cells of a cluster (involving passage of dye into up to 10 cells or more).
2. No transfer of tracer from epitheloid into fibroblastoid cells.
3. No transfer of tracer as far as noticeable from a fibroblastoid cell into adjacent fibroblastoid cells.

The results of observation in these cells are listed in Table II, see also Fig. 5. Oligomycin was added as an inhibitor of mitochondrial energy transfer, on the basis of studies by Politoff et al. (68) suggesting a dependence of junctional communication on energy metabolism (e.g. depression of junctional permeability by cooling, cyanide, dinitrophenol, oligomycin and other inhibitors of ATP synthesis). In islet cells, oligomycin (10^{-6}M) did not seem to affect the intercellular transfer of fluorescein.

Table II

Intercellular transit times (τ) in cultivated pancreatic islet cells. The intercellular flow per unit concentration difference between injected cell and adjacent neighbours (F) is also indicated, as approximated. It can be seen that τ remains near constant ($\sim$ 1 sec) in a variety of experimental conditions.
n = number of observations

SPECIAL GROWTH CONDITION	ADDITION TO THE EXPERIMENTAL MEDIUM OR SPECIAL CONDITIONS	MICROINJECTION	n	τ (sec)	F ($10^4 \times \mu^3$/min)
24 H. IN 16.7 mM GLUCOSE	GLUCOSE-FREE, 10^{-5}M Ca^{++}	FLUORESCEIN + GLUCOSE-6-P	4	0.8	(60)
24 H. IN 16.7 mM GLUCOSE	GLUCOSE-FREE, 10^{-5}M Ca^{++} 10^{-6}M OLIGOMYCIN (5-25 MIN)	FLUORESCEIN + GLUCOSE-6-P	4	0.7	(90)
24 H. IN 16.7 mM GLUCOSE	GLUCOSE-FREE, 10^{-3}M Ca	CARBOXY-FLUORESCEIN (0.2M) + GLUCOSE-6-P (1M)	2	0.8	(198)
GROWN IN 5.6 mM GLUCOSE	GLUCOSE-FREE, 10^{-3}M Ca	CARBOXY-FLUORESCEIN (0.2M) + GLUCOSE-6-P (1M)	5	0.6	(12)
GROWN IN 16.7 mM GLUCOSE	GLUCOSE-FREE, 10^{-3}M Ca	FLUORESCEIN ISOTHIOCYANATE GLUTAMIC + GLUCOSE-6-P	4	1.0	(8)

A numerical example can be provided on the intercellular transfer of fluorescein into adjacent islet cells. In Figure 5 a cluster of 5 cells are seen. Injection of fluorescein into cell A results in transfer to cells B,C,D,E.

Three of these cells A,B,C are viewed by detector channels from 217 to 320. A maximum fluorescence is reached in cell B about 6 scans after the maximum in cell A, and a maximum in cell C is reached about 17 scans after the maximum in cell B. Thus, the intercellular transit time from cell A to B is 0.4 sec and the intercellular transit time from cell B to C is about 1.1 sec.

It may seem _a priori_ strange that the fluorescence intensity in cell C would rise at levels over that in A or B. However, the fluorescence intensity measured at the center of each cell is a function not only of concentration but also of cell thickness. Thus at equal concentration, the intensity will appear higher in the thicker cell, also cell B is not fully covered by the viewing channels.

Using the same example, a computation of intercellular flux F can be also attempted from the decay of fluorescence curves.

The fluorescence intensities at the center of cells A,B,C at the instant t are $I_A(t)$, $I_B(t)$ and $I_C(t)$ (Fig. 5). When plotted on a semilogarithmic scale the time courses of $I_A(t)$ and $I_B(t)$ are linear, which indicates that the drop of fluorescence in cells A and B, i.e. $I_A(t)$ and $I_B(t)$ follows an exponential equation:

On this basis F_A and F_B out of cells A and B can be calculated (see formula in Methods).

To minimize the error the times t_1 and t2 should be such, that:

1. $I_A(t_1) \gg I_B(t_1)$ for transfer from A into B
2. $I_B(t_2) \gg I_C(t)$ for transfer from B into C

The above inequalities are easily satisfied for t_1, but not for t_2, as until $I_B(t)$, the fluorescence at the center of cell B starts decreasing, $I_C(t)$, the fluorescence at the center of cell C has already increased considerably. It is possible to move to a point earlier in time and to consider the rising branch of the intensity in cell C ($I_C(t)$), since the flux into cell C will be about one-third of the flux out of cell B (going into cells C,D,E), but correction is also needed for the fact that cell B is not completely covered by the viewing channels.

Intercellular Transfer of Metabolites

The cell-to-cell transfer of glucose-6-P or a catabolite along the glycolytic or pentose shunt pathways, was investigated via the resulting NAD(P)H transients in EL2 ascites cells, NCTC 8739, L cells and cultivated pancreatic islet cells. In EL2 cells, a long established undifferentiated malignant line, metabolic transients were limited to the cell injected with glucose-6-P. However, in NCTC 8739 cells, showing *in vitro* acquired ability to generate sarcoma in syngenic mice, the cell-to-cell transfer of glucose-6-P was observable.

In atractylate-grown L cells (see Table I) NAD(P)H transients were observable in secondary or tertiary cells adjacent to a primary cell microinjected with glucose-6-P. No cell-to-cell transfer of glucose-6-P was observable in L cells treated with oligomycin or under partial anaerobiosis.

In the cultivated pancreatic islet cells, it was possible to observe negative transients (NAD(P)H $\rightarrow$ NAD(P)) instead of NAD(P) $\rightleftarrows$ NAD(P)H) both in injected cell and neighbor. Such transients are observable upon acceleration of lower, post-NADH steps in glycolysis. A positive transient was occasionally observable in injected cell, with a negative transient in adjacent cell, suggesting different velocities of NAD reducing and reoxidizing steps in the two neighbors.

DISCUSSION AND CONCLUSIONS

The main potential of multichannel microfluorometry lies in rapid determinations of cell-to-cell transit times for various tracers or metabolites within the dynamic context of intra and intercellular metabolic interactions. On the basis of observations to this date the following conclusions may be drawn:

1. The definition of communicating territories in cell clusters.

While the unilateral scan of the detector (silicon intensifier tube) restricts microfluorometric observations to two or three adjacent cells (6 in new design), using the special dichroic mirror, the cells of an entire cluster are visualized. The intercellular spread of fluorescent tracers into 2-10 cells or more, is obvious. While only 2-3 cells are quantitatively evaluated by microfluorometry, their neighbors into which the fluorescent tracer will have spread exhibit visually fluorescence intensities comparable to those of the cells being measured, but other neighboring cells into which the tracer did not spread show much weaker fluorescence. Such visual definition of a communicating territory is observable in cultures of rat liver cells, most strikingly in cultivated pancreatic islet cells where large territories are involved (almost an entire cluster).

Whether a functional meaning can be attributed to such "communicating territories", should await further studies, e.g. morphometric analysis (P. Meda, personal communication) of intercellular junctions (size of junctional complexes and junctional area per unit intercellular area)by freeze-etching EM technique. Cultivated pancreatic islets, may be a suitable material for such functional analysis in view of their mixed population (α ,β and Δ cells). The β cells form the large majority ($\sim$ 80%). Until now there has been some hesitation as to whether β cells are indeed functionally communicant, e.g. to explain their concerted response. Each time an intercellular transfer of tracer (or possible metabolites?) is detected in islet cultures there is 6.4 over 10 chance that it will be a β to β communication (43-45), but identification of α and Δ cells by EM (61,62) or immunofluorescence (69) is required to elucidate other alternatives.

2. Observations in malignant cells.

There is now a body of evidence that certain but not all malignant cells have altered communication. The present experiments allow to add 2-3 new malignant cell types which exhibit altered communication to those already known as such. No intercellular transfer of tracer or metabolite was detectable in the highly malignant EL2 ascites or HPM melanoma cell. Also there was no or little intercellular transfer of fluorescein in the NCTC 8739 sarcoma (33). However, intercellular transfer of glucose-6-P or a catabolite was observable in the NCTC 8739 sarcoma (33), by reduction of NAD in cells adjacent to injected cell.

3. Intercellular transfer of metabolites.

The ultimate purpose of these studies refers to the intercellular transfer of metabolites. Fluorescent tracers can provide some information on intercellular fluxes and molecular sizes involved, but not on the real functional mechanisms unless putative messenger molecules could be labelled as tracers. The sensitivity of multichannel microfluorometry allows, the analysis of in situ metabolic events and metabolic cooperation, via intracellular fluorescent coenzymes. Thus, the reciprocal interactions of intracellular compartments and intercellular transfers can be studied as a whole system.

However, the cells in which the largest metabolic transients (NAD(P) $\rightleftarrows$ NAD(P)H) are observed are generally non-communicant (e.g. EL2 cells). The cells in which the highest fluxes of the intercellular transfer are observed (e.g. cultivated pancreatic islets liver) exhibit very weak transients. This could be due to spread of injected metabolites over a network of communicating cells, thus decreasing considerably the concentration of metabolite per unit volume. This is indeed possible, since a transient will last 20-60 sec, while intercellular transfer may be completed within a sec, so there is no time to observe the transient while the local concentration of metabolites would be still high near the site of injection. In such case the difficulty to observe metabolite-induced transients in these cells would be another sign of intercellular communication (i.e. intercellular transfer of microinjected metabolites) in these cells.

It is also possible to observe in communicating cell lines NAD(P)H $\rightarrow$NAD(P) transients, i.e. reverse transients which are harder to evaluate and understand. While the positive transients have been extensively studied in EL2 cells by microfluorometry, so far the reverse transients have been studied mainly in cell-free preparations of yeast (70). They are due to an acceleration of the post NADH lower part of the glycolysis (70). This leads to a decrease in the intracellular levels of glycolytic intermediates and NADH. It is possible that the rate-limiting enzyme step is at the level of pyruvate kinase (70). Thus, the key to a better evaluation of intercellular metabolic cooperation, may reside at the intracellular level in better understanding of the controls affecting this enzyme. The study of intercellular metabolite transfer should be facilitated if negative transients could be somehow reconverted to positive transients. A slower activity of pyruvate kinase may lead to a reversal from negative to positive NADH transients in communicating cells.

Studies initiated with fluorescent tracers, and pursued with metabolites, affecting the oxidation-reduction of cell coenzymes, when directed to suitable models (e.g. cultivated islet cells) can help to unravel directly in situ the mechanisms involved in intercellular functional regulation and metabolic cooperation.

ACKNOWLEDGEMENTS

The authors acknowledge thankfully the stimulating discussions with Professor W.K. Loewenstein, Drs. B. Rose, I. Simpson (Dept. Phys. Sch. Med. Univ. Miami). The useful comments of Drs. D.L. Wilson, R. Wagner (Dept. Phys. Schl. Univ. Miami) and Professor B. Thorell (Dept. Pathol. Karolinska Inst. Stockholm, Sweden) are also acknowledged. The authors are thankful to P.R. Bartick for the error computation in the estimate of intercellular flow. The authors are also indebted to Dr. K.K Sanford, National Cancer Institute, Bethesda, Maryland, for the supply of NCTC 8739 cells to Professor D. Schachtschabel, Physiological Chemistry, Philipps Univ.-Marburg, G.F.R. for the supply of HPM melanoma cultures, to Drs. D.H. Mintz, A. Rabinovitch and Mrs. C. Quigley, Division of Endocrinology, Department of Medicine, University of Miami, for the monolayer cultures of pancreatic islet cells. The useful discussions with Drs. H. Mintz and A. Rabinovitch on studies referring to these monolayer cultures are thankfully acknowledged. This work was supported by Grant No. 2R01 GM20866-02 through 04 awarded by the National Institute of General Medical Sciences, DHEW and Grant No. 1R01 AM21330-01 BBCA awarded by the National Institute of Arthritis Metabolism and Digestive Diseases.

REFERENCES

1. W.R. Loewenstein, Ann. N.Y. Acad. Sci. 137(1966)441-472.

2. W.R. Loewenstein,Devel. Biol. 19, Sup. 2 (1968) 151-183.

3. W.R. Loewenstein, Perspectives Biol. Med., 11 (1968) 260-272.

4. W.R. Loewenstein, Fed. Proc. 32 (1973) 60-64.

5. Y. Kanno and W.R. Loewenstein, Nature, 212 (1966) 629-631.

6. W.R. Loewenstein and Y. Kanno, J. Cell. Biol. 22 (1964)565-568.

7. S. Ito and N. Hori, J. Gen. Phys. 49 (1966)1019-1027.

8. E.J. Furshpan and D.D. Potter In Current Topics in Dev. Biol., A.A. Moscona, ed., Academic Press, New York (1968) 95-127.

9. D.D. Potter, E.J. Furshpan and E.S. Lennox, Proc. Nat. Acad. Sci., 55 (1966) 328-336.

10. Y. Kanno and W.R. Loewenstein, J. Cell. Biol., 33 (1967) 225-234.

11. Y. Kanno and W.R. Loewenstein, J. Cell. Biol., 33 (1967) 235-242.

12. C. Borek, S. Higashino and W.R. Loewenstein, J. Membrane Biol., 1 (1969) 274-293.

13. R. Azarnia, W.R. Loewenstein, J. Membr. Biol. 6 (1971) 368-385.

14. R. Azarnia, W. Larsen and W.R. Loewenstein, Proc. Nat. Acad. Sci. USA 71(1974) 880-884.

15. R. Azarnia, W. Michalke and W.R. Loewenstein, J. Memb. Biol. 10 (1972) 247-258.

16. R. Azarnia and W.R. Loewenstein, Nature, 241 (1973) 455-457.

17. R. Azarnia and W.R. Loewenstein, J. Memb. Biol. 30 (1976) 175-186.

18. I. Simpson, B. Rose and W.R. Loewenstein, Science, 195 (1977) 294-296.

19. B. Rose, I. Simpson, and W.R. Loewenstein, Nature, 267 (1977) 625-627.

20. W.R. Loewenstein In Cellular Membranes and Tumor Cell Behavior, J. MacCay, ed., Williams and Wilkins, Baltimore, Mds., (1975) 239-248.

21. T. Caspersson, J. Roy Microsc. Soc., 60 (1940) 8-25.

22. B. Chance and B. Thorell, J. Biol. Chem. 234 (1959) 3044-3050.

23. B. Chance, Science, 137 (1962) 499-508.

24. F.F. Jobsis, V. Legallais and M. O'Connor, I.E.E.E.B.M.E. 13 (1966) 93-99.

25. E. Mills and F.F. Jobsis, J. Neurophysiol. 35 (1972) 405-428.

26. F.F. Jobsis, M. O'Connor, A. Vitale and H. Vreman, J. Neurophysiol. 34 (1971) 735-749.

27. A. Lahti and L. Saxen, Exptl. Cell. Res., 44 (1966) 563-571.

28. L. Saxen, O. Koskimies, A. Lahti, H. Miettinen, J. Rapola & J. Wartiowaara, Adv. Morphogenesis 7 (1968) 251-293.

29. L. Saxen and J. Kohonen, Intern. Rev. Exptl. Pathol., 8 (1969) 57-128.

30. P. Sundelin, J. Wartiowaara, L. Saxen and B. Thorell, Exptl. Cell Res., 54 (1969) 347-352.

31. L. Saxen, In Control Mechanisms of Growth and Differentiation. Symp. Soc. Exp. Biol. Vol. 25, D.D. Davies and M. Balls, Academic Press, New York (1971) 207-221.

32. G.H. Pollack, J. Physiol., 255 (1976) 275-298.

33. E. Kohen and C. Kohen, Exptl. Cell Res., 107 (1977) 261-268.

34. B. Chance, The Harvey Lectures, 49 (1955) 145-175.

35. E. Kohen, C. Kohen and B. Thorell, Exptl. Cell Res., 101, (1976) 47-54.

36. D.E. Atkinson, Science, 150 (1965) 851-857.

37. E. Kohen, B. Thorell and C. Kohen In Advances in Biological and Medical Physics, Vol. 15, J.H. Lawrence, J.W. Gofman and T.L. Hayes, ed., Academic Press, New York (1974) 271-297.

38. W.R. Loewenstein, S.J. Socolar, S. Higashino, Y. Kanno, N. Davidson, Science, 149 (1965) 295-298.

39. W. Michalke and W.R. Loewenstein, Nature, 232 (1971) 121-122.

40. E. Kohen, J.G. Hirschberg, A.W. Wouters, C. Kohen, H. Pearson, A. Salmon and J.M. Thorell, Biochim. Biophys. Acta, 396 (1975) 149-154.

41. R. Lim, D.E. Turriff, S.S. Troy, B.W. Moore, L.F. Eng, Science, 195 (1977) 195-196.

42. A.E. Lambert, B. Blondel, Y. Kanazawa, L. Orci and A.E. Renold, Endocrinology, 90 (1972) 239-248.

43. L. Orci, A.A. Like, M. Amherdt, B. Blondel, Y. Kanazawa, E.B. Marliss, A.E. Lambert, C.B. Wollheim and A.E. Renold, J. Ultrastruct. Res. 43 (1973) 270-297.

44. L. Orci, R.H. Unger, A.E. Renold, Experientia, 29 (1973) 1015-1018.

45. L. Orci, F. Malaisse-Lagae, M. Amherdt, M. Ravazzola, A. Weisswange, R. Dobbs, A. Perrelet and R. Unger, J. Clin. Endocrinology and Metabolism, 41 (1975) 841-844.

46. L. Orci, F. Malaisse-Lagae, M. Ravazzola, D. Rouiller, A.E. Renold, A. Perrelet and R. Unger, J. Clin. Investigation 56 (1975) 1066-1070.

47. H.P. Meissner, J. Physiol., Paris, 72 (1976) 757-767.

48. N.B. Gilula, O.R. Reeves and A. Steinbach, Nature, 235 (1972) 262-265.

49. R.P. Cox, M.R. Krauss, M.E. Baliss, J. Dancis, Exptl. Cell Res., 74 (1972) 251-268.

50. J.D. Pitts In 3rd Lepetit Colloquium on Cell Interactions, L.G. Silvestri, ed., American Elsevier Press, N.Y. (1972) 277-285.

51. C.M. Corsaro and B.R. Migeon, Proc. Natl. Acad. Sci., USA 74 (1977) 4476-4480.

52. D.O. Schachtschabel, Virchows Arch. Abt. B. Zellpath. 7 (1971) 27-36.

53. V.J. Evans and W.F. Andresen, J. Natl. Cancer Inst., 37 (1966) 247-249.

54. V.J. Evans, J.C. Bryant, H.A. Keer and E.L. Schilling, Exptl. Cell Res., 36 (1964) 439-474.

55. R.S. Chang, Proc. Soc. Exp. Biol. Med., 87 (1954) 440-443.

56. P. Biberfeld, J.L.E. Ericsson, P. Perlmann and M. Raftell, Z. Zellforsch, 71 (1966) 153-168.

57. L. Pasteur, C.R. Acad. Sci. (Paris) 80 (1875)452-457.

58. M. Klingenberg and E. Pfaff In Regulation of Metabolic Processes in Mitochondria, J.M. Tager, S. Papa, E. Quaglieriello and E.I. Slater, eds., Elsevier, Amsterdam (1966) 180-201.

59. H.A. Lardy, P. Witonsky and D. Johnson, Biochemistry, 4 (1965) 552-554.

60. A. Anderson and C. Hellerstrom, Diabetes, 21 (1972) 546-554.

61. J.I. Braaten, M.J. Lee, A. Shenke and D.H. Mintz, Biochem. Biophys. Res. Commun., 61 (1974) 426-432.

62. J.I. Braaten, U. Jarlfors, D. Smith and D.H. Mintz, Tissue and Cell, 7 (1975) 747-762.

63. P.E. Lacy, E.H. Finke, S. Conant and S. Naber, Diabetes 25 (1976) 485-493.

64. W.L. Nastuk, Fed. Proc., 12 (1953) 102.

65. D.R. Curtis In Physical Techniques in Biological Research, Vol. 5, Pt. A, W.L. Nastuk, ed., Academic press, New York (1964) 144-190.

66. E. Kohen, C. Kohen, G. Bengtsson and J.M. Salmon, IEEE Transac. BME, 22 (1975) 424-426.

67. E. Kohen, C. Kohen and M. Michaelis, Exptl. Cell Res. 77 (1973) 195-206.

68. A.L. Politoff, S.J. Socolar and W.R. Loewenstein, Biochim. Biophys. Acta, 135 (1967) 791-793.

69. J.T. Braaten, M.H. Greider, J.E. McGuigan and D.H. Mintz, Endocrinology, 99 (1976) 684-691.

70. B. Hess In Rate of Control of Biological Process, D.D. Davies, ed., Symp. Soc. Exptl. Biol., Cambridge University, 27 (1973) 105-131.

MUTANT MICE FROM MUTAGENIZED TERATOCARCINOMA CELLS

BEATRICE MINTZ
Institute for Cancer Research
Fox Chase Cancer Center
Philadelphia, Pennsylvania 19111 U.S.A.

Abstract: From recent experimental evidence, the malignancy of mouse teratocarcinoma cells is nonmutational in origin. It is apparently due to aberrant gene expression in primitive, developmentally totipotent stem cells. The unrestrained proliferation and chaotic behavior of the malignant cells is permanently supplanted by an orderly developmental sequence of genetic programming if they are placed in a normal embryo (blastocyst) environment. There they are able to contribute functionally mature cells in all somatic tissues and in the germ line. The malignant stem cells may therefore be utilized as vehicles for the deliberate introduction of particular mutant genes (or foreign genes) into mice, by blastocyst injection, following mutagenesis and selection for specific mutant cells *in vitro*. This new experimental system provides many hitherto unavailable approaches to the experimental study of mammalian differentiation. It also permits the creation of mouse models of many human genetic diseases. An example is Lesch-Nyhan disease, caused by a deficiency of the enzyme hypoxanthine phosphoribosyltransferase (HPRT). Mosaic animals have been successfully produced from blastocysts injected with teratocarcinoma cells preselected in culture for this deficiency. The unique value of the mosaic mice lies in the fact that some have $HPRT^-$ cells in only one or a few tissues, whose role in the disease syndrome can now be analyzed. When progeny, with all-mutant cells, are obtained, they will be models of the human disease.

ISBN 0-12-045450-5

INTRODUCTION

There is increasing evidence that neoplasia is an aberration of differentiation and not merely of cell proliferation (1, 2). All of normal development may be thought of as a hierarchical succession of progressively more restricted stem cells, i.e., of generative cells that function as precursors of one or more specialized cell types (Fig. 1). Such stem cells have been extensively studied only in the "renewal" tissues in which cell loss and replacement continue throughout life, e.g., in hematopoiesis. It is likely that all tissues arise in a similar manner, although the stem cell pool may disappear relatively early, as is the case in brain.

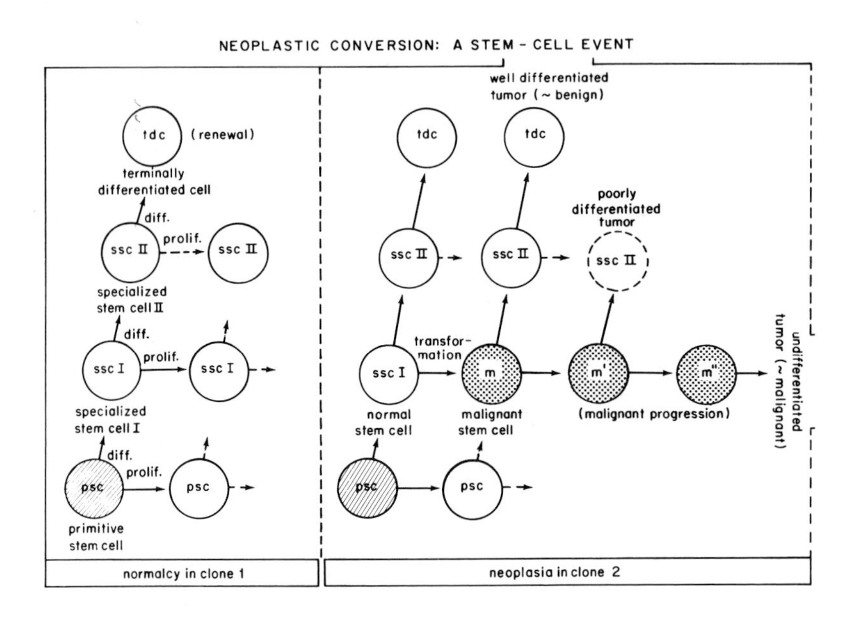

Fig. 1. In normal development (left), a given kind of stem cell produces mitotic progeny like itself, with the capacity to continue proliferation (prolif.) or to differentiate further (diff.), in a hierarchical succession. In neoplasia, any target stem cell that is transformed undergoes reduced differentiation but not reduced proliferation. If differentiation is reduced only slightly, the tumor is relatively benign. If malignant progression occurs, with increasing selection for more proliferative variants, a highly malignant (anaplastic) tumor results. (From Mintz, ref. 1.)

The most reasonable explanation for malignancy, and one which is consistent with all available facts, is that normal stem cells are the targets in neoplastic conversion. The transformed stem cell then yields the generative cell population of the tumor. The tumor tissue usually also comprises some relatively differentiated cells (often abnormal) that develop from the stem cells.

According to this view, the most developmentally primitive normal stem cells (psc, in Fig. 1) are capable ultimately of giving rise to all the more restricted stem cells in normal development; the latter eventually yield terminally differentiated, non-dividing cells (tdc, in Fig. 1). In short, only the least specialized stem cells in the normal hierarchy are developmentally totipotent.

TERATOCARCINOMA: A MALIGNANCY OF DEVELOPMENTALLY TOTIPOTENT STEM CELLS

Teratocarcinomas are unique tumors insofar as they contain many tissues unrelated to their site in the body. From this fact, it seemed possible that they might represent malignant aberrations of developmentally totipotent stem cells. This likelihood was increased by the demonstration that a single proliferative cell from a mouse teratocarcinoma could, upon transplantation, produce a tumor with a variety of tissues (3). Nevertheless, the tumors invariably lack a number of tissues and exhibit only immature versions of most others (4). For this reason, the full developmental potentialities of the stem cells remained unknown.

I became interested, a few years ago, in learning the answer to this question, for two reasons: The first was that a successful demonstration of totipotency of the stem cells, through their orderly differentiation and maturation into _all_ tissues, would require that all genes be present. Such a result would, in effect, mean that genes had neither been deleted nor mutated during carcinogenesis; thus the malignancy would have arisen without mutational change. The second reason for my interest in this question was that if the tumor stem cells could be made to form all tissues, they could be regarded as "surrogate eggs" through which specific mutant genes could be established in mice. This will be discussed presently in greater detail.

The initial experiments involved teratocarcinoma cells that had been grown only _in vivo_, as a transplant line, for as long as 8 years. This tumor (OTT 6050) had been induced by grafting a young embryo to an ectopic location under the testis capsule (4). It was chosen for our earliest tests partly because of the greater likelihood that euploidy might have been retained in an in vivo transplant line than in lines adapted to culture in vitro. The tumor did indeed prove to be karyotypically entirely normal and of X/Y male sex chromosome constitution (C. Cronmiller and B. Mintz, manuscript in preparation). It should be emphasized that this tumor, as well as the others used in our subsequent work, are highly malignant by pathologists' criteria: they invade, metastasize, and grow rapidly. The OTT 6050 tumor had been propagated in the peritoneal cavity as a modified ascites of multicellular "embryoid bodies". These, when still small, consist only of a "core" of stem cells surrounded by a "rind" of yolk-sac-like epithelial cells.

If the teratocarcinoma stem cells were in fact totipotent, the most promising place in which they might actually realize their maximum developmental potentialities seemed to be in the company of entirely normal early stem cells, hence in an early embryo. The contemplated experimental arrangement would therefore be comparable to what had already been carried out here some time ago, by associating normal early mouse embryo cells of two separate genotypes (5), except that tumor cells would be utilized in place of one of the donor embryos. Cleavage-stage blastomeres were found not to adhere well to teratocarcinoma cells; therefore the latter were microinjected into the cavity of slightly older (blastocyst-stage) embryos; there they were entrapped (6). Since the embryo-forming region of the blastocyst probably comprises only about 3 cells (5), comparable small cell numbers, i.e., 1-5 tumor stem cells, were introduced. In choosing the blastocyst strain so as to distinguish, by genetic markers, its contributions from those of the tumor, the alleles of the inbred 129 strain in which the tumor had arisen were tentatively assumed to be present, although most genes in the tumor cells had not been expressed for 8 years.

The results of the experiment were striking (6, 7). Many healthy tumor-free mice were obtained although the genetic markers specific for each strain clearly demonstrated that normal, mature cells derived from the tumor now coexisted with embryo-derived cells. The two cell populations remain separate, except in the formation of skeletal muscle, which is known (8) to occur by myoblast fusion. The tumor-lineage cells produced numerous tissue-specific proteins (e.g., hemoglobin, melanins, immunoglobulins) normally found in the 129 inbred strain, and coded for by genes that had given no evidence of function during all the years of tumor-cell transplantation. Moreover, the tumor-lineage cells remained normal, even when grafted subcutaneously to recipient mice.

Comparable results were obtained in blastocyst injections with the stem cells of a spontaneous ovarian teratocarcinoma of the LT strain (B. Mintz and K. Illmensee, manuscript in preparation). This tumor, which had also been maintained by serial transplantation in syngeneic hosts, was chromosomally X/X female and karyotypically normal (C. Cronmiller and B. Mintz, manuscript in preparation).

For both the euploid X/Y and X/X tumors discussed above, an observation even more striking than somatic-tissue formation was their capacity to contribute to the germ line in the mosaic animals. The X/Y tumor lineage gave rise to functional sperms and to many progeny whose "father", genetically speaking, had been a tumor (Mintz and Illmensee, 1975). The X/X tumor lineage gave rise to functional eggs and to progeny whose "mother" had been a tumor (B. Mintz and K. Illmensee, manuscript in preparation). (Earlier studies, reviewed in ref. 5, have shown that allophenic mice, including X/X ⟷ X/Y sex chromosome mosaics, produce functional gametes only from the germ-cell strain of the same sex type as the host's sex phenotype.) These F_1 progeny, and their F_2 offspring as well, are healthy, and they document transmission of the tumor-strain genes.

The results most critical for tests of stem cell capacities were those from injections of single teratocarcinoma cells (of the OTT 6050 tumor) into blastocysts (9). As shown in Fig. 2, even a single stem cell could give rise to contributions in all of the many tissues tested. This provides conclusive evidence for the developmental totipotency of the donor cell.

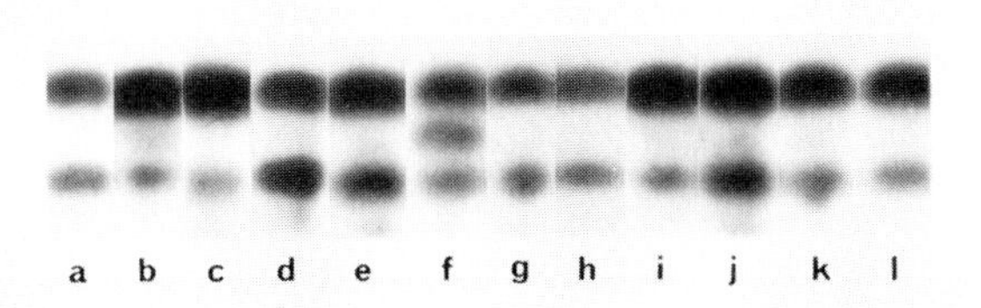

Fig. 2. Starch gel electrophoretic tests of tissue homogenates from a mouse obtained by injecting one teratocarcinoma cell into a blastocyst. The tumor-strain type of glucosephosphate isomerase (slow-migrating, lowermost) occurs in all tissues. Slot (a) a control mixture; (b-l) tissues from the experimental mouse, including blood, brain,

spleen, heart, skeletal muscle (with a hybrid band due to myoblast fusion), kidney, reproductive tract, liver, pooled gut and pancreas, thymus, and lungs. (From Illmensee and Mintz, ref. 9.)

The single-cell injections were also particularly instructive with regard to the stability of "normalization" of the once-malignant cells (9, and further unpublished data from R. P. Custer and B. Mintz): Of the 18 animals with tumor-strain cells in any or all tissues, none had tumors; in 5 mice with tumors (all teratocarcinomas), tumor-strain cells were found only in their tumors, not in any of their other tissues. (The "pancreatic adenocarcinoma" preliminarily diagnosed as such in one case [9] proved, upon further study, to be merely a teratocarcinoma.) Thus, the injected malignant cell in each blastocyst appears to have followed only one of two mutually exclusive paths after it was introduced into the blastocyst: If it was integrated into the embryo, it participated in embryogenesis and was permanently normalized; if it failed to become integrated, it remained a malignant teratocarcinoma, as if in a culture situation.

These results furnish the first unambiguous example of complete reversal of an animal malignancy. The experiments strongly imply that the initial conversion to malignancy (induced, in one case; spontaneous, in another) came about through changes in gene function rather than through true mutation. Once placed in a more appropriate environment, the stem cells changed from a program of little differentiation and much proliferation to an orderly series of normal gene expressions indistinguishable from the genetic events in early normal embryogenesis. The mechanism whereby the change is effected is still unknown. Nevertheless, the possibility of cure-by-differentiation rather than cure-by-killing is an attractive one, in view of the clinical risks to normal cells in conventional therapy directed at destruction of the stem cells.

It seems possible that the nonmutational basis for this malignancy may not be unique and that it may offer a model for some other nonmutational tumors. There is good evidence (10, 11), even if indirect, that mutation is sometimes (and perhaps usually) involved in tumorigenesis. Thus, there seem to be two classes of tumors: those of mutational origin and others that are nonmutational in origin. In <u>both</u> classes, malignancy may be viewed as a failure of, or diminution in, the differentiation of a previously normal stem cell (Fig. 1)

(1). Whether a tumor could be "cured" by restoration of normal stem cell differentiation would depend not only on whether a significant primary mutation had occurred, but also on whether secondary errors had intruded during malignant progression. Laboratory tests of reversal, as in the blastocyst injections, do not offer any clinical application, but are nonetheless worth pursuing with some other candidate tumors. Where the relevant stem cell is of a more specialized type (e.g., as in a hepatoma), the appropriate test environment might then be in the corresponding early organ primordium, rather than in a blastocyst.

TERATOCARCINOMA CELLS AS VEHICLES FOR MUTANT GENES

The demonstrated developmental totipotency of teratocarcinoma stem cells makes available a unique material for "directed" mammalian mutagenesis (1, 6, 12). These cells may be grown in large numbers in culture, where each may in a sense be regarded as a potential mouse (if aneuploidy has been avoided). By techniques familiar to somatic cell geneticists, the cells may be mutagenized, and selected or screened *in vitro* for a particular mutation of interest, and then placed in a genetically marked blastocyst for further differentiation and for germ-line transmission. While the basic *in vitro* methods of somatic cell genetics have in fact been available for some time, the cells to which such methods have hitherto been applied (e.g., fibroblasts) are incapable of any substantial degree of differentiation. Genes of other species may also be introduced into teratocarcinoma cells in culture, for example, by cell hybridization or chromosome transfer, and thereby into mice (12). Cells bearing mutations that would ordinarily be lethal to the embryo could be "rescued" by the accompanying normal cells of blastocyst provenance.

Many problems of genetics, differentiation, and metabolism might be elucidated by this scheme. One of the most promising and important possibilities lies in the creation of laboratory animal models of human genetic diseases. While there are numerous human hereditary maladies that are due to known biochemical lesions, no precise animal model counterpart is yet available for any of these. Until now, mutant mice have been produced fortuitously, without predetermination or choice of the mutation. In most instances, the mutant animal is recognized only by gross phenotypic changes (e.g., behavioral defects) and the primary biochemical lesion has remained unidentified. In those human genetic diseases where the biochemical defect is known, the clinical syndrome is usually very complex and the chain of events from the metabolic lesion to the syndrome is still mysterious. Prenatal changes and cascades of errors are doubtless often involved and cannot easily be traced in such patients. Nor can cures readily be attempted on an experimental basis.

An example of such a hereditary disorder is the Lesch-Nyhan syndrome (13). It is characterized by excessive synthesis of purines, and excretion of uric acid, by spasticity, and by mental retardation and compulsive self-mutilation. The disease is known to arise from a severe deficiency of the purine salvage enzyme hypoxanthine phosphoribosyltransferase (HPRT) (14). The gene is an X-linked recessive and the disease is generally fatal in males at a young age. The availability of an <u>in vitro</u> medium for selection of HPRT-deficient cells made this defect favorable for the first attempt at applying this experimental approach in mice.

The experiment (15) is diagrammed in Fig. 3.

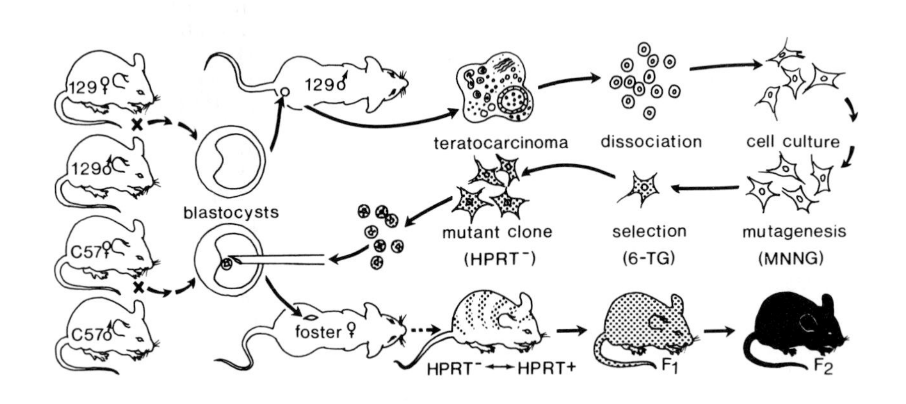

Fig. 3. A teratocarcinoma was first induced by grafting a blastocyst from a pair of 129 strain parents (upper left) under the testis capsule of a male of the same strain. A malignant teratocarcinoma was formed. The cells were dissociated and explanted and a stem-cell culture line established. After exposure to the mutagen <u>N</u>-methyl-<u>N</u>'-nitro-<u>N</u>-nitrosoguanidine (MNNG), $HPRT^-$ cells (stippled) were selected for their resistance to the purine base analog 6-thioguanine (6-TG). Cells from a resistant clone were then injected into genetically marked blastocysts (C57BL/6 strain). The injected embryos were transferred to the uterus of a pseudopregnant foster mother (mated to a sterile

vasectomized male). Among the live mice that were born, some had $HPRT^-$ cells from the teratocarcinoma lineage, along with blastocyst-derived cells, in their coats and/or other tissues. In future germ-line transmission, chromosomally female teratocarcinoma cells (X/X or X/0) could yield affected "Lesch-Nyhan" males in the F_1 generation; male (X/Y) cells would yield affected male progeny in the F_2. (From Dewey *et al.*, ref. 15.)

As shown in Fig. 3, the mutagenized stem cells of this cultured teratocarcinoma line were selected for resistance to the purine base analog 6-thioguanine. The cells of a resistant clone, known as NG 2, were found to be completely deficient in HPRT activity. Following injection into genetically marked blastocysts, the previously malignant cells participated effectively in embryogenesis and viable mice free of tumors were obtained. Cells of tumor-strain provenance were identified by strain markers in virtually all tissues of some individuals. Mature cellular function was seen by their tissue-type products, such as melanins and liver proteins. These cultured tumor stem cells are therefore developmentally totipotent, as was true for the *in vivo* tumor transplant lines previously described. The fact that NG 2 is a *clonal* line dispels any possibility that it comprised malignant and non-malignant cells. The test grafts of NG 2 cells in subcutaneous sites, rather than blastocysts, yielded typical malignant tumors.

The teratocarcinoma cell line employed here was chromosomally quasi-normal. The only two anomalies were X/0 female sex chromosomal constitution and trisomy of chromosome 6 (C. Cronmiller and B. Mintz, manuscript in preparation). The former is not a disadvantage, as X/0 mice are fertile females (16). Trisomy-6 is ordinarily lethal at 12-14 days of gestation (17), but the cells were "rescued" in the present experiment by coexistence with wild-type cells.

Retention of the severe HPRT deficiency after cellular differentiation *in vivo* was documented in two ways: First (Table 1), extracts of mosaic tissues were assayed for HPRT activity, in relation to adenine phosphoribosyltransferase (APRT), which is unaffected in Lesch-Nyhan patients. When appreciable populations of NG 2-derived cells were present (e.g., 50-55% in heart or kidney of one animal), there was a marked depression of HPRT activity. However, when only minor amounts occurred (e.g., 5-10% in brain or pancreas of the same mosaic individual), there was no significant lowering of the HPRT/APRT ratio.

TABLE 1

HPRT specific activity in tissues of mosaic mice partially derived from HPRT-deficient teratocarcinoma cells

	HPRT/APRT			% 129 type
Tissue	C57 control	129 control	Exp.	GPI*
Heart	3.08	1.90	0.50	55
Kidney	0.79	1.01	0.48	50
Brain	1.58	1.79	1.81	10
Pancreas	0.56	0.58	0.48	5

*The teratocarcinoma strain contribution, based on the glucosephosphate isomerase isozyme type (The table is from Dewey _et al_., ref. 15.)

Additional evidence was obtained by autoradiographic visualization of $HPRT^-$ colonies in cultures of subcutaneous connective tissue. After the cells were plated at low density and incubated in [^{3}H]hypoxanthine, two kinds of colonies were seen: Some had many silver grains over the cells, as in wild-type ($HPRT^+$) controls; others were devoid of grains, as in known $HPRT^-$ controls (15).

It should be emphasized that in our previous experiments with injected teratocarcinoma cells, as well as in the experiment involving $HPRT^-$ mutant cells, the injected blastocysts gave rise to some mice with tumor-lineage cells in all tissues, to some with such cells in only a few or even only one tissue, and to some with no vestige of that strain. This variability is presumably due in part to delayed or haphazard integration of the injected cells.

This variability in tissue derivatives among individuals

becomes an especially useful tool, as in the present case of mutant teratocarcinoma cells carrying a disease-producing gene. While the actual model of Lesch-Nyhan disease, an F_1 "Lesch-Nyhan" male mouse (Fig. 3), has not yet been obtained, the mosaic animals themselves offer exceptional possibilities for conducting causal analyses, precisely because of the variable tissue distribution of their $HPRT^-$ cells (15). The metabolic and clinical characteristics of an individual with "Lesch-Nyhan" cells in, say, only its brain, or only its liver, could help to pinpoint the tissue(s) in which the complex syndrome may originate. The mosaics thus far have no obvious behavioral defects, even though some have $HPRT^-$ cells in their brains. These and other features are being followed up for possible clues.

A notable observation is that there appears in the mosaic mice to be particularly strong selection against HPRT-deficient cells in the red blood cell population (15). A similar trend has been reported in human heterozygous female carriers of the disease (18). This parallelism is an encouraging sign that the mosaic mice and, ultimately, their Lesch-Nyhan sons, will offer relevant information for examining the basis for the complex human disease.

This is merely the first among many possibilities for utilizing the malignant stem cells of mouse teratocarcinomas as a means of "synthesizing" mutant mice with defects comparable to those in human genetic diseases.

REFERENCES

(1) B. Mintz. Harvey Society Lectures, Series 71 (1976). (Academic Press, New York) In press.

(2) G. B. Pierce, in: Developmental Aspects of Carcinogenesis and Immunity, 32nd Symposium of the Society for Developmental Biology, ed. T. J. King (Academic Press, New York, 1974) p. 3.

(3) L. J. Kleinsmith and G. B. Pierce, Jr. Cancer Res. 24 (1964) 1544

(4) L. C. Stevens. Dev. Biol. 21 (1970) 264.

(5) B. Mintz. Annu. Rev. Genet. 8 (1974) 411.

(6) B. Mintz, K. Illmensee and J. D. Gearhart, in: Symposium on Teratomas and Differentiation, eds. M. I. Sherman and D. Solter (Academic Press, New York,

1975) p. 59

(7) B. Mintz and K. Illmensee. Proc. Nat. Acad. Sci. USA 72 (1975) 3585.

(8) B. Mintz and W. W. Baker. Proc. Nat. Acad. Sci. USA 58 (1967) 592.

(9) K. Illmensee and B. Mintz. Proc. Nat. Acad. Sci. USA 73 (1976) 549.

(10) J. McCann and B. N. Ames. Proc. Nat. Acad. Sci. USA 73 (1976) 950.

(11) A. G. Knudson, Jr. Advan. Cancer Res. 17 (1973) 317.

(12) B. Mintz, in: Genetic Interaction and Gene Transfer, Brookhaven Symposia in Biology 29, ed. C. W. Anderson. In press.

(13) M. Lesch and W. L. Nyhan. Am. J. Med. 36 (1964) 561.

(14) J. E. Seegmiller, F. M. Rosenbloom and W. N. Kelly. Science 155 (1967) 1682.

(15) M. J. Dewey, D. W. Martin, Jr., G. R. Martin and B. Mintz. Proc. Nat. Acad. Sci. USA 74 (1977) 5564.

(16) W. J. Welshons and L. B. Russell. Proc. Nat. Acad. Sci. USA 45 (1959) 560.

(17) A. Gropp. Clin. Genet. 8 (1975) 389.

(18) W. L. Nyhan, B. Bakay, J. D. Connor, J. F. Marks and D. K. Keele. Proc. Nat. Acad. Sci. USA 65 (1970) 214.

These investigations were supported by United States Public Health Service Grant Nos. HD-01646, CA-06927, and RR-05539, and by an appropriation from the Commonwealth of Pennsylvania.

DISCUSSION

R. OSHIMA: Since your studies dramatically demonstrate that malignant embryonal carcinoma cells can take part in normal development of a mouse embryo, can you tell us if they can completely replace the inner cell mass of a mouse embryo as can be done with the inner cell mass of other embryos?

B. MINTZ: Dr. Oshima is pointing out that in the recipient embryo in which the malignant cells are introduced the only portion of the embryo that will actually form a mouse is in the area known as the inner cell mass. And I believe from my earlier experiments that only a very small number of those cells are ever going to form an embryo; that number may be as small as three. The question that is raised therefore is whether cells other than those of the future embryo can influence the malignant cells to behave normally. We did in fact surgically remove the inner cell mass of some blastocysts, having only the trophoblast and stuffed these with embryonal carcinoma cells. After placing them in the uterus of an appropriate recipient, such "blastocysts" could implant, but the tumor cells continued to be malignant.

R. OSHIMA: In light of those results, would you speculate about the nature of the difference between the embryonal carcinoma cell and the inner cell mass cells? Since the malignant cells can be propagated through more than 200 transplant generations and there are also strain differences, concerning the spontaneous appearance of testicular teratocarcinomas, what would be the nature of the difference that would give rise to a malignant cell?

B. MINTZ: So far as we know, there do not seem to be strain differences in the capacity for normal embryo cells to influence these malignant cells. As to what normal stem cell counterparts there might be to these malignant cells, this may be the stage shortly after implantation, not in the blastocyst.

R.C. LEIF: In the teratomas what is the efficiency of producing a mouse when you implant as compared with your previous experiments with normal blastocyst, because you may have some degree of selection, for those cells that are capable of producing a take?

B. MINTZ: I am happy to tell you that it is not much below other experiments. An even more critical answer is the fact that if we inject only one stem cell as opposed to five, we do not diminish the takes.

R.C. LEIF: With Don Rounds' microbeam technique you could literally map all the chromosomes by deleting specific areas (Ohnuki, Y., Rounds, D.E., Olson, R.S., and Berns, M.W., Exp. Cell Res., 71:132-144 (1972)).

B. MINTZ: We are interested in other ways of getting at this. One can introduce parts of chromosomes, or even chromosomes of other species including man.

R.G. KALLEN: Have you tried to transplant adult tissue or embryonic tissue from your teratocarcinoma derived mice, and do they have an increased neoplastic potential?

B. MINTZ: The answer is no. If we take differentiated tissues from the mosaic animals and transplant them, they do not become malignant.

R.G. KALLEN: Embryonic tissues as well?

B. MINTZ: We do not sacrifice many embryos because we like to wait and see what they are going to turn into.

I.B. FRITZ: Are there HPRT-deficient cells in the gonads and have you observed any HPRT-deficient germ cells?

B. MINTZ: We have seen the presence of such cells in the gonads, but we are looking at a homogenate with isozyme markers and therefore cannot be sure whether the cells are in the germ line or in the somatic cells. We can only show them to be in the germ cells if there are progeny with the marker.

I.B. FRITZ: Do you have a male yet?

B. MINTZ: No, not yet. We may have to get a better cell line for that.

C. WEST: Do you have any experience in introducing cells of other types into the blastocyst? How about other embryonal carcinoma lines which seem to have a restrictive capacity for differention?

B. MINTZ: I am not greatly interested in them.

C. WEST: Have you had any experience in introducing normal embryonic cells other than blastomeres into the blastocysts?

B. MINTZ: Other people have done that experiment - so far unsuccessfully.

H.V. RICKENBERG: You suggest that the teratocarcinoma was of non-mutational origin. An alternative explanation could be that in fact a mutation did occur, but that the mutation affected the synthesis of a presumptive regulatory protein, and that the function of this protein, the regulatory function, could be taken over by the surrounding cells. A probable test for this hypothesis would be to re-isolate the teratocarcinoma derived cells from the animal and to reinject them, not into the blastocyst, but rather into other tissue to see whether they become malignant.

B. MINTZ: The answer to the last part of your comment, is, as I stated earlier, that if such cells are transferred they do not become malignant again. The fact that the company of normal embryo cells is required for these cells to behave normally might actually mean that there has been a mutation that was critical in the malignancy which is now being compensated for by some product from the normal cells. I think that is a possibility and it cannot be ruled out. The most probable interpretation is the one that I presented, namely that these are non-mutational in origin.

S. WEINHOUSE: To follow up on that - I know the mutation theory is going to die hard.

B. MINTZ: I hope not.

S. WEINHOUSE: I am wondering if there is a mutation in a regulatory gene which might be passed on to the progeny; and if so, whether cells such as fibroblasts that you would isolate from these mice might be more susceptible to transformation, either by viruses or by chemicals?

B. MINTZ: Well, I would be glad to give to anybody a culture of such fibroblasts if they wished to investigate the question. We have not done that. I certainly hope that the mutation theory will not die hard and will not die at all, because although we seem to have the first example of a non-mutational tumor, and there may indeed be others, I think it highly unlikely that all tumors arise non-mutationally. I am much persuaded by the kind of evidence that Bruce Ames and others have been propounding. I think it is altogether reasonable that there may be tumors of mutational origin, and that may even comprise the majority of tumors. It is not necessary to postulate a single origin of tumors. I do not believe in the sanctity of unifying theories, and having said that, I will add that I think the only unifying theory about cancer is that I believe all cancers are developmental imperfections of stem cells, whether they arise mutationally or not.

W. CIEPLINSKI: Are both of your strains H2 histocompatible?

B. MINTZ: No, we deliberately choose this as another marker, such that in some of our experiments they are different. But in the early experiments that I have done making mosaic mice out of normal embryo cells of two different strains, we have always found that differences in histocompatibility type between the two strains made absolutely no difference, and this is just exactly what I would expect. That is to say that those animals are permanently tolerant of their two immunogenetic strains. This is presumably related to the normal mechanism of self-tolerance and possibly to the fact that the antigenic differences or immunological differences are present before the immune system is present, so that the latter does not recognize them as foreign.

W. CIEPLINSKI: Have you looked for the histocompatibility at the H_2 locus at the time you inject cells into the blastocyst and whether it alters the rate of takes?

B. MINTZ: We have not systematically investigated that question.

W.T. McALLISTER: Will the mature mice derived from the tumor germ line support development of teratocarcioma?

B. MINTZ: We did not do that experiment but I think that they would behave no differently than ordinary mice.

NUCLEAR TRANSPLANTATION AND GENE INJECTION IN THE ANALYSIS OF AMPHIBIAN DEVELOPMENT

J.B. GURDON, E.M. DE ROBERTIS, R.A. LASKEY, and A.H. WYLLIE
MRC Laboratory of Molecular Biology,
Hills Road, Cambridge CB2 2QH,
England.

Abstract: We summarize recent results from this laboratory involving the transplantation of nuclei and the injection of DNA into eggs and oocytes of *Xenopus laevis*. Nuclear transplantation experiments indicate the presence of gene-controlling substances in the cytoplasm of oocytes and eggs. Purified DNA is converted into a stable nucleoprotein complex and correctly transcribed if injected into the nucleus of an oocyte. A similar nucleoprotein complex can also be formed *in vitro* with components of egg cytoplasm.

INTRODUCTION

The eventual aim of the work summarized below is to understand the cytoplasmic control of gene activity in early development. We have tried to exploit the large size and ready availability of eggs and oocytes of the South African clawed frog *Xenopus laevis*. Their large size ($>$ 1 mm diameter) makes them particularly suitable for microinjection and for the biochemical analysis of injected cells.

In this article, we first summarize our reason for believing that gene-controlling substances exist in the cytoplasm of eggs and oocytes of this species. We then describe experiments involving the injection of purified genes, as DNA, into oocytes and show that correct transcription can be obtained from purified genes. The fate of injected DNA has established that circular molecules are converted into a nucleoprotein complex, which is very stable. The distribution of oocyte nuclear proteins injected into oocytes suggests that it may be possible to do meaningful experiments to test the activity of chromosomal proteins from specialized cells. Lastly we summarize progress with a

ISBN 0-12-045450-5

cell-free system in which a nucleoprotein complex can be formed from purified DNA and components of Xenopus eggs. The structures formed in vitro closely resemble those that are recovered from DNA-injected oocytes.

The transplantation of nuclei to enucleated eggs.

It has been known for some years that the cytoplasm of amphibian eggs or embryos must contain components which can control gene activity. The most direct demonstration of this has come from experiments in which nuclei from adult skin cells which have undergone the early stages of keratinization were transplanted to enucleated Xenopus eggs (1). Tadpoles containing many different cell-types, including muscle, nerve, pigment and lens cells, were formed as direct mitotic products of a transplanted skin cell nucleus and egg cytoplasm. The lens, muscle, and nerve cell genes were not expressed in the adult skin cells, but were re-activated after nuclear transplantation. It is not certain in these experiments whether the reactivation took place immediately after transplantation or at some later stage in development, since the reactivated genes were not expressed until tadpoles had been formed (2 - 3 days after nuclear transplantation). To pursue this question, nuclei have been transplanted to oocytes which are growing egg cells taken from the ovary of a female frog.

The transplantation of nuclei to oocytes.

Nuclei transplanted to oocytes do not divide or initiate new rounds of DNA synthesis. This is in contrast to similar nuclei injected into eggs, as a result of which they always synthesize DNA and undergo mitotic division within 1 - 2 hours. Nuclei in oocytes may continue to synthesize RNA for as long as oocytes remain viable in culture (up to 4 weeks (2)). This greatly exceeds the time (2 - 3 days) needed for eggs and transplanted nuclei (or their mitotic products) to form tadpoles. Therefore we have looked for changes in gene activity in nuclei transplanted to oocytes. Injected oocytes were cultured for the desired amount of time and then incubated in medium containing ^{14}C-amino acids for periods of about 6 hours after which they were frozen. Proteins were extracted, fractionated by 2-D gel electrophoresis, and fluorography of gels was used to locate labelled proteins.

Several series of experiments were carried out with HeLa nuclei. As a result it was established, using α-amanitin and various other controls, that the transcriptional activity of genes in nuclei could be seen after transplantation to oocytes (2, 3). It was found that, of the 25 HeLa-coded proteins readily detected on a background of *Xenopus* oocyte proteins, 22 were not expressed, and only 3 continued to be active after injection; in addition a few other proteins not readily identifiable as being HeLa or *Xenopus* proteins, were synthesized. This result suggests that components of oocytes may cause a selective switching off and switching on of genes in transplanted nuclei. However, these experiments must be interpreted with caution, because human nuclei were combined with amphibian cytoplasm.

Direct evidence for selective gene expression by nuclei transplanted to oocytes has come from experiments in which cultured kidney cell nuclei of *Xenopus* were transplanted to oocytes of *Pleurodeles*, another amphibian species (4). Oocytes of *Pleurodeles* were chosen so that proteins characteristically synthesized by *Xenopus* oocytes could be seen. Three to five days after injecting *Xenopus* cultured kidney cell nuclei into *Pleurodeles* oocytes, labelled proteins were analyzed. None of the proteins characteristic of *Xenopus* kidney cells was seen. However, 3 proteins characteristic of *Xenopus* oocytes were seen, and these represent the activation of oocyte-active genes in the cultured kidney cell nuclei. From this experiment it can be concluded that oocyte cytoplasm has components which can activate some genes and inhibit the expression of others.

In other experiments using two species of Axolotl, Etkin (5) has injected liver nuclei into oocytes. He observed the non-expression of liver-specific alcohol dehydrogenase, but the continued expression of oocyte- and liver-active lactic dehydrogenase.

The transcription of purified DNA injected into oocytes.

A detailed analysis of gene-controlling substances in oocytes is difficult if not impossible to undertake using whole nuclei in which numerous different genes are undergoing changes in activity. Analysis would be enormously facilitated if it were possible to do transplantation experiments with single kinds of genes. Our first experiments made use of SV40 DNA (6) which contains approximately five genes. It soon became clear that transcripts of SV40 DNA were obtained so long as DNA was injected into the nucleus of an oocyte (6). Some of these transcripts are able to code for SV40 virion proteins (7). Since these early experiments, many different kinds of DNA have been injected, and all are found to be transcribed, with one exception, namely mouse satellite DNA. This is the only naturally untranscribed type of DNA which we have tested. All other kinds of DNA probably include promoters. Therefore it is possible, though not yet proved, that the transcription of injected DNA is promoter-dependent.

A particularly advantageous aspect of the transcription of injected DNA is its fidelity. For example, if purified 5S genes are transcribed *in vitro* with *E.coli* polymerase, the transcripts obtained are commonly from the wrong strand and from spacer regions (8). But the same kind of DNA injected into oocytes is transcribed almost entirely from the naturally transcribed strand and mainly from the gene rather than spacer region (9). It is possible that fidelity of transcription is greater for polymerase III transcribed genes than for other genes (mostly polymerase II), though genes transcribed by polymerase I and polymerase II have not yet been adequately tested.

The amount of transcription from injected DNA is substantial. For example, 25% of the total RNA synthesized by a group of injected oocytes is transcribed by injected 5S genes (9). Since DNA is successfully deposited in the nucleus of, on average, only half of all injected oocytes, this value represents 50% for successfully injected oocytes, or approximately a 200-fold increase in the endogenous rate of 5S RNA synthesis. Even then, the number of 5S genes (about 10^9) injected is considerably greater than a 200-fold excess of the endogenous number (c. 100,000 in a tetraploid oocyte (10)). Therefore injected genes must be transcribed at a lower rate than endogenous genes, or perhaps more likely, a small proportion of them are transcribed at the normal rate. The most important point, however, is that the transcription of one type of gene is increased enormously above the natural rate, thereby making it possible, in effect, to transplant one type of gene into an oocyte, and to analyze its transcriptional products. It is fortunate that injected genes seem either to be transcribed correctly or not at all; we have not yet been able to detect incorrect transcripts from 5S genes (9).

The fate of injected DNA

To study the control of transcription of a gene in detail,

it will be necessary to identify the control region - i.e. the promoter sequence at which RNA polymerase starts to transcribe. This is the part of a gene which we presume to be regulated, and to which repressors and activators are presumed to bind. The most direct means of identifying a promoter region of a DNA molecule would be to cut away increasing amounts of a molecule until the gene it contains is no longer transcribed. Before attempting to do this, we must be sure that very short lengths of DNA can be meaningfully transcribed. So far all types of DNA which have given correct transcription after injection have been long linear molecules (e.g. genomic 5S DNA) or circular molecules (e.g. SV40 or plasmids with and without insertions of other genes). We have therefore surveyed the fate of injected DNA molecules of different configurations, using ^{3}H-SV40 (11). The results are the following. All forms of SV40 DNA are degraded when deposited in oocyte cytoplasm. Supercoiled and open circles are cut to linear molecules, which are then progressively degraded to smaller fragments. In contrast, any circular molecules deposited in the nucleus are repaired if they were nicked and are converted into covalently closed supercoiled molecules which are indefinately stable. On the other hand, linear molecules in the nucleus are slowly degraded by exonuclease activity. Thus the only stable form of DNA for injection is circular. It is unimportant whether it is supercoiled, since supercoiled molecules are immediately relaxed within a few minutes of injection and then slowly converted back into a supercoiled state.

It has also been possible to follow the conversion of free DNA into a nucleoprotein complex (11). After appropriate extraction procedures, about half of the ^{3}H-SV40 injected is recovered from a sucrose gradient as a 55S complex (free SV40 DNA sedimenting at about 21S). This complex has been cross-linked with formalin and banded on a CsCl gradient. Labelled DNA bands at a density of about 1.46, a result expected if it has associated with it an equal weight of protein. Material from the CsCl gradient contains beaded circles, resembling the "minichromosome" (12) of SV40.

From these results we conclude that oocyte nuclei contain components which can convert DNA into a stable nucleoprotein structure. It is known from other work (13) that this effect is indeed due to the presence in the oocyte nucleus of stabilizing components and not to the exclusion from the nucleus of a cytoplasmic nuclease. Since oocyte nuclei evidently contain an exonuclease activity, adverse effects are likely to be particularly severe when very short DNA molecules are injected. Therefore we consider that the use of injected oocytes for identifying promoter regions in DNA molecules will best be achieved by circularizing restriction segments before injection.

The intracellular distribution of proteins injected into oocytes.

We have discussed the injection of DNA into oocytes with the implication that injected genes will be subjected to the same types of regulatory controls as apply to whole transplanted nuclei. We have not yet been able to test this supposition because we have not had available in purified form a gene of the type that is believed to be transcriptionally *inactive* in oocytes. Thus chromosomal 5S genes and histone genes are naturally transcribed in oocytes. SV40 and various plasmid and prokaryote virus DNAs used for our experiments are never normally present in *Xenopus* oocytes and are presumably not subject to regulation if injected.

It is conceivable, if a generalization is based on the results of Perlman et al. (14), that all genes are transcribed in oocytes. In this case, the non-expression of many genes in the transplanted nuclei referred to above must take place at a post-transcriptional level. If so, oocytes would have to be injected not only with genes whose activity or non-activity is being tested, but also with other molecules presumed to carry out these functions in specialized cells. For example, chromosomal proteins from different kinds of differentiated cells could be injected into oocytes containing purified genes which are active or inactive in these cell types. This approach can be undertaken (i) if segments of chromosomal DNA containing such genes are available, and (ii) if nuclear proteins are able to move freely within injected oocytes. Substantial advances have recently been made in the rapid selection of pieces of chromosomal DNA containing known genes. Some indication that injected nuclear proteins are able to diffuse freely within injected oocytes and therefore to associate with or dissociate from injected genes comes from several investigations.

It was observed some years ago that labelled histones are rapidly concentrated in the nuclei of injected oocytes, whereas bovine serum albumin and other non-nuclear proteins were not (15). Bonner (16) showed that non-histone nuclear proteins also accumulate in the nuclei of oocytes when injected into the cytoplasm. De Robertis et al. (17) have followed individually the compartmentation of several nuclear and cytoplasmic proteins in injected oocytes. They have found that each protein will distribute itself in injected oocytes so as to take up its normal distribution between nucleus and cytoplasm. These experiments were carried out with nuclear proteins from *Xenopus* oocytes, but they encourage us to believe that it will also be possible to use injected oocytes for testing the biological activity of chromosomal proteins or other kinds of molecules from specialized cells.

The *in vitro* assembly of chromatin from components of *Xenopus* eggs.

The detailed analysis of how proteins or other molecules interact with DNA injected into oocytes would be greatly facilitated if some of the events we have described for injected DNA could be reproduced in a cell-free system. During the last two years, Laskey and colleagues have developed a cell-free system from *Xenopus* eggs which is able to assemble SV40 DNA into chromatin under physiological conditions.

It was first established (18) that a 145,000 g supernatant fraction from eggs is able to convert SV40 DNA into a nucleoprotein structure which closely resembles naturally-occurring SV40 chromatin, and which is not distinguishable from the SV40 nucleoprotein complex formed in injected oocytes. Assembly requires histones, a nicking-closing enzyme activity, and an additional factor present in egg supernatant. It was subsequently found (19) that the supernatant factor is required for the ordered interaction of histones with DNA. This factor is thought not to contain small molecules, such as ATP, since it is excluded by Sephadex G-25, and can be recovered as a sharp peak of material which sediments at 11S in a sucrose gradient. It is thought that the factor contains histones combined with one or more other components. For example, it has a net negative charge, which is not true of histones or histone aggregates. It behaves as if it can catalyze the assembly of histones and DNA into nucleosomes. It is unlikely to contain RNA or DNA.

Although this *in vitro* assembly system is derived from *Xenopus* eggs, it is likely to create the same complex as is formed in injected oocyte nuclei. It may permit a detailed analysis of the structure of complexes formed by injected DNA and may also be able to provide nucleoprotein structures whose composition is known and whose transcriptional activity can be assayed in living cells by injection.

CONCLUSIONS

There are several respects in which *Xenopus* eggs and oocytes have proved favourable for the aim of our work. Cytoplasmic components evidently exist in oocytes, which can regulate gene expression. Experiments using purified genes as free DNA have been successful in achieving correct transcription. As purified genes become available, there is a reasonably good prospect of being able to study the control of transcription of individual genes in injected oocytes. Substantial progress has been made in the development of a cell-free system, based on components of *Xenopus* eggs, which can assemble purified DNA into complexes similar to those formed in injected oocytes.

REFERENCES

(1) J.B. Gurdon, R.A. Laskey and O.R. Reeves, J. Embryol. exp. Morph., 34 (1975) 93.

(2) J.B. Gurdon, E.M. De Robertis and G.A. Partington, Nature, London, 260 (1976) 116.

(3) E.M. De Robertis, G.A. Partington, R.F. Longthorne and J.B. Gurdon, J. Embryol. exp. Morph. 40, (1977) 199.

(4) E.M. De Robertis and J.B. Gurdon, Proc. Nat. Acad. Sci. US 74 (1977) 2470.

(5) L.D. Etkin, Devel. Biol., 52 (1976) 201.

(6) J.E. Mertz and J.B. Gurdon, Proc. Nat. Acad. Sci. US., 74 (1977) 1502.

(7) E.M. De Robertis and J.E. Mertz, Cell, 12, (1977) 175.

(8) R.H. Reeder and D.D. Brown, J. Mol. Biol., 51 (1970) 361; and unpublished results of D.D. Brown.

(9) D.D. Brown and J.B. Gurdon, Proc. Nat. Acad. Sci. US, 74 (1977) 2064.

(10) D.D. Brown, P.C. Wensink and E. Jordan, Proc. Nat. Acad. Sci. US, 68 (1971) 3175.

(11) A.H. Wyllie, R.A. Laskey, J. Finch and J.B. Gurdon, (1978) submitted for publication.

(12) J.D. Griffith, Science 187 (1975) 1202.

(13) A.H. Wyllie, J.B. Gurdon and J. Price, Nature, 268 (1977) 150.

(14) S.M. Perlman, P.J. Ford and M.M. Rosbash, Proc. Nat. Acad. Sci. US, 74 (1977) 3835.

(15) J.B. Gurdon, Proc. Roy. Soc. London B, 176 (1970) 303.

(16) W.M. Bonner, J. Cell Biol., 64, (1975) 431.

(17) E.M. De Robertis, R.F. Longthorne and J.B. Gurdon, (1978) submitted for publication.

(18) R.A. Laskey, A.D. Mills and N.R. Morris, Cell, 10 (1977) 237.

(19) R.A. Laskey, B.M. Honda, A.D. Mills, N.R. Morris, A.H. Wyllie, J.E. Mertz, E.M. De Robertis and J.B. Gurdon, Cold Spring Harbor Symposium (1978) in press.

DISCUSSION

C.C. LIEW: If you analyze this 55S protein complex, does the complex contain both histones and other proteins?

J.B. GURDON: The way we are analyzing these complexes is to remove the protein from the complexes and iodinate the proteins. Analysis is by conventional methods and, at present, shows that there is an excess amount of material in addition to the gene-protein complexes that we have isolated. Almost certainly the excess of the material makes the analysis imprecise. It probably consists of ribosome subunits which tend to come out in the same fractions as the complex; so I am afraid the answer to your question is that we are not in a position to say which particular kinds of proteins are present in these complexes just yet.

I.C. LO: Have you ever recomplexed SV40 DNA with protein and then tested its activity by injection?

J.B. GURDON: This sort of experiment is being done by my collegue Ron Laskey who has a particularly favorable way of reconstituting SV40 into chromatin. We have not done that particular experiment but the one that we are most looking forward to doing is one where individual proteins, for example particular histone fractions, are recomplexed into SV40 as a nucleo-protein complex which is then injected for testing both the transcription and the persistence of the individual molecules.

I.C. LO: After you inject a larger amount of exogenous DNA do you find an increase in histone synthesis by the host cell?

J.B. GURDON: No, there is no apparent increased synthesis of histones. This is not surprising because there is an enormous pool of histones in the oocyte; one of the characteristics of oocytes is that they contain most of the components needed for embryos to develop to the blastula stage (which has 30,000 cells). So I think we are making use of this enormous developmental reserve which is brought into use for the genes we introduce. That also applies for RNA polymerases.

S. YUSPA: Does SV40 DNA get replicated in the oocyte?

J.B. GURDON: There is no replication of any kind of injected DNA though it does seem that repair synthesis might occur.

INCREASE IN FATTY ACID SYNTHETASE CONTENT DURING THE DIFFERENTIATION OF 3T3 FIBROBLASTS INTO ADIPOSE CELLS

Patricia M. Ahmad, Thomas R. Russell[†] and Fazal Ahmad, Papanicolaou Cancer Research Institute, Miami, Florida 33123 and [†]Department of Biochemistry University of Miami School of Medicine, Miami, Florida 33152.

Mouse 3T3-L fibroblasts isolated from the original stock of Swiss 3T3 cells differentiate into adipocytes when the cell cultures are maintained in the confluent state for prolonged periods of time (1). Treatment of the confluent cultures either with prostaglandin $F_{2\alpha}$ ($PGF_{2\alpha}$) or 1-methyl-3-isobutyl xanthine (MIX) potentiates the process of differentiation of the pre-adipocytes to adipose cells (2). Previous studies have shown that the increased lipid accumulation in cells undergoing the adipose conversion is due to an increase in the activities of enzymes known to be involved in the biosynthesis of triglycerides (3,4).

Acetyl CoA carboxylase and fatty acid synthetase (FAS) catalyze the initial steps of lipogenesis by converting acyl CoA derivatives to long chain fatty acids. The rate of *in vivo* fatty acid synthesis is considered to be regulated by changes in quantity as well as in catalytic efficiency of these enzymes. To determine which of these factors regulate FAS during the adipose conversion of 3T3-L cells that had previously been induced with $PGF_{2\alpha}$ or MIX, antibodies raised against purified rat mammary gland FAS were employed. The data so far obtained are consistent with the suggestion that like acetyl CoA carboxylase (3) the FAS content of these cells increases markedly during the conversion to adipocytes.

Supported by USPHS grants CA-15196 (F.A.),RR-05690 to PCRI and The Juvenile Diabetes Research Foundation of Florida (T.R.).

1. H. Green and M. Meuth, Cell 3 (1974) 127.
2. T.R. Russell and R.-J. Ho, Proc. Natl. Acad. Sci. USA 73 (1976) 4516.
3. J.C. Mackall, A.D. Student, S.E. Polakis and M.D. Lane, J. Biol. Chem. 251 (1976) 6462.
4. W. Kuri-Harcuch and H. Green, J. Biol. Chem. 252 (1977) 2158.

ISBN 0-12-045450-5

EFFECT OF SHORT CHAIN FATTY ACIDS ON THE MEMBRANE MICROVISCOSITY OF HELA CELLS AS MEASURED BY FLUORESCENCE ANISOTROPY

E. Esat Atikkan and Peter H. Fishman
Laboratory of Molecular Biology and the Developmental and Metabolic Neurology Branch, NINCDS, NIH, Bethesda, MD 20014.

Straight chain fatty acids inhibit proliferation in HeLa cells; butyrate, and to a lesser degree pentanoate and propionate induce morphological and biochemical changes(1). Since microviscosity and differentiation was correlated by others(2) we examined the effect of butyrate and other fatty acids on the fluorescence anisotropy of HeLa cells using 1,6-diphenyl-1,3,5-hexatriene (DPH) as the probe(3).

We found that C_2-C_7 straight chain, saturated, unsubstituted fatty acids decreased membrane microviscosity in whole cells; butyrate, hexanoate and heptanoate were the most effective. In isolated plasma membranes the order of effectiveness was: butyrate>hexanoate=propionate>pentanoate. The results with whole cells and isolated membranes can not be directly correlated because of dye internalization in whole cells.

The effect of butyrate was both time and concentration dependent; the fluorescence anisotropy reached a minimum after 14h exposure to 5mM butyrate. As the decrease occured over a finite time and since butyrate did not change the microviscosity of a butyrate resistant HeLa mutant that is deficient in butyrate incorporation, the microviscosity change can not be explained only as the partition of butyrate in the membrane. In addition, methoxyacetate, a non-metabolizable analog of butyrate, did not alter membrane microviscosity.

The time course of the microviscosity change does not correlate with the differentiating effects of butyrate. The pronounced effect of hexanoate (a non-differentiating homolog) in decreasing microviscosity, and the limited decrease produced by pentanoate, a homolog that mimics butyrate, make it difficult to correlate differentiation with changes in membrane microviscosity.

(1) Henneberry, R.C. and Fishman, P.H., Exp. Cell Res., 103, 55-62 (1976).
(2) Arndt-Jovin, D.J., Ostertag, W., Eisen, H., Klimek, F. and Jovin, T.M., J. Histochem. Cytochem., 24, 332-347 (1976).
(3) Shinitzky, M. and Barenholz, Y., J. Biol. Chem., 249, 2652-2657 (1974).

DIFFERENTIAL SYNTHESIS OF RABBIT AORTA CONTRACTING SUBSTANCE (THROMBOXANE A_2) BY HUMAN LUNG VERSUS SKIN FIBROBLASTS IN TISSUE CULTURE

J.M. Bailey, R.W. Bryant, B. Brynelson and S. Feinmark
Department of Biochemistry, School of Medicine, George Washington University, Washington, D.C.

Guinea pig lungs, during anaphylaxis, or when perfused *in vitro* with arachidonic acid, release a highly vasoactive labile material (RCS) which contracts rabbit aorta and induces platelet aggregation. This material has been identified as thromboxane A_2 (1).

Cultures of 20th-25th passage human diploid lung fibroblasts (strain WI-38) were grown for 4 days to confluency. When washed cell monolayers were exposed for brief periods of 1 to 15 minutes to ^{14}C-arachidonic acid, material was released which contracted rabbit aorta and had a half-life at 25° characteristic of thromboxane A_2. The medium was analyzed by two-dimensional thin-layer and gas-liquid radiochromatography. Radioactive thromboxane B_2 and prostaglandin E_2 were the major products, and were identified by selected ion monitoring mass spectrometry. Synthesis of thromboxane B_2 and prostaglandin by lung fibroblasts was blocked by the anti-inflammatory drugs aspirin or indomethacin.

Human diploid skin fibroblasts similarly treated with ^{14}C-arachidonate released only prostaglandin E_2 and the extracts displayed no RCS activity. In addition, lung fibroblasts transformed with SV40 virus, lost the ability to synthesize thromboxane A_2 in this system, demonstrating a pattern of arachidonic acid metabolism similar to that of skin fibroblasts.

These results in which differential synthesis of vasoactive thromboxane A_2 has been observed indicate that fibroblasts in tissue culture can display biochemically differentiated functions characteristic of their tissue of origin.

Supported by research grants from USPHS and NSF.

(1) Svensson, J., Hamberg, M., and Samuelsson, B. *Acta. Physiol. Scand.* *94*, 222-228 (1975).

ISBN 0-12-045450-5

MEMBRANE LIPID CHANGES OF STARFISH OOCYTES AFTER 1-METHYL ADENINE TREATMENT

Mary Lee Barber and Foster R. Montalbano
Department of Biology, California State University, Northridge, CA 91330

Five separable changes in the cell membrane of oocytes have been experimentally demonstrated following treatment with 1-methyladenine of immature starfish oocytes of several species: 1) electrophysiological and permeability changes (1); 2) capacity for egg membrane fusion with sperm (2); 3) capacity for cortical granule response (2); 4) capacity for reaction preventing polyspermy (2); 5) reduction or loss of surface microvilli (3).

Many of these investigations have suggested a membrane surface receptor for 1-MA, an involvement of Ca^{++}, and a change in surface properties which are usually accompanied by correlate changes in the content and arrangement of lipids, proteins, and other molecular components of the membrane. These experiments investigated the qualitative and quantitative representation of cell membrane lipid fractions before and after induction of the maturation response induced by 1-MA treatment of *Patiria miniata* oocytes. Previously published methods were used (4).

Immature and mature egg membranes differ in that the mature egg membranes contain lower relative concentrations of triglycerides (TG) and phosphatidyl serine-inositol (PS) but more cholesterol (Ch) and phosphatidyl choline (PC). Profiles of the fatty acid methyl and esters made from seven lipid fractions also differ between the two types of membranes in terms of saturation and chain length. The proportion of cholesterol/P-lipid is greater in the mature egg membranes.

Such lipid changes in other cell or artificial membrane systems have been shown to alter membrane properties such as those altered by 1-MA.

Supported by CSUN Foundation grant #4.263.01
(1) Shen, S. and R. Steinhardt, Devel. Biol. 48 148-162 (1976)
(2) Hirai, S., Bull. Marine Biol. Sta. Asamushi 15 165-71 (1976)
(3) Hirai, S, Kubota, J., Kanatani, H., Exp. Cell Res 68 137-43 1971.
(4) Barber, M.L., and Mead, J.F. Wilhelm Roux' Archiv 177 19-27 (1975).

ISBN 0-12-045450-5

GENETIC ANALYSIS OF THE CYCLIC AMP CHEMOSENSORY SYSTEM OF *DICTYOSTELIUM DISCOIDEUM* AND ITS ROLE IN MORPHOGENESIS

Stephen L. Barclay and Ellen J. Henderson, Department of Chemistry, 18-049, Massachusetts Institute of Technology, Cambridge, MA 02139

Development of *D. discoideum* during the multicellular stages of its life cycle is organized about a single axis and leads to two terminal cell types. Indirect evidence suggests that cyclic AMP released by the cells of the organizer controls the pattern of cellular differentiation. To test this hypothesis, cAMP chemosensory system mutants are being studied for the effect of their mutations on morphogenesis.

Sensory system mutants are being isolated by their failure to migrate through Millipore filters toward a source of cAMP. Thermo-sensitive and absolutely defective mutants have been found to be altered in (a) the reception of a cAMP signal, (b) the transduction of the signal to a form that activates the contractile system, or (c) the response of the contractile system.

The properties of three mutants indicate that the cAMP sensory system is essential for all stages of development. One mutant is thermo-sensitive during aggregation and later developmental stages and lacks surface cAMP binding activity at the restrictive temperature. Multicellular aggregates formed at the permissive temperature do not form the organizer at the restrictive temperature and cease development. A second mutant forms multicellular aggregates of normal size which then develop several organizers and subdivide into smaller morphogenetic fields. This mutant has elevated surface cAMP receptor activity. Another thermo-sensitive mutant fails to aggregate at the restrictive temperature although it shows chemotaxis toward cAMP. This mutant, which may fail to synthesize or release cAMP, is also thermo-sensitive during subsequent stages of development.

Supported by an NIH Postdoctoral Fellowship to S. L. B. and by NIH Grant GM21828 to E. J. H.

ISBN 0-12-045450-5

INDUCTION OF CADMIUM-THIONEIN IN ISOLATED LIVER AND TUMOR CELLS.

S. E. Bryan, Vasantha Koppa and Humberto A. Hidalgo.
Department of Biological Sciences, University of New Orleans, New Orleans, La.

Sulfur-rich metal-binding proteins (metallothioneins), which have been identified in a variety of organisms and tissues are induced in response to sublethal doses of certain heavy metals. Information on the effects of cadmium and other metals has been largely obtained by studying the response of a given tissue following whole body exposure to the metals. This method provides no control over either synergistic or antagonistic effects between different tissues, or the metal concentrations which finally reach the target organ. Hence, there are few studies on the effect of cadmium on cells which synthesize cadmium-thionein, in the absence of variables. The isolated single cell system described in this communication which allows one to control certain of these variables, is applied to normal liver and tumor cells.

The uptake of cadmium by isolated rat liver cell was linearly related to the cadmium concentration the cells were exposed to in the medium. Cadmium-treated cells synthesized de novo proteins with the characteristics of cadmium-thionein induced in the liver of cadmium-treated animals. Thionein from liver cells incorporated cadmium and ^{35}S-cysteine, had a Ve/Vo (Sephadex G-50) of 1.8-1.9, and was separated into two subfractions by DEAE-cellulose ion-exchange chromatography. Cycloheximide and actinomycin D when added after a cadmium exposure prevented the synthesis of thionein. However, addition of actinomycin D after synthesis had started, only reduced the total amount of thionein synthesized.

When primary cultures of mouse tumor cells were exposed to cadmium (0.25 μg/ml), the cells incorporated the metal into a fraction identified as cadmium-thionein by gel filtration and ion-exchange chromatography. The appearance of cadmium-thionein in tumor cells was accompanied by the incorporation of radioactive cysteine into this protein indicating de novo synthesis of thionein. Thus, tumor cells derived from mouse hepatoma have retained the capacity to synthesize thionein.

ISBN 0-12-045450-5

CHANGE IN PHENOTYPIC EXPRESSION INDUCED BY CYCLIC AMP IN DIFFERENTIATION-ARRESTED MURINE LYMPHOMAS

Scott W. Burchiel and Noel L. Warner, Immunobiology Laboratories, University of New Mexico School of Medicine, Albuquerque, NM 87131

During recent years, the great diversity in the development of various populations of lymphoid cells has become particularly appreciated. Although it has become clear that lymphoid cells become differentiated to perform specific immunological functions under the influence of various environmental and hormonal stimuli, the exact nature of these processes are only now beginning to be elucidated.

The work described in this report represents ongoing research on the characterization of cultured murine lymphoma cells, with the belief that such cells represent a clonal population of lymphocytes that are arrested in a particular stage of differentiation, and that such cells can be utilized to investigate the modulation of lymphocyte differentiation by various pharmacological and hormonal agents. This concept is supported by all available data that shows that lymphoma cells express particular cell surface components that are specific for various lymphocyte lineages during a particular stage of development. As such, various murine lymphoma cells can be examined for the induction of differentiated cell function, or the expression of cell surface components that are characteristic for particular stages of differentiation, following treatment with specific pharmacological or hormonal agents.

The results of initial studies show that treatment of the pre- B cell lymphoma ABE-8 with various agents that elevate intracellular levels of cyclic AMP produces an increase (3 fold) in the percentage of cells expressing Fc receptor (as measured by the EA-rosette method using rabbit IgG). The induction of Fc receptor expression is both dose and time dependent. Preliminary studies (A. Harris and N. Warner) also have shown that cyclic AMP elevation results in an increased percentage of cells expressing membrane immunoglobulin (mIg). Because the Fc receptor and mIg are characteristic for a more differentiated B cell, these results suggest that cyclic AMP can induce differentiation in ABE-8. Further experiments are being conducted to determine the stability of the expression of induced phenotypes (such as Fc receptor and mIg) following removal of drugs, and quantitation of these and other determinants.

ISBN 0-12-045450-5

CHROMOSOMAL PROTEINS FROM THE EMBRYOID BODY/TERATOCARCINOMA SYSTEM

Don B. Carter and Steve E. Harris. National Institute of Environmental Health Sciences, Research Triangle Park, North Carolina

The chromosomal proteins were isolated from simple embryoid bodies and teratomas, derivatives of the transplantable mouse testicular teratoma, OTT-6050 of 129 mouse origin. Comparison was made with analogous proteins from mouse liver, brain and thymus. We have extracted the histones and low salt 5 M urea soluble nonhistones from embryoid bodies, small and large teratomas which are grown *in vivo*. Use of long Laemmli (28 cm) discontinuous gradient gels enables detailed comparison of the developmental changes in the 5 M urea soluble chromosomal proteins in the embryoid body, small attached teratoma (<0.2 cm diameter) and large attached teratoma (<0.5 cm). Electrophoretic patterns of histones from three stages of development indicate that histone 4 (H4) from the teratomas is modified (acetylated) more than H4 from the embryoid bodies. The other four histones show no detectable differences during differentiation of embryoid bodies to teratoma tumors. However, the 5 M urea extractable chromatin proteins show a significant difference during this growth and differentiation process. The 5 M urea soluble proteins above 30,000 daltons from embryoid body chromatin resolve into approximately 68 stained bands and the corresponding group of proteins from small and large teratoma tumors resolve into about 70 distinct bands. Both quantitative and qualitative differences at a minimum of 14 positions are detectable by visual inspection of the gels. On the other hand, there are only 6 visible differences between similar gels of small and large teratomas. Thus, the differentiation process from an embryoid body containing only endoderm and embryocarcinoma cells to teratomas containing a predominance of neuroepithelial cells appears to involve a large variation in synthesis of 5 M urea soluble chromosomal proteins. Further study of this class of chromatin proteins seems warranted based on the observation (1) that the 5 M urea soluble proteins of chick oviduct chromatin may be involved in the expression of the specific gene for ovalbumin.

(1) Tsai, S.Y., Harris, S.E., Tsai, M. and O'Malley, B.W., *J. Biol. Chem.* *25* 4713-4721 (1976).

IDENTIFICATION OF NONHISTONE CHROMATIN PROTEINS IN THE CHROMATIN SUBUNIT DEVOID OF HISTONE H_1

P.K. Chan and C.C. Liew
Dept. of Clinical Biochemistry, University of Toronto, Toronto, Canada.

Rat liver chromatin was prepared by the method of Reeder (1) and nuclease digestion was carried out as described by Axel (2). Chromatin subunits (nucleosomes or nu bodies) were isolated by sucrose density gradient and subsequently fractionated by 6% polyacrylamide gel electrophoresis (3) into two major components. One component (N_1) of the chromatin subunits had a higher mobility and contained four histones but no histone H_1, while the other (N_2) contained all five histones. Both chromatin subunits (N_1 and N_2) were found to contain nonhistone chromosomal proteins (NHCP) which constituted about 10% of the total chromosomal proteins in the subunits. By electrophoresis in 15% SDS-polyacrylamide gel, eight major bands of NHCP were identified in component (N_1); there were more bands in component (N_2). We have observed that treatment of chromatin subunits with 0.6M NaCl removes most of these NHCP.

We had previously identified a highly phosphorylated protein fraction among the NHCP in the chromatin subunit (4). The NHCP have now been isolated from total chromatin subunits by phenol extraction and fractionated by DNA-CM-cellulose affinity chromatography. After the unbound fraction of NHCP was removed, a fraction containing the specific phosphoproteins that were identified previously, was eluted at salt concentrations greater than 0.4M. We conclude that these NHCP are intimately associated with the chromatin subunits.

Supported by the Medical Research Council, Ontario Heart Foundation and the Muscular Dystrophy Association of Canada.

(1) Reeder, R.H., J. Mol. Biol. 80, 229-241(1973)
(2) Axel, R., Biochem. 14, 2921-2925 (1975).
(3) Vanshavsky, A.J., Bakayer, V.V. & Georgiev, G.P. Nucleic Acid Research, 3, 477-492 (1976).
(4) Chan, P.K. & Liew, C.C., Can. J. Biochem. 55, 847-855 (1977).

ISBN 0-12-045450-5

TERMINAL DIFFERENTIATION OF THE CARDIAC MUSCLE CELL

William C. Claycomb Department of Biochemistry,
LSU School of Medicine, New Orleans, La. 70112

During the terminal stages of differentiation, mammalian cardiac muscle cells completely lose the ability to replicate their DNA, undergo mitosis and proliferate. These cells never reenter the cell cycle for the remaining life of the animal. Terminal differentiation of these cells also involves the loss of enzymes which are needed to replicate DNA, presumably by selective and programmed gene inactivation. This aspect of differentiation of cardiac muscle cells can be viewed as the turning off of the DNA replication process which was turned on when the egg was fertilized.

A serious problem which is encountered when studying cardiac muscle differentiation is that the muscle tissue is heterogeneous as to cell type. Only approximately 55% of the total cell population of ventricular cardiac muscle of the 1-day old rat are actually myocytes; the remaining 45% are fibroblasts, phagocytes, circulating erthrocytes and cells comprising nerve and vascular tissue. Myocytes comprise only 20-25% of the total cell population of muscle tissue of the adult. Experiments were carried out to characterize a system which could be used to study differentiation of the cardiac myocyte. Cells from cardiac muscle of rats at various ages of development were dissociated using 0.1% collagenase and 0.1% hyaluronidase and the myocytes were purified to homogeneity on a Ficoll-BSA gradient. *In vivo* and *in vitro* experiments determined that these cells cease DNA synthesis and lose the activity of several enzymes needed to replicate DNA by the middle of the third week of postnatal development. After this period of development acquisition and assembly of the contractile proteins increases greatly and the cells enlarge approximately 10-fold. These studies define a system which can be used to study terminal differentiation of the cardiac muscle cell.

Supported by the American Heart Association

ISBN 0-12-045450-5

POSSIBLE ROLE OF METABOLIC OSCILLATIONS IN THE INDUCTION OF cAMP SIGNALLING IN DICTYOSTELIUM DISCOIDEUM.

E.L. Coe, and W.-J.K. Chung, Dept. of Biochemistry, School of Medicine, Northwestern Univ., Chicago, Ill. 60611.

The early differentiation in the cellular slime mold, *D. discoideum* is characterized by aggregation of free-living amoebae into multicellular bodies. A chemotactic agent, 3',5' cAMP, is released in a pulsatile fashion from the forming aggregation centers and the pulses are relayed through populations of amoebae by a delayed synthesis and release of new cAMP in cells exposed to extracellular cAMP. This spontaneous pulsing and the relaying system develop after 4-6 hrs of starvation, in association with an increase in cAMP binding sites and cAMP phosphodiesterase on the cell surface. The chemotactic response to cAMP develops before the spontaneous pulsing appears. The period of starvation somehow brings about these changes, but the mechanisms are unclear.

Metabolic oscillations in nucleotides and glycolytic intermediates are relatively easy to produce in cell-free systems from yeast and mammalian cells, but intact cells tend to approach stable states exponentially and avoid oscillating. However, intact cells may be forced to oscillate under conditions of stress. Ascites tumor cells, for example, will develop glycolytic oscillations after prolonged starvation.

We have now demonstrated that the spontaneous cAMP pulsing in suspended *D. discoideum* (NC-4) amoebae is correlated with oscillations in ATP, ADP, pyruvate, and glucose-6-P. These metabolite oscillations are also correlated with waves of decreased turbidity and are well developed even when the cAMP pulses are small. An elicited pulse of cAMP, obtained by adding extracellar cAMP, occurs about 90 sec after the addition and is preceded by a positive wave in ADP which coincides with a negative wave in ATP. This suggests that a transient period of increased ATP hydrolysis resulting from the immediate chemotactic response to added cAMP may serve to activate generation of intracellular cAMP. It is, therefore, conceivable that spontaneous oscillations in metabolism appear first in response to starvation stress, and that these oscillations induce development of the spontaneous cAMP pulsing mechanism.

Supported by Fluid Research Funds from Northwestern University Medical School.

ISBN 0-12-045450-5

DIFFERENTIATION OF CYTOTOXIC T LYMPHOCYTES MAY REQUIRE INDUCTION ON THE T CELL MEMBRANE OF A FREE RADICAL PRODUCING MECHANISM

R. G. Devlin, C. S. Lin and H. Dougherty. Merck Institute for Therapeutic Research, Rahway, New Jersey 07065

The cytotoxic thymus dependent lymphocyte is a presumably end stage effector cell which differentiates *in vivo* or *in vitro* upon sensitization with cells bearing foreign histocompatibility antigens on their surface membranes. Differentiated lymphocytes kill target cells upon contact by a process that does not require protein synthesis[1], DNA synthesis[2], or even viability[3], since dead cytotoxic cells are capable of killing. There thus appears to be a destructive molecular apparatus on the killer cell membrane which develops after sensitization and which needs only to be triggered by the appropriate specific antigen in order to carry out its destructive function. We have hypothesized that the destructive mechanism responsible for the death of target cells involves the production of oxygen-centered free radicals induced by the recognition of target cell alloantigens by killer lymphocytes. We have found that the killing process is inhibited by scavengers of superoxide radicals (O_2^-) and peroxy radicals (ROO·) of the appropriate molecular size (e.g., tiron and phenol, respectively). This inhibition appears to be independent of effects on prostaglandin biosynthesis and is not related to simple toxic effects of the scavengers. In addition, we have found that O_2^- generated *in vitro* by the reaction of KO_2^- and culture medium is cytotoxic to tumor cells. The cytotoxicity of chemically generated O_2^- for target cells is inhibited by tiron, as is the cytotoxicity of killer lymphocytes for target cells. These data demonstrate that oxygen-centered free radicals are capable of killing target cells and provide additional evidence that these radicals may be involved in the molecular mechanism of target cell destruction by sensitized lymphocytes.

(1) Thorn, R. M. and Henney, C. S., *J. Immunol.* *115* 145, (1975).

(2) Mauel, J., Rudolf, H., Chapuis, B. and Brunner, A. T., *Immunology* *18* 517, (1970).

(3) Sanderson, C. J. and Taylor, G. A., *Cell Tissue Kinet.* *8* 23.

ISBN 0-12-045450-5

THE MACROPHAGE SURFACE: SPECIFIC CONSTITUENTS RELATED TO CELL TYPE AND CELL SHAPE

S.R. Dienstman, E. Pearlstein, and V. Defendi.
Dept. of Pathology, N.Y.U. Med. Sch., NYC, NY 10016

Macrophage substrate adhesion is distinguished by: 1. failure to form monolayers of coherent cells, 2. once attached to glass the resistance to trypsin detachement, 3. 2 forms of adhesion - (I) as round cells and (II) as spread cells (flat, motile, ruffled membranes).

"Activated" mouse peritoneal macrophages (STpMAC) were obtained from exudates after 2% starch injection. STpMAC spread within 10 min of incubation (Form II). Non-activated MAC (NpMAC), from unstimulated mice, show Form I in 10 min, but few cells spread in 12 hr in vitro. NpMAC show Form II in 24 hr in serum containing medium. ST and NpMAC have been subjected to lactoperoxidase catalysed 125-Iodination. Cell lysates were run on SDS polyacrylamide gels with dithiothreitol and labeled (external) bands visualized on X-ray film. STpMAC always showed a major band, "A" (195K est. MW), that was not displayed by cultured mouse embryo cells (MEFs) or by freshly isolated NpMAC. NpMAC always showed a major band, "B" (180K est. MW) not found on MEFs or minimally on fresh STpMAC. By 24 hr in vitro NpMAC displayed A and B. MEFs had a major band, "CAF" (=LETS, 250K est. MW) not found on ST or NpMAC.

Iodinated cells were trypsin treated. A and B of ST or NpMAC were protease resistant. However, the entire pattern from MEF controls was sensitive including CAF. Treating with the local anaesthetic lidocaine resulted in: Most NpMAC detaching, with their band pattern unchanged, 2. most STpMAC changing shape (II to I), many detaching, with their pattern stable except for a reduction in A.

Conclusions: The labeling of A is a characteristic feature of activated pMAC. Labeling of A and B is relative to the degree of spreading. MAC protease resistance for surface constituents and adhesion correlate. Lidocaine sensitivity in adhesion, shape and display of A correlate.

ISBN 0-12-045450-5

THYROID HORMONE RECEPTOR: INTERACTION WITH HISTONE H_1 AND IMPLICATIONS FOR THE MECHANISM OF REGULATION OF TRANSCRIPTION.

Norman L. Eberhardt, Lorin K. Johnson, Keith R. Latham and John D. Baxter
Metabolic Research Unit, Departments of Medicine and Biochemistry and Biophysics, University of California, San Francisco, California 94143

Thyroid hormones appear to regulate specific mRNAs through interactions with chromatin receptors. The isoelectric point of the partially purified nuclear thyroid hormone receptor from rat liver was determined to be 5.85, demonstrating that the receptor is an acidic, non-histone chromosomal protein. When a crude nuclear extract was focused under identical conditions receptor-bound hormone migrated in two peaks with the bulk of the activity migrating at pH 7.9. These results suggested that the receptor could be interacting with basic proteins, e.g. histones, in the crude extract.

Hormone binding studies with the receptor which had been partially purified by Sephadex G-100 chromatography indicated that purified H_1 and whole histone, but not ovalbumin, lysozyme or poly-L-lysine resulted in a stimulation of the receptor binding activity. H_1 stimulated triiodothyronine (T_3) binding at concentrations well below those of whole histone suggesting that the effect was specific for H_1. Scatchard analysis of this effect indicated that T_3 and thyroxine (T_4) binding were both influenced; the concentration of binding sites was doubled in the presence of histones.

These results inidicate that H_1 is capable of interacting with and influencing the binding properties of the receptor. Thus, the histone may be important for receptor function and the data also suggest possible mechanisms whereby the receptor may be involved in regulation of transcription. For instance, H_1 may affect higher order chromatin structures; the receptor could participate with H_1 in modulation of chromatin conformation and activity.

Supported by NIH Grant No. 1-RO1-AM18878-02

ISBN 0-12-045450-5

INFORMOSOMAL AND POLYSOMAL MESSENGER RNA: DIFFERENTIAL KINETICS OF POLYADENYLATION AND NUCLEOCYTOPLASMIC TRANSPORT IN CHINESE HAMSTER CELLS

M. Duane Enger, Helen L. Barrington, and John L. Hanners. Cellular and Molecular Biology Group, Los Alamos Scientific Laboratory, University of California, Los Alamos, New Mexico

Cytoplasmic messenger RNA exists ribosome-bound (polysomal, PRNA) and not bound to ribosomes (informosomal, IRNA). In cultured mammalian cells, about one-half of cytoplasmic mRNA is in each form. Although informosomal mRNA is known to contain sequences both present in and absent from polysomal mRNA, the role of this population as possible or obligate precursor to polysomal mRNA is not defined. This study follows the cytoplasmic appearance and relative polyadenylation of these RNAs. The data obtained are discussed in terms of limits placed on the temporal and quantitative relationship of informosomal to polysomal messenger. Kinetic analysis of the appearance of uridine-labeled informosomal RNA shows that it enters the cytoplasm within 5 min after synthesis. This compares with 20 min for polysomal mRNA. If IRNA is precursor to PRNA, then a period of cytoplasmic processing is required before IRNA can associate with ribosomes. Label in PRNA enters the cytoplasm 15 min after labeling with adenosine (5 min before uridine). This accords with the fact that polyadenylation is a post-transcriptional event. Nascent IRNA contains less poly A than PRNA. The ratio of incorporated adenosine/uridine in IRNA is 66% and 77% that in PRNA after 30 min and 60 min of incorporation, respectively. The difference in polyadenylation that maintains throughout the putative cytoplasmic IRNA processing period indicates that, if most or all of the PRNA is derived from IRNA, additional adenylation of IRNA occurs immediately before association of IRNA with polysomes. However, this would suggest coordinate appearance of adenosine and uridine into PRNA, which is not consistent with the 5-min lag (uridine following adenosine) observed. It is concluded that the IRNA population cannot be the sole source of PRNA.

This work was performed under the auspices of the Division of Biomedical and Environmental Research of the U. S. Energy Research and Development Administration.

ISBN 0-12-045450-5

RELATIONSHIPS BETWEEN THE APPEARANCE OF T CELL SUBSETS AND SURGICAL INTERVENTION.

Robert S. Epstein, Diana M. Lopez and M. Michael Sigel. Department of Microbiology, School of Medicine, University of Miami, Miami, Fla.

The emergence of novel T cells during tumorigenesis has been previously described. T cells with surface markers usually associated with B cells, i.e. Fc receptors (FcR) and complement receptors (CR), appear in spleens of tumor bearing mice. We now report on the fate of these cells following surgical procedures. Both surgical removal of tumor and sham surgery were performed. The representation of CR^+ T cells was not affected by either procedure. In contrast both surgery to remove tumor and sham surgery caused a delay and suppression in the appearance of FcR^+ T cells. Mere surgical trauma rather than tumor removal appeared to be the affecting agent in the depression of these cells' appearance. Since we have shown that FcR^+ T cells contribute significantly to elevated antibody dependent cellular cytotoxicity observed during tumor progression, a suppression in the level of FcR^+ cells as a result of surgical trauma may indicate an improved prognosis for the tumor bearing host.

The behavior of these subpopulations of lymphocytes represent a form of differentiation of T cells apparently under the influence of tumor growth.

Suppoated by Contract NO1-CP-53532 within the Virus Cancer Program of the National Cancer Institute, NIH, PHS.

ISBN 0-12-045450-5

ENZYMATIC ACETYLATION OF HISTONES IN VITRO

Mark S. Fischer. Department of Biochemistry, University of Massachusetts, Amherst, Massachusetts

Histone acetyltransferase (HAT) has been partially purified from calf thymus by batchwise techniques which include DEAE-Sephadex, CM-Sephadex and precipitation of chromatin by dilution of salt with salt-free buffer. Enzyme so prepared is substantially free of endogenous histone and DNA.

Histone has been isolated as a soluble denatured substrate by repeated extractions of calf thymus with 0.25 N HCl. Nucleosomes have been isolated from calf thymus nuclei by digestion with staphylococcal nuclease at 37° and removal of nonsolubilized nucleohistone by centrifugation. Soluble nucleosomes thus isolated are comprised of DNA and histone in equal amounts and very small amounts of nonhistone nuclear protein.

Soluble denatured histone is readily acetylated by HAT with [^{3}H]-acetyl coenzyme A as donor. Equal amounts of exogenous calf thymus DNA added at low ionic strength decrease the ability of the histone to be acylated by more than 90%. Similarly, histone complexed with DNA in nucleosomes is a very poor substrate for HAT. Factors which regulate enzymatic acylation of histone will be discussed.

Supported by USPHS Grant #HD 08341, the Massachusetts Division of the American Cancer Society and Faculty Research Grant Career Development Award from the University of Massachusetts.

ISBN 0-12-045450-5

ORIGIN OF STROMA IN HUMAN BREAST CARCINOMA

LUNA GHOSH, BIMAL C. GHOSH, TAPAS K. DAS GUPTA. DIVISION OF SURGICAL ONCOLOGY, THE ABRAHAM LINCOLN SCHOOL OF MEDICINE, CHICAGO, ILLINOIS.

THIS INVESTIGATION IS DESIGN TO STUDY THE ORIGIN OF STROMAL COMPONENT OF HUMAN MAMMARY CARCINOMA. TISSUES WERE OBTAINED FROM THIRTY FIVE PROVEN BREAST CARCINOMA, TEN FIBRO ADENOMA, TEN FIBROCYSTIC DISEASE. FIVE NORMAL BREAST TISSUE WERE TAKEN AS CONTROL. UNDER LIGHT MICROSCOPY WITH SPECIAL STAINS AN INCREASED NUMBER OF CONNECTIVE TISSUE COMPONENT WAS SEEN IN CARCINOMA. GLUTERALDEHYDE FIXED TISSUE WERE EXAMINED UNDER ELECTRON MICROSCOPE. IN CARCINOMA, THE STROMA APPEARED TO BE MORE CELLULAR AND THE CELLS WERE COMPOSED OF HISTIOCYTIC AND A FEW FIBROBLASTIC, MORE CLOSE TO PRIMITIVE MESENCHYMA. IN SOME AREAS THE ELASTIC FIBERS WERE INTIMATELY INTERMINGLED WITH COLLAGEN AND RETICULUM AND WERE NOT ASSOCIATED WITH FIBROBLASTS. THE ELASTIC FIBERS WERE DIFFERENT THAN WHAT IS SEEN ON THE NORMAL BLOOD VESSEL WALL AS THEY WERE SMALLER IN DIAMETER, NOT ARRANGED IN AGGREGATE AND PRESENCE OF A CENTRAL TRANSLUCENT ZONE. THEY ARE FRAGMENTED AND APPEARED TO BE CONTINUOUS MATRIX RATHER THAN AS INDIVIDUALLY IDENTIFIABLE FIBERS IN CARCINOMA AND IS SEEN INSIDE THE CANCER CELLS OFTEN WITH DISCONTINUOUS CYTOPLASMIC MEMBRANES AS IF THEY ARE FORMED BY THE CANCER CELLS.

THIS OBSERVATION SUGGESTS THAT BREAST CARCINOMA CELLS TAKES AN IMPORTANT ROLE IN THE PRODUCTION OF STROMAL COMPONENT RATHER THAN THE GENERAL BELIEF THAT A PROLIFERATION OF THE PRE-EXISTING CONNECTIVE TISSUE.

ISBN 0-12-045450-5

DIFFERENTIATION OF CERVICAL CARCINOMA CELLS MEDIATED BY INHIBITORS OF DNA REPLICATION. Nimai K. Ghosh and Rody P. Cox. Division of Human Genetics, Departments of Medicine and Pharmacology, New York University Medical Center, New York, N.Y. 10016.

Phorbol diesters, promoters of tumor growth inhibit differentiation of erythroleukemia cells. The reverse process is demonstrated in the present study by showing that sodium butyrate and certain other inhibitors of DNA replication that are used as antitumor drugs e.g. hydroxyurea, methotrexate and 1-β-D-arabinofuranosyl cytosine, are inducers of morphologic and biosynthetic differentiations of two clones of human cervical carcinoma cells, $HeLa_{65}$ and $HeLa_{71}$. HeLa cells produce five different placental peptides and hormones - choriogonadotropin (HCG), choriofollitropin (FSH), trophoblastic alkaline phosphatase, placental aldolase and placental progesterone. Inhibitors of DNA synthesis cause these cells to assume a fibroblastic morphology and to exhibit 300-fold induction of HCG[1,2], FSH[3] and perhaps lutropin (LH). The capacities of four aliphatic monocarboxylates (C_3 to C_6) to induce peptide hormone synthesis, as measured by radioimmunoassays (RIA), are proportional to their abilities to inhibit DNA synthesis in HeLa cells. Inducers of glycotropic hormones cause marked inhibition of cell growth and DNA replication. Synthesis of RNA, however, is either unaltered or increased during biosynthetic differentiation, suggesting transcriptional control of ectopic peptide hormone production. Morphologic differentiation induced by sodium butyrate is accompanied by a slight reduction in ^{3}H-Concanavalin A binding of HeLa cells, indicating perturbation of their membrane structures.

Of particular interest is approximately 100-fold (0.1 to 10.5 ng/10^6 cells) stimulation of placental progesterone production (quantitated by RIA) during increased synthesis of placental peptide hormones in HeLa cells grown with sodium butyrate and hydroxyurea. Inhibitors of DNA synthesis can thus regulate expression of differentiated functions along with derepression of trophoblast-specific embryonic genes in cultured neoplastic cells.

Supported by USPHS, NIH Grant GM-15510 and American Cancer Society Institutional Grant IN-14-R.

1. Ghosh, N.K. and Cox, R.P. Nature, 259 416 (1976).
2. Ghosh, N.K., Rukenstein, A. and Cox, R.P. Biochem. J. 166 265 (1977).
3. Ghosh, N.K. and Cox, R.P. Nature, 267 435 (1977).

ISBN 0-12-045450-5

Substances controlling morphogenesis in Hydra.

C.J.P. Grimmelikhuijzen, T. Schmidt and
H.C. Schaller.
European Molecular Biology Laboratory,
Postfach 10.2209, 6900 Heidelberg, Germany.

Four morphogenetic substances have been isolated from *Hydra attenuata* : an activator[1] and an inhibitor[2] of head and bud formation and an activator[3] and an inhibitor[4] of foot formation. The head factors are distributed in the animal as descending gradients from head to foot, the foot factors as descending gradients from foot to head. When tested in physiological concentrations, the foot factors do not influence head and bud formation and *vice versa* the head factors do not influence foot formation. All four factors are present in hydra in two forms, an inactive structure-bound form and an active free form. All the factors have molecular weights around or lower than 1,000 and are therefore, when released, able to diffuse well.
The four substances can be extracted from hydra tissue with methanol and separated from each other by chromatography on Sephadex G-10 and Sephadex DEAE A-25 with high yield and considerable purification.

1. Schaller, H.C. (1973). J. Embryol. Exp. Morph. 29, 27-38.
2. Berking, S. (1977). Wilhelm Roux's Archives 181, 215-225.
3. Grimmelikhuijzen, C.J.P. and Schaller, H.C. (1977). Submitted for publication.
4. Schmidt, T. and Schaller, H.C. (1976) Cell Diff. 5, 151-159.

ISBN 0-12-045450-5

EFFECTS OF ELEVATED CYCLIC AMP LEVELS IN CULTURED MOUSE HEPATOMA CELLS

Helen C. Hamman and Barry E. Ledford. Dept. of Biochemistry, Medical University of South Carolina, Charleston, S.C., 29403

The rate of cell proliferation and the expression of tissue specific functions in eukaryotic cells has been related to cellular levels of either cyclic AMP or cyclic GMP (1,2). The purpose of these studies was to determine the effect of altered cyclic AMP levels in a cell line which retains the ability to synthesize serum proteins.

The cell line used in these studies, Hepa, was derived from the mouse hepatoma BW 7756 (Jackson Laboratory) and secretes the serum proteins albumin, alpha-fetoprotein and transferrin into the culture medium (3). Addition of dibutyryl cyclic AMP (1mM), cholera enterotoxin (1 μg/ml) or methyl isobutyl xanthine (0.1 mM) to the culture medium alters the cellular morphology. Within one hour, cells become rounded and long processes are apparent. Scanning electron micrographs reveal an increased number of blebs and microvilli. Cyclic AMP and its effectors also reduce the rate of proliferation and the rate of secretion of proteins. Rates of protein secretion increased 2-4 fold within 24 hours after addition of cyclic AMP or its effectors. This effect represents a specific stimulation in the synthesis of each of the secreted proteins relative to total cellular protein. This stimulation is accompanied by elevated medium levels of cyclic AMP for both methyl isobutyl xanthine and cholera enterotoxin. Removal of cyclic AMP or its effectors results in a decreased synthetic rate, approaching that of untreated cultures.

These studies indicate that cyclic AMP levels specifically influence the expression of a major liver specific function, the secretion of serum proteins, while also effecting the rate of cellular proliferation.

This investigation was supported by Grant Number CA 17037, awarded by the National Cancer Institute, DHEW.

(1) Koyama, H., Tomida, M. and Ono, T. (1976) J. Cell Physiol. 87, 189-198.
(2) Pastan, I., Johnson, G.B., and Anderson, W. (1975) Ann. Rev. Biochem. 44, 491-522.
(3) Bernhard, H.P., Darlington, G.J., and Ruddle, F.H. (1973) Dev. Biol. 35, 83-96.

ISBN 0-12-045450-5

ISOLATION OF THE BRAIN SPECIFIC LIKE PROTEIN S-100 FROM TERATOCARCINOMAS OTT 6050 AND PARTIAL PURIFICATION OF S-100 LIKE mRNA

Stephen E. Harris, Sandra Gipson and Alan B. Silverberg. Molecular Embryology Workgroup, NIEHS, LET, Research Triangle Park, North Carolina 27709.

Simple embryoid bodies 100μ (EB) consisiting of an outer ring of endoderm and a core of embryocarcinoma cells, are capable of attaching *in vivo* to a surface such as a spleen, mesentary, or liver, or *in vitro* to a collagen coated tissue culture dish, and initiating a differentiation process leading to teratocarcinomas (0.2-1.5 cm) containing a high concentration of neuroepithelial cells. In our initial attempts at defining some of the molecular parameters which appear in the teratocarcinomas (T), we noticed on SDS-acrylamide gels of 100,000 x g supernatants a band of protein(s) of molecular weight 7,000 in the T but absent in EB. Subsequently, we have shown that this protein(s) is soluble in 80% ammonium sulfate and precipitated at 100% ammonium sulfate titrated to pH 4.0. The material binds tightly to DEAE-Sephadex and is therefore acidic in nature. We presume this teratocarcinoma (T) specific protein to be S-100 although this has not been proven conclusively.

We are preparing antibodies to this protein isolated from T. This antibody will then be used in the purification of S-100 Like mRNA using Sepharose 4B fractionation and formamide sucrose gradient fractionation of total poly(A)-containing RNA and subsequent analysis using the *in vitro* wheat germ translation system.

ISBN 0-12-045450-5

STUDIES OF A HEME REQUIREMENT IN THE DIFFERENTIATION OF FRIEND ERYTHROLEUKEMIA CELLS

Leslie M. Hoffman and Jeffrey Ross
McArdle Laboratory for Cancer Research, University of Wisconsin, Madison, Wisconsin 53706

Cultured murine erythroleukemia cells of the T3-Cl_2 line differentiate when an inducer such as dimethylsulfoxide (DMSO) or hemin is added to the culture medium (1). The extent of differentiation has been quantitated by measuring intracellular heme, hemoglobin, and globin mRNA concentrations in induced cells.

We wished to determine if elevated intracellular heme concentrations could trigger the maturation of T3-Cl_2 cells. If so, then factors which decrease *de novo* heme synthesis should also retard the differentiation process. To test this hypothesis, T3-Cl_2 cells were cultured in the presence of the heme synthesis inhibitor isonicotinic acid hydrazide (INH) (2) at concentrations that did not affect the growth rate. The incorporation of ^{59}Fe into heme was 7-fold greater in 1.5% DMSO-treated cells than in untreated controls. In contrast, ^{59}Fe incorporation in 1.5% DMSO plus 4 mM INH-treated cells was about one-third that of cells receiving DMSO alone. Cells treated with DMSO and INH accumulated 10-fold less hemoglobin and 5-fold less globin mRNA than DMSO-treated cells.

Cells receiving 10^{-4}M hemin or hemin plus INH showed an increase in benzidine staining, but no significant ^{59}Fe incorporation into heme was detected. The steady-state quantity of hemoglobin or of globin mRNA in hemin or hemin plus INH-treated cells was about 6 times that of uninduced cells. Thus, INH did not prevent T3-Cl_2 cells from taking up exogeneous hemin and accumulating hemoglobin and globin mRNA. Cells treated with hemin plus DMSO plus INH contained higher steady-state quantities of hemoglobin and globin mRNA than cells treated with hemin, hemin plus INH, or DMSO plus INH.

The results indicate that heme might play a role in the differentiation of immature erythroblasts, perhaps by controlling the expression of the globin genes.

Supported by USPHS grant CA-07175.

(1) Ross, J. and Sautner, D., Cell 8, 513-520 (1976).
(2) Ponka, P. and Neuwirt, J., Blood 33, 690-707 (1969).

ISBN 0-12-045450-5

ROLE OF LIPIDS IN MORPHOGENETIC PROLIFERATION OF MOUSE MAMMARY EPITHELIUM

H. L. Hosick and M. E. Anderson, Dept. Zoology, Washington State University, Pullman, WA 99164.

Mammary epithelium of the mouse is embedded in a matrix of adipose tissue; it will not proliferate if transplanted to other sites in the body. We are employing cell-culture techniques to analyze the nature of the adipose/epithelial cell interaction. We hypothesized that adipose cells liberate free fatty acids which are then utilized by midpregnant epithelium to support its growth. However, because of the inconvenient orientation of the mammary epithelial cells in culture, with their basal surface inaccessible against the culture dish and the impermeable apical surface upward, we had to first devise methods using liposomes to efficiently insert free fatty acids directly into the epithelial-cell cytoplasm.

Primary monolayer cultures of mammary epithelial cells from midpregnant mice were maintained in a culture medium containing 10% fetal calf serum and 5 μg/ml insulin. These cells grew poorly (as assayed by total DNA accumulation) when supported with medium containing delipidated serum. Growth could be increased when free fatty acids were inserted into the cells using liposomes but the cells responded in this way only if an adrenal steroid was added to the culture fluid. The free fatty acids were metabolized rapidly to phospholipids and triglycerides. Other cell types and mammary tumor cells were distinguishable from normal epithelium in the degree of their responses to free fatty acids inserted into cells and in the pattern of metabolism of free fatty acids. We conclude that normal mammary epithelial cells are particularly senstive to fatty acids to which they respond by growing, and that the culture and liposome technologies can be utilized together to explore this response in detail.

Supported by NIH grant CA-16392 and contract No1-CB-63986.

ISBN 0-12-045450-5

ANALYSIS OF RAT DNA PURIFIED BY NONHISTONE CHROMOSOMAL PROTEIN-DNA INTERACTIONS

L. L. Jagodzinski, C. Castro, D. Lee, and J. S. Sevall.
Department of Chemistry, Texas Tech Univ., Lubbock, Texas 79409

Nonhistone proteins (NHP) have been purified from rat liver nuclei by their affinity for phosphocellulose at physiological ionic strengths. Constituting 0.8 percent of the total chromosomal proteins, this class of DNA-binding NHP consists of a herterogeneous mixture of polypeptides which are histospecific.

The DNA-binding-NHP can fractionate the rat genome into bound and unbound DNA sequences. To demonstrate the presense of high affinity DNA-binding sites, 5 percent randomly sheared 2700 base pair (bp) DNA was isolated as protein bound DNA (BDNA). Reassociation kinetics of BDNA sheared to 400 bp showed that equal amounts of repetitive and nonrepetitive sequence elements are present. The non-repetitive element represents 1.29 percent of the total rat genome compared to 1.3 percent for the repetitive element. Both kinetic components exhibit lower kinetic complexities than the rat genome and each reassociate with a rate constant twenty fold faster than the rat genome. Self reassociation of the 2700 nucleotide BDNA fragments demonstrates that each BDNA fragment contains a repetitive sequence element. Qualitative studies on sequence distribution indicate that we have isolated a fraction or subclass of the interspersed sequences of the rat genome. RNA excess hybridization with trace amounts of nonrepetitive BDNA indicates a subset of the low abundant class of mRNA drive the hybridization. Fourteen point nine percent of the total mRNA drives the hybridization of total non-repetitive sequences; whereas, 1.86 percent of the total mRNA drives the hybridization of nonrepetitive BDNA.

Reassociation and hybridization data indicate that a subset of repetitive and nonrepetitive sequences is observed via NHP-DNA interactions. This subset represents a subclass of the interspersed repetitive sequence region and may contain a fraction of the abundant class of mRNA molecules.

Supported by: NSF BMS-75-09232, and NIH-GM- 22653.
L.L.J. is a Welch Fellow.

ISBN 0-12-045450-5

ACCESSIBILITY OF DNA IN CHROMATIN TO THE COVALENT BINDING OF A CHEMICAL CARCINOGEN Carolyn L. Jahn and Gary W. Litman. Cornell Univ. Graduate School of Medical Sciences and Sloan-Kettering Institute, New York, New York, 10021

As a step in determining whether specificity of binding of chemical carcinogens is involved in their mechanism of action, we have analyzed the accessibility of DNA in chromatin to the covalent binding of the polycyclic aromatic hydrocarbon, benzo(a)pyrene (BP). The low extent of binding (1 per 10^4 bases) has required the use of an *in vitro* system utilizing rat liver microsomes for the metabolic activation and subsequent binding of (^{3}H)BP to calf thymus nuclei. Quantitation has been made on the basis of specific activity (CPM/OD_{260}) after extensive extraction, since binding to proteins and RNA and non-covalent interactions with DNA are known to occur. The distribution of covalently bound BP has been examined by analyzing the extent of binding relative to digestibility by Staphylococcal nuclease or DNase I.

The specific activity of undigested DNA determined as a function of percent digestion differs for the two enzymes. Staphylococcal nuclease results in a linear decrease in specific activity leading to a 40-60% reduction at 50% digestion; while DNase I results in a slight linear increase in specific activity of 8-12% by 50% digestion (continuing to approximately 20% by maximal digestion).

The preferential removal of carcinogen by Staphylococcal nuclease suggested that there is more extensive binding in the spacer region. This was further investigated by isolation of monomer, dimer, and trimer particles from sucrose gradients after limited digestion and analysis of the binding to each size DNA repeat. These show a decrease in specific activity with increasing digestion which parallels the total sample. In addition, at a single time point, the monomer, dimer, and trimer differ in specific activity by amounts expected if a single spacer (60 bases) were being removed from increasing size DNA repeats. Further fractionation of monomers by precipitation with KCl yields "cores" or non-H1-containing monomers of lower specific activity and H1-containing monomers of higher specific activity.

The distribution of BP binding resembles the binding of ethidium bromide (EB), which is known to have higher affinity for DNA associated with H1. BP and EB share certain structural features: thus their intercalation mechanism may result in a similar binding pattern. (Supported by NCI Grants CA 17085, CA 08748, and USPHS Grant GM 01918).

ISBN 0-12-045450-5

PROPERTIES OF MESSENGER RIBONUCLEOPROTEIN PARTICLES OF EMBRYONIC MUSCLE CELLS AND THEIR ROLE IN TRANSLATIONAL CONTROL.

S.K. Jain, M.G. Pluskal and S. Sarkar. Dept. of Muscle Res., Boston Biomed. Res. Inst.; and Dept. of Neurol., Harvard Med. School, Boston, MA, 02114

A key question in the regulation of eukaryotic mRNA translation during differentiation is how the mRNA-associated proteins affect the biological activity of the mRNAs. We have approached this problem by studying the isolation and characterization of both cytoplasmic free (CmRNP) and polysome-derived (PmRNP) mRNP particles from 12-14 day-old chick embryonic muscles which are at a stage of development when synthesis of muscle-specific proteins is accelerated. Poly A-containing mRNP particles are isolated in pure and nondenatured form by a novel method of affinity chromatography on columns of oligo-dT-cellulose. The properties of mRNP particles indicate: (i) the CmRNP are protein rich particles (RNA/protein ratio 1:3) containing about 9-10 polypeptides, while PmRNP are relatively protein deficient (RNA/protein ratio 1:1) and contain only two major proteins of 52,000 and 78,000 dalton, which are common to both classes of mRNPs; (ii) two polyA-associated proteins of mol. wt. 78,000 and 64,000 are present in both classes of mRNP particles; these are bound to mRNA in a mutually exclusive manner; (iii) not every PmRNP particle contains one mole of 78,000 dalton and one mole of 52,000 dalton protein as previously believed; and (iv) the two types of mRNP have similar size PolyA segments (mean length 120-140 A residues) suggesting that the CmRNP are not obligatory precursors of PmRNP. The protein components of CmRNP particles isolated from liver, reticulocyte and muscle are very similar, suggesting that these mRNA-associated specific proteins have evolved in a wide variety of tissues. Since a CmRNP particle containing myosin heavy chain mRNA and the protein-free mRNA obtained from this particle are translated *in vitro* with equal efficiency; and the protein moieties of CmRNP particles are released in the cytoplasm when their mRNA moieties are transferred to polysomes in an *in vitro* translation system it is concluded that the proteins of CmRNP do not exert a stringent negative control on the translation of the mRNA *per se*. These results support a model of translational control in which the proteins which are present only in CmRNP are involved in the *in vivo* regulation of the entry of the mRNAs from the nonpolysomal pool of the cytoplasm to the polysomes.

ISBN 0-12-045450-5

EVIDENCE FOR A CATALYTIC MECHANISM INVOLVED IN THE ACTIONS OF GLUCOCORTICOID RECEPTORS ON CHROMATIN

Lorin K. Johnson, Nancy C.Y. Lan and John D. Baxter
Metabolic Research Unit and Departments of Medicine and Biochemistry and Biophysics, University of California, San Francisco, California 94143

Glucocorticoid hormones specifically increase growth hormone (GH) mRNA in cultured rat pituitary tumor (GC) cells without major effects on total RNA synthesis. Under GH stimulating conditions (log phase of cell growth) chromatin from dexamethasone treated cells (1 hr) displayed a major (50%) increase in cell-free template capacity when assayed with *E. coli* RNA polymerase. Examination of the rapidly initiating (rifampicin resistant) polymerase binding sites showed no change in the affinity ($K_d \sim 5$ nM), the kinetics of enzyme binding, chain elongation, or RNA chain size. Instead, the 50,000 glucocorticoid receptors that bound to the nucleus increased the number of initiation sites by 500,000 per cell.

Under conditions where growth hormone is not induced (confluency) the glucocorticoid surprisingly decreased the number of such rapidly initiating sites by 750,000 per cell. The effect followed receptor binding to the nucleus by about 10 min, and it disappeared in parallel with receptor dissociation from chromatin once the glucocorticoid was removed ($t_{1/2}$=20 min). With mouse lymphoma (S49) cells that are killed by glucocorticoids, the steroid also decreased the number of polymerase initiation sites. This effect was not elicited in the mutant (r^-) line which lacks a glucocorticoid receptor or the NTi line which has a defective receptor. Thus, this receptor-mediated effect on the number of RNA polymerase "initiation" sites greatly exceeds the number of receptors (10:1), and can be either positive or negative in the same cell. Further, the direction of the effect parallels the specific biological response. Therefore, it is unlikely that a stoichiometric mechanism explains the effect of the receptor. Rather the receptor may function catalytically at many sites on a reaction that can proceed in two directions depending on the metabolic state of the cell. It is hypothesized that chromatin differentiation distal to the receptor allows only a subset of the expressed gene to respond to such actions.

Supported by NIH Grant 1-RO1-AM-18878-01

ISBN 0-12-045450-5

CONVERSION OF NON-MUSCLE CELLS INTO FUNCTIONAL STRIATED MYOTUBES BY 5-AZACYTIDINE

Peter A. Jones, Philip G. Constantinides and Shirley Taylor
Division of Hematology-Oncology, Childrens Hospital of Los Angeles, Los Angeles, California 90027; and Department of Medical Biochemistry, University of Stellenbosch Medical School, Box 63, Tygerberg 7505, South Africa.

The C3H/10T½CL8 line of mouse embryo cells was derived for studies on chemical oncogenesis *in vitro* (1). It has none of the characteristics of myoblasts yet we have found that functional striated muscle cells are induced in the line 9-10 days after exposure to micromolar concentrations of 5-azacytidine (2).

Further characterization of these myotubes revealed that they arise from the fusion of mononucleated precursors and not as a result of endoreplication. They accumulate histochemically demonstrable myosin ATPase activity as well as acetylcholine receptors capable of binding α-bungarotoxin. The deoxy analog, 5-aza-2'-deoxycytidine, induced myogenic conversion at one-tenth of the maximally effective concentration of 5-azacytidine. The ability of both analogs to induce myotube formation and to cause cytotoxicity was strongly influenced by co-treatment with certain pyrimidine nucleosides. These effects were consistent with a requirement for metabolism of both aza compounds to phosphorylated derivatives and with a mechanism based on their incorporation into DNA. Experiments with synchronized cells showed that myotubes were only induced in cells which had been exposed to 5-azacytidine in early S phase. These analogs may therefore be useful in studies on the molecular mechanisms of cellular differentiation.

Supported by the South African Medical Research Council and the National Cancer Association of South Africa.

(1) Reznikoff, C.A., Brankow, D.W., and Heidelberger, C., *Cancer Res.*, *33*, 3231-3238 (1973).

(2) Constantinides, P.G., Jones, P.A., and Gevers, W., *Nature*, *267*, 364-366 (1977).

ISBN 0-12-045450-5

A STUDY OF ESTROGEN RECEPTOR-DNA INTERACTION BY A COMPETITION METHOD.

J. Kallos and V. P. Hollander, Research Institute of the Hospital for Joint Diseases & Medical Center, Mount Sinai School of Medicine, 1919 Madison Ave., New York, N.Y. 10035

We have used DNA-cellulose competition experiments to examine whether rabbit uterus estradiol receptor (1) discriminates among DNA sequences, (2) unwinds the DNA double helix and (3) recognises DNA conformational alteration induced by intercalating drugs.

Increasing amounts of competing soluble DNA was mixed with fixed amounts of DNA-cellulose and a partially purified uterine cytosol preloded with ^{3}H-estradiol (^{3}H-E labeled 105,000 g supernatant fractions, 3 mg protein/ml in 50 mM KCl, 20 mM mercaptoethanol, ammonium sulfate precipitated and filtered through Sephadex G-25) was added. Afterward, the DNA bound-radioactivity in the pellet was determined. The receptor complex indiscriminately binds to mammalian and bacterial DNAs, but binds preferentially to Poly dAT (AT containing synthetic DNA). The receptor binding is drastically reduced upon DNA denaturation and fully recovered upon renaturation, suggesting that the receptor is probably not an unwinding protein. Finally, the receptor-DNA interaction is inhibited by the intercalating drug, actinomycin D, usggesting that the intact conformation of the DNA double helix is required for optimum binding.

This investigation was supported by Grant Numbers CA 10064 and P30 14194, awarded by the National Cancer Institute, DHEW.

ISBN 0-12-045450-5

THE EFFECT OF ETHIONINE ON LYMPHOCYTE ACTIVATION AND ASSOCIATED MACROMOLECULAR SYNTHESIS

Rochelle Seide-Kehoe,* Peter Zabos,+ David Kyner,+ Judith Christman,+ and George Acs+
*Northeastern Ohio Universities College of Medicine, Rootstown, Ohio and +Mt. Sinai School of Medicine, New York, N.Y.

The effect of the hepatocarcinogen, ethionine, on events associated with mitogen induced activation of lymphocytes has been studied. DNA synthesis (as measured by ^{3}H-thymidine or ^{32}P incorporation) by mitogen stimulated cells was totally, but reversibly, inhibited by 4-8 mM DL-ethionine. DNA replication commenced within 4-6 hours of ethionine removal even when ethionine had been present in the culture for up to 60 hours.

Experiments using Con A and α-methyl mannose indicated that "commitment" of lymphocytes to DNA synthesis was not affected by the presence of ethionine in the culture medium. Early reactions associated with lymphocyte activation take place in the presence of ethionine. The rates of both RNA and protein synthesis in cells stimulated in the presence of ethionine were higher than those seen in unstimulated cells although they were approximately 50% of the rates observed in cells stimulated with mitogen alone.

Further studies have shown that the synthesis of 4, 18 and 28S RNA occurs in the presence of ethionine although at slower rates than in cells stimulated under normal conditions. In addition the presence of ethionine decreased the methylation of these various RNA species in mitogen stimulated cells.

Experiments utilizing actinomycin D during a pulse label suggested that previous ethionine exposure leads to the synthesis of undermethylated tRNA molecules, which can be rapidly methylated once the ethionine is removed.

Taken together, the results indicate that ethionine allows the progression of stimulated lymphocytes from the G_0 to the G_1 phases of the cell cycle but blocks their entry into the S phase. The exact nature of this block is currently unknown.

(Supported by NIH grant CA16890 and American Cancer Soc. grant NP-36

ISBN 0-12-045450-5

COMPETITIVE LABELLING ANALYSIS OF CHROMATIN STRUCTURE.
B. Malchy and H. Kaplan, Departments of Biochemistry, Queen's University and University of Ottawa.

Competitive labelling is a method which measures the rate of reaction of reactive groups in proteins relative to a primary standard. The method has been applied to the analysis of chromatin structure using ^{3}H-acetic anhydride as the labelling reagent. When applied to chromatin isolated from either calf thymus or rat liver the procedure confirms that histone ε-amino groups are not available to solvent. Both pH and ionic strength perturbations have been used in order to determine under what conditions the ε-amino group interactions become disrupted. It has been observed that in .01M veronal-phosphate buffer histone amino groups become selectively exposed above pH 11 and that histone I is more easily exposed than H2B, and H2A which are in turn more easily exposed than H3 and H4. Raising the ionic strength also raises the degree of exposure of the histone amino groups. Histone H2B because of its N-terminal proline is characteristically more reactive than the others. This proline is observed to have approximately the reactivity of an exposed proline and an estimated pK of 9.7. Lysines 5, 8, 12 and 16 of H4 have been found to have a reactivity index (R.I.) of 0.3 in 0.25M NaCl and are thus substantially buried. However, under conditions in which H4 has an R.I. around 0.5 (0.9M NaCl pH 8 and 0.25M NaCl pH 10) the R.I of these lysines is approximately 1, demonstrating that this highly charged region of H4 can become exposed to solvent at least as easily as other parts of H4.

ISBN 0-12-045450-5

INHIBITION OF ACCUMULATION OF MESSENGER RNA FOR A BRAIN-SPECIFIC PROTEIN IN RAT GLIAL CELLS BY COLCHICINE

A. Marks, J.B. Mahony, and I.R. Brown
Banting and Best Department of Medical Research, and Department of Zoology, University of Toronto, Toronto, Canada.

The synthesis and accumulation of the brain-specific S100 protein increase when clonal rat glial cells, C6, progress from logarithmic to stationary growth in monolayer culture. The induction of S100 protein may be mediated by intercellular contact and is dependent on an accumulation of S100 protein mRNA (1). The synthesis of S100 protein in logarithmic cultures is increased by succinyl Con A and inhibited in stationary cultures by drugs which disrupt the microtubular network (2). Total cellular RNA was prepared from stationary cultures of C6 cells, before and after treatment with colchicine (1 μM) for 24 hours, and enriched for poly A- containing RNA by affinity chromotography on oligo-dT cellulose. The capacity of this RNA fraction to direct the synthesis of S100 protein was assayed in a cell-free system derived from wheat embryos (1). The inhibition of S100 protein synthesis by colchicine was paralleled by a decrease in the activity of S100 protein mRNA, suggesting that the disruption of the microtubular network interrupts the signals which maintain the induction of S100 protein in stationary cultures of C6 cells.

REFERENCES

1. G. Labourdette, J.B. Mahony, I.R. Brown, and A. Marks (1977). Eur. J. Biochem, in press.

2. A. Marks and G. Labourdette (1977). Proc. Nat. Acad. Sci. U.S.A. 74, 3855-3858.

Supported by NCI and MRC, Canada.

ISBN 0-12-045450-5

THE THREE DIMENSIONAL STRUCTURE OF A DNA UNWINDING PROTEIN: THE GENE 5 PRODUCT OF fd BACTERIOPHAGE

Alexander McPherson and Frances A. Jurnak
Dept. Biochemistry, Milton Hershey Medical Center, Hershey, PA

Andrew Wang and Alexander Rich
Dept. Biology, M.I.T., Cambridge, MA

Ian Molineux
Imperial Cancer Research Fund, Mill Hill Laboratories, London, England

The three dimensional structure of the gene 5 product of Fd bacteriophage, a DNA unwinding protein of 9800 molecular weight has been determined to 2.4 Å resolution by single crystal X-ray diffraction analysis. The 88 amino acids are organized almost entirely in a pattern of β structure and the location of a cleft in the protein, with associated residues implied in function, suggests the region of DNA binding. A three dimensional model of the protein will be presented as well as the electron density map from which it is derived.

Crystals have also been grown of the DNA unwinding protein complexed with a series of deoxy-oligonucleotides and these suggest the nature of the protein DNA complex and its possible analogy to that which may exist in chromosomal material. Preliminary data on these complex crystals will be presented.

ISBN 0-12-045450-5

MODULATION OF GRANULOPOIESIS BY PROSTAGLANDINS AND 1-METHYL 3-ISOBUTYL XANTHINE.

Alan M. Miller, Thomas R. Russell and Adel A. Yunis. Departments of Medicine and Biochemistry, University of Miami School of Medicine and the Howard Hughes Medical Institute, Miami, Florida

The role of prostaglandins (PG) in cell proliferation and differentiation has recently received wide interest. To explore a possible role for PG in granulopoiesis we have examined the effects of PGE and PGF on mouse myeloid colony (CFU-C) growth *in vitro*. The number of colonies was determined on days 4 and 6 of culture and morphology observed at day 7.

PGE_1 and PGE_2 were strongly inhibitory with both causing approximately 40% inhibition of colony formation at 10^{-8} M concentrations. In contrast $PGF_{1\alpha}$ at concentrations as low as 10^{-10} M and $PGF_{2\alpha}$ in the range of $0.5-2 \times 10^{-9}$ M caused a 30-50% increase in the number of colonies above that seen with colony stimulating factor alone. PGE caused a reduction in the number of both macrophage and granulocyte containing colonies, while PGF stimulation was specific for granulocytic colonies.

The compound 1-methyl 3-lisobutyl xanthine (MIX) has been shown to act like $PGF_{2\alpha}$ in triggering the conversion of 3T3 fibroblasts into adipocytes (1). In the range of $0.2-1 \times 10^{-5}$ M MIX augmented CFU-C growth and differentiation by 46-66%. MIX stimulation was also specific for granulocytic colonies.

PGF stimulation of colony formation could be diminished by increasing the levels of PGE, and conversely inhibition of PGE can be reversed by increasing the concentration of PGF.

The opposing effects of PGE and PGF on CFU-C growth *in vitro* and the known transformation of PGE_2 to $PGF_{2\alpha}$ by the PGE_2 9-keto reductase suggest that the PGF/PGE ratio may play an important role in the control of granulopoiesis *in vivo*.

1. T.R. Russell and R.J. Ho, Proc. Natl. Acad. Sci., USA 73 (1976) 4516.

ISBN 0-12-045450-5

REGULATION OF GLUTAMINE SYNTHETASE IN 3T3-L1 ADIPOCYTES

Richard E. Miller*, James F. Jorkasky*, and Howard Gershman+
Departments of Medicine* and Biochemistry+, Case Western Reserve University, and Department of Medicine, Cleveland Veterans Administration Hospital*, Cleveland, Ohio 44106

The 3T3-L1 fibroblast cell line, when maintained at confluence, develops morphological and biochemical characteristics of adipocytes. This conversion from preadipocytes to adipocytes is accelerated by insulin (Green, H. and Kehinde, O. (1975) Cell 5, 19-27).

We find that glutamine synthetase (GS) specific activity increases >100-fold in 3T3-L1 cells during the insulin-mediated adipocyte conversion. In contrast, glucose-6-P dehydrogenase (G6PD) and hexokinase (HK) specific activities increase only 1.5- to 3-fold, and culture protein increases only 4- to 5-fold. The increase in GS specific activity precedes the morphological changes and the increases in G6PD, HK, and protein. A half-maximal increase in GS specific activity occurs at a culture medium insulin concentration of 10 ng/ml, while a half-maximal increase in G6PD and HK specific activities and culture protein requires medium insulin concentrations between 250 and 1000 ng/ml. Incubation of 3T3-L1 cells in the presence of 1μg/ml hydrocortisone for 2 to 3 days increases GS specific activity 2-fold in 3T3-L1 adipocytes (maintained after confluence in the presence of insulin) and >20-fold in 3T3-L1 preadipocytes (maintained after confluence in the absence of added insulin). In contrast, incubation of the adipocytes with hydrocortisone yields<1.5-fold increase in G6PD and HK specific activities and no change in culture protein; incubation of the preadipocytes with hydrocortisone results in no change in G6PD, HK, or protein. Incubation of 3T3-L1 adipocytes in the presence of dibutyryl cyclic AMP(1mM) plus theophylline (1mM) decreases GS specific activity >70% in 24 hr. During the same incubation period there is little or no effect of dibutyryl cyclic AMP plus theophylline on G6PD or HK specific activity or on culture protein. These data indicate that glutamine synthetase activity is hormonally regulated in 3T3-L1 cells. Since the increase in glutamine synthetase precedes other insulin-mediated changes in 3T3-L1 cells, our data also suggest that glutamine synthetase is required for differentiation of 3T3-L1 cells into adipocytes.

Supported by the Diabetes Association of Greater Cleveland and the Medical Research Service of the Veterans Administration.

ISBN 0-12-045450-5

THE EFFECT OF 5 -BROMODEOXYURIDINE ON THE INDUCTION OF CYCLIC AMP PHOSPHODIESTERASE IN 3T3-L FIBROBLASTS

Thomas Murray and Thomas R. Russell.
Department of Biochemistry, University of Miami School of Medicine, Miami, Florida 33152

The thymidine analog, 5-bromodeoxyuridine (BrdU), has been employed in studies concerned with the regulation of cAMP phosphodiesterase (PDE) activity in mouse 3T3 fibroblasts grown in cell culture. After plating cells were grown in the presence of BrdU until a confluent monolayer was attained and cell division ceased. Experiments were conducted on the confluent monolayers to examine the effects of BrdU incorporation into DNA on the levels of cAMP PDE activity in the presence and absence of a potent inducer of the enzyme, 1-methyl-3-isobutyl xanthine (MIX).

Monolayer cultures treated with MIX (0.5mM) exhibit a continual increase in cAMP PDE activity over a 48 hour period, culminating in a 3 to 4 fold elevation. This inducibility can be inhibited by prior growth of the cells in the presence of BrdU. Concentrations of BrdU from 10^{-7} to 10^{-5}M show progressively greater inhibition of the MIX dependent induction of cAMP PDE. None of the concentrations of BrdU tested had any significant effect on basal levels of the enzyme. Competition experiments with excess thymidine suggest that BrdU is functioning via incorporation into DNA. Since the incorporation of BrdU into DNA can inhibit cAMP PDE induction, these results suggest that induction of the enzyme involves a DNA linked event. This contention is supported by experiments that show actinomycin D and cycloheximide block induction of the enzyme.

Supported by the Juvenile Diabetes Research Foundation of Florida.

ISBN 0-12-045450-5

MILK PROTEINS AND THEIR mRNA's DURING THE LACTOGENIC DIFFERENTIATION OF RAT MAMMARY GLAND. Hira L. Nakhasi and Pradman K. Qasba, Laboratory of Pathophysiology, NCI, National Institutes of Health, Bethesda, Md. 20014 U.S.A.

Rat mammary gland specific proteins - α-lactalbumin and three species of caseins, 42K, 29K and 25K-, and their mRNA's have been purified[1,2,3,4]. Using radioimmunoassays[5], and cDNA probe for α-LA, the levels of these proteins and of α-LA mRNA have been quantitated in the rat mammary gland during gestation and lactation. The results show that the mammary gland undergoes first phase of lactogenesis at the beginning of gestation, with a 3-4 fold increase in the levels of α-LA and caseins in the gland within 2-4 days as compared to the virgin. Increase in the levels of α-LA mRNA within 2-5 days of gestation, measured directly with the cDNA probe, concurs with the increase in α-LA protein levels. These levels recede by the end of the week and so does the levels of α-LA and casein proteins, reaching a value comparable to virgin mammary gland. α-LA mRNA levels thereafter again steadily increase till two days before paturation. Total casein, α-LA and other mRNA activities measured by the wheat germ translational system increase from the 7th day of gestation, with the appearance of two peaks of activities one corresponding to the 12th day and the other the 17th day of gestation. Protein levels of α-LA and caseins in the gland also increase during the second and third week of gestation and the proportion of these remain as 42K>29K>25K>α-LA until paturation. After paturation α-LA mRNA levels increase about 10 fold till the 8th day of lactation thereafter declining to the virgin levels by the 21st day of lactation. The proportion of these proteins remain as 42K>25K>29K>α-LA throughout lactation. The mRNA activities also increase by 10 fold during lactation and appear sequentially; α-LA activity plateaus by the 4th day, caseins by the 8th day and the other mRNA activities by the 14th day. Thereafter these activities decline sequentially. These results show: (a) the phasic expression of differentiated functions during gestation and (b) sequential expression of the milk protein mRNA's and other mRNA's during lactation.

1. Qasba, P.K. and Chakrabartty, P.K., J.Biol.Chem.,in press.
2. Chakrabartty, P.K. and Qasba, P.K., Nucl.Acid Res., 4: 2065-2074 (1977).
3. Rosen, J.M., Biochem., 15: 5263-5271 (1976).
4. Rosen, J.M. and Barker, S.W., Biochem., 15: 5272-5279 (1976).
5. Qasba, P.K. and Gullino, P.M.,Cancer Res.,37: 3792-3795 (1977).

PARTIAL PURIFICATION OF PROLACTIN RNA FROM BOVINE ANTERIOR PITUITARY GLANDS

J. Nilson, E. Convey and F. Rottman.
Departments of Biochemistry and Dairy Physiology, Michigan State University, E. Lansing, MI 48824

Steroids and some polypeptide hormones have been shown to stimulate transcription of specific mRNAs in their respective target tissues. The possibility also exists that hormones regulate mRNA metabolism at other secondary points i.e., by post-transcriptional structural alterations that could lead to changes in mRNA processing, stability or translational efficiency. Our goal is to use a cDNA probe for prolactin (PRL) mRNA to detect, isolate, and characterize PRLmRNA in cultures of bovine pituitary cells treated with steroid and peptide hormones. In this particular study we report the partial purification of PRLmRNA from bovine anterior pituitary glands and its potential use as a template in the synthesis of cDNA.

When unfractionated cytoplasmic RNA was translated in the wheat germ cell-free system, two major labeled proteins were detected by SDS-gel electrophoresis and fluorography. The two proteins migrated slightly behind PRL and growth hormone (GH) markers. The slower migrating protein band could be immunoprecipitated with PRL antibody. Furthermore, if the wheat germ lysates which had been directed by unfractionated cytoplasmic RNA were analyzed by radioimmunoassay (RIA), both GH and PRL were detected. We interpret these results to mean that PRLmRNA and GHmRNA are the primary mRNA species in anterior bovine pituitary glands. Poly(A+) mRNA was subsequently isolated from the cytoplasm of pituitary cells by oligo(dT) cellulose chromatography and resolved by electrophoresis in denaturing agarose-urea gels. A major mRNA peak was detected (∿15S) that migrated behind a much smaller secondary peak. The RNA from the major peak was eluted from the gel onto an oligo(dT) column, recovered, and then translated in the wheat germ cell-free system. Only one labeled protein band could be detected after analysis of the cell-free lysates by electrophoresis/fluorography. RIA of the translation products with GH and PRL antibodies revealed only the presence of PRL. We calculate the MW of this gel-purified RNA to be approximately 3.9×10^5 daltons. This purified PRLmRNA serves as an efficient template for the synthesis of cDNA, using AMV reverse transcriptase.

ISBN 0-12-045450-5

QUANTITATIVE DIFFERENTIATION *IN VITRO* OF EXPONENTIALLY GROWING EMBRYONAL CARCINOMA CELLS AT LOW DENSITY.
Robert Oshima. Dept. of Pediatrics, Univ. of Calif., San Diego, La Jolla, California.

Extraembryonic endoderm is the first identifiable cell type to appear when cultured mouse embryonal carcinoma (ec) cells, such as the PSA1 cell line, differentiate *in vitro*. Endoderm can be distinguished biochemically at the single cell level from its undifferentiated parental cell by its low activity of alkaline phosphotase (AP^-), secretion of plasminogen activator (PA^+) and synthesis of a basement membrane-like material called Reichart's membrane (RM^+). F9 is an ec cell line which differentiates to only a very limited extent either *in vivo* or *in vitro*. In order to investigate what regulatory changes have occured in the F9 cell line which result in the loss of its ability to differentiate substantially, techniques have been developed to measure the frequency at which embryonal carcinoma cells differentiate spontaneously to an identifiable differentiated cell type. The fibrin-plasminogen overlay system of Beers was utilized to measure the spontaneous appearance of cells secreting PA in exponentially growing cultures of PSA1 and F9 at low cell density. PA^+ colonies appear at an average frequency of $0.5 \pm 0.10/10^4$ cells for F9 and three subclones derived from it. A recently cloned culture of PSA1 produced PA^+ colonies at a frequency of $51 \pm 6.6/10^4$ cells. The relative frequency at which these cell lines spontaneously differentiate is a stable characteristic since subclones of both cell lines reisolated from tumors arising from injections of the parental lines into 129SJ mice give similar results.

Histochemical localization of alkaline phosphotase and immunoflourescent staining of cells synthesizing Reichart's membrane support the contention that the cells secreting plasminogen activator are extraembryonic endoderm. The increasing size of the lysis zones formed over undissociated cultures overlaid for 4 hours at daily intervals suggests that proliferation of the PA^+ cells follows the initial differentiative transition event. Determination of the total number of individual PA^+ cells in single cell suspensions of dissociated cultures support this view. This is the first quantative cellular measurement of the differentiation of clonal ec cell lines *in vitro*. In addition, it indicates that aggregation is not a strict requirement for differentiation to endoderm.

--Supported by a Basil O'Connor Starter Research Grant from the National Foundation.

ISBN 0-12-045450-5

INDUCTION OF SURFACE IMMUNOGLOBULIN (sIg) EXPRESSION ON A MURINE B-CELL LEUKEMIA BY LIPOPOLYSACCHARIDE (LPS).

Christopher J. Paige, Paul W. Kincade, and Peter Ralph.
Laboratory of Hematopoietic Development, Sloan-Kettering Institute, Rye, New York.

A chemically induced leukemia of BDF_1 mice, designated 702/3 (donated by Dr. P. Baines, Manchester, England), was adapted to tissue culture in our laboratory. A small percentage (<5%) of these cells displayed sIg as detected by immunofluorescence with rhodamine-labeled class (IgM) specific antibodies. Addition of the B-cell mitogen, LPS, to the cultures induced sIg expression on all of these cells. The kinetics of this transition were dependent on the dose of LPS. As little as 0.1 μgm/ml induced sIg on >97% of the cells within 36 hours. Other mitogens (Con A, PPD) or inducing agents (DMSO, dimethyl formamide, butyric acid), tested over a wide range of concentrations, failed to induce sIg expression.

The cells bear H-2 antigens but lack IgD, IgG, IgA, Ia, Thy 1.1, and receptors for the Fc portion of immunoglobulin. Exposure to LPS had no effect on the presence or absence of these structures. A small percentage of cells were positive for complement receptors. Cytoplasmic IgM was detectable within all of the cells and constitutive production of small quantities of IgM was confirmed by SDS-polyacrylamide gel electrophoresis of serological precipitates of cell lysates after pulsing with radioactive amino acids.

This cell line has properties similar to cytoplasmic Ig^+, surface Ig^- cells found in immature tissues and in bone marrow of mice and humans which are thought to be immediate precursors of sIg^+ B lymphocytes. It may provide a model for studying the mechanism of LPS activation and the molecular events associated with externalization of cell surface receptors.

Supported by NIH grant AI-12741 and NSF grant BMS 75-19734.

ISBN 0-12-045450-5

TEMPORAL AND SPACIAL DISTRIBUTION OF MESSENGER RNA IN PRIMATE SPERMATOGENIC CELLS

R. L. Pardue, D. G. Capco, L. E. Franklin, and W. R. Jeffery*.
University of Houston, Department of Biology, Houston, Texas, and University of Texas, Department of Zoology, Austin, Texas*

The temporal and spacial distribution of messenger RNA (poly(A)+RNA) during the spermatogenic cycle of the bush baby, Galago senegalensis, was analyzed by in situ hybridization of testes sections with (^{3}H)-poly(U) (1). The location of (^{3}H)-poly(U):poly(A) complexes was detected by autoradiography at the light and electron microscope levels. Alterations in the density and cellular distribution of grains were found to accompany the progressive differentiation of spermatids from the spermatogonial cells in the semeniferous tubules. In spermatogonia, grain density is initially elevated in the nucleus (stages I-III), followed by a substantial increase in the cytoplasm (stages IV-VI). Later during spermatogonial development (stages VII-XII) nuclear and cytoplasmic grain density markedly decreased. In spermatocytes, although grain density above the nucleus becomes maximal during the leptotene-zygotene transition (stages VII-IX), no concommitant increase in cytoplasmic labeling was observed. The spermatids exhibited a progressive decrease in nuclear and cytoplasmic grain density (stages I-IX), until the pre-spermiogenic period (stages X-XII) when labeling was again elevated. Electron microscopic localization of poly(A)+RNA indicated a random distribution of grains in the nucleoplasm and cytoplasm.

The data are consistent with the occurrence of three periods of poly(A)+RNA accumulation during the spermatogenic cycle; in the primary spermatogonial cells, during the leptotene-zygotene interpahse in spermatocytes, and during the pre-spermiogenic period of spermatids. The first two periods correspond to intervals of intense DNA and RNA synthesis in other mammalian testes (2). The pre-spermiogenic period increase in the present report is proposed to be a necessary prelude to the sperm maturation.

Supported by American Cancer Society Grant NP188a.

(1) Capco, D. G. and Jeffery, W. R., Development(submitted).

(2) Monesi, V., Experimental Cell Research 39:197-224,1965.

ISBN 0-12-045450-5

GENETIC CONTROL OF MILK ZINC AVAILABILITY

J. E. Piletz and R. E. Ganschow.
Institute for Developmental Research, Children's Hospital Research Foundation, Cincinnati, Ohio.

Mice homozygous for the recessive mutation lethal milk (lm) produce milk which is 34% deficient in zinc content (1). This mutation results in symptoms of dietary zinc deficiency and eventual death for all nursing pups. Since even wild-type pups die when nursed on milk of mutant mice, the defect resides in the milk. Moreover, lmlm pups develop normally if foster-nursed on a normal dam. Administration of zinc to pups nursing on lmlm dams corrects mortality and morbidity.

The human disease acrodermatitis enteropathica (AE), like lethal milk, involves the availability of milk zinc and is controlled by a recessive genetic determinant. AE is expressed as acute dermatitis and eventual death when genetically-predisposed infants are weaned to cow's milk. Successful treatment has been achieved with oral zinc administration. Recent evidence suggests that the symptoms of AE result from the absence in bovine milk of a low-molecular-weight, zinc-binding ligand (ZBL) necessary for zinc absorption and normally present in human milk (2). Moreover, the ZBL is only present in maternal milk until mid-lactation when a similar ZBL develops in the infant's intestinal mucosa (2). Thus the presence of ZBL in maternal milk appears to be temporally co-ordinated with the production of intestinal ZBL in infants. Presumably AE patients fail to produce intestinal ZBL.

Preliminary evidence with the lethal milk mutant suggests that this milk zinc deficiency may also involve a defective zinc transport ligand and thus may be an animal model of AE.

Supported by USPHS Grant AM14770 and HD05221.

(1) Piletz, J.E. and Ganschow, R. E., Science, in press.

(2) Hurley, L.S., Duncan, J.R., Sloan, M.V., and Eckhert, C.D., Proc. Nat. Acad. Sci., 74 3547-3549 (1977).

ISBN 0-12-045450-5

STUTTERING: HIGH LEVEL MISTRANSLATION IN ANIMAL AND BACTERIAL CELLS

J. W. Pollard[†], J. Parker[*], J. Friesen[*], and C. P. Stanners[†]

[†]Ontario Cancer Institute, 500 Sherbourne St., Toronto, Ontario, Canada and [*]Dept. of Biology, York University, 4700 Keele St., Downsview, Ontario, Canada.

The fidelity of translation by animal cells has been of considerable interest since the suggestion that errors in translation should be autocatalytic and could thus be responsible for aging(1).

We have demonstrated that extreme starvation for certain amino acids in both bacterial and mammalian cells resulted in translational errors which could be detected by two dimensional polyacrylamide gel electrophoresis. The faulty proteins were detected as a trail of spots with similar molecular weights to the authentic proteins but separated from them in the isoelectric focussing dimension. We suggest that this charge displacement is a consequence of amino acid substitution and call this phenomenon, stuttering.

The direction of displacement in the isoelectric focussing dimension could be predicted by misreading pyrimidine for purines at the third position of the codon (2). It was predicted that upon His starvation, Gln would be substituted, resulting in a charge shift towards the acidic end of the gel. Similarly upon Asn starvation it was predicted that Lys would be substituted with a resultant shift towards the basic end of the gel. These predictions were investigated in CHO cells by starvation for Asn with a _ts_ asparagyl tRNA synthetase mutant, and for His by treatment with histidinol (3) in the absence of His. The results confirmed the predicted direction of displacement.

It is expected that this two dimensional assay will provide a rapid means of measuring the fidelity of the translational machinery from cell type to cell type. Experiments are underway to investigate the relationship of mistranslation with aging and tumor progression.

This work was supported by grants from the NCI, NRC (NRC 57 34), MRC (MT 1877) of Canada and the NIH(NIH-NCI Cont. N01-CP-4331)USA. J.W.P. is a Research Fellow of the NCI of Canada.

(1) Orgel, L.E. _Proc. Nat. Acad. Sci._ _49_ 517-521 (1963).
(2) Woese, C.R. in _Prog. Nucleic Acid Res. Mol. Biol._ Eds. Davidson, J.N. and Cohen, W.E., Academic Press, Vol. 7, 107-172 (1967).
(3) Hansen, B.S., Vaughan, M.H., and Wang, L.J. _J. Biol. Chem._ _247_ 3854-3857 (1972).

ISBN 0-12-045450-5

EFFECT OF CYTOCHALASIN B ON THE EXPRESSION OF MITOGEN RECEPTORS IN CULTURED CHICK EMBRYO FIBROBLASTS (CEF)

M.K. Raizada and R.E. Fellows, Department of Physiology and Biophysics, University of Iowa, Iowa City, IA 52242.

Previous studies utilizing ^{125}I-insulin as a model mitogen have demonstrated a positive correlation between mitogen-receptor interactions and pleiotypic events associated with cell multiplication (1). In uninfected confluent cultures of CEF, mild trypsin treatment and serum starvation have been observed to regulate expression of mitogen receptors, while in Rous Sarcoma Virus (RSV)-transformed CEF, with 2-4 times more mitogen receptors, expression is independent of these controls (2). To investigate the role of microfilaments in mitogen receptor expression, cytochalasin B (CB) has been incubated with confluent cultures of CEF. This results in a time- and dose-dependent enhancement of ^{125}I-insulin binding with 2-4 fold increase after 24 hours of treatment with 10 μg/ml CB. Scatchard analysis reveals an increase in the number of insulin binding sites, from 9×10^3 in untreated cultures to 20×10^3 in CB-treated cultures, without significant change in K_a. This is accompanied by an increase in insulin-stimulated ^{3}H-thymidine incorporation. Removal of CB results in a time-dependent decrease in ^{125}I-insulin binding, with return to pretreatment levels by 24 hours. CB also has an additive effect on the increase in binding induced by serum starvation. However, trypsin treatment, which disrupts microfilaments as well as increases mitogen receptor numbers, does not result in a further increase in ^{125}I-insulin binding by CB-treated CEF. Scanning electron microscopic examination of serum-starved and unstarved CEF after CB treatment shows that in both, cells retract from the plate and develop a large number of surface folds and protrusions. By contrast, CB treatment does not alter ^{125}I-insulin binding of RSV-transformed CEF, which already have increased surface folds and protrusions, depolymerized microfilaments, and increased numbers of mitogen receptors. These results suggest that microfilaments may be an important structural component in mitogen receptor expression.

Supported by The Milheim Foundation and AM 19901.

(1) Raizada, M.K. & Perdue, J.F., J. Biol. Chem. 250, 6445.

(2) Perdue, J.F. & Raizada, M.K., Prog. Clin. Biol. Res. 9 (V. Marchasi, ed) Alan Liss Press, NY, P. 49 (1976).

ISBN 0-12-045450-5

DEVELOPMENT OF ADULT MUSCLES IN THE ABDOMEN OF DROSOPHILA

Christoph A. Reinhardt Center for Pathobiology, University of California, Irvine, California 92717.

Adult muscles in the thorax of the fruit fly are known to develop from adepithelial cells of the imaginal discs during metamorphosis (1). By analogy, we expected to find adepithelial cells in the abdominal imaginal primordia, the histoblast nests.

Using SEM and TEM we discovered small clusters of adepithelial cells attached to the center of each histoblast nest in 18 h and 24 h old pupae. Younger pupae and larvae exhibit one single layer of epithelial cells, all of which seem to be able to contribute to larval or pupal cuticle; adepithelial cells are absent at these stages. In 36 h old pupae the adepithelial cells start to differentiate into muscle cells by stretching in the appropriate directions and by accumulating microtubules and myofilaments in their cytoplasm.

We present some morphological evidence that the adepithelial cells derive from histoblasts, which is consistent with histological observations on adepithelial cells of imaginal discs (2). Between 10 h and 24 h after pupariation adepithelial cells appear at the base of the epithelium (see fig. below) which gradually detach from the epithelial sheet by loosing their apical cell junctions and their contact to the cuticle. In any stage the ultrastructure of the adepithelial cells allows a clear distinction from blood cells.

This leads to the conclusion that muscle elements derive from ectodermal tissue a long time after embryogenesis is completed.

(1) Reed, C.T., Murphy, C. and Fristrom, D., Wilh. Roux Arch. 178 285-302 (1975).
(2) Madhavan, M. and Schneiderman, H. A., Wilh. Roux Arch. (in press).

Fig.: Development of a histoblast nest in the early pupa (h= histoblasts, L= larval cell, a= adepithelial cells c= cuticle)

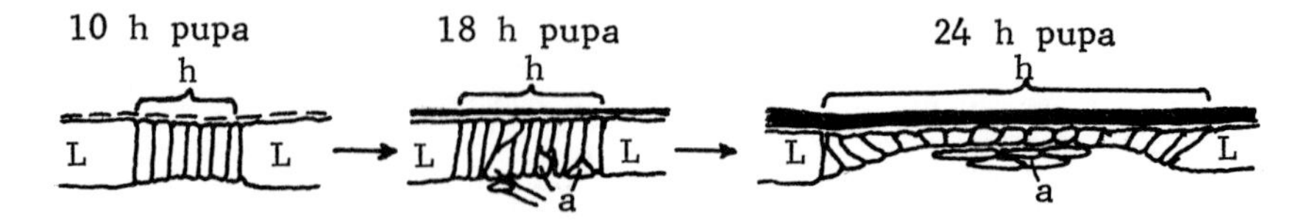

ISBN 0-12-045450-5

SEQUENCE AND STRUCTURE OF THE ACTH/LPH PRECURSOR PROTEIN AND mRNA

James L. Roberts, Peter Seeburg, John Shine, Howard Goodman, John Baxter, and Edward Herbert
Departments of Medicine and Biochemistry, Univ. of California, San Francisco, Ca. 94143 and Department of Chemistry, Univ. of Oregon, Eugene, Oregon, 97403

Using RNA isolated from the AtT20-D_{16v} mouse tumor cell line, the mRNA coding for the ACTH/LPH precursor has been isolated and its primary translation product identified as a 28,500 dalton polypeptide. Poly A enriched RNA was fractionated by formamide-sucrose gradient centrifugation and urea-agarose gel electrophoresis and the ACTH/LPH mRNA was identified as having a sedimentation coefficient of 16S. A highly enriched ACTH/LPH mRNA fraction was used as the template for single stranded DNA synthesis using reverse transcriptase. Double stranded ACTH/LPH DNA was also synthesized from this single stranded cDNA. These DNA's were digested with a variety of different restriction endonucleases and the resulting fragments were isolated by slab acrylamide gel electrophoresis. Several of these DNA fragments have been sequenced. Tryptic peptide and cyanogen bromide peptide analysis of the ACTH/LPH mRNA primary translation product labeled with 17 different radioactive amino acids was also performed. By comparing the nucleotide sequence results with the peptide structure data and the known portion of the ACTH/LPH precursor protein (1), the reconstructed sequence of the ACTH/LPH mRNA should be obtained and hence, the primary sequence of the precursor protein.

Supported by NIH grants AM 16879 and AM 19997-01.

(1) Roberts, J.L., and Herbert, E. *Proc. Nat. Acad. Sci.*, in press.

ISBN 0-12-045450-5

APPEARANCE OF A SPECIES-SPECIFIC SPERM NUCLEAR PROTEIN IN THE SPERMATOGENIC SEQUENCE DEMONSTRATED BY CYTO-IMMUNOFLUORESCENCE.

Toby C. Rodman* **, Stephen D. Litwin* and Giorgio Vidali**
Cornell University Medical College* and The Rockefeller University**, New York, N.Y.

Spermatozoa were collected from mouse vasa deferentia and caudae epididymides and the basic proteins were obtained from a fraction of sperm heads. A protein migrating as a single band in acetic acid-urea gel, distinct from histones, was isolated. Antibodies to the protein, induced in female rabbits by a series of injections of the protein (0.2 mg./dose) emulsified with complete Freund's adjuvant, were used for detection of the protein by an indirect immunofluorescence method.

When the method was applied to cytologic preparations of mature and immature spermatogenic cells, the following observations were made: (1) Smears of untreated or methanol-acetic acid fixed, but otherwise untreated, spermatozoa were negative. When similar smear preparations were treated with a solution of guanidinium chloride-beta-mercaptoethanol in Tris buffer at pH 8.5 the sperm heads swelled and were brilliantly positive. The intensity of the fluorescence was related to the degree of swelling. (2) In untreated smears of a suspension of cells from teased mouse testes, the mature spermatozoa were negative, primary spermatocytes were negative and spermatids of all stages were positive. After treatment with the guanidinium chloride solution, only primary spermatocytes were negative. (3) All parallel control preparations of mouse sperm and testes, utilizing non-immune rabbit sera, were negative. (4) Similar tests on rat and rabbit spermatogenic cells were negative with both immune and non-immune rabbit sera.

The cytologic observations reported here are consistent with previously reported biochemical data[1] that mammalian sperm proteins are species-specific. These observations also indicate that the unique sperm protein of the mouse appears as early as the beginning of spermiogenesis and its antigenic determinants are present, but masked, in mature spermatozoa.

Supported by grants from the Harry Winston Foundation and The Rockefeller Program for Reproductive Biology.

1. Bellve, A.R. et al (1975) Dev. Biol. 47, 349-365.

ISBN 0-12-045450-5

EVIDENCE FOR AN INTRINSIC DEVELOPMENTAL PROGRAM DURING GROWTH AND DIFFERENTIATION OF SKELETAL MUSCLE FIBER

R.K. Roy, F.A. Sreter and S. Sarkar. Dept. Muscle Research, Boston Biomedical Research Institute; and Dept. Neurology, Harvard Medical School, Boston, MA 02114

Using tropomyosin (TM) subunits and myosin light chains as muscle-specific markers, we have looked for differential gene expression in skeletal muscles during development. Both α and β subunits of TM are present in embryonic skeletal muscles of chicken and rabbit, the β TM being the major species (α/β = 20/80) in all embryonic fibers. With development, the β TM gradually decreases and the subunit pattern switches either to the single α TM, as in the case of adult chicken m. pectoralis, or to a predominantly α type, as in the case of rabbit skeletal fast muscle (α/β = 80/20). In contrast, the cardiac muscles of both chicken and rabbit consist of only the α subunit which remains invariant during development. The adult chicken breast TM, which gives a single band of α TM in SDS gels, shows two distinct components in acidic urea gel electrophoresis, one of which is identical in mobility to the single band of chicken cardiac TM indicating the presence of distinct multiple forms of α TM in the same species. In rabbit muscles the speed of contraction (slowness) seems to be correlated with an increased β TM content. However, the α/β ratio of 55/45 is virtually identical in the slow ant. and fast lat. dorsi of chicken suggesting that the variation in the relative amounts of TM subunits in avian muscles cannot be explained solely on the basis of fiber types.

In terms of myosin light chains, embryonic myosin clearly resembles the fast myosin light chains except that LC_3 light chain is present in smaller amounts in an embryonic stage. It is concluded that the high β TM content and the fast myosin light chains LC_1 and LC_2 observed in all embryonic skeletal fibers regardless of their future fiber type are due to an intrinsic developmental program of the fiber. Subsequent events such as the appearance of slow myosin light chains in adult slow muscles, the increased accumulation of LC_3 light chain in adult fast muscle, and the changes in the TM subunit ratio represent differential gene expression during maturation of the fiber, probably in response to exogenous stimuli such as neural influence and activity pattern.

ISBN 0-12-045450-5

HYPOSTOMAL DOMINANCE-A ROLE FOR NERVE CELLS

Peter G. Sacks and Lowell E. Davis. Biology Department,
Syracuse University, Syracuse, N.Y. 13210

The freshwater hydra exhibits polarized form and regulative control over its pattern formation. The role of nerve cells in hypostomal dominance has been investigated using animals lacking nerve cells (1), and a new type of nerveless animal which we developed from hydroxyurea(HU) treatments (2). Normal animals were compared to HU (2) and colchicine(C) nerveless animals (1).

Control animals averaged 6.1 tentacles with a range from 4 to 8 tentacles. Nerveless clones (HU and C) average from 7 to 8.6 tentacles per clone with some individual animals having as many as 12 tentacles. Nerveless clones also contain significant numbers of aberrant animals possessing two or more hypostomes. In regeneration experiments, nerveless animals average 5.6 tentacles with a range from 0 to 14, whereas control average 4.8 with a range from 3 to 8 tentacles. These increases in mean and variance are statistically significant. Normal form is altered as shown by the abnormal animals and in increased tentacle number.

For grafting, hydra was divided into H1234B56F regions and homo- and heterographs were made (3). Nerveless -12- pieces posses as much and perhaps a higher inhibitory level than normal -12- pieces. Nerveless 1-F pieces form significantly more structures at the graft border than normal pieces. They do not respond to normal control from -12- pieces and this loss of control may explain the many aberrant forms present in clones.

These results show that nerveless animals can establish true dominant regions but that the levels of dominance and control over pattern formation are altered. It is suggested that nerve cells may function in regulating pattern established by epitheliomuscular cells.

(1) Campbell, R.D. J. Cell Sci. 21 1-13 (1976).
(2) Sacks, P.G. and Davis, L.E. J. Cell. Bio. in press.
(3) Wolpert, L. and Hornbruch, A. and Clarke, M.R.B. Amer. Zool. 14 647-663 (1974).

ISBN 0-12-045450-5

SEQUENCE COMPLEXITY STUDIES WITH POLY(A)-MRNA FROM MURINE NEUROBLASTOMA CELLS IN SUSPENSION CULTURE.

Bruce K. Schrier, Lawrence D. Grouse and Carol Letendre
Lab. of Devel. Neurobiol., NICHD, NIH, Bethesda, Md. 20014

Poly(A)-containing mRNAs were isolated from NS20Y cloned neuroblastoma cells grown in suspension culture. The weight-average size of the mRNAs was determined, by sedimentation through sucrose gradients under denaturing conditions, to be 1800 nucleotides. Complementary DNAs (cDNAs) were prepared from these mRNAs by use of AMV reverse transcriptase. The mean length of the cDNA molecules was 300 nucleotides, as determined on alkaline sucrose gradients. Hybridization of tracer amounts of cDNA with a vast excess of poly(A)-mRNA was performed and assayed for DNA-RNA hybrid by S_1 nuclease digestion. Eighty-five percent of the cDNA was capable of reacting with RNA in the cDNA-mRNA reactions. The most rapid component (15% of the cDNA, $R_0t_{1/2}$ = 0.006 had a sequence complexity of 1700 nucleotides, indicating that it contained a high concentration of only one or two messenger sequences. Middle abundance sequences (35% of the cDNA, $R_0t_{1/2}$ = 5.3) had a sequence complexity of 1.5×10^6 nucleotides; late-reacting species (50% of the cDNA, $R_0t_{1/2}$ = 93) had a complexity of 2.1×10^7 nucleotides. The data showed that the poly(A)-mRNA represented 1.35% of the single copy sequences in the mouse genome, which was similar to data obtained previously by us with another neuroblastoma clone (Schrier, et al., submitted for publication). Our goal is to examine cDNA-mRNA reaction kinetics in neuroblastoma cells under a variety of growth conditions.

SITES IN SIMIAN VIRUS 40 CHROMATIN WHICH ARE PREFERENTIALLY CLEAVED BY ENDONUCLEASES

Walter A. Scott and Dianne J. Wigmore. Department of Biochemistry, University of Miami School of Medicine, Miami, Florida 33152.

Deoxyribonucleoprotein complex isolated from Simian Virus 40-infected BSC-1 cells (SV40 chromatin) has been analyzed for sites preferentially cleaved by endonucleases. Digest conditions were chosen to introduce one double-strand cut into most of the viral genomes present in a chromatin preparation. After deproteinization, full-length linear viral genomes were isolated by electrophoresis on agarose gel slabs and the initial cleavage site was mapped by restriction enzyme digestion. The initial cut was introduced either by DNase I or by an endonuclease present endogenously in infected BSC-1 cell nuclei. With either of these endonucleases, the initial cleavage site mapped at one of a limited set of discrete sites located in a limited region of the viral genome. Major cutting sites for the endonuclease endogenous to BSC-1 cell nuclei are indicated on the map of SV40 shown below.

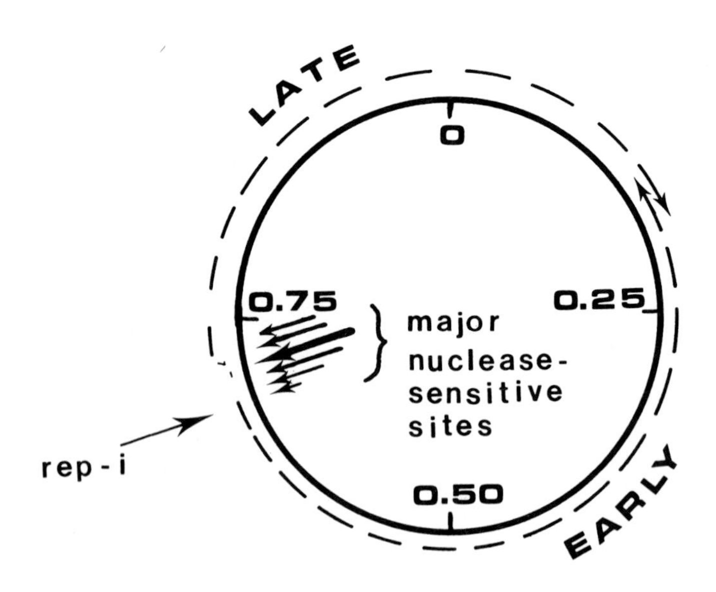

ISBN 0-12-045450-5

EMBRYONIC GENE REGULATION: ROLE OF A SPECIFIC INDUCER RNA IN CONTROL OF HEART-LIKE DIFFERENTIATION IN EARLY CHICK EMBRYO

M.A.Q. Siddiqui, Hans-Henning Arnold, Amrut K. Deshpande, and Sonia B. Jakowlew
Roche Institute of Molecular Biology, Nutley, N.J.

Models of mechanisms of eukaryotic gene regulation during embryogenesis have been proposed in which Both RNA and proteins are implicated to exercise regulatory functions affecting the pattern of gene activity (1). Although the control of gene expression by cytoplasmic and extracellular factors in general has been well documented, there is no clear and direct experimental evidence yet in support of the main theme of the model. Embryonic systems which permit identification and functional analysis of putative regulatory macromolecules would, therefore, be highly suitable for elucidation of the underlying mechanism(s).

In the stage 4 chick blastoderm, an area located 0.6 mm posterior to Hensen's node, the post-nodal piece (PNP), consists of an undifferentiated population of cells, since the explants when cultivated *in vitro* in a variety of media or as chorioallantoic transplants do not develop into any histologically identifiable structures. However, addition of a specific low molecular weight RNA isolated in our laboratory (2) from the 16-day old chick embryonic heart promotes the appearance of a distinct mode of morphological and biochemical changes that is similar to that of embryonic cardiogenic differentiation (3). In the absence of RNA the embryonic cells remain undifferentiated. RNA from a variety of other sources isolated under identical conditions are not effective. The inducer RNA, which was rich in polyadenylate residues, does not appear to possess mRNA like properties.

Thus, the dependence of a distinct mode of differentiation in the embryonic chick cells upon a specific RNA offers an opportunity in which the role of a presumptive regulatory molecule in control of embryonic development can be tested in an experimentally analyzable biological system.

(1) Davidson, E.H. and Britten, R.J., *Quart. Rev. Biol.* *48*, 565-613 (1973).
(2) Deshpande, A.K., Jakowlew, S.B., Arnold, H-H., Crawford, P.A. and Siddiqui, M.A.Q. *J. Biol. Chem.* *252*, 6521-6527 (1977).
(3) Deshpande, A.K. and Siddiqui, M.A.Q., *Develop. Biol.* *58*, 230-247 (1977).

ISBN 0-12-045450-5

PATTERNS OF PROTEIN SYNTHESIS AS MEASURED *IN VIVO* IN SIMPLE EMBRYOID BODIES AND TERATOCARCINOMAS

Alan B. Silverberg, William D. Willis, Doug Tully and Stephen E. Harris. Molecular Embryology Workgroup, NIEHS, LET, Research Triangle Park, North Carolina 27709.

The mouse teratocarcinoma can be used as a model system for early mammalian differentiation and development. In our effort to study gene expression during the process of differentiation from simple embryoid bodies (EB) to predominately neuroepithelial teratomas (T), we first looked at *in vivo* patterns of protein synthesis in these two states.

Five 129/Sv-Sl male mice, in which the EB and T are maintained as ascitic and solid differentiating tumors respectively, were injected IP with 1 mCi of ^{3}H-leucine (spec. act.: 60 Ci/mmol) 48 hours prior to sacrifice. EB were removed by lavage of the peritoneal cavity with phosphate buffered saline, and the T were excised and immediately frozen on dry ice. SDS-polyacrylamide slab gel electrophoresis was done on 105,000 xg supernatants and the gels were processed for scintillation fluorography.

Histologically the T have been shown to contain a high concentration of neuroepithelial cells. At least two neural specific proteins, glial fibrillar acidic protein (GFAP) and S-100, appear to be present in the T and not in the EB. GFAP has previously been demonstrated immunohistochemically *in vivo* as well as in differentiating cultures of EB *in vitro* (1). Alpha-fetoprotein (AFP) has been reported to be present in many human teratomas. We have noted a band corresponding to AFP in the T and EB, but the amount of material is much less in the EB. Other differences in the stained and fluorographic patterns exist, but overall the T and EB were similar.

Poly(A)-containing messenger RNA from EB and T have been isolated, and studies on the *in vitro* translation are being carried out in the nuclease-treated rabbit reticulocyte lysate *in vitro* translation system. This assay is being used to establish the presence and initiate the isolation of the messenger RNA's for AFP, S-100, and GFAP.

(1) VandenBerg, S.R., Herman, M., Ludwin, S., and Bignami, A., *Am. J Pathol.* *79* 147-168 (1975).

ISBN 0-12-045450-5

CONTINUED EXPRESSION OF DOPA OXIDASE IN CHICK MELANOCYTE X CHICK FIBROBLAST HETEROKARYONS

Marlene S. Skulsky and John A. Brumbaugh.
School of Life Sciences, University of Nebraska, Lincoln, Nebraska.

Primary cultures of somite-derived chick embryo melanocytes, wild-type with respect to eumelanin production, were fused with primary cultures of chick embryo fibroblasts of the same genotype. The fibroblasts were grown in media with 2 μ latex beads. Multinucleate cells containing both pigment granules and latex beads were recognized as heterokaryons.

48 hours post-fusion the cells were fixed for TEM and incubated in 5 mM DOPA to test for DOPA oxidase. The presence of active DOPA oxidase results in an electron opaque, golgi-related end product. Cells were scored as either DOPA positive or DOPA negative. Preliminary ultrastructural observations have shown no difference in DOPA oxidase activity between control melanocytes and heterokaryons. However, treatment with α-amanitin or cycloheximide dramatically reduced the percentage of DOPA oxidase positive control cells.

Davidson et al (1) showed DOPA oxidase was "extinguished" in hamster melanoma X L cell hybrids. In this heterokaryon system, however, extinction does not occur during the first 48 hours. In other heterokaryon systems extinction of the differentiated function has occurred within the first 48 hours (2, 3).

Supported by NIH grant #GM 18969.

(1) Davidson, R.L., Ephrussi, B. and Yamamoto, K., Proc. Nat. Acad. Sci. 56 1437-1440 (1966).

(2) Thompson, E.B. and Gelehrter, T.D., Proc. Nat. Acad. Sci. 68 2589-2593 (1971).

(3) Zeuthen, J., Stenman, S., Fabricius, H.A. and Nilsson, K., Cell Differ. 4 369-383 (1976).

ISBN 0-12-045450-5

Association of Cyclic GMP with Gene Expression in Polytene Chromosomes of *Drosophila*.

W. Austin Spruill*, David R. Hurwitz**, John C. Lucchesi**, and Alton L. Steiner*.
*Departments of Medicine and Pharmacology and **Curriculum of Genetics, University of North Carolina, Chapel Hill, North Carolina, 27514.

The distribution of cyclic nucleotides on polytene chromosomes of *D. melanogaster* was examined using indirect immunofluorescence. Salivary glands are dissected out in Ringers, fixed with 2% formaldehyde in a phosphate buffer containing salts and triton, and squashed between a slide and coverslip in a drop of 5% acetic acid-2% formaldehyde. After the coverslip is removed the slides are placed in phosphate buffered saline until application of antibodies. The chromosome preparations are then treated with rabbit antibodies directed against either cGMP or cAMP, followed by exposure to fluorescein isothiocyanate-conjugated goat anti-rabbit antibodies. Slides are examined by dark field fluorescence microscopy and later by bright field microscopy following staining with aceto-orcein.

This technique allows the observation of cGMP but not cAMP along the chromosomal arms. The subchromosomal distribution of cGMP correlates with genetically active sites on the chromosomes (i.e. puffs, interbands and diffuse bands but not highly condensed dark bands). Following heat shock treatment the intensity of cGMP fluorescence is markedly enhanced in specific loci on the chromosomes with locus 93D as the most intensely fluorescent locus. Autoradiographic analysis, using 3H-uridine, revealed that 93D is the most transcriptionally active locus within a particular nucleus. In addition, the subchromosomal distribution of cGMP correlates with the subchromosomal distribution of RNA polymerase II as determined immunocytochemically by Jamrich, Greenleaf and Bautz (Proc. Nat. Acad. Sci. (USA) 74, 2079-2083 (1977). These observations suggest that cGMP may participate in processes associated with transcription on polytene chromosomes.

Supported by NIH grants AM05330, AM283, GM-15691 and T01-GM-0685.

ISBN 0-12-045450-5

BIOCHEMICAL MARKERS FOR HEMATOPOIETIC CELL DIFFERENTIATION AND LEUKEMIAS

B. I. Sahai Srivastava, Department of Experimental Therapeutics, Roswell Park Memorial Inst., Buffalo, New York 14263

Terminal deoxynucleotidyl transferase (TDT), a serine protease and a 3-4S deoxyribonuclease (DNase), found predominantly associated with the chromatin and DNA polymerase (DP) α and β activity were examined in normal and leukemic human leukocytes. High activity of DP and TDT and little or no activity of the protease and DNase was found in untreated or relapsed leukemic patients with T- or non-T/non-B cell acute lymphocytic leukemia (ALL). Normal leukocytes, leukocytes from ALL patients in remission, or from patients with chronic lymphocytic or chronic myelocytic leukemias had low DP and TDT but high protease and DNase activity, although the activity of the latter two enzyme was much higher in myeloid cells. On the basis of these markers, acute myelocytic leukemia (AML) and blastic phase chronic myelocytic leukemia could be divided into two categories: a major one in which cells have high DP, protease and DNase and low TDT and a minor category characterized by high DP, and intermediate to high TDT, protease and DNase activity. Alterations in the above enzyme profiles could be useful for the study of differentiation of hematopoietic cells, since maturation of cells along both lymphocytic and granulocytic series seems to lead to a decline in DP and TDT and an increase of protease and DNase. Furthermore, two new markers, protease and DNase, together with TDT could be useful for the differential diagnosis of leukemias, and the serine protease and DNase especially for discrimination of ALL from AML which cannot be done unambiguously by the TDT assay. Supported by USPHS Grants No. CA-17140, and CA-5834.

ISBN 0-12-045450-5

DIFFERENTIAL EFFECT OF DIMETHYL SULFOXIDE (DMSO) ON CLONED CELL STRAINS OF NEURAL ORIGIN

Allen C. Stoolmiller and Helen O. Hincman.
E.K. Shriver Center, Waltham, MA 02154

Established tumorigenic animal cell strains of neural origin were cultured in medium containing up to 2% DMSO and changes in cell morphology, proliferation and macromolecular synthesis were studied.

Rat C6 astrocytes underwent a marked morphological change within 24 h after exposure to low concentrations of DMSO. Low density cultures incubated for 5 days in medium containing 0.4-0.6% DMSO yielded 40-80% more cells than control cultures. Higher concentrations of DMSO (0.8-2.0%) caused progressive inhibition of cell growth with 50% inhibition at 1.4%. The effect of DMSO on the proliferation of C6 cells was paralleled by either increased or decreased synthesis of DNA, protein, glycoprotein, glycosaminoglycans, glycolipids and lipids. Even after maintaining C6 cells in medium containing 2% DMSO for 5 weeks, normal growth and morphology were restored within 3 days after omitting DMSO from the medium.

DMSO inhibited the growth of both mouse Schwannoma cells (G26-20 and G26-24) and mouse 2a neuroblastoma cells at all concentrations tested (0.4-4.0%); 50% inhibition was observed at DMSO concentrations of 0.8% or less. While the synthesis of most macromolecular products was decreased in proportion to the reduction in cell proliferation, the production of sulfated glycosaminoglycans by 2a cells treated with 0.4 to 0.8% DMSO was stimulated 2- to 4-fold over controls.

These results suggest that DMSO will be a valuable agent for *in vitro* studies on the differentiated characteristics of tumorigenic cells of glial and neuronal origin.

Supported by U.S.P.H.S. grants HD-05515, HD-04147 and U.S.D.A. Cooperative Agreement No. 12-14-5001-269.

ISBN 0-12-045450-5

REGULATION OF DNA-DEPENDENT: RNA POLYMERASE I ACTIVITY IN RAT LIVER NUCLEI

John A. Todhunter, Michael Tainsky, Herbert Weissbach, and Nathan Brot. Department of Biochemistry, Roche Institute of Molecular Biology, Nutley, New Jersey.

The hypothesis that the activity of RNA polymerase I is controlled by an interconversion between bound (chromatin active) and unbound (chromatin inactive) forms of the enzyme (1,2) has been examined. Stimulation of the RNA polymerase I activity following *in vivo* administration of hydrocortisone or methylisobutylxanthine appears to be mediated by a labile factor. This factor which is sensitive to trypsin and sonication is tightly bound to the transcriptional complex since it is possible to isolate nucleoli with elevated levels of polymerase activity. Conversion of unbound to bound polymerase does not appear to be involved in the stimulation of polymerase activity. Synthesis of new enzyme is also ruled out by the finding that after proteolysis or sonication, the level of polymerase activity decays to the stable level found in control liver. The endogenous template also appeared to be unaltered in stimulated and control livers.

The decrease of RNA polymerase I activity by the *in vivo* administration of cycloheximide results in a level of activity lower than that of control livers but stable to proteolysis or sonication. This decrease in activity was also not due to a conversion of bound to unbound polymerase. Indeed it was found that the amount of unbound polymerase may be related to the procedure used to prepare the nuclei.

It has been suggested that ornithine decarboxylase is a regulator of RNA polymerase I activity (3) and may be the labile component associated with stimulated polymerase. At present we have found no evidence to support this view but further studies are in progress on the nature of the stimulated enzyme.

(1) Yu, F. and Feigelson, P.; PNAS 69 2833 (1972).

(2) Lampert, A. and Feigelson, P.; BBRC 58 1030 (1974).

(3) Manen, C.A., and Russell, D.H.; Life Sciences 17 1769 (1976).

ISBN 0-12-045450-5

PANCREAS HISTONE 1 PHOSPHORYLATION IN DEVELOPMENT, REGENERATION AND CANCER

F. Varricchio, F. Sharkey, D. Kim, and P. J. Fitzgerald.
Memorial Sloan-Kettering Cancer Center, New York, New York and the Life Sciences Center, Nova University, Fort Lauderdale, Florida.

It has been suggested that the phosphorylation of histones at different sites serves different functions in DNA replication and mitosis. *In vitro* a direct correlation between the rate of DNA synthesis and histone phosphorylation has been shown. H1 is the most phosphorylated histone and H1 has shown the most change in degree of phosphorylation in previous studies.

One hour after adult rats were injected i.p. with ^{32}P the specific activity of histone H1 was highest in spleen, lower and about the same in liver, pancreas and kidney, and lowest in lung. The specific activity of H1 was increased 2-3 fold in embryonic and regenerating rat pancreas, and in rat pancreas cancers grown in nude mice.

Analysis of histones by polyacrylamide gel electrophoresis showed phosphorylated H1 species in regenerating pancreas and in a rat acinar cell pancreas tumor grown in a nude mouse. In addition, embryonic pancreas and a rat acinar cell cancer grown in nude mice have decreased levels of the minor histone $H1^{o}$ band.

^{32}P labeled H1 from control rat pancreas, embryonic and regenerating rat pancreas, and a rat pancreas tumor grown in a nude mouse were digested with trypsin. The tryptic peptides were separated in 2 dimensions by electrophoresis and chromatography and the phosphopeptides located by autoradiography. The phosphopeptide map showed that the phosphorylation state of H1 is much more complex and varied than can be shown by polyacrylamide gel electrophoresis. Five to ten phosphopeptides were found which vary in occurence and intensity in development, regeneration and tumors. In particular in developing and regenerating pancreas as well as a rat pancreas tumor grown in a nude mouse that was a very basic phosphopeptide not found in controls. This phosphopeptide may be involved in DNA synthesis.

Supported in part by USPHS Grant #CA19584.

ISBN 0-12-045450-5

REGULATION OF DIHYDROFOLATE REDUCTASE GENE EXPRESSION IN 3T6 CELLS DURING TRANSITION FROM THE RESTING TO GROWING STATE

Leanne Wiedemann, Carol Fuhrman and Lee Johnson.
Department of Biochemistry, The Ohio State University, Columbus, Ohio 43210

The rate of synthesis of dihydrofolate reductase was studied in resting, growing and serum stimulated mouse 3T6 cells by first exposing the cells to 10^{-6} M methotrexate (MTX) for 20 min to specifically and irreversibly inactivate the pre-existing enzyme, then determining the rate of recovery of reductase activity after removal of MTX. Reductase was quantitated by measuring the ability of a cell extract to reduce ^{3}H-folic acid or to bind ^{3}H-MTX. In all cases, recovery of enzyme activity was inhibited by cycloheximide, confirming that the recovery was due to *de novo* synthesis of reductase.

We found that the rate of synthesis of reductase was quite high in exponentially growing cells, as expected. Similar studies with resting 3T6 cells showed that the rate of synthesis of reductase was extremely low in these cells. To determine when (following growth stimulation) the reductase gene was expressed, resting 3T6 cells were serum stimulated and the amount of reductase was determined at various times thereafter. We found that the amount of reductase began to increase linearly beginning about 10 hrs after serum stimulation (slightly preceding the onset of DNA synthesis). The increase in reductase activity was not affected by the presence of hydroxyurea or cytosine arabinoside, indicating that there is no direct coordination between DNA replication and reductase gene expression. The increase in reductase was inhibited by Actinomycin (5 μg/ml) if the drug was added any time between 0 and 8 hrs after stimulation but was not inhibited if the drug was added after 12 hrs of stimulation. This is consistent with the idea that translatable reductase mRNA is synthesized only during this interval.

Supported by Grant No. 5 P30 CA 16058 0351 awarded by the National Cancer Institute.

ISBN 0-12-045450-5

INCREASED NUCLEASE-SUSCEPTIBILITY OF ACTIVE CHROMATIN AT HEAT-SHOCK LOCI IN DROSOPHILA.

Carl Wu, Ken Livak and Sarah Elgin.
The Biological Laboratories, Harvard University, Cambridge, Massachusetts 02138

An alteration in chromatin structure following gene activation as evidenced by increased susceptibility to DNase I digestion was initially reported by Weintraub and Groudine for the chick globin sequence (1). We have probed the chromatin structure of sequences coding for the 910,000 dalton major heat-shock mRNA of *Drosophila melanogaster* using the nucleolytic enzymes Micrococcal Nuclease and DNase I.

At various extents of digestion, the size distribution of all nuclease-resistant DNA fragments was determined by ethidium bromide fluorescence after gel electrophoresis. To selectively visualize those sequences complementary to the major heat-shock mRNA, the DNA fragments were then transferred by the Southern technique (2) from gel to nitrocellulose and hybrdized with ^{32}P-labelled DNA from plasmid pPW229.1. This plasmid is a subclone of pPW229, screened in the laboratories of Pieter Wensink and Matthew Meselson, and contains an 840 base-pair insert complementary to sequences totally internal to the major heat-shock mRNA. These sequences exhibit an increase in susceptibility to both nucleases in nuclei isolated from heat-shocked embryos of *Drosophila melanogaster*. Non-heat-shocked embryos do not show this phenomenon.

(1) Weintraub, H. and Groudine, M., *Science* *193* 848-856 (1976).
(2) Southern, E.M., *J. Mol. Biol.* *98* 503-517 (1975).

ISBN 0-12-045450-5

Keratin Production as a Marker for Normal Differentiation in Mouse Epidermal Cell Cultures. Stuart H. Yuspa, Miriam C. Poirier and Peter Steinert. Experimental Pathology Branch and Dermatology Branch, National Cancer Institute, Bethesda, Maryland 20014.

Mouse epidermal cell cultures have been a useful in vitro system for studies of chemical carcinogenesis and cellular differentiation. In all previous reports the determinants for normal differentiation have been the histochemical, morphological or non-specific chemical nature of the cells or cell products. This study was designed to demonstrate that biochemically and immunologically characterizable keratin is produced by these cells in vitro.

When mouse stratum corneum or isolated mouse epidermal cells are extracted with 8M urea buffer and electrophoresed on SDS polyacrylamide gels, two proteins of M.W. 68,000 (K_1) and 60,000 (K_2) are isolated in equimolar amounts. These proteins comprise more than 70% of the extractable material from stratum corneum or the differentiated fraction of the isolated epidermal cells. When epidermal cells are cultured for 2 weeks, K_2 consistently comprises about 20% of the urea extractable protein in the attached cells, while K_1 is initially low, increases in content to be equimolar to K_2 at 48 hours and then gradually decreases to a low (5%) stable value. When urea extracts of the cellular material sloughed into the medium at any time in culture are examined by SDS PAGE, K_1 and K_2 comprise more than 70% of the protein and are equimolar. Urea extracts of attached or sloughed cells from dermal fibroblast cultures do not contain K_1 or K_2. Purified K_1 and K_2, isolated from stratum corneum or cells in culture, recombine in vitro to produce filaments of 70-80Å in diameter and up to 10 µm in length which display an α-type x-ray diffraction pattern and contain K_1 and K_2 in a molar ratio of 1:2 or 2:1. Antibodies formed in rabbits against purified K_1 or K_2 produce immunoprecipitin bands when reacted against purified K_1 or K_2 or urea extracts of epidermal cells (both attached and sloughed) during 6 days in culture.

These results demonstrate that mouse epidermal cells in culture produce keratin proteins identical to stratum corneum in vivo. Furthermore the concomitant production of K_1 and K_2 leads to irreversible differentiation and sloughing of cells in vitro, analogous to keratin production in vivo. Therefore these two major proteins should be useful markers for further studies on the control of differentiation in vitro.

EARLY VOLUME CHANGES DURING DIFFERENTIATION OF FRIEND LEUKEMIC CELLS

R.M. Zucker, N.C. Wu and A. Mitrani, Papanicolaou Cancer Research Institute, Miami, Florida

This investigation involves the study of the early volume changes which occur during the differentiation of Friend leukemic cells (FLC, clone 745).[1] The kinetics of the early induction process were studied with a Coulter H_4 Channelyzer by measuring cell volume, cell growth and nuclei volume. The statistical parameters derived from these data include mean, mode, skewness and coefficient of variation.

The comparison of the inducing agents, DMSO (1.5%) and butyric acid (1 mM), on FLC revealed different rates in the reduction of cell volume. The DMSO reduced the cell volume by 5% within 3 hours, while the butyric acid appeared to have no effect. However, by 24 hours the butyric acid reduced the mean cell volume by 31% in contrast to 12% for DMSO. This volume decrease was produced primarily by the reduction in the proportion of large sized cells. The volume decrease was observed to be concentration dependent for both inducers with the variations being greater for changes in butyric acid concentrations (.5 mM - 2 mM).

The mean and modal cell volume decrease was correlated to a change in the coefficient of variation (C.V.) and skewness. These functions represent the degree of homogeneity of the cell populations. A maximum skewness for DMSO was observed at 16.5 hrs and for butyric acid at 19 hrs. A minimum C.V. was observed at 19 hrs for DMSO and 23 hrs for butyric acid. From these data it appears DMSO reacts quicker than butyric acid in the differentiation process but the reduction in cell volume is greater with the butyric acid. The nuclei volume distributions support the cell volume data in demonstrating a build up of small nuclei and a decrease in larger sizes during the 12-24 hr time period. During the 24-40 hr time period the populations of cells and nuclei show small changes in C.V. skewness, mean and mode. This study has shown the early differentiation effects of Friend leukemic cells involve a volume decrease which is concentration dependent, agent specific and might be related to a redistribution of cell cycle stages.

1. Loritz, F., Bernstein, A., Miller, R.G. J. Cell Physiology 91;423-437, 1977

Supported in part by NCI contract N01CB33861 and NIH grant HL-15999

ISBN 0-12-045450-5

THE ORIGIN AND SIGNIFICANCE OF TISSUE-SPECIFIC HISTONE VARIANT PATTERNS IN MAMMALS.

Alfred Zweidler, Michael Urban and Patricia Goldman.
The Institute for Cancer Research, Philadelphia, Pa.

We have previously shown that by electrophoresis in presence of nonionic detergents (1), the mammalian histones 2A, 2B and 3 can be resolved into nonallelic primary structure variants which occur in different relative amounts in different tissues (2). We now present evidence that the tissue-specific histone variant patterns are the result of differential activity of at least two independent histone operons which contain different sets of variants.

During embryonic growth all cell types appear to have the same histone variant pattern. This embryonic pattern is maintained in adult tissues capable of rapid growth (hemopoetic tissues, mammary glands, placenta, etc.) but changes dramatically in tissues with limited growth potential (e.g. liver, kidney, lung, thyroid) throughout life. In all cases, the predominant embryonic variants are gradually replaced by less "hydrophobic" variants. In mouse liver, for instance, the H2A.2:H2A.1 and H3.3:H3.1 ratios change from 1:3 in the embryo to 1:1 at ca. 20 days, 2:1 at ca. 100 days, 4:1 at ca. 200 days of age. The kinetics of this change are different for different tissues. Both the variant patterns and the tissue-specific rates of change appear to be similar in all mammals.

Our results indicate that the tissue-specific histone variant patterns are related to cell growth patterns rather than cell differentiation. Since histone complexes containing different histone variants have distinct physical properties (3), the structure of the chromatin of most mammalian tissues is changing continuously with age. This may lead to general changes in the functional properties of the genome and could contribute to the phenomenon of aging.

Supported by NIH grants #CA 15135, #CA 06927, #RR 05539 and an appropriation from the Commonwealth of Pennsylvania.

(1) Franklin, S. and Zweidler, A., Nature 266 273 (1977).

(2) Zweidler, A., Methods in Cell Biology 18 (1977) in press.

(3) Zweidler, A. et al., J. Cell Biol. 70 254 (1976).

ISBN 0-12-045450-5

THE OVALBUMIN STRUCTURAL GENE SEQUENCES IN NATIVE CHICK DNA ARE NOT CONTIGUOUS

Savio L.C. Woo, Eugene C. Lai, Achilles Dugaiczyk, James F. Catterall and Bert W. O'Malley.
Baylor College of Medicine, Houston, Texas.

Plasmid pOV230, a chimera previously constructed in our laboratory using pMB9 and full-length double-stranded ovalbumin DNA synthesized from purified ovalbumin messenger RNA, was cleaved with *Hae* III and *Hind* III to yield two distinct DNA fragments containing the left and right halves of the ovalbumin DNA insert. The two fragments were separated by gel electrophoresis, labeled with (32p)dGTP by nick translation and employed as specific hybrization probes to study the organization of the ovalbumin gene in native chick DNA. Total chick DNA was digested witha variety of restriction endonucleases and transferred to nitrocellulose filters using the method of Southern after agarose gel electrophoresis. The filters were then hybridized with the radioactive probes containing the left and right halves of the structural ovalbumin gene followed by radioautography. While Hae III generated only one hybridizable band with OV_R as expected, it generated two bands with OV_L, indicating the presence of a *Hae* III site in native chick DNA that is absent in $cDNA_{ov}$. Furthermore, *Eco*RI produced 3 discrete fragments although it does not cleave the structural ovalbumin gene. While the 2.4 and 1.8 Kb fragments hybridized only with OV_L, the 9.5 Kb fragment formed a hybrid only with OV_R. The fidelity of these bands was established by the fact that they could be enriched by "R-loop" formation using highly purified ovalbumin mRNA followed by oligo dT-cellulose column chromatography. These results are consistent with the interpretation that one or more non-structural sequences may exist in the native ovalbumin structural gene that were cleave by *Hae* III and *Eco*RI. A rather detailed model for the structure of the native ovalbumin gene was thus constructed.

ISBN 0-12-045450-5

DNA SEQUENCE ANALYSIS OF THE CHICKEN OVALBUMIN GENE

Daniel Kuebbing, Charles D, Liarakos, and Bert W. O'Malley.
Department of Cell Biology, Baylor College of Medicine,
Houston, Texas.

The DNA sequence of the ovalbumin structural gene around the initiation codon has been determined by the method of Maxam and Gilbert. The recombinant plasmid pOV230 that contains an insert of ovalbumin cDNA was used as starting material. This plasmid was cleaved with either of two restriction endonucleases - HinfI or SstI. The products were separated by electrophoresis on polyacrylamide gels. After elution from the gels, the individual fragments were terminally labeled with polynucleotide kinase and ATP-γ-^{32}P. The fragments were secondarily digested with the other restriction endonuclease (i.e., Sst I ur Hinf I) and the products were separated on a polyacrylamide gel. These terminally labeled fragments were sequenced by the specific chemical degradation procedures of Maxam and Gilbert. The results indicate that the AUG initiation codon is located very close to the 5' end of the ovalbumin mRNA, probably within the first 50-60 nucleotides. Based on the known size of this mRNA (1850 N), the 3' untranslated region must be very large (~600 N).

A
B
C 8
D 9
E 0
F 1
G 2
H 3
I 4
J 5

ISBN 0-12-045450-5